PLANT MATERIAL OF AGRICULTURAL IMPORTANCE

IN TEMPERATE CLIMATES

M.A. Farragher

University College Dublin Press

Preas Choláiste Ollscoile Bhaile Átha Cliath

First published 1996 by University College Dublin Press,
Newman House, St Stephen's Green, Dublin 2, Ireland

ISBN 1 900621 00 2

Cataloguing in Publication data available from the British Library

Printed in Ireland by Colour Books, Dublin

1 NOMENCLATURE AND CLASSIFICATION

Mention of the word species conjures up visions of countless plants which look very similar in appearance. Historically, the preoccupation with the recognition of species is possibly as old as mankind itself. From the earliest of times it was important to know which species were useful for any given purpose. Defining species solely on the basis of the possession of certain characteristics, and the absence of others, has, many obvious limitations. One of the few agreements between the earliest botanists was that individuals of one species were in some way isolated or protected from individuals of another species, otherwise the species would lose the distinctive characteristics by which they could be identified.

Morphological species

John Ray, a seventeenth century English naturalist, was the first to propose that species be recognised on the basis of differences in structure that were transmitted from generation to generation. This, the *Morphological Species Concept*, was favoured by other early naturalists, including Linnaeus, at least until shortly before his death. The latter believed that different organisms owed their origin to special creation, and that entities or species were incapable of change or intergradation by hybridisation or any other means. In other words, variation within species, and the possibility of environmental influence, was not considered.

Strict adherence to this concept would cause many obvious problems in relation to some well known wild and cultivated plants. As regards the different cole plants, all would be deemed species in their own right! Thus, cauliflowers, Brussels sprouts, cabbage, rape, etc., would belong to separate species. The same rationale, if applied to *Primula vulgaris* which displays floral dimorphism, would necessitate the recognition of two species. On the basis that plants which looked morphologically similar being treated under the same species, serious problems would arise in relation to some brome grasses. *Bromus commutatus* and *B. racemosus* are very alike, in terms of morphological characteristics. Should these bromes belong to the same species? It has long been accepted that the environment can influence the development of a plant. A good example would be *B. hordeaceus*. Under very dry conditions, the inflorescence is often represented by one spikelet. Growing on damp heavy clays, the inflorescence has numerous spikelets. The two phenotypes are startling in their contrast. Should they be treated as separate species?

Biological species

Apart from the difficulties already mentioned in connection with the morphological concept, another major objection is that it does not show to advantage the genetic relationship between members of a breeding population, as well as continuity in time. The theory of evolution has been with us for some time. It was even suggested or hinted at by Linnaeus shortly before his death, nearly a century before Darwin's (1859) *Origin of Species*. Yet, it was not until 1937 that a taxonomist by the name of Dobzhansky put forward another concept completely divorced from morphological characteristics. This latest concept, which has been somewhat refined by later workers, has become known as the *Biological Species Concept*. Using this concept, species would be defined as *groups of actually or potentially interbreeding natural populations which are reproductively isolated from other such groups*. According to the biological concept, species are classified on the basis of one criterion, *viz.* reproductive isolation. Put in simple terms, plants which are cross-fertile, belong to the same species, those which are cross-sterile, with minor exceptions, belong to different species.

In reality, both concepts are used simultaneously, in that species which look alike tend to be cross-fertile. Under the biological concept, the cole plants mentioned earlier, on the basis of cross-fertile, are classified as a single species. As regards the two phenotypes of soft brome, again, because of cross-fertility, belong to the same species. The same rationale applies to the primrose. Meadow brome and smooth brome, already referred to, are cross-sterile, and, consequently, are best considered in separate species.

NOMENCLATURE

The use of the vernacular is totally unsuited to biological nomenclature. Apart from the multiplicity of languages, the same common name has often been applied to different organisms. One example is the use of the common term 'foxtails' for grasses of the genera *Alopecurus* and *Setaria*. Apart from the application of the same common name to different groups of plants, some plants may have a multiplicity of common names. *Senecio jacobaea*, for example, is said to have two hundred and forty eight common names, worldwide. To avoid any confusion, a plant must have only one name, *viz.* one valid scientific name.

The study of the naming of plants, and the various *taxa* (singular *taxon*) or *units* of classification, is known as *nomenclature*. The earliest nomenclature systems involved the use of lengthy *Latinised* descriptions, sometimes using up to thirty, or more, words! These lengthy descriptions were known as *polynomial* systems. A less verbose nomenclature system, involving the use of twelve words, was introduced by Linnaeus. He also coined a shortened form consisting of two words which was eventually accepted as the official scientific name. This latter system, *i.e.* involving use of two words, is now known as the *binary, binominal* or *binomial* system. Subsequent to Linnaeus, a proliferation of binomial names took place as new species were established and higher taxonomic categories were formed. This led, in the nineteenth century, to much confusion in the nomenclature of many groups of organisms. In the twentieth century, however, the establishment of rules by international committees in the fields of zoology, botany, bacteriology and virology has done much to improve the situation. One improvement was the formulation of three *Codes of Nomenclature* for the following:

1. Plants, including fungi -   **International Code of Botanical Nomenclature (ICBN);**
2. Animals -   **International Code of Zoological Nomenclature (ICZN);**
3. Bacteria, including actinomycetes - **International Code of Nomenclature of Bacteria (ICNB).**

Subsequent to the formulation of the above Codes, a fourth, the *International Code of Nomenclature of Cultivated Plants*, was established in 1953, and revised in 1969. This covers the nomenclature of cultivated forms of Agricultural, Horticultural and Silvicultural plants. It follows closely the rules and recommendations of the Botanical Code, but the emphasis is on the nomenclature and legitimacy of the 'cultivated variety' or *cultivar*, which is discussed later.

Scientific names

The Code of Botanical Nomenclature requires that the scientific names of all taxa be *Latin* in form, written in the Latin alphabet and subject to the rules of Latin grammar. The scientific names are therefore Latin or treated as Latin, even if, as is often the case, they are derived from other languages. In the interests of uniformity, the Code also lays down a number of conventions which must be observed.

Names of species

Species names, properly written consist of two terms and are, therefore, called *binary, binomial* or *binominal*. In addition, they are *always* followed immediately by what is known as the *'Authority'*. The name of a species consists of the name of the *genus* in which the species is classified; this is followed by a second term which is peculiar to the *species*. Thus, the scientific name of red fescue is - **Festuca rubra L.**

The scientific name, when first written in a text, must be cited in full. For example, the name for red fescue should appear as *Festuca rubra* L. Thereafter, it may be abbreviated to *F. rubra* L. Also, in type-written or handwritten manuscripts, the name should be underlined; in printed documents it should be *italicised*. The second term of the binary name, the *specific epithet*, may not be used on its own.

Type specimen:

The application of scientific names is determined by *nomenclature types*. Every scientific name, be it that of a family, genus or species, is based on a *type specimen*. The determination of the exact organism designated by a particular name usually requires more than the mere reading of the description of the taxon to which the name applies. Circumstances could change since the the descriptions were first written. New forms, with different traits, may be discovered; or later workers may discover, by the inspection of the original material, that the first authors inadvertently confused two or more types. Validation, therefore, of a name can only be determined from an examination of the original specimen. In the case of binomials, this can only be based on an actual plant specimen. Therefore, when a new name is coined, a single *type specimen* must be designated, and placed in a reliable public institution where it can be properly cared for and made available to all taxonomists wishing to inspect it.

Apart from binomials, typification is also necessary for the naming of other taxa. In the case of a family, or lower taxon above the rank of genus, the type of the name is a genus. The type of the name of a genus, or lower taxon above the rank of a species, is a species. As stated earlier, the type of a name of a species, or infraspecific taxon, is an actual plant specimen.

Authority

The scientific names of plants, when fully written, are always followed by one or more personal names. Sometimes, if the personal name is long, it may be abbreviated. Thus, the name for red fescue may be written as *Festuca rubra* Linnaeus or *Festuca rubra* L. The personal names are the *authority citations* for the names they follow. The *authority* of a binomial is the name of the worker who has published it validly. For a new binomial to be valid, there are three requirements which must be fulfilled. Firstly, the name must not be applied to another plant; secondly, it must be accompanied by a written description, always in Latin, and thirdly, published in works that are printed, reasonably permanent and generally available.

Sometimes authority citations consist of two names separated by the prepositions *in* or *ex*. A hypothetical example would be *Lolium maximum* Farragher in Reilly. This means that Farragher was responsible for the valid publication of the name *L. maximum* in a paper edited or otherwise written by Reilly. The last part, 'in Reilly', may, in the interest of brevity, be omitted. The name - *Lolium maximum* Farragher ex Reilly - means that Reilly was the first to validly publish the name *L. maximum* which was originally coined by Farragher. The latter, for a number of possible reasons, failed to do so in accordance with the rules of the Botanical Code.

Frequently, one encounters the use of the double citation with one name in parentheses. Under the Botanical code, use of the double citation is necessary when a taxon below the rank of genus is transferred to another taxon, or when a genus or taxon of lower rank is altered in rank but retains its original epithet. The use of the double citation, therefore, signifies that there has been a revision of taxonomic status. Thus, the common plant, lesser swine-cress, was originally named by Linnaeus as *Lepidium didymum* L. Seeman transferred it, on the basis of new taxonomic information, to another genus but retained the specific epithet. Its new valid name is now *Coronopus didymus* (L.) Sm. There are many examples of plants, once considered varieties of a species, being raised to the rank of species. Here, viviparous fescue was once classified as a variety of sheep's fescue by Linnaeus, and given the name *Festuca ovina* var. *viviparous*. Seeman raised it, again on the basis of additional taxonomic information, to the rank of species and gave it the name *Festuca vivipara* (L.) Sm.

Synonyms

Because of ongoing taxonomic research which often results in name changes, a review of botanical literature frequently shows two or mose scientific names for the same plant. Under the principle of priority, however, there can be only one valid scientific name. Names other than the valid name are known as *synonyms*. In general, the older or oldest, is the legitimate or valid name, and all others are referred to as synonyms. The common pineappleweed may be taken as an example. Its valid name is *Matricaria matricarioides* (Less.) Porter. A previous name, *Chamomilla suaveolens* (Pursh) Rydb., is its synonym.

Names of genera

These are always uninomial. Generic names are singular nouns in the nominative case. They may be descriptive, an original or ancient term referring to a type of plant, or a noun adopted from the name of a person. The initial letter is written in the *higher* case.

Names of families

The names of all taxonomic groupings or taxa above the rank of genus are all of one term and, therefore, are called *uninomial, uninominal* or *unitary*. They are either *plural* noun or adjectives used as nouns. The initial letter is always written in the higher case. As for family names, originally these were coined in a most haphazard way, the name was often based on some floral feature. A good example would be the turnip family, Cruciferae, where the name was based on the 'crucifix-like' arrangement of the petals. Under the Code, names must now end with the suffix 'aceae', and must be based on the *type genus*. Taking the Buttercup family as an example, the type genus is *Ranunculus*. The stem of the name is derived by dropping the last syllable. The full name is obtained by adding the suffix 'aceae' to the stem. Thus, the name of the Buttercup family is obtained from the genus *Ranunculus*, by dropping the last syllable *us*, and adding the suffix 'aceae', *i.e.* Ranunculaceae.

Generally, each family should have only one scientific name. Eight families, however, are exempted in that each has an alternative name; both the original and the alternative are legitimate under the Botanical Code. One, the new name, conforms to the above rule and has the recognised ending 'aceae'. The second, the older, is the exception and is sanctioned by long usage.

The eight exceptions, with the older names given in parentheses, are as follows: Arecaceae (Palmatae), Brassicaceae (Cruciferae), Lamiaceae (Labiatae), Fabaceae (Leguminosae), Clusiaceae (Guttiferae), Poaceae (Gramineae), Apiaceae (Umbelliferae) and Asteraceae (Compositae).

CULTIVATED PLANTS

Strictly speaking, the nomenclature of cultivated plants is governed by the Botanical Code. In 1969, however, the *International code of Nomenclature of Cultivated Plants* was established. Since then, it has been revised to make it, as far as is possible, complete in itself, so that consultation of the Botanical Code should only be necessary under exceptional circumstances. In essence, it concentrates on rules and recommendations governing the 'cultivated variety' or *cultivar*.

The naming of cultivated plants is governed by the Botanical Code when the names refer to taxa such as species. To this extent, they do not differ from plants occurring in the wild. However, cultivated plants are markedly different in their origin, history and biological significance. In reality, they occur as artificial populations, maintained and propagated through the intervention of man. For these reasons, the Botanical Code, which is concerned with the various taxa of wild plants, is not appropriate. The International Code of Nomenclature of Cultivated Plants has a more limited scope in that it governs only one infraspecific unit of classification, namely the 'cultivated variety'

CONTENTS

The term cultivar is derived from *cultivated variety*, or its etymological equivalent in other languages. It is now the internationally recognised term denoting an assemblage of plants which is clearly distinguishable by morphological, physiological, cytological, chemical or other characters, and which, when reproduced either sexually or asexually, retains its distinguishing characters. They are often incorrectly called 'varieties' in practically all seed catalogues. The term variety is derived from *varietas*, a latin word referring to an infraspecific taxon attributable only to wild plants. True varieties, in the strict botanical sense, are more or less interbreeding natural populations occurring in the wild, and are capable of self perpetuation. Cultivars are neither. Each cultivar owes its distinction to one or more characters that are usually maintained in the population *artificially,* by asexual propagation, by controlled breeding, or by the continued selection to an accepted standard. Sometimes cultivars are separated by rather insignificant features. For example, some wheat cultivars can only be distinguished by differing iso-enzymes of alpha-amylase. In any event, no new cultivar can be accepted unless it can be distinguished from all other existing cultivars. Clearly, since cultivars differ in their mode of reproduction, there are different types:

1. Clones: A clone is a genetically uniform assemblage of individuals, derived originally from a *single* plant by asexual propagation. There are many examples, such as populations derived from cuttings, divisions, grafts or those propagated by obligate apomixis. In addition, individuals propagated from a distinguishable bud-mutation form a cultivar distinct from the parent plant.

2. Lines: A line is a population of plants propagated by breeding known parental stocks. Stability is maintained by selection to a recognised standard. Plants in this category are normally self-fertilising. Examples would be most cereals, most vegetables and some garden annual such as petunias.

3. Hybrids: Another kind is an assemblage of individuals reconstituted on each occasion by crossing. These include single-crosses, double-crosses, three-way crosses, top-crosses and intervarietal hybrids. Some wheat and maize cultivars fall into this category.

4. Strains: Some cultivars cannot be separated morphologically, yet they are quite distinct. They are generally distinguished by some intrinsic characteristic such as earliness or lateness of flowering. Within *Lolium perenne*, for example, there are hay and pasture types. Each type constitutes a different strain.

Names of cultivars

Any cultivar name published on or after the 1st of January, 1959, must, other than in exceptional cases, be a 'fancy' name, *i.e.* one markedly different from a botanical name in Latin form. Those in Latin form, given prior to the above date, may not be rejected. For example, the name *Hibiscus syriacus* 'Totus Albus', may be retained. Another exception is permitted when a botanical epithet, published in conformity with the Botanical Code, before, on, or after the same date, for a plant subsequently considered to be a cultivar, unless it duplicates an existing cultivar of the species concerned. Each name should preferably consist of one or two words, or at most, three words. There can be only one correct name, the name by which it is internationally known. Names are subject to the ordinary operation of priority and synonymy, and a cultivar may one or more legitimate synonyms. Should there be a change in the binomial of the species to which the cultivar belongs, the cultivar name remains constant. New names must not be the same as the botanical or common name of the genus or the common name of the species to which it belongs, if confusion might be caused. Furthermore, no name is deemed legitimate until it has been published. Publication is achieved by distribution to the public of a printed or similarly duplicated description.

Under the International Code of Nomenclature of Cultivated Plants, the names of all cultivars are written with the initial letter in the *higher* case. They are preceded by the abbreviation *cv.* or placed in quotation marks. They may be used with the generic, specific or common name if these are unambiguous. Thus, in the case of barley, the following example may be cited in relation to the cultivated form, Chariot:

Hordeum sativum cv. Chariot > *Hordeum* cv. Chariot > *Hordeum sativum* 'Chariot' > Barley 'Chariot'

CLASSIFICATION

Classification is the placing of similar things together into groups or categories. No classificatory system is definitive, as circumstances can change with the advent of new data. Any system involves the grouping of plants into a taxonomic hierarchy. Each level of the hierarchy is called a *rank, category* or *taxon*, such as family, genus or species. The basic rank of the hierarchy is the *species*. Species are grouped into *genera* (singular - *genus*). A genus may be described as a group of species thought to have a common ancestor, but which do not interbreed together, or if they do, as is very rarely the case, form sterile hybrids. Next, the genera are grouped into *families*. Often large families are separated into *tribes*. Some families, such as Brassicaceae (Cruciferae) and Gramineae are very natural. In these instances, the names are based on some distinctive feature. Others are more mixed. Thereafter, families are grouped into *orders*. At this level of classification degrees of relationship become more difficult to establish, as the common ancestor to the group becomes more obscure. In deed, it is not uncommon for families to be moved from one order to another, or for families to be grouped into different orders in different classificatory systems, as new taxonomic data become available. The next taxon in the hierarchy is the *class*. Orders are grouped into these. Here the common characteristics are even fewer. With recent data from many fields, many long established classes have been split up. For example, in the fungi, the class *Phycomycetes* has been divided into six. Following the class, the *division* or *phylum*, forms the next category. These contain the classes. The highest level of all is the *kingdom*.

There may be additional categories between some of the above ranks. These are invariably prefaced with the prefix 'sub', as, for example, *subdivision, subfamily, subtribe* and *subspecies*. The last three categories are generally not necessary for small uniform families. Finally, some authors recognise two additional infraspecific ranks such as *variety* and *form*.

Originally there were two kingdoms, Plant and Animal. The position of the *prokaryotes* (bacteria and blue-green algae) and *eukaryotes* (all other organisms other common characters) is under discussion. These two distinct groups may, in time, be afforded kingdom status. The position and nomenclature of some other long established hierarchal taxa is also in a state of revision. There are as many suggestions, as to change, as there are authors on the subject. Anyone interested in pursuing this topic further should consult *Outline of Plant Classification*, by Holmes (1983).

So that the rank of a taxon may be readily apparent from its name, the Code of Botanical Nomenclature stipulates that some, particularly the lower, should have a distinctive ending. A required ending has been discussed already for the rank of family. Other obligatory endings are as follows: subfamily -oideae, tribe -eae and subtribe -inae. An example of a full classification scheme is given hereunder for cultivated wheat. The system adopted follows the old classical format.

Kingdom*	Plantae
Division*	Embryophyta
Subdivision	Phanerogamia
Branch	Angiospermae
Class*	Monocotyledoneae
Order*	Poales
Family*	Poaceae
Subfamily	Festucoideae
Tribe	Triticeae
Subtribe	Triticinae
Genus*	*Triticum*
Species*	*aestivum*
Subspecies	not applicable
Variety	*aestivum*

* denotes important ranks of the taxonomic hierarchy.

2 MISCELLANEOUS FAMILIES

WEED PLANTS

Floral Phases

General: Use of the word 'weeds' immediately conjures up visions of hosts of troublesome plants growing in all kinds of situations. Indeed weeds are variously described as troublesome plants, or plants with a negative value, or plants which compete with man for soil. Possibly the shortest most succinct definition of a weed is *a plant out of place*. Using this broad definition, crop plants may be regarded as weeds when they contaminate other crop plants. For example, barley plants growing in a crop of wheat would qualify as weeds. In general, however, such contaminants lack the aggression and persistence over generations normally associated with more pernicious plants that are usually classified as weeds.

Weeds, on the other hand, have an abundance of characteristics which enables them to persist under the most adverse conditions. One characteristic common to all is their excellent adaptation to the disturbed environment in which they grow. Many have the so-called weed characteristics, such as abundant seed production, dormancy, ability to survive unfavourable growing conditions, aggressive competitive ability and ability to spread vegetatively. Weed plants are are essentially usurpers of essential plant nutrients and squatters on valuable ground that would be better utilised by economic plants. They are many and varied, and are undesirable for a variety of reasons.

Classification: Weeds have been classified in various ways. One classificatory system involves separation into parasitic and non-parasitic plants. The vast majority of weeds belong to the latter. Members of the first group are of no great significance in the British Isles, though 3 families, Convolvulaceae, Scrophulariaceae and Orobanchaceae, are represented. The first family contains the most important member, *Cuscuta epithymum* (common dodder), which parasitises woody plants but can also effect others, including crop plants. The second family is represented by several species of 3 genera, *Pedicularis*, *Euphrasia* and *Rhinanthus*, the most widespread being *R. minor* (yellow-rattle). All genera of the third family are parasitic.

Apart from the classical treatment, given later, weed plants are frequently classified as annuals, biennials and perennials. Annuals, by their very definition, are quick growers, and complete their life-cycle within 1 year from germination to seed production and then die. Ephemerals, a subdivision, complete their life-cycle in a few weeks. *Senecio vulgaris, Stellaria media, Veronica hederifolia, Mercurialis annua* and *Cardamine hirsuta* are good examples. Other subdivisions include: (i) summer annuals, and (ii) winter annuals. Summer annuals, such as *Chenopodium album* and *Avena fatua*, germinate in the spring, make most of their growth during summer, and tend to be more serious in most crops grown in the British Isles. Winter annuals, which, among others, include *Capsella bursa-pastoris*, are troublesome mostly in winter grown cereal crops. A biennial plant lives for more than 1, but less than 2, years. All members of the Apiaceae (carrot family), *Senecio jacobaea* and *Cirsium vulgare*, and many more, are biennial in longevity. Apparently a cold period or vernalisation is necessary for flower initiation and, consequently, biennials occur characteristically in temperate areas. Perennial plants live for more than 2 years, and may live almost indefinitely. These are subdivided into: (i) simple and (ii) creeping perennials. The former spread only by seed, whereas the latter spread mainly by stolons, rhizomes, tubers or creeping roots. *Plantago* species and *Cirsium vulgare*, respectively, are examples of this final group. Creeping perennials are possibly the most difficult to control, because regeneration can occur from underground parts.

Dissemination: Weeds may be scattered in various ways. Some of the well-known methods by which this occurs include contaminants in crop seed, grain feed, hay and straw, seed carried by the wind, water, animals and farm machinery. Plants differ in their ability to produce seed, some do so in abundance, whereas others are meagre seeders. Dormancy plays an important role in the perpetuation of many weeds. It is essentially a state of suspended development which allows for germination over a long period of time and, possibly, when conditions are more favourable. The hard seed coat phenomenon of most legume seeds may be considered an advantage in that germination is also spread over a longer period of time. Many perennial weeds are propagated asexually. Plants with underground stems, such as *Elymus repens*, and others with stolons or tubers, are readily propagated from cuttings brought about by farm machinery. Such plants are difficult to eradicate.

Weed Damage: Essentially weeds effect losses by competing with crops for essential nutrients, light, water and space. Reduced light and space, with a concomitant reduction in photosynthesis, lead to direct crop losses. The presence of weeds can reduce land values, render cultivation more difficult and expensive, make crop handling and harvesting more difficult, interfere with the efficient working of machinery, increase overall production costs, and reduce crop values substantially. Many have poisonous properties which can cause death to man and livestock, or cause serious illness, or taint or impart a distasteful smell to animal products. Furthermore, many weeds act as hosts for numerous insects, fungi and viruses which can seriously damage crop plants. The need for weed control, therefore, is obvious. Effective control involves the recognition of the weed plants present and the adoption of appropriate control techniques.

Plant identification: The first step towards eradication is identification. The 'key' presented later provides for the identification of a wide range of plants (excluding grasses) which are common or frequent as weeds of tillage fields, rubbish heaps, waste places, and poorly managed grasslands. Use of the 'key' may be facilitated by the following brief review of the terminology used in plant description. The accompanying illustrations (Figs 01.01-01.60) are only diagrammatic representations of the most important features of each species; they are not intended as complete illustrations.

Leaves: Identification is based on leaf, flower and fruit features; growth-habit is also an important consideration. Leaves, which are produced on the stems of most flowering plants, vary greatly from the minute scale-like structures of *Equisetum* species (Figs 02.01-04) to the broad types associated with *Rumex obtusifolius* (Fig. 05.12). The form and arrangement are often characteristic of a species, enabling one to make identification by the leaves alone. They may be *simple* or *compound, entire, toothed* or *lobed, amplexicaul, sessile* or *stalked*, and *pubescent* or *glabrous*. Close examination is required to determine the presence or absence of hairs; these are often short and can be easily overlooked. In a few species they are confined to a particular part of the plant (*e.g.* Fig. 12.23). Pubescence or hairiness, *per se,* is not a very dependable feature.

Blade outline, or leaflet outline in the case of compound leaves, is a most useful characteristic. The shape of the leaf-base is invariably consistent for plants of any given species. The blade may be gradually or abruptly tapered towards the leaf-stalk, cut-off squarely, or 2-lobed; the lobes may be pointed or rounded. The manner in which the leaves are attached to the stem, or *phyllotaxy*, is an indispensable criterion; little variation is shown by this feature. Thus they are *alternate* - 1 per stem node and alternating from 1 side to the next, *opposite* - 2 per node, or *whorled* - 3 or more per node. Some plants, for example, *Bellis perennis*, only produce *radical* leaves, *i.e.* leaves which appear to emanate from the top of the root. Such leaves form a rosette.

The presence or absence of *stipules* - mostly leaf-like outgrowths form the base of the petiole - provides additional information. They are mostly present or absent at family level and differ widely in texture, shape, size and outline. Often an examination of the stipules is sufficient for identification. The Polygonaceae, for example, are characterised by their possession of sheathing stipules. Consistent variation is also found in leaf margins and venation. Continuous or unbroken outlines give entire blades; they are *toothed* when the divisions are confined to a peripheral zone. Teeth show distinct patterns. Deep divisions give *lobed* blades. Entire, toothed and lobed types constitute simple leaves; there are many different

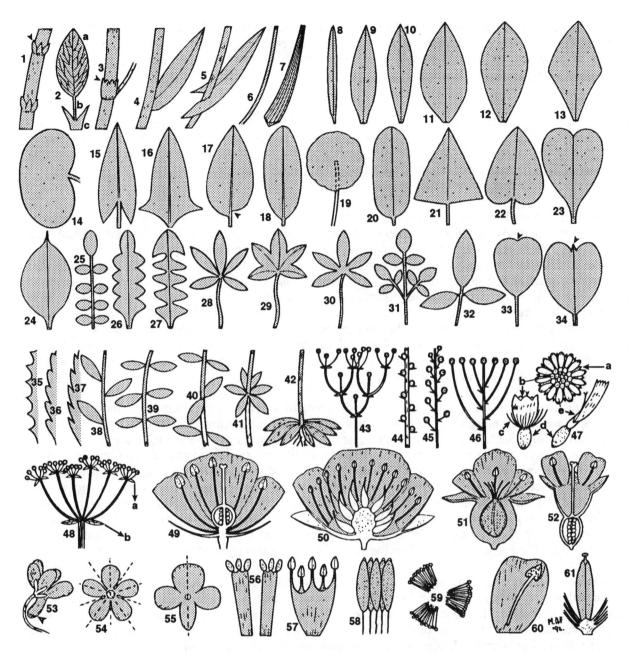

Legend: 1, scale-leaves; **2,** simple leaf showing (a) blade with reticulate venation, (b) petiole and (c) stipule; **3,** sheating stipule; **4,** sessile leaf; **5,** amplexicaul leaf (auriculate); **6,** bristle-like leaf; **7,** grass-like with 'parallel' venation; **8,** linear; **9,** lanceolate; **10,** oblanlceolate; **11,** ovate; **12,** obovate; **13,** rhomboid with cuneate base; **14,** reniform; **15,** sagittate; **16,** hastate; **17,** obligue; **18,** oval; **19,** peltate; **20,** oblong; **21,** deltoid or triangular; **22,** cordate base; **23,** obcordate; **24,** cuspidate; **25,** pinnnte; **26,** pinnately lobed; **27,** lyrate and pinnatifid; **28,** palmate; **29,** palmately lobed; **30,** palmatifid; **31,** biternate; **32,** trifoliate; **33,** emarginate apex; **34,** mucronate apex; **35,** dentate; **36,** crenate; **37,** serrate; **38,** alternate leaves; **39,** opposite leaves; **40,** opposite and decussate; **41,** whorled leaves; **42,** radical leaves; **43,** cyme; **44,** spike; **45,** raceme; **46,** corymb; **47,** capitulum (head) showing (a) an involucre, (b) tubular florets, (c) pappus, (d) ovaries and (e) ligulate florets; **48,** compound umbel showing (a) bracteoles and (b) bracts; **49,** hypogynous arrangement with superior ovary; **50,** perigynous arrangement, apocarpous pistil, showing superior ovaries; **51,** epigynous arrangement with inferior ovary; **52,** epigynous arrangement with an epigynous hypanthium; **53,** spurred flower; **54,** regular flower (actinomorphic); **55,** irregular flower (zygomorphic); **56,** diadelphous stamens; **57,** monadelphous stamens; **58,** syngenesious stamens; **59,** fascicled stamens; **60,** epipetalous stamens; **61,** syncarpous pistil.   N.B. DRAWINGS NOT TO SCALE

Fig. 01.00

types. Divisions which extend to the mid-ribs form *compound* leaves of different types. The principal types are described as being *pinnate*, *i.e.* divided in a feather-like manner, *palmate* or *digitate* when the leaflets are arranged like the fingers on a hand, *trifoliate* where the blade has 3 leaflets, or *ternate* when the leaf as a whole is divided into 3 main parts, which are often further divided.

Flowers: A simple flower has 4 parts - *sepals, petals, stamens* and 1 or more *carpels*. The first 2 are protective organs and, collectively, form the *perianth*. Ordinarily the sepals and petals are clearly differentiated but in some families, *e.g. Polygonaceae,* they are undifferentiated. One or both may be absent, reduced in size or modified, in some families, genera or species. Wind pollinated flowers are frequently *apetalous*. Finally, the number of each, their colour, shape, size, texture and whether they are united or free, provide indispensable taxonomic criteria.

Stamens, collectively the *androecium*, and carpels, collectively the *gynaeceum* or *pistil*, constitute the essential organs. Again, wide variation in the number of each. They may be united or free and their mode of attachment can provide further important information. When both are present in the same flower the latter is described as being *bisexual* or *hermaphrodite*; one or other gives *unisexual* forms. *Dioecious* species have male an female flowers on separate plants; *monoecious* forms have both types on the same plant. Stamens may be free or united in different ways: united by their anthers only - *syngenesious*; united by their filaments to form a tubular structure - *monadelphous*; united in 2 bundles - *diadelphous*, or occurring in several bundles - *fascicled*. In some plants, *e.g.* all *Veronica* species, they are attached to the petals, to give the *epipetalous* condition. Free independent carpels form an *apocarpous gynaeceum* or *pistil*; united carpels give a *syncarpous gynaeceum* or pistil. The syncarpous condition is considered to be more evolved than the apocarpous arrangement.

The relative arrangement of the floral parts differs with different plant groups. The most primitive - the *hypogynous* arrangement - has the ovary attached to the pedicel at a higher level than the other 3 parts. More evolved flowers show a *perigynous* arrangement, *i.e.* the sides of the receptacle are expanded in a saucer-like manner and the floral parts, other than the pistil, are attached to the lip. The most advanced condition - *epigynous* - is characterised by having the 'lip of the saucer' folded around the ovaries. The first 2 arrangements have *superior* ovaries, while the last shows an *inferior* ovary. Flowers with a multiplicity of floral parts indicate a primitive structure. Another feature of these is the hypogynous arrangement. The trend in plant evolution is towards a reduction in floral parts. The epigynous condition is associated with this trend. The most evolved flower is shown by the *Euphorbia helioscopia*. Here the male flower is reduced to a single stamen, while the female is represented by a solitary pistil.

Floral symmetry is an important consideration in plant identification. A flower is *regular* or *actinomorphic* when it is possible to define 2 or more planes of symmetry, and irregular or *zygomorphic* when only 1 plane is possible. In the interests of expediency only the petals need be examined. Flowers with petals of unequal size, except those of the *Veronica* species, are invariably regular, whereas those of equal size are regular. Furthermore, flowers with an unequal number of spurs or projections, such as those of the genera *Viola* and *Fumaria,* are also irregular.

An inflorescence, which is the arrangement of the flowers on a plant, may be of different types. Thus the flowers may be attached directly to the central stem to form a *spike* (*e.g. Plantago lanceolata*), by primary branches giving a *raceme* (*e.g. Sinapis arvensis*), or by secondary and tertiary branches to give a *panicle* (*e.g. Avena fatua*). In 2 taxa, the Asteraceae and Dipsacaceae, the flowers are attached directly but to an expanded central stem. This is the *capitulum* or compact 'head' and it is always subtended by 1 or more whorls of *bracts*, collectively known as the *involucre*. In the Asteraceae the flowers are either *tubular* or *ligulate*, and a capitulum may have one or other, or both, flower types. The Dipsacaceae have all tubular florets; in addition, each flower has a standard type calyx and an *epicalyx*. In another, the Apiaceae, carried on stalks which appear to radiate from a common level giving the *umbel* type inflorescence. Most members of this taxon have have compound umbels. Here, also, *bracts* and *bracteoles*, terms decided by their level of attachment, may be present or absent. The final racemose type inflorescence is the *spadix*, which is represented in only 1 family - the Araceae. Here the flowers are attached directly a fleshy axis which is enveloped in a *spathe* or leaf-like bract.

All the foregoing types are indeterminate inflorescences. Determinate or cymose types are more complex and difficult to identify. The essential characteristic of this group is the termination of each stem in a flower. They occur in several families, particularly in the Caryophyllaceae and the Solanaceae; the best example of a simple cyme is shown by *Stellaria holostea* (Fig. 12.26). Leaf and inflorescence features are illustrated in Figure 01.00.

KEYS TO FAMILIES, GENERA AND SPECIES

1. Plant *flower-bearing, i.e.* plants *other than* horsetails and ferns .... **3 (page 12)**
   Plant *not* flower-bearing, *i.e. horsetails* and *ferns* **2**

2. True leaves *absent;* minute scale-leaves present, whorled .... **Equisetaceae**
   True leaves, entire or pinnately divided, *present* .... **Polypodiaceae**

EQUISETACEAE

HORSETAIL FAMILY: Perennial herbs with creeping rhizomes which give rise to aerial stems at intervals. Stems either all alike, green and assimilating, or of 2 kinds, green and assimilating sterile stems and fertile stems without chlorophyll; all ridged or grooved. Branches whorled, ridged, readily disjoining at the nodes, resembling the stems or with whorls of slender green branches from the nodes; traversed by a central cavity with a ring of smaller cavities within the cortex (Clapham, Tutin and Warburg, 1962). True leaves absent. Scale-leaves very small, whorled, usually not green, combined into sheaths above the nodes, the former ending in free teeth; sheaths of the branches much smaller and with fewer teeth. Spores all alike, overlaid by 2 spiral bands which show hygroscopic movement. *Equisetum* is the only genus; there are 4 common species in the British Isles (Figs 02.01-02.04).

   a. Sterile stems *thin* and *green* .... b
      Sterile stems *robust* and *dirty white*, finely grooved, with numerous spreading branches; the latter 4-grooved. Fertile shoots unbranched. Rhizomes *pubescent*, often with pyriform tubers (Fig. 02.01). *Common in ditches, streams and other wet areas.* **Equisetum telmateia**

   b. Stem-cavity *one-half* or *less* of overall diameter c
      Stem-cavity large, at least *two-thirds* of overall diameter; the sterile and fertile stems similar, with or without branches, with numerous, scarcely perceptible grooves, smooth; rhizomes *glabrous* (Fig. 02.02). *Common in ditches, streamsides, rivers, etc.* **Equisetum fluviatile**

   c. Branches *hollow*, mostly erect, rarely absent (Fig.02.03). Stems 8.0-70.0 cm, erect or decumbent, with 4-8 deep grooves; the central cavity about the same size as the outer. Rhizomes *glabrous. Common; bogs; ditches; streams; marshes; shallow water.* **Equisetum palustre**
      Branches *solid*, numerous, spreading (Fig. 02.04). Stems 15-80 cm, erect or decumbent, with 6-18 grooves; central cavity about *twice* the size of the outer. Rhizomes *pubescent, tuberous. Common in tilled fields, hedgerows, roadsides, waste places, etc.* **Equisetum arvense**

POLYPODIACEAE

BRACKEN FAMILY: Leaves variable; entire, pinnate, bipinnate, tripinnate or pinnatifid; sporangia, which are *all alike,* present on abaxial surface of some leaves. Plants rhizomatous or tufted. This family contains, worldwide, about 170 genera 7000 and species. Two are of some significance in the British Isles, principally because of their poisonous properties (*see poisonous plants*). One, *Pteridium aquilinum*, is very common on dry sandy soils, while the second - *Dryopteris filix-mas* - is usually found in shaded habitats. There are 5 other frequent or common species (Figs 02.05-02.12) throughout the British Isles (for an alternative classificatory system for many of the species included in this family, see Stace, 1992).

a. Leaves *divided* .... b
Leaves *entire* (Fig. 02.05); the latter cordate basally, light green, glossy. Rhizomes densely covered with *brown scales. Common in moorlands, woods, etc.* **Phyllitis scolopendrium**

b. Leaves *pinnate* with *divided* pinnae .... e
Leaves *pinnatifid*, or *pinnate* but with *undivided* pinnae c

c. Lowest pinnae *slightly* shorter, *equal* to, or *larger*, than those in the middle .... d
Lowest pinnae *clearly* decreasing in size basally (Fig. 02.06). Leaves 40-100 cm, *bipinnate* or *tripinnate*, petioles and mid-ribs moderately covered with brown scales; pinnules undivided, oblong, rounded apically, more or less toothed laterally and apically; sori large, 3-6 on each pinnule. *Abundant; shaded hedgerows, woodlands, etc.* **Dryopteris filix-mas**

d. Leaves *3.0-15 cm* (Fig. 02.07); the latter tufted, mostly pinnate, *dull green* throughout except for the petiole base; petiole 1-2 times as long as blade; pinnae 3-5 each side; sori confluent to cover the entire lower surface. *Common in rock crevices.* **Asplenium ruta-muraria**
Leaves up to 200.0 cm (Fig. 02.08); the latter mostly *tripinnate*, arising from a pubescent rhizome, light *glossy* green; petiole pubescent when young, later glabrous; pinnules narrow-oblong, numerous and 'parallel', entire; sori continuous on segment margins. *Abundant; poor pastures, hedgerows, woods, especially those on sandy soils.* **Pteridium aquilinum**

e. Abaxial surface *sparsely* covered with, or *lacking*, scales .... f
Abaxial surface *completely* and *densely* covered with scales (Fig. 02.09); pinnae alternate, short, broad, entire or crenate, obtuse. *Frequent; mortared walls.* **Ceterach officinarum**

f. Petioles and mid-rib *neither* black, *nor* contrasting in colour with the pinnae .... g
Petioles and mid-rib *black, shiny,* and contrasting in colour with the pinnae (Fig. 02.10); pinnae oval to oblong, bluntly toothed. *Common; walls and rocks.* **Asplenium trichomanes**

g. Sori *continuous* on each side of the pinnae mid-rib (Fig. 02.11); leaves numerous, tufted, of 2 types; sterile with flat oblong, slightly curved pinnae; the fertile with narrower, longer, more widely spaced pinnae. *Common; woodlands, heaths, moorlands.* **Blechnum spicant**
Sori in *1* row on each side of the pinnae mid-rib (Fig. 02.12); leaves *solitary*, pinnatifid nearly to the mid-rib, dull green; pinnae oblong, entire or finely toothed, and *nearly equal* in length. Rhizomes creeping. *Frequent on walls, buildings, trees, etc.* **Polypodium vulgare**

3. Inflorescence *other than* a spadix .... **4** (page 13)
Inflorescence *a spadix* (Fig. 02.13), enclosed in a large *green, leaf-like sheath* .... **Araceae**

ARACEAE

ARUM FAMILY: Herbs, frequently with tuberous or elongated rhizomes, rarely woody and climbing. Plants often with a milky or watery sap, calcium oxalate crystals, or sharply pungent. Inflorescence a *spadix*. Flowers generally small, but *crowded* and *sessile* on a fleshy stalk; the latter mostly enclosed in a *spathe* or large leaf-like bract, and either hermaphrodite or *unisexual*. Unisexual flowers usually on *monoecious* plants. Perianth present in hermaphrodite flowers but *absent* from the unisexual. Stamens 2-8. Leaves alternate, simple, linear to ovate and then often cordate to sagittate basally. Worldwide, there are about 100 genera and 1500 species. There is 1 common species in the British Isles, and it is of importance because of its known poisonous properties (Fig. 02.13).

Leaves mostly *triangular-hastate* (Fig. 02.13); the latter long-stalked, and frequently spotted with purple; dark, shiny. Flowers unisexual; perianth *absent;* upper part of the axis *barren* and bright purple; male florets massed above the female; fruits of bright red berries. Plant tuberous. *A frequent perennial of woods and shaded hedgerows.* **Arum maculatum**

4. Leaves *not* grass-like, or very rarely so but corolla *coloured*, or *sepaloid* and *4-lobed* .... **7** (p 18)
  Leaves *somewhat* grass-like, *i.e. long* and *narrow,* 'parallel-veined', or *bristle-like* or *terete*, variously arranged but *not* whorled; sometimes blades absent and only *sheaths* present **5**

5. Flowers arranged in racemes .... **Juncaginaceae**
  Flowers arranged *other than* in racemes **6**

JUNCAGINACEAE

ARROWGRASS FAMILY: Annual or perennial marsh or aquatic herbs, *rush-like*, rhizomatous or with tuberous roots. Leaves *radical*, ligulate, linear. Inflorescence a few-flowered spike or *raceme*. Flowers small, hermaphrodite or unisexual, mostly *actinomorphic*. Perianth of *6 sepaloid* segments. Androecium of 6 stamens. Gynaeceum *syncarpous*, separating in fruit; ovary *superior*. Fruit a *follicle*. Worldwide, there are 3 genera and 12 species. There are 2 frequent species (Figs 02.14-02.15).

  Leaves deeply *furrowed adaxially* towards the base (Fig. 02.14); the latter *flat* or *convex* abaxially. The raceme slender, elongating after flowering; the flowers small, greenish. Rhizomes *long, slender. Very frequent plant; marshes and wet pastures.* **Triglochin palustris**
  Leaves *unfurrowed* adaxially (Fig. 02.15); the latter linear, *flat* adaxially and *convex* abaxially, somewhat fleshy. Racemes *short*, scarcely elongating; flowers greenish. Rhizomes *short, stout. Very frequent plant of salt marshes and muddy shores.* **Triglochin maritima**

6. Perianth segments *absent* or *bristle-like*; stamens usually *3* .... **Cyperaceae**
  Perianth segments *present*, dry and membranous; stamens *3-6* .... **Juncaceae**

CYPERACEAE

SEDGE FAMILY: Usually perennial, often rhizomatous. Leaves linear, *grass-like,* some or all often reduced to sheaths, often ligulate. Stems mostly solid, *triquetrous* or *terete*. Inflorescence mostly a 1-many-flowered *spike* or *spikelet,* rarely an umbel-like spike. Flowers hermaphrodite or unisexual, subtended by bracts or glumes. Spikes or spikelets solitary, terminal or grouped. Perianth of 1-many *bristles* or *scales,* or *absent.* Androecium usually of *3,* stamens. Gynaeceum *syncarpous,* often enclosed in an *utricle.* The fruit is a *nut.* The family, worldwide, contains about 100 genera and 3000 species. There are about 19 frequent or common species (Figs 02.16-03.06).

  a. Flowers *unisexual* .... h
    Flowers *hermaphrodite* .... b

  b. Bristles *6* or less, *shorter* than the glumes, and *not* silky .... d
    Bristles *numerous*, long and silky c

  c. Spikes *2* or more (Fig. 02.16); the latter nodding, many-flowered; bracts several; glumes lanceolate, acuminate; bristles up to 4 cm, whitish in fruit. Leaves 3-6 mm, *channelled*. Rhizomes very extensive. *Common perennial in moors and bogs.* **Eriophorum angustifolium**
    Spikes *solitary* (Fig. 02.17); the latter terminal, ovoid or globular; the bracts few or absent; glumes ovate-lanceolate, acuminate; bristles about 2.0 cm, white in fruit. Leaves 1-2 mm, *setaceous. Tufted-rhizomatous plant, common; moors, bogs, etc.* **Eriophorum vaginatum**

  d. Sheaths *tapered* to the blade, the latter long or very much reduced .... e
    Sheaths *truncate* apically (Fig. 02.18); the latter yellowish-brown. Leaves reduced to membranous sheaths at the stem bases. Spikes *solitary*, on stiff, wiry, angled, long peduncles; glumes several, ovate-oblong, obtuse, reddish-brown, with membranous margins. *Densely tufted, frequent perennial, of marshy ground and shallow water.* **Eleocharis palustris**

e. Spikes, all, *sessile* .... f

Spikes, at least some, clearly *branched* (Fig. 02.19); the latter 3 or more, subtended by a few rather short bracts. The leaves variable, but mostly reduced to sheaths. Glumes broadly ovate, often *fringed, emarginate* with rounded lateral lobes and a short awn. *Frequent rhizomatous plant of shallow water and neighbouring wet ground. **Schoenoplectus lacustris***

f. Sheaths *few* or *absent* entirely .... g

Sheaths *numerous* and crowded at the base of the plant (Fig. 02.20); the latter constricted and short-pointed apically, overlapping, shiny; leaves otherwise absent. Spikes solitary, terminal, brown, subtended by shorter, green bracts, with leafy tips; peduncles long, stiff. *A common densely tufted perennial of bogs, moors, heaths, etc. **Trichophorum cespitosum***

g. Bracts *2* (Fig. 02.21); the latter unequal, at least 1 with a leafy tip. The spikes *solitary*. Spikelets inflated, flattened, short-stalked; lower glumes asexual. Stems smooth, wiry, cylindrical. Leaves narrow, rigid, involute; the basal sheaths dark reddish-brown or almost black, shiny. Plant *densely* tufted. *A common perennial of bogs, heaths, etc. **Schoenus nigricans***

Bracts *solitary* (Fig. 02.22). Spikes mostly *2-3*, rarely solitary. Spikelets sessile; glumes mucronate; the bristles *absent*. Stems slender, short. Leaves short, narrow, tapered from base to a fine point. *A tufted annual plant of damp sandy or gravelly places. **Isolepis setacea***

h. Spikes *2* or more per stem i

Spikes *solitary* (Fig. 02.23). Stems very slender, rough, often curved. Leaves somewhat flat, narrow, tapered, dark green, shiny, mostly basal. Stigmas 2; utricles winged to base serrate above, plano-convex. *Frequent; bogs, heaths and mountain pasture. **Carex pulicaris***

i. Spikes clearly of *2* types l

Spikes more or less all *alike* in appearance .... j

j. Lowermost bract *4 cm* or less .... k

Lowermost bract *6 cm* or more (Fig. 02.24). The stems slender, weak, tufted. Leaves narrow, channelled, tapered to a long, pendulous point, mid-green; ligule blunt; sheaths persistent Spikes sessile, contiguous at top, remote in lower part; stigmas 2; beak very short, broad, split. *A very frequent plant in woods, hedgerows and damp shaded places. **Carex remota***

k. Utricle *narrowly* winged on its *upper* half (Fig. 02.25); the latter plano-convex, prominently nerved, bifid; beak about 1 mm, rough, bifid; stigmas 3; female glume broadly lanceolate, margins hyaline. Spikes ovoid, brown. Leaves soft, tapered to a *trigonous* point, mid to dark green; ligule blunt; sheaths persistent. *Common in wet upland pastures. **Carex ovalis***

Utricle *broadly* winged on its upper half, *narrowly winged* basally (Fig. 02.26), and distinctly nerved; wing serrate on upper half; beak 1-1.5 mm; female glume lanceolate, *broadly* hyaline. Spikes crowded. Leaves dark green, mostly flat, rigid, thick, tapered to a *trigonous* point; rhizomes extensive. *A frequent plant on sand-hills and sea shores. **Carex arenaria***

l. Stigmas *3* .... m

Stigmas *2* (Fig. 02.27). Female spikes oblong-cylindrical; the male slender, solitary; the male glumes *obovate*; female oval; all blunt, dark brown or *black*; utricle *truncate*, beak *absent*. Leaves *glaucous*, flat, tapered. *Very common; marshes and other wet areas. **Carex nigra***

m. Utricle *glabrous* .... p

Utricle *pubescent* or *papillose* .... n

n. Sheaths *glabrous* .... o

Sheaths *pubescent* (Fig. 02.28); leaves pubescent, dull green. Female spikes erect, stalked; utricle ending abruptly in a long, deeply bifid beak; female glume abruptly tapered from mid-way into a ciliate awn. *Very frequent in damp grasslands, marshes, etc. **Carex hirta***

N.B. DRAWINGS NOT TO SCALE

Fig. 02.00

o. Leaves *dark green* (Fig. 03.01); the latter flat, 1.5-2.5 mm, abruptly *trigonous*. Male spikes solitary, obovate; female 1-3, sessile or short-stalked; utricle *clearly pubescent*, slightly notched, *trigonous*, obovoid; beak short; female glumes broadly ovate, with an excurrent green mid-rib. *Very frequent plant in dry pastures and grassy heaths.* **Carex caryophyllea**
Leaves very *glaucous* (Fig. 03.02); the latter flat, gradually tapered. The male spikes oblong, mostly 2, dark brown; female stalked, cylindrical, mostly drooping; female glumes dark, oblong-ovate, dark, with excurrent green mid-rib; the utricle minutely *papillose, truncate* apically, short beaked. *Very common in pastures, heaths and grassy places.* **Carex flacca**

p. Female glume width *equalling*, or *nearly* equalling, that of the utricle .... q
Female glume much *narrower* than the utricle (Fig. 03.03); the former blunt; utricle inflated, faintly nerved, yellow-green, ovoid; beak bifid, smooth; stigmas 3. Female spikes dense, 2-5, suberect, cylindric, short-stalked. Male spikes 2-4, narrowly cylindrical. The leaves rough, glaucous adaxially, dull green abaxially. *Common in marshy areas.* **Carex rostrata**

q. Female spikes *erect* .... r
Female spikes *pendulous* (Fig.03.04); the latter very slender, rather lax, long-stalked; female glume ovate-lanceolate, tapered, mid-rib green and hyaline margins; the utricle obovoid-trigonous, with 2 prominent nerves; the beak long, bifid. Leaves yellowish-green, tapered to a sharp point, soft. *Very frequent in woods and other shaded places.* **Carex sylvatica**

r. Utricle with *prominent* lateral nerves (Fig. 03.05); the latter broadly elliptic, purple-brown, rarely green; the beak long, rough, deeply bifid; female glume short-awned. Female spikes well spaced, stalked, cylindric, mostly erect. Leaves up to 6 mm wide, abruptly and finely tapered. The stem trigonous-terete. *Common; upland moors and heaths.* **Carex binervis**
Utricle *indistinctly* nerved, trigonous-ellipsoid and rounded basally (Fig. 03.06); the beak distinct, bifid. Female spikes well spaced, stalked, erect; female glume ovate-oblong, brown with greenish mid-rib, tapered to an acute point. Leaves up to 6 mm wide, flat, tapered to a fine point, grey-green. *Frequent in salt marshes and damp coastal areas.* **Carex distans**

JUNCACEAE

RUSH FAMILY: Annual to perennial herbs, glabrous or sparsely hairy, frequently tufted or with rhizomes. Leaves mostly cylindrical and *terete*, rarely *grass-like*, with sheathing bases, alternate, exstipulate, and with or without internal partitions; or terminal portions absent and sheaths only present. Inflorescence *panicles, corymbs, cymes,* or *heads*, or of solitary flowers. Flowers hermaphrodite, with the hypogynous arrangement, actinomorphic. Perianth *undifferentiated* with 6 greenish, brownish or membranous *tepals*, in 2 whorls of 3. Androecium usually of *6* stamens. Gynaeceum *syncarpous;* ovary *superior.* Fruit a *capsule*; seeds 3-many, very small, often with appendages. There are about 8 genera and 350 species, worldwide. The following species are frequent or common (Figs 03.07-03.19).

a. Plant *glabrous* .... d
Plant *pubescent* .... b

b. Leaves *6* mm wide or less c
Leaves *10* mm or more wide (Fig. 03.07); the latter *grass-like,* mostly radical, up to 30 cm x 20 mm, sparsely long-haired, glossy, broadly linear, gradually tapered to an acute point. Plant tufted. Flowering stems up to 80.0 cm; flowers brown, numerous, in a loose spreading panicle; branches spreading. *Common in woods, open moorlands, etc.* **Luzula sylvatica**

c. Plant shortly *stoloniferous* (Fig. 03.08). Leaves *grass-like*, linear, usually 2.0-4.0 mm wide, sparsely covered with long whitish hairs. Flowers small, brown, 3-12 together, in 1 sessile and 3-6 spherical-obovate clusters; the latter in a lax or sometimes drooping panicle; filaments *shorter* than anthers. *Common in heaths, open moorlands, etc.* **Luzula campestris**

Plant *cespitose* (Fig. 03.09). Leaves *grass-like*, 3-6 mm wide, sparsely covered with numerous, long, whitish hairs. Flowers small, brown, 8-16 together, in ovate or elongate clusters; the latter on erect slender branches, mostly in umbel-like arrangements; filaments as long as the anthers. *Very frequent in moorlands and upland heaths.* **Luzula multiflora**

d. Leaf-blades *present* .... g
    Leaf-blades *absent* .... e

e. Stems with a *continuous* core of pith .... f
    Stems with an *interrupted core* of pith (Fig. 03.10); the former in a dense tuft, mostly *glaucous*, ribbed, wiry, stiff. Leaf-sheaths brown or blackish, glossy. Flowers brown, small, in *1-sided*, loose cymes. Rhizomes short. *Common on wet calcareous soils.* **Juncus inflexus**

f. Stems, when fresh, *smooth* or *scarcely ribbed* (Fig. 03.11), yellowish green or often bright green, rather soft, glossy, densely tufted, stiffly erect. The leaf-blades absent, sheaths reddish to dark brown. Inflorescence apparently lateral, well down from stem apex, 1-sided, mostly lax, cyme; capsule blunt. *A common perennial in wet acid soils.* **Juncus effusus**
    Stems, when fresh, especially below the inflorescence, strongly and coarsely *ribbed* (Fig. 03. 12); the former bright to greyish-green, dull. Leaf-sheaths brown or blackish in colour, and width expanded basally. Flowers in a *dense, compact* 1-sided cluster; capsule shortly mucronate. *Frequent in moorlands and damp upland grasslands.* **Juncus conglomeratus**

g. Leaf-blades *without* septa .... j
    Leaf-blades *with* distinct or *scarcely discernable* septa .... h

h. Flowering stems with *numerous* flower clusters in a *well-branched* cyme i
    Flowering stems with a *few* sessile or *short-stalked* clusters (Fig. 03.13); the former about 15 cm, very slender, often slightly swollen basally. Leaves narrow, *needle-like*, bases sheathing. *Common in upland bogs and wet heaths, especially on acid soils.* **Juncus bulbosus**

i. Leaves *strongly* laterally compressed (Fig. 03.14); the latter hollow, deep green, with up to 25 *inconspicuous* septa. The stems variable; rhizomes slender. Flowers in little-branched panicles; perianth segments lanceolate, outer *acute*, inner *obtuse*; capsule *abruptly* short-pointed. *Common in marshy ground, especially those with acid soils.* **Juncus articulatus**
    Leaves *slightly* compressed and subterete (Fig. 03.15); the latter deep green, with up to 25 *conspicuous* septa. The stems stiffly erect; rhizomes extensive. Flowers in panicles, well-branched; the perianth segments *gradually* tapered to fine apices; capsule also *gradually* tapered. *Common in marshy ground, especially those with acid soils.* **Juncus acutiflorus**

j. Inflorescence *confined* to the upper parts of the stems .... k
    Inflorescence *scattered* over most of the plant (Fig. 03.16); the latter of many small, leafy, sessile or short-stalked, few-flowered clusters. The stems many, tufted, slender. Leaves bristle-like, mostly all basal, *channelled. Common in wet muddy places.* **Juncus bufonius**

k. Leaves *few*, at least some cauline l
    Leaves *numerous*, all radical (Fig. 03.17); blades of the latter *stiff*, channelled, glossy, very narrow, in a dense basal tuft; sheaths broad. Stems stiff, slender, erect. Flowers in narrow, somewhat lax terminal cymes. *Common in moors and upland heaths.* **Juncus squarrosus**

l. Leaves and bracts *ending* in needle-like points (Fig. 03.18); the former similar to the stems but shorter. The stems tufted, very stiff, wiry, ending in sharp points. Flowers in a 1-sided, rather loose panicle. *Very frequent in salt marshes and sandy shores.* **Juncus maritimus**
    Leaves *not* ending in needle-like points (Fig. 03.19); the former narrow, dark green, usually channelled, mostly basal. The stems erect, slender, arising from a slender extensive rhizome. Flowers in loose terminal panicles. *Frequent; wet coastal areas.* **Juncus gerardii**

7. Flowers arranged *other than* in capitula .... **9** (page 25)
    Flowers arranged *in* capitula which are subtended with bracts **8**

8. Epicalyx *present;* anthers *free* .... **Dipsacaceae**
    Epicalyx *absent;* anthers *united* .... **Asteraceae**

DIPSACACEAE

TEASEL FAMILY: Annual, biennial or perennial herbs, rarely shrubs. Leaves opposite, rarely whorled, *exstipulate*. Inflorescence usually a *capitulum*, subtended by a common *involucre* of bracts. Flowers *hermaphrodite,* slightly *zygomorphic, epigynous*, each surrounded by an *epicalyx*, or 'involucel' of united bracteoles. Corolla *gamopetalous*, with the tube often curved, and the 4-5 lobes more or less equal or forming 2 lips. Calyx small, variable, often cup-shaped or more or less deeply cut into 4-5 segments, or into numerous teeth or hairs. Androecium of 4 or 2 epipetalous stamens, which alternate with the corolla lobes. Gynaeceum *syncarpous*; ovary *inferior*. Fruit an *achene*. Worldwide, there are about 9 genera and 160 species. There are 2 common or frequent species (Figs 03.20-03.21).

> Leaves *pinnatifid* or pinnate (Fig. 03.20); the latter lanceolate, densely covered with reflexed bulbous hairs, opposite, connate. Capitula long-stalked; corolla *blue* or *pale lilac*; those of the outer flowers with larger, spreading, and more irregular lobes; the calyx of 8 ciliate teeth; stamens 4. *Frequent: dry pastures and dry gravelly waste places.* ***Knautia arvensis***
> Leaves *mostly* entire (Fig. 03.21); the latter mostly radical, oblanceolate, tapered into long stalks, narrowed apically, glabrous or scarcely pubescent; cauline few, narrower. Capitula long-stalked; the corolla *purplish-blue* or rarely white or pink; the calyx of 4 bristle-like teeth. *Abundant: marshes, upland pastures, heaths and cut-away bogs.* ***Succisa pratensis***

ASTERACEAE

DAISY FAMILY: Herbaceous or sometimes woody plants of diverse habit, often with oil-canals or latex. Leaves *exstipulate;* alternate, opposite or rarely whorled, frequently in a basal rosette, compound or simple, or pinnately or palmately lobed. Inflorescence usually a *capitulum*, subtended by an *involucre* of 1 or more whorls of bracts. Corolla *gamopetalous*, of 5 petals, and of 3 main types: (i) *tubular, actinomorphic*, the corolla-tube being surmounted by 5 more or less equal teeth; (ii) *tubular, zygomorphic*, the upper portion 2-lipped; (iii) *ligulate, zygomorphic*, the upper portion long, narrow and *strap-shaped*. Composite heads consisting of either tubular or ligulate, or both, floret types. Calyx modified into a tuft of numerous, simple or feathery, silky hairs (pappus) which may be present or absent; rarely represented by a ring of short teeth or bristles. Androecium of 5 epipetalous, *syngenesious* stamens. Gynaeceum of 1 carpel; ovary *inferior,* 1-celled. Fruit an *achene,* mostly *ribbed.* This is a very large family with over 900 genera and 14000 species, worldwide; it is divided into a number of subfamilies. An alternative legitimate name, Compositae, may also be used for this family. There are about 40 common or frequent species throughout the British Isles (Figs 03.22-05.06).

> (1). Leaves *not* grass-like; venation *reticulate* .... (3)
> Leaves somewhat *grass-like, i.e.* long, narrow, entire, and 'parallel-veined' .... (2)
>
> (2). Heads *bicolour*, corymbose (Fig. 03.22). The ligulate florets spreading, *blue-purple,* rarely *whitish* or *absent*; tubular florets *yellow.* Basal leaves oblanceolate to obovate, tapering into a long petiole; cauline leaves narrowly oblong to linear, tapered apically; all mostly entire, *fleshy*, glabrous, and *3-veined. Common perennial; salt-marshes.* **Aster tripolium**
> Heads *monocolour*, large, solitary on stem and branches (Fig.03.23). Stem with latex. Tubular florets *absent*, ligulate *yellow.* The basal leaves linear-lanceolate, tapered apically into a long point, widened and somewhat sheathing at the base, clearly veined, glabrous; cauline semi-amplexicaul. *A frequent perennial; grassy places.* **Tragopogon pratensis**

N.B. DRAWINGS NOT TO SCALE

Fig. 03.00

(3). Leaves *entire* or variously divided, but *segments not linear* .... (6)

Leaves finely *divided* into *linear segments*; pappus *not* of a tuft of hairs .... (4)

(4). Ray florets *present* .... (5)

Ray florets *absent* (Fig. 03.24); tubular florets *greenish-yellow*, arranged in *cone-shaped heads;* the latter short-stalked; the bracts with membranous margins and apices. Pappus represented by an *obscure membranous, toothed rim*. The achenes marked with 4 inconspicuous ribs. Stems glabrous, green, often purplish basally. The leaves alternate, *broad and flat* basally, *bi-* or *tripinnate;* the ultimate segments *linear, bristle-pointed. An aromatic annual, very common in farm-yards, waste places, etc. **Matricaria matricarioides**

(5). Heads large, *solitary,* on long terminal peduncles (Fig. 03.25). The ray florets *numerous, white, spreading;* tubular *yellow, several.* Pappus a *narrow,* nearly *entire rim.* Achenes with *3 wide ribs* on the inner face, and *2 dark oil-glands* on the outer face. Stems erect, spreading or prostrate. The leaves spirally arranged, *bi-* or *tripinnate,* segments *linear,* mostly sessile, glabrous. *Common biennial; waste places.* **Tripleurospermum inodorum**

Heads small, *numerous,* in *dense terminal corymbs* (Fig. 03.26). Stems erect, ribbed, somewhat woolly. Ray florets usually *5,* mostly *white,* rarely mauve, pink or reddish; tubular florets *white* or *creamy-white.* Basal leaves stalked; cauline sessile; all *bi-*or *tripinnate,* more or less woolly, the ultimate segments *linear* and *pointed. A very common stoloniferous, aromatic plant of roadsides, old pastures, waste places, etc.* **Achillea millefolium**

(6). Flowers *yellow, yellow* and *white* or *greenish-white* .... (16)

Flowers *purple, reddish-purple, light-purple* or *pinkish-mauve* .... (7)

(7). Leaves very *prickly* .... (12)

Leaves *smooth* or *roughly hairy* .... (8)

(8). Leaves *alternate* or *spirally* arranged .... (11)

Leaves *basal* or *opposite* .... (9)

(9). Leaves, all, *basal* .... (10)

Leaves, at least some, *opposite* (Fig.03.27). Basal leaves oblanceolate, stalked. Stem leaves, some, divided into 3-5 toothed, glandular segments. The flowers in dense compound clusters, and either pink, reddish or rarely grey-white; all tubular. *A frequent, scarcely branched perennial, of wet marshy ground and rocky places.* **Eupatorium cannabinum**

(10). Flowers *lilac* or *pale pinkish-mauve,* and carried in short racemes (Fig. 03.28); flowering stems up to 30 cm. Leaves somewhat rounded, cordate basally, long-stalked, not angled, serrate, 10-15 cm, rarely up to 20.0 cm across, green adaxially and abaxially; somewhat downy abaxially. *Rhizomatous perennial, common in many habitats.* **Petasites fragrans**

Flowers *dull purple* or *reddish-violet,* and carried in long dense racemes (Fig. 04.01); flowering stems up to 80.0 cm. Leaves similar to the above but much larger, up to 100.0 cm, mostly radical, long-stalked, roundish, deeply cordate, downy; petioles stout, hollow. *A frequent perennial of riversides, streamsides and other wet places.* **Petasites hybridus**

(11). Heads in *racemes* or small *clusters* (Fig.04.02). Bracts ending in *hook-shaped bristles.* Florets tubular, reddish-purple, the wider upper part of the corolla about equalling the filiform lower part. Basal leaves ovate-oblong, entire or distantly toothed, nearly glabrous adaxially, cottony abaxially. *A common biennial plant of waste places.* **Arctium minus**

Heads *solitary,* subsessile (Fig. 04.03). Florets all tubular, and purple in colour. Bracts *not* ending in hook-shaped bristles; the latter, at least the outer, comb-like. Stems grooved and somewhat hairy.The basal leaves petiolate, entire, sinuate-toothed or somewhat pinnatifid; cauline leaves sessile or with a few teeth basally; softly or roughly pubescent. *A perennial, common in pastures, meadows and waste grassy places.* **Centaurea nigra**

(12). Pappus *branched* .... (13)

Pappus *unbranched*, though toothed (Fig. 04.04). Heads cylindrical, sessile, in dense terminal clusters. Florets all tubular, pale purple-red. Stems continuously spinous-winged. All basal leaves oblanceolate, blunt, and narrowed basally; cauline leaves decurrent; all sinuate-pinnatifid with prickly margins, and somewhat cottony abaxially. *A frequent perennial of sandy ground and dry banks, especially near the sea.* **Carduus tenuiflorus**

(13). Blades, adaxial surfaces, *smooth* .... (14)

Blades, adaxial surfaces, *rough* or *prickly* (Fig.04.05). Basal leaves obovate-lanceolate, tapered into a short stalk-like base, often deeply pinnatifid, and with segments usually 2-5 lobed; the upper lobe toothed towards the base, the lower entire; cauline leaves similar but all sessile and decurrent, and with a *spear-like terminal lobe;* all with prickly margins. Heads solitary or in clusters of 2-3; florets tubular, pale red-purple. *Biennial, it is a common plant in many meadows, pastures, gardens and waste areas.Cirsium vulgare.*

(14). Flower heads *clustered* .... (15)

Flower heads *solitary* (Fig. 04.06). Florets all tubular, dark or red-purple. Stems unwinged, cottony, with a few bract-like leaves on the upper halves. Basal leaves elliptical lanceolate, long stalked, sinuate-toothed or somewhat pinnatifid; cauline leaves few, oblong-lanceolate, and semi-amplexicaul; all green adaxially, white cottony abaxially, and with soft marginal prickles. Plant shortly stoloniferous with obliquely ascending stock; stems simple, unwinged. *A perennial, frequent in wet or peaty pastures.* **Cirsium dissectum**

(15). Stems *continuously* spiny-winged (Fig. 04.07); the latter cottony, and with few ascending branches. Heads short-stalked, and clustered at the end of the main stem and branches. Florets all tubular, and mostly dark red-purple. Basal leaves narrowly oblanceolate, petiolate, pinnatifid; the lobes shallow; stem leaves sessile, long-decurrent, and deeply pinnatifid and undulate; all leaves with long marginal spines, pubescent adaxially, slightly cottony abaxially. *A common biennial in marshes and wet grasslands.* **Cirsium palustre**

Stems *not* continuously winged (Fig. 04.08). Rhizomatous plant with whitish lateral roots, and numerous adventitious non-flowering and flowering unwinged, or sparingly winged stems. Heads stalked, and mostly in terminal clusters. Florets all tubular, and dull pale purple. The basal leaves in a compact rosette, oblong-lanceolate, narrowed basally into a short stalk, mostly pinnatifid; the lobes undulating, triangular; the cauline leaves similar but sessile and semi-amplexicaul. *Perennial, abundant in all habitats.* **Cirsium arvense**

(16). Leaves, at least some, *cauline* .... (23)

Leaves, all, *radical* .... (17)

(17). Ray florets *yellow* or *absent* entirely .... (18)

Ray florets *white*, often *tinged* with pink (Fig. 04.09). Tubular florets bright *yellow*. Flowering stems (scapes) up to 15.0 cm, rarely more, leafless, pubescent, and carrying solitary heads. Leaves in a basal rosette, spatulate-obovate, broad and rounded apically, tapered basally, pubescent, toothed. *A perennial with a short erect rootstock and stout fibrous roots; regenerates easily after spraying; common; many habitats.* **Bellis perennis**

(18). Stolons *absent* .... (19)

Stolons, long and pubescent, *present* and terminating in rosettes (Fig. 04.10). The flowering stems (scapes) leafless, with blackish glandular hairs. Heads solitary. Florets ligulate, yellow. Leaves entire, densely covered with long white stellate hairs, and mostly oblanceolate in general outline. Stolons numerous, bearing small distant leaves, terminating in rosettes. *Frequent perennial of wall-tops and other dry habitats.* **Hieracium pilosella**

(19). Heads *2* or more per flowering stem .... (22)

Heads *1* per flowering stem .... (20)

(20). Blades *not* cordate .... (21)

Blades *cordate* basally (Fig. 04.11); the latter petiolate, densely covered with white cottony hairs abaxially, more or less rounded but shallowly lobed; petioles grooved, and lobes toothed. The heads solitary, carried on scaly stems; the florets, both ligulate and tubular, yellow. *Common perennial of tillage fields, waste places, roadsides.* **Tussilago farfara**

(21). Bracts, outer, *reflexed* or *spreading* (Fig.04.12). The scapes smooth, erect, hollow, and containing latex. Heads solitary; florets ligulate, and yellow; the achenes prickly distally, and long-beaked. Leaves variable, but mostly lyrate, rarely shallowly toothed or nearly entire; mostly glabrous. *Ubiquitous perennial plant.* **Taraxacum officinale, sensu lato**

Bracts, outer, *neither* reflexed *nor* spreading (Fig. 04.13). Scapes ascending from a decumbent base, leafless, glabrous or sparsely covered with bifid hairs; latex absent; heads solitary; the florets ligulate, yellow. Leaves mostly oblanceolate, tapered basally, unevenly pinnately lobed or nearly entire, and mostly with dense bifid hairs; or rarely nearly glabrous. *Frequent perennial; gravelly pastures and sandy places.* **Leontodon taraxacoides**

(22). Scales *present* between the florets (Fig. 04.14). The pappus hairs in *2 rows*, the inner *feathery*, the outer *shorter* and *simple*. Flowering stems usually branched, with a few small scales, and often thickened apically. Florets ligulate, yellow. Blades oblong-lanceolate, tapered basally, sinuate-toothed to shallowly pinnatifid, dull green adaxially, and covered with short stiff hairs. *Abundant perennial; most grasslands.* **Hypochoeris radicata**

Scales *not* present between the florets (Fig. 04.15). Pappus hairs in a *single row* of *feathery* hairs. The flowering stems usually branched, ascending, with small scale-like bracts, and either glabrous or somewhat pubescent. Flowers ligulate, golden yellow, the outer red-streaked abaxially. Leaves either distantly toothed or deeply pinnatifid, glabrous or simple-haired. *A common perennial of many damp grasslands.* **Leontodon autumnalis**

(23). Tubular florets *absent* .... (33)

Tubular florets *present* .... (24)

(24). Ray florets some colour *other than* white, or absent entirely .... (25)

Ray florets *white* and tubular florets *yellow* (Fig. 04.16). Flowering stems erect, simple or branched, sparsely pubescent or glabrous. Heads solitary, long-stalked. Basal leaves oblanceolate to spatulate, crenate-dentate, long-stalked; the upper small, distantly toothed, very rarely nearly entire; sessile and semi-amplexicaul; all dark green and sparsely pubescent. *Perennial, abundant in several dry grassy situations.* **Leucanthemum vulgare**

(25). Pappus *present* .... (28)

Pappus *absent* or else consisting of *4 short teeth* or *bristles* .... (26)

(26). Pappus *absent* .... (27)

Pappus represented by *4 short bristles* (Fig. 04.17). Flowering stems erect. Heads solitary, long-stalked. Ray florets usually absent, the tubular florets yellow; bracts unequal, 3-6, leaf-like, short stiff hairs on margins; others small, in 2 rows. The leaves sessile, roughly serrate, glabrous or pubescent. *A frequent annual of marshy places.* **Bidens cernua**

(27). Heads very large, *few* (Fig. 04.18). Ligulate florets *present* and *conspicuous*; tubular florets also yellow. Flowering stems erect, simple or branched; bracts with broad membranous margins. Basal leaves coarsely toothed or pinnatifid, tapered basally; upper sessile, oblong, toothed or nearly entire, and semi-amplexicaul; all *glaucous, glabrous,* and somewhat *fleshy. Calcifuge; annual; frequent in broken soils, etc.* **Chrysanthemum segetum**

Heads small, *numerous*, and arranged in *panicles* or *racemes* (Fig. 04.19). Ligulate florets *absent*. Tubular florets yellow and tinged with red. Stems erect, glabrescent or sparsely hairy, grooved, reddish, angled. Basal leaves short-stalked, tomentose abaxially, deeply divided, segments narrow; cauline sessile. *Perennial; waste places.* **Artemesia vulgaris**

Fig. 04.00

N.B. DRAWINGS NOT TO SCALE

(28). Leaves *not* prickly .... (29)
Leaves very *prickly* (Fig. 04.20). Basal leaves oblong-lanceolate, acute, tapered to a sessile base, cottony especially abaxially; cauline leaves semi-amplexicaul, broader-based and shorter; all pinnately lobed with spinous lobes. Heads 2-5, rarely solitary; inner bracts *like* ray florets, many, long-linear, straw-coloured, spreading when dry; florets yellowish. Stems stiffly erect, purplish, cottony. *A biennial of gravelly places.* **Carlina vulgaris**

(29). Leaves *deeply* pinnatifid or clearly lobed .... (30)
Leaves *distantly* toothed or nearly entire (Fig. 04.21). Basal blades tapered basally; cauline blades strongly amplexicaul, and oblong-lanceolate; all densely covered with soft hairs. Flowering stems erect, sparsely pubescent. Heads mostly in a loose corymb. Ray florets numerous, linear, yellow. *A frequent perennial of damp habitats.* **Pulicaria dysenterica**

(30). Ray florets *conspicuous* .... (32)
Ray florets *inconspicuous* or *absent entirely* .... (31)

(31). Ray florets *absent* (Fig. 04.22); tubular *yellow.* Heads, at least some, *short-stalked*, in terminal and axillary clusters. Stems erect or ascending, succulent, grooved, glabrous or thinly clothed with rather long cottony hairs. Blades *somewhat glaucous, glabrous* or *cottony*; pinnatifid, with distant, oblong, blunt, irregularly toothed lobes; lower short-stalked, upper semi-amplexicaul. *An ubiquitous annual of broken soils.* **Senecio vulgaris**
Ray florets *inconspicuous, revolute* (Fig. 04.23); florets yellow. Heads, all, *clearly stalked*, and in a terminal flat-topped corymb. The stems erect, furrowed and mostly pubescent. Blades *yellowish-green, woolly* at first; the upper sessile or semi-amplexicaul; all deeply and irregularly pinnatifid; toothed. *Frequent annual of dry acid soils.* **Senecio sylvaticus**

(32). Heads in *irregular loose corymbs* (Fig.04.24); the latter 2.0-3.0 cm in diameter, and carried on relatively long branches. Ray florets very conspicuous; all florets yellow. Flowering stems erect, up to 90.0 cm, and glabrous or cottony distally. Basal leaves petiolate, not divided to lyrate-pinnatifid, with large terminal lobe; laterals smaller; cauline leaves, at least some, semi-amplexicaul; margins crenate to coarsely serrate, and more or less glabrous. *Poisonous; common biennial plant; wet grasslands, ditches.* **Senecio aquaticus**
Heads in *dense flat-topped corymbs* (Fig. 04.25); the latter 0.5-1.0 cm in diameter, and carried on relatively short branches. Ray florets conspicuous; all florets yellow. Flowering stems erect, up to 150 cm, and glabrous or cottony. Basal leaves in a dense rosette, petiolate, lyrate-pinnatifid, terminal lobe large, laterals smaller; cauline pinnatifid to bipinnatifid, petiolate or semi-amplexicaul; dark green, sinuate-toothed, glabrous to sparsely cottony abaxially. *Poisonous; ubiquitous biennial of many habitats.* **Senecio jacobaea**

(33). Pappus *present* .... (34)
Pappus *absent* (Fig.04.26). Stems erect, with spreading hairs basally, more or less glabrous distally. Basal leaves long-stalked, lyrate-pinnatifid, with a large, ovate-cordate, toothed terminal, and smaller lateral lobes; the cauline short-stalked, toothed or entire; all pubescent. Heads many, in a corymbose panicle. Florets all ligulate and yellow; the pappus absent. *Common annual; tillage fields, waste ground, gardens, etc.* **Lapsanna communis**

(34). Stems *not* angled; latex *absent* .... (37)
Stems *angled*, hollow, succulent; latex usually *present* .... (35)

(35). Bracts and peduncles *not* densely covered with glandular hairs .... (36)
Bracts and peduncles *densely* covered with yellowish glandular hairs (Fig. 05.01). Stems angular, hollow, with glandular hairs distally. Basal leaves tapered basally, pinnatifid; lobes triangular, spinous-ciliate; cauline leaves similar but sessile, less well divided, and with *rounded, appressed auricles.* The florets all ligulate, yellow. *A rhizomatous perennial, common; tillage fields, waste places, hedgerows, roadsides, etc.* **Sonchus arvensis**

(36). Blades *prickly* (Fig. 05.02); the basal leaves pinnatifid, petiolate, usually in a rosette, some-what spatulate; the cauline similar but sessile, and with *rounded, appressed auricles;* all dark glossy-green adaxially, often tinged with purple, *crisped* and *sharply spinous cili-ate* on the margins. Stems angular, succulent, with a milk-like juice, glabrous. Florets all yellow. *A common annual plant of cultivated ground, waste places*, etc. **Sonchus asper**

Blades *not* prickly (Fig. 05.03); the basal leaves petiolate, irregularly pinnatifid, usually in a rosette, with a large terminal lobe; cauline similar but with *spreading auricles*; all var-iable, more or less glaucous, with finely toothed lobes, never spinous, mostly glabrous or cottony when young. Stems angular, hollow, succulent, often purplish.Florets yellow. *A very common annual plant of cultivated ground, waste places, etc.* **Sonchus oleraceus**

(37). Florets *10 to many* .... (38)

Florets *few*, mostly *4-5* (Fig. 05.04); the latter ligulate, yellow. Heads in a large open pani-cle. The stems erect, slender, glabrous, often covered with *'bloom'*. Basal leaves on long-winged stalks, lyrate-pinnatifid; terminal lobes large; the cauline amplexicaul; all gla-brous, with toothed lobes. *Frequent biennial of tilled fields, wall-tops.* **Mycelis muralis**

(38). Stems, branches, peduncles and leaves *densely* covered with *short, dark, stiff hairs* (Fig. 05. 05). Stems erect, ribbed, branched. Basal leaves petiolate, lyrate-pinnatifid; lobes vari-able; cauline pinnatifid or nearly entire; some amplexicaul. Heads 1.5-2.5 cm, *many*, in corymbs; florets ligulate, yellow; styles *brownish-green;* the outer involucral bracts fre-quently *spreading horizontally. Biennial, common in dry grassy places.* **Crepis vesicaria**

Stems, branches, peduncles and leaves *glabrous* or only *sparsely* hairy (Fig. 05.06). The former erect or ascending; branches *few*. Basal leaves lyrate-pinnatifid, lobes toothed, mostly tapered basally; middle cauline lanceolate, amplexicaul. Heads 1-1.5 cm, *few*, in a *lax* corymb; florets all ligulate, yellow; styles *yellow*; the outer involucral bracts most-ly *appressed. A common annual or biennial of dry grassy places, etc.* **Crepis capillaris**

9. Stipules *absent*, or *minute* but then leaves *entire* and either *opposite* or *whorled* .... **17** (page 33)

Stipules *present* **10**

10. Corolla *funnel-shaped* .... **17** (page 33)

Corolla *absent*, or present but *not* funnel-shaped **11**

11. Perianth segments *3* or *absent* .... **17** (page 33)

Perianth segments *more* than 3 in number **12**

12. Perianth *differentiated* into petals and sepals .... **14** (page 27)

Perianth *not differentiated* into petals and sepals **13**

13. Leaves *opposite* .... **Urticaceae**

Leaves *alternate* .... **Polygonaceae**

URTICACEAE

NETTLE FAMILY: Herbs, small shrubs or rarely soft-wooded trees. Leaves simple, opposite or alter-nate, mostly with small *stipules*, toothed and usually with *stinging* hairs. Inflorescence mostly *cymes*, rarely modified into a head by loss of pedicels, or reduced to a single flower. Flowers small, *actino-morphic*, with the *hypogynous* arrangement, generally unisexual. Perianth *undifferentiated*. Corolla *absent*. Calyx *gamosepalous* of 4 green, rarely 5, sepals. Androecium of male flowers with 4-5 stamens; rudimentary ovary usually present. Gynaeceum of female flowers with a 1-celled, *superior* ovary; small staminodes often present; style simple; ovule solitary. Fruit an *achene*, rarely a drupe. Plants *monoecious* or *dioecious*. Worldwide, the family contains about 45 genera and 600 species. Two species are very common throughout the British Isles (Figs 05.07-05.08).

Leaves, at least some, *cordate* basally (Fig. 05.07); the latter acuminate, crenate-serrate, petiolate, opposite, covered with *stinging hairs;* stipules *small, narrow.* Flowers in axillary
panicles, small, green; corolla *absent;* calyx of *4 obovate* sepals. Plant *dioecious.* Stems
up to 1.0 m, or more on rich soils or shaded places, erect, 4-angled, covered with stinging
bristly hairs. *Rhizomatous, abundant; pastures, roadsides, waste places, etc.***Urtica dioica**
Leaves *not cordate* basally (Fig. 05.08); the latter ovate to elliptic, obtuse to acuminate, with
stinging hairs, crenate-serrate, petiolate, *3-nerved*; stipules *small, narrow.* Flowers green,
small, in *axillary clusters*; corolla *absent;* the calyx of *4 obovate* sepals; the latter in the
female flowers always *unequal,* often with bristle-hairs dorsally. Plant *monoecious.* Stems
4-angled. *Common creeping annual of cultivated ground, waste places, etc.* **Urtica urens**

POLYGONACEAE

DOCK FAMILY: Herbaceous annuals, shrubs, woody climbers, or rarely trees, sometimes twining;
stems often with swollen nodes, occasionally geniculate. Leaves alternate, very rarely opposite or subopposite, invariably simple, usually entire, with characteristic membranous, *sheathing* stipules (*ocreae*).
Inflorescence very variable, either in *cymes, racemes, panicles, spikes* or *heads.* Flowers hermaphrodite or
unisexual, or variable, *actinomorphic*, and with the *hypogynous* arrangement. Perianth segments *undifferentiated*, in 1 or 2 whorls, Corolla *absent.* Calyx *polysepalous*, rarely gamosepalous, sometimes *petal-
like*, mostly of 5-6, rarely 3 or 4, sepals. Androecium usually of 5-9 stamens. Gynaeceum *syncarpous*;
ovary *superior,* 1-celled; stigmas 2-3, sessile or on styles, capitate to finely divided. Fruit a *nut,* hard, *trigonous* or rarely lenticular; ovule solitary, basal. The family contains over 30 genera and 800 species,
worldwide. There are about 10 important species throughout the British Isles (Figs 05.09-05.18).

a. Leaves *other than* sagittate-cordate *or* hastate .... d
    Leaves *sagittate-cordate* or *hastate* .... b

b. Leaves *sagittate* or *sagittate-cordate* c
    Leaves *hastate* (Fig. 05.09); the latter petiolate, narrowly lanceolate to oblanceolate, except
    for basal lobes or auricles which are spreading at right angles to the mid-rib, covered with
    minute tubercular protuberances. The stems erect. Flowers small, brown, clustered in leaf-
    less, loose, slender, terminal panicles; mostly unisexual and then dioecious; sepals small,
    glabrous. *Small perennial; calcifuge; common in heathlands, bogs, etc.***Rumex acetosella**

c. Perianth segments *6* (Fig. 05.10); the latter dark brown; the inner broadly oval andconcealing
    the fruit; outer reflexed and appressed to the pedicel after flowering. Stems erect. Flow-
    ers clustered in long, lax, terminal panicles; dioecious. The leaves sagittate with pointed
    lobes; entire; upper short-stalked and and clasping the stem; stipule fringed. *A very com-
    mon perennial of pastures, shaded places, especially those on acid soils.* **Rumex acetosa**
    Perianth segments *5* (Fig. 05.11); the latter glabrous, green, with white margins; the outer 3
    large and keeled. Flowers small, clustered in *axillary* panicles. Stems angular or ribbed,
    slender, *climbing* or trailing on ground. Leaves deeply cordate-sagittate at the base, ab-
    ruptly long-pointed apically or obtuse, usually entire, longer than its petiole, usually gla-
    brous. *Common annual plant; cultivated ground and waste places.* **Fallopia convolvulus**

d. Perianth segments *5* .... h
    Perianth segments *6*, 3 large and 3 small .... e

e. Large perianth segments *entire* .... f
    Large perianth segments clearly *toothed* (Fig.05.12); the latter triangular, dark brown; 1, rare-
    ly all 3, with a prominent *tubercle.* Flowers numerous, in whorls, and in a terminal, leafy
    branched panicle. Lower leaves large, broadly elliptic, rounded or shallowly cordate ba-
    sally, gradually and bluntly pointed, dull green, mostly entire, glabrous; cauline narrower;
    *A very common perennial of roadsides, grasslands, waste places, etc.***Rumex obtusifolius**

f.　Leaf margins *not* wavy...g

　　Leaf margins very *wavy* (Fig. 05.13); the latter lanceolate or oblong-lanceolate, somewhat rounded or abruptly constricted basally, tapered from about the middle to an obtuse point, dull green, glabrous. Flowers in whorls, or in leafy, sparingly branched panicles; perianth segments broadly ovate-cordate, usually *all 3* with tubercles; 1 tubercle often larger than other 2. *A common perennial of roadsides, grasslands, waste places, etc.* **Rumex crispus**

g.　Large segments, *all 3*, tuberculate (Fig. 05.14); the latter ovate to oblong, dull brown. Flowers in whorls, in *leafy*, well-branched, panicles; branches *spreading*. Lower leaves ovate-lanceolate, rounded or slightly cordate basally, acute, entire or finely toothed; upper narrower. *Frequent perennial; woods, grassy and damp waste places.***Rumex conglomeratus**

　　Large segments, *1*, tuberculate (Fig. 05.15); the latter oblong, obtuse, dull brown. Flowers in distant whorls, in *mostly leafless*, well-branched, terminal panicles. Lower leaves ovate-lanceolate, rounded or slightly cordate basally, acute, glabrous, mostly entire; cauline narrower. *Frequent perennial of woods, grassy and damp waste places.* **Rumex sanguineus**

h.　Flowers in *conspicuous* spike-like racemes i

　　Flowers in very small *inconspicuous* axillary clusters (Fig. 05.16); the latter almost sessile; sepals united basally, green with pink or white margins. The stems prostrate, spreading. Leaves very variable in size, up to 2.5-5 x 0.5-2 cm, those on the main stem larger than those on the branches, narrowed basally, ovate-lanceolate, sessile or short-stalked, entire, glabrous. *Abundant annual of cultivated ground and waste places.* **Polygonum aviculare**

i.　Leaves *tapered* to the petiole (Fig. 05.17); the former entire, lanceolate, short-stalked, up to 12 x 5 cm, glabrous or pubescent abaxially and often with a dark blotch adaxially; stipule fringed. The stems erect, pinkish, and swollen above the nodes. Flowers in erect,compact, terminal, axillary, cylindrical spike-like panicles; the sepals petaloid, mostly *pink*, rarely white. *An abundant annual of cultivated ground and waste places.* **Polygonum persicaria**

　　Leaves *cut off squarely* at the base (Fig. 05.18); the latter mostly rounded-triangular, large, mostly entire and glabrous, short-stalked. Stems erect, stout, coarse, and arising from an extensive rhizome. Flowers in axillary spike-like panicles; green; 3 of the 5 sepals keeled or winged. *A frequent perennial of roadsides and waste places, etc.* **Reynoutria japonica**

14.　Flowers *zygomorphic* .... **16** (page 30)

　　　Flowers *actinomorphic* **15**

15.　Leaves *alternate* .... **Rosaceae**

　　　Leaves *opposite* .... **Geraniaceae**

ROSACEAE

ROSE FAMILY: Trees, shrubs or herbs. Features very variable; leaves petiolate, mostly alternate, very rarely subopposite, and either simple, pinnate or palmate; the number of leaflets, their size, the presence of secondary leaflets, varies. *stipules conspicuous*. Inflorescence of *various* determinate and indeterminate types. Flowers usually *actinomorphic, hermaphrodite*, and with the *epigynous* or *perigynous* arrangement. Corolla *polypetalous*, mostly of 5 petals; the latter rarely present. Calyx *polysepalous*, usually of 5 sepals; epicalyx often present. Androecium of 2, 3, or 4 times as many stamens as sepals, very rarely 1-5 only. Gynaeceum *apocarpous* or *syncarpous*, of 1-many carpels; ovary *superior* or *inferior;* ovules mostly 2, rarely 1 or more. Fruits either *achenes, drupes, follicles* or *pomes*. This is a large family of great economic importance, with about 150 genera and 3000 species, worldwide (see page 261). It is divided into 6 subfamilies. There are 10 common or frequent species (Figs 05.19-05.28).

a.　Epicalyx *present* c

　　Epicalyx *absent* .... b

b. Petals *creamy-white*, mostly obovate, clawed (Fig. 05.19); sepals usually 5, triangular-ovate, reflexed, pubescent. Flowers numerous, small, in irregular *corymbose cymes* or panicles Leaves large, pinnate; leaflets *large* and *small,* ovate, sharply doubly crenate, dark green, glabrous adaxially, white-tomentose abaxially; rarely glabrous; terminal leaflet largest, 3-lobed; stipules leafy. *A very common perennial; damp grassy places.* **Filipendula ulmaria**
Petals *bright yellow* (Fig. 05.20). Flowers in *very long terminal spikes*; the calyx tube *fringed* with a ring of hooked bristles. Leaves irregularly pinnate, softly *pubescent*; leaflets *large* and small, alternating, deeply and coarsely serrate-crenate; stipules large, deeply toothed
A frequent perennial of roadsides and other dry grassy places, etc. **Agrimonia eupatoria**

c. Stamens *numerous* .... d
Stamens *1-2* (Fig. 05.21). Vestigial stamens present; flowers minute, in sessile axillary clusters and hidden by large leafy stipules; petals green; epicalyx *small.* Leaves small, very short-stalked, fan-shaped, mostly 3-lobed, the lobes further divided, thinly pubescent. *A small annual, frequent in cultivated ground, dry roadsides, walls, etc.* **Aphanes arvensis**

d. Flowers *yellow* or *purple* .... f
Flowers *white* .... e

e. Hairs on leaves *appressed* (Fig. 05.22); the latter ternate in a basal rosette; the leaflets ovate, obovate or oblong, coarsely serrate-dentate, pubescent, bright green adaxially, pale and somewhat glaucous abaxially; the petioles long. Flowers in few-flowered irregular cymes; petals white, *obovate*, blunt apically. Fruits *fleshy.* Stock thick and woody; runners slender, arching, extensive. *A small perennial, common in dry shaded places.* **Fragaria vesca**
Hairs on leaves *exserted* (Fig. 05.23); the latter ternate, on long petioles, densely pubescent adaxially and abaxially; leaflets coarsely crenate-serrate, bluish-green adaxially, paler abaxially, broadly obovate, blunt apically, short-stalked. Flowers *solitary*, on long slender pedicels; the petals white, somewhat *obcordate.* Fruits *dry.* Stock thick, somewhat woody, Runners slender, short. *A small common perennial in dry shaded places.* **Potentilla sterilis**

f. Corolla *yellow* .... g
Corolla *dark purple* (Fig. 05.24); the latter of 5 rather small, lanceolate petals; sepals sharply pointed, *bigger* than the petals; the flowers in loose cymes. Stems rooting basally, upper part erect. Leaves petiolate, pinnate; leaflets oblong, toothed, scarcely pubescent; stipules adnate to petioles. *Frequent perennial; bogs and marshy ground.* **Potentilla palustris**

g. Leaves, at least *some*, cauline i
Leaves, *all*, arising at ground level .... h

h. Leaves *palmate* (Fig. 05.25); the latter on long slender petioles, and mostly of 5 obovate or oblanceolate, sparingly pubescent, coarsely serrate leaflets; stipules lanceolate, entire or finely toothed. Runners extensive and rooting at the nodes. Flowers *solitary*, axillary, on long slender pedicels. Corolla of 5 slightly notched petals; calyx and epicalyx pubescent. *An abundant perennial of poor pastures, roadsides, waste places, etc.* **Potentilla reptans**
Leaves *pinnate* (Fig. 05.26); the latter on short petioles, *silvery with silky hairs,* particularly abaxially; leaflets alternating large and small, oblanceolate, deeply and regularly serrate; stipules entire or finely toothed, adnate to the petioles. The runners slender, short. Flowers *solitary*, axillary; pedicels long, slender; corolla of 5 obovate petals; calyx and epicalyx hairy. *An abundant perennial of cultivated ground, waste places, etc.* **Potentilla anserina**

i. Petals *5* (Fig. 05.27); the latter obovate or oblong, apex blunt; sepals 5, united basally. Flowers solitary or in cymes, axillary. Radical leaves long-stalked, with 2-3 pairs of unequal-sized leaflets, and a larger, mostly 3-lobed terminal leaflet; the cauline, at least some, *ternate* or 3-lobed; all leaflets hairy and toothed; stipules large and *leaflet-like.* Stems erect. Fruits *prickly. Common perennial of hedges, woods and shaded places.* **Geum urbanum**

Fig. 05.00

N.B. DRAWINGS NOT TO SCALE

Petals *4* (Fig. 05.28); the latter cuneiform and *emarginate*; the sepals 4. Flowers *solitary* or in cymes; pedicels long. Stems erect, rarely trailing. Leaves ternate, rarely some palmate; pubescent or glabrous; leaflets mainly cuneiform, coarsely serrate apically, entire basally; stipules adnate to petioles, small. *Common perennial; heaths, bogs, etc.* **Potentilla erecta**

GERANIACEAE

CRANE'S-BILL FAMILY: Plants are mostly herbaceous, very rarely undershrubs or trees. Leaves simple or compound, opposite or alternate, always *stipulate, palmately* or *pinnately* lobed, and mostly pubescent. Inflorescence *cymose*, or flowers solitary or in pairs. Flowers mostly hermaphrodite, very rarely dioecious, *actinomorphic*, very rarely slightly zygomorphic, and with the *hypogynous* arrangement. Corolla *polypetalous*, of 5 petals. Calyx *polysepalous*, of 5 sepals, the latter often *aristate*. Androecium of 5 or 10 stamens, rarely more; filaments often somewhat connate basally. Gynaeceum *syncarpous*; ovary *superior*. Fruit *schizocarpic*, separating into 5 1-seed units. This is a moderate size family with about 11 genera and 900 species, worldwide. There are 4 common or frequent species (Figs 06.01-06.04).

a. Stamens *10* .... b
Stamens *5* (Fig. 06.01); flowers umbel-like, on long axillary peduncles; petals *purple* or *pink*, sometimes with a blackish spot on the base of the upper 2; sepals bristly, *aristate*. Stem leaves opposite, each pair unequal; basal long-stalked; all *pinnate*, with *pinnately* divided leaflets; sparingly pubescent. *A frequent annual of dry sandy places.* **Erodium cicutarium**

b. Petals *emarginate* apically c
Petals *rounded* or *blunt* apically (Fig. 06.02); the latter *pink*, rarely white, spoon-shaped; sepals pubescent, *aristate*; flowers solitary or in pairs; peduncles long. Leaves *palmate*, divided into 3 main lobes which are further lobed; softly pubescent; stipules small. Stems erect or straggling, pinkish. *Abundant annual of shaded places.* **Geranium robertianum**

c. Leaves divided *half-way* to the base, or slightly beyond, into 5-9 lobes (Fig. 06.03); the former pubescent, long-stalked; lobes deeply toothed; the stipules narrow. Flowers mostly in pairs; peduncles short; petals *rose-purple*, rarely white, *notched*; sepals *shortly-aristate*, pubescent. *Common annual of dry pasture, roadsides, waste places, etc.* **Geranium molle**
Leaves divided *close* to the base (Fig. 06.04); the latter pubescent, long-stalked; lobes cuneiform and further divided into narrow segments; stipules narrow. Flowers mostly in pairs, on short peduncles; petals obovate, *reddish-purple, emarginate*; sepals *glandular-hairy, aristate. Common annual; dry pasture, roadsides, waste places, etc.* **Geranium dissectum**

16. Flowers *spurred* basally .... **Violaceae**
Flowers *not* spurred basally .... **Fabaceae**

VIOLACEAE

VIOLET FAMILY: Herbaceous annuals or perennials, shrubs or trees. Leaves usually simple, petiolate, alternate and *stipulate*. Inflorescence racemose or cymose. Flowers often *solitary*, mostly *zygomorphic*, rarely *actinomorphic*, hermaphrodite, and with the *hypogynous* arrangement. Corolla of 5 sepals, continued into a *spur* beyond point of insertion when flowers are zygomorphic. Calyx *polysepalous*, of 5 sepals, continued beyond point of insertion. Androecium of 5 free stamens, alternate with the petals, the 2 lower spurred in zygomorphic flowers. Gynaeceum *syncarpous*, of 3-5 carpels; ovary *superior*, unilocular; style simple, often curved or thickened. Fruit a *capsule* or *berry*. Worldwide, there are about 16 genera and 900 species. Six species are common or frequent (Figs 06.5-06.10).

a. Stipules *small*, entire or toothed but *not* lobed c
Stipules *large*, leafy, *divided* into *narrow* lobes .... b

b.   Corolla *shorter* than the calyx (Fig. 06.05); the former *blue* tinged with white or yellow, spur
     sometimes violet; sepals linear-lanceolate, acute. Flowers on long pedicels. Leaves vari-
     able, *mostly* oval to oblanceolate and glabrous, crenate; the terminal stipule lobe *ovate* and
     *toothed. A frequent annual of cultivated ground, and dry gravelly places. **Viola arvensis**
     Corolla *longer* than the calyx (Fig. 06.06); the former usually a mixture of *purple, white* and
     yellow; sepals triangular to linear-lanceolate acute. Flowers on long pedicels. Leaves var-
     iable, mostly oval to oblanceolate, *mostly* glabrous, crenate; terminal stipule lobe *entire*.
     A frequent annual of dry sandy cultivated ground and other sandy places.   **Viola tricolor**

c.   Style *hooked* apically .... d
     Style *straight* or *oblique* apically (Fig. 06.07). The flowers on long pedicels; the latter longer
     than the peduncles; the corolla pale *bluish-violet* with *dark* streaks; sepals oblong, obtuse.
     Leaves orbicular or reniform, cordate basally, blunt apically, obscurely toothed, glabrous;
     stipules narrow, fringed with glands. *A frequent perennial of wet ground.* **Viola palustris**

d.   Stipules *fringed* with hairs .... e
     Stipules *not* fringed (Fig. 06.08); the former entire or toothed, linear. Radical leaves cordate
     basally, somewhat ovate, blunt, shallowly crenate or crenate-serrate, scarcely pubescent,
     sometimes broader than the upper. Flowers on long pedicels; corolla *clear blue*, spur *yel-
     lowish*; sepals lanceolate, entire. *A frequent perennial of mostly dry habitats.***Viola canina**

e.   Spur *slender, deep* violet (Fig. 06.09); flowers on long pedicels; petals *lilac* or *reddish-lilac*,
     rarely pink or white, rather narrow; sepals lanceolate, rounded basally. Leaves orbicular-
     ovate, cordate basally, blunt, scarcely pubescent, crenate; petioles variable; stipules with
     *long comb-like teeth. Frequent perennial in woods; shaded areas.***Viola reichenbachiana**
     Spur *stoutish, pale* violet or whitish (Fig.06.10); corolla mostly *blue-violet*, but variable; sep-
     als pointed, notched basally; the flowers on long pedicels. Leaves orbicular-ovate, cordate
     basally, crenate, scarcely pubescent; petioles variable; stipules fringed with *short comb-
     like teeth. Common perennial; upland pastures, banks and sandy places.* **Viola riviniana**

FABACEAE

LEGUME FAMILY: Herbs, less frequently shrubs or trees. Leaves usually *compound*, rarely simple, *stipulate*, mostly *alternate*; often with *tendrils*. Inflorescence mostly a *raceme*, rarely a contracted raceme,
or more rarely as few-flowered clusters or as solitary flowers. Flowers hermaphrodite, with the *hypogynous* arrangement, *zygomorphic*; actinomorphic in 1 exotic subfamily. Corolla of 5 petals, *polypetalous*
or, frequently, with the 2 anterior petals connate basally forming the *keel* petals; 1 petal, *i.e.* the *standard*
or *vexillum* clearly larger than anyone of the remaining 4. Calyx *gamosepalous*, of 5 sepals. Androecium
mostly of 10 *monadelphous* or *diadelphous* stamens. Gynaeceum *syncarpous*, of 5 carpels; ovary *superi-
or*. Fruit is usually a *legume*. This is a large family of worldwide importance. It contains over 750 genera
and 16-19000 species. Three subfamilies are recognised (see page 241); most of the plants in the temperate zones belong to the subfamily Faboideae. The Leguminosae is an alternative legitimate family name.
Another, Papilionaceae, is obsolete. There are many common species (Figs 06.11-07.01).

a.   Leaves, all, ending in a *leaflet* .... h
     Leaves, at least some, ending in a *tendril* or short *point* .... b

b.   Leaf ending in a *tendril* .... d
     Leaf ending in a *short point* c

c.   Corolla *yellow* (Fig. 06.13); the standard petal recurved; calyx teeth triangular, of equal size.
     Flowers in a loose raceme, on long peduncle. The leaflets 2, *sagittate* basally, lanceolate,
     acute, 'parallel-veined', mostly hairy; stipules leaflet-like, *sagittate*. Stems *sharply angu-
     lar. A common perennial of damp meadows, hedgerows and ditches.* **Lathyrus pratensis**

Corolla *reddish-purple* (Fig. 06.12). Flowers few; calyx teeth *unequal*. The stems *angular*, winged, *glabrous*, simple; the rhizomes *creeping, tuberous.*Leaflets 'parallel-veined', few, lanceolate, acute or obtuse, mucronate, glabrous; stipules mostly *unequally 2-lobed*, often finely toothed. *A frequent perennial of heaths and upland pastures.* ***Lathyrus montanus***

d. Corolla *blue, lilac* or *purple* .... e
Corolla *yellow* (Fig. 06.13); the standard petal recurved; calyx teeth triangular, of equal size. Flowers in a loose raceme, on long peduncle. The leaflets *2, sagittate* basally, lanceolate, acute, 'parallel-veined', mostly hairy; stipules leaflet-like, *sagittate*. Stems *sharply angular. A common perennial of damp meadows, hedgerows and ditches.* ***Lathyrus pratensis***

e. Racemes *1-5-flowered* .... f
Racemes *10-flowered* or *more* (Fig. 06.14); the latter dense; peduncles long; flowers usually reflexed; corolla *blue-purple*; the calyx teeth *uneven*. Leaflets many; the latter lanceolate-linear, acute or mucronate, pubescent or nearly glabrous; stipules inconspicuous; tendrils branched. *An abundant perennial of hedgerows and waste grassy places, etc.****Vicia cracca***

f. Calyx teeth *unequal* .... g
Calyx teeth *equal* (Fig. 06.15); the latter pubescent; flowers very small, few, on a relatively long peduncle; corolla *pale blue* or *whitish*; fruit usually 2-seeded, hairy. Leaflets many, linear-oblong, apices blunt to emarginate or mucronate, mostly hairy; tendrils branched; stipules often 4-lobed. *A frequent annual of dry grassy, waste places, etc.* ***Vicia hirsuta***

g. Leaflets *narrow* (Fig. 06.16); the latter linear to obovate, pointed, or emarginate and mucronate, scarcely hairy; stipules half-sagittate, often with a dark central spot. Flowers solitary or in pairs, 10-16.0 mm long, short-stalked; the corolla *light-purple*; calyx hairy, teeth *unequal*, as *long* as the tube. *Common annual of dry gravelly places.* ***Vicia sativa*** ssp. ***nigra***
Leaflets *broad* (Fig. 06.17); the latter *ovate*, rounded basally, rounded and mucronate apically, glabrous or scarcely hairy; stipules half-sagittate, often toothed; the tendrils branched. Flowers 2-5, short-stalked; the corolla *dull purple*; calyx hairy, teeth *unequal,* lower much *shorter* than tube. *Common annual of hedges, roadsides, waste places, etc.* ***Vicia sepium***

h. Flowers in contracted *racemes* or *clusters* i
Flowers *solitary* (Fig. 06.18); the latter axillary, short-stalked; corolla *pink*, standard streaked with a deeper shade; calyx deeply 5-toothed, densely glandular. Leaflets mostly 1, rarely 3, toothed; stipules *adnate* to the short petioles, with *sticky glandular* hairs. Stems spreading, densely *glandular* hairy. *A frequent perennial of dry sandy places, etc.****Ononis repens***

i. Corolla *blue, purple, pink* or *white* .... o
Corolla *yellow* .... j

j. Flowers in compact *heads* or *clusters* .... k
Flowers in *racemes* (Fig. 06.19); the latter slender, compact, with numerous reflexed flowers; wing, keel, and standard petals *about equal*; calyx with 5 *subulate* teeth. Leaves *trifoliate*; short-stalked, leaflets oblong to obovate, toothed, mostly glabrous; the stipules *subulate* to *setaceous. A frequent perennial of dry sandy coastal areas.* ***Melilotus altissima***

k. Leaflets *3* l
Leaflets *7* or more (Fig. 06.20); leaves *imparipinnate*; lateral leaflets ovate-elliptic, terminal largest and lanceolate; all acute and mostly pubescent; the stipules entire and *leaflet-like*. Flowers in compact clusters; the corolla *yellow*; calyx *inflated*, somewhat *2-lipped*, longer than the fruit. *A common perennial of dry banks and sandy pastures.* ***Anthyllis vulneraria***

l. Stipules *not* leaflet-like .... m
Stipules *leaflet-like* .... n

m. Stipules *entire* (Fig. 06.21); the latter small, broadly ovate, acuminate; leaves short-stalked; leaflets cuneiform, toothed and emarginate apically, and pubescent or glabrous. Flowers in small compact axillary clusters; peduncles longer than petioles; corolla *yellow*. Stems prostrate or erect. *Common small annual of pastures and waste places.* **Trifolium dubium**
Stipules *toothed* (Fig. 06.22); the latter lanceolate; leaves short-stalked; the leaflets obovate, toothed, *mucronate*, glabrous or pubescent. Flowers in small axillary clusters; peduncles slender and longer than the petioles; corolla *yellow*; fruit *black, coiled*. Stems procumbent or ascending. *A common small annual of pastures and waste places.* **Medicago lupulina**

n. Stems *solid* (Fig. 06.23); the latter decumbent or ascending. Leaves *trifoliate*, short-stalked; leaflets obovate to ovate, obtuse or apiculate, mostly *glabrous* or slightly pubescent; stipules large and *leaflet-like*. Flowers few, in loose axillary clusters; peduncles stout, longer than leaves; corolla *yellow*, often streaked or tipped with red; calyx teeth narrow, hairy, equal. *A common perennial of dry pastures, roadsides, sand-hills, etc.* **Lotus corniculatus**
Stems *hollow* (Fig. 06.24); the latter erect or ascending; rootstock slender. Leaves *trifoliate*, short-stalked; leaflets obovate, often obliquely so, obtuse or mucronate, lower pair ovate; mostly *pubescent;* stipules large and *leaflet-like*. The flowers in lax axillary clusters; peduncles short; corolla *yellow*, streaked with red; the calyx teeth narrow, spreading, equal. *Very common perennial plant of damp grasslands, marshy ground, etc.* **Lotus uliginosus**

o. Plant *pubescent* .... q
Plant *glabrous* .... p

p. Growth-habit *stoloniferous* (Fig.06.25); stems long, wiry, spreading and rooting at the nodes. Leaves *trifoliate, long-stalked;* leaflets very variable in size and shape, often obovate or nearly rounded, often *emarginate* apically, *toothed, glabrous*, mostly with *white* crescent-shaped markings; stipules large, entire, acuminate. Flowers in compact clusters; petioles *much longer* than the peduncles; corolla *white*, often tinged with pink; the calyx teeth unequal. *An ubiquitous perennial of pastures, meadows, waste places, etc.* **Trifolium repens**
Growth-habit *erect* (Fig. 06.26); stems erect, glabrous. Leaves *trifoliate, long-stalked*; leaflets broadly obovate, toothed, *without* white markings, generally *glabrous*; stipules wide, acute. Flowers in compact clusters; peduncles long; corolla *white* or *pink*; calyx teeth narrow, nearly equal. *A perennial, now mostly as a relic of cultivation.* **Trifolium hybridum**

q. Calyx *shorter* than the corolla .... r
Calyx *much longer* than the corolla (Fig. 06.27); the former with *long-ciliate* teeth, campanulate, feebly ribbed; corolla small, *white* or *pale pink*; flower heads *cylindrical*; peduncles thin, equalling or exceeding the leaves. The latter *trifoliate*, short-stalked; leaflets narrow, oblong-ovate, pubescent. *Common annual plant; coastal sandy places.* **Trifolium arvense**

r. Flower-heads *globose* (Fig. 06.28); the latter becoming *sessile*, and subtended by 1 pair of short-stalked leaves; the corolla *purplish-red*, rarely whitish; calyx teeth unequal, 1 longer than other 4, often hairy. Leaves *trifoliate*, lower petioles long; leaflets variable, often obovate, acute or emarginate, and frequently with *white crescent-shaped* markings; stipules large, acuminate. *A common perennial of grassy and waste places.* **Trifolium pratense**
Flower-heads *cylindrical* (Fig. 07.01); the latter sessile, subtended by 1 leaf; corolla *crimson*, rarely pink or pale cream, exceeding the calyx; the latter ribbed, with pubescent, bristle-like teeth. Leaves *trifoliate*, long-stalked; leaflets obovate to obcordate, blunt or emarginate; stipules mostly obtuse. *Perennial; mostly relic of cultivation.* **Trifolium incarnatum**

17. Gynaeceum *syncarpous* or flowers staminate .... **19** (page 36)
Gynaeceum *apocarpous* **18**

18. Stamens *10* or less .... **Crassulaceae**
Stamens *12* to many .... **Ranunculaceae**

CRASSULACEAE

STONECROP FAMILY: Annual or perennial herbs or under shrubs. Leaves opposite, whorled or alternate, usually *succulent, exstipulate*, simple and entire. Inflorescence mostly *cymes*. Flowers hermaphrodite, *actinomorphic*, rarely unisexual when the plants are then dioecious, and with the *hypogynous* arrangement. Corolla *polypetalous* or *gamopetalous*, of 4-30 petals. Calyx *polysepalous* or *gamosepalous*, of 4-30 sepals. Androecium with a stamen number once or twice that of the petals, in 2 whorls. Gynaeceum mostly *apocarpous*, with the carpel number equalling that of the petals; ovary *superior*. Fruit a bunch of *follicles*, or a *capsule*. Worldwide, there are about 33 genera and 500 species (estimates of the latter vary widely). There are 3 frequent or common species (Figs 07.02-07.04).

a. Leaves *sessile* .... b
    Leaves *long-stalked* (Fig. 07.02); the latter *peltate*, fleshy, slightly crenate, decreasing in size upwards, glabrous. The flowers in a long terminal raceme; corolla 5-lobed, whitish-green; calyx small, 5-lobed. *A frequent perennial of rock and wall crevices.* **Umbilicus rupestris**

b. Corolla *white*, rarely pinkish (Fig. 07.03); flowers in a flat-topped, terminal cyme; sepals 5, shortly united basally, fleshy. The leaves crowded, alternate, obovoid, terete or somewhat flattened, entire, succulent. *A frequent perennial of wall and rock crevices.* **Sedum album**
    Corolla *yellow* (Fig.07.04); flowers crowded at the tops of short shoots, in short small; calyx of 5 fleshy sepals. The leaves crowded, ovoid, very fleshy, entire, and *not* spurred at the base; taste sharp and peppery. *A frequent perennial of wall and rock crevices.***Sedum acre**

RANUNCULACEAE

BUTTERCUP FAMILY: Annual or biennial herbs, occasionally woody climbers; very rarely trees. Leaves mostly spirally or alternately arranged, rarely opposite, *exstipulate*, simple or compound; often palmately lobed. Inflorescence *cymes, racemes, panicles* or of solitary flowers. Flowers hermaphrodite, rarely monoecious or dioecious, *actinomorphic*, and with the *hypogynous* arrangement. Corolla *polypetalous*, of 3-8 petals, rarely more, or absent entirely. Calyx *polysepalous*, of 3-8 sepals. Androecium of *numerous*, spirally arranged, stamens. Gynaeceum *apocarpous*, of *numerous* carpels; ovary *superior*. Fruit mostly an *achene*, rarely a follicle. This is a moderately large family with up to 35 genera and 1500 species, worldwide. Many are deemed poisonous because they contain volatile oils and acrid substances. There are 9 frequent or common species (Figs 07.05-07.13).

a. Leaves deeply *lobed* .... f
    Leaves *entire, toothed* or very *shallowly* lobed .... b

b. Leaves *lanceolate* or *broadly-lanceolate* .... e
    Leaves *cordate, reniform* or somewhat *triangular* c

c. Sepals *green* or *greenish* .... d
    Sepals *petal-like* (Fig. 07.05); the latter golden yellow. The flowers large and showy, on long pedicels, axillary, solitary or 2-3 together. Petals *absent*. Fruit a *follicle*. Leaves large, up to 12 x 12 cm, *cordate* or *reniform*, blunt apically, regularly crenate, glabrous; the lower long-stalked; upper sessile. *A common perennial of wet marshy ground.* **Caltha palustris**

d. Sepals *3*, rarely more (Fig. 07.06). Flowers mostly solitary; pedicels long; petals 8-12, yellow, narrow. Leaves small, 1-4.0 cm, long-stalked, broadly triangular to cordate, entire to distantly and finely toothed, glabrous; the lower in a rosette; upper in unequal-sized pairs. plant with many *'tubers'*. *Common perennial of damp shaded places.* **Ranunculus ficaria**
    Sepals *5* (Fig. 07.07); the latter green. Flowers solitary; petals 5, pale yellow. Leaves opposite or alternate, *reniform, shallowly* lobed, more or less glabrous, all floating. *A very frequent, prostrate annual: shallow waters and wet muddy places.* **Ranunculus hederaceus**

Fig. 06.00

N.B. DRAWINGS NOT TO SCALE

e.	Leaves with clear *reticulate* venation (Fig. 07.08); the lower leaves long-stalked, mostly lan-
	ceolate; sheaths broad; cauline becoming sessile; all entire or distantly and finely toothed,
	and more or less glabrous. Flowers few, in a loose corymb; sepals 5, hairy; petals 5, pale
	yellow. *A very common perennial of ditches and marshy ground.* **Ranunculus flammula**
	Leaves with 'parallel' venation (Fig. 07.09); lower stalked, sheaths broad, almost circling the
	stems, sheaths margins membranous; cauline mostly sessile or half-clasping the stems; all
	lanceolate, long-pointed, entire or remotely toothed. Flowers large, on long pedicels; sep-
	als 5; petals 5, bright yellow. *A frequent perennial in marshy places.* **Ranunculus lingua**

f.	Leaves deeply *pinnately* lobed .... g
	Leaves deeply or shallowly *palmately* lobed .... h

g.	Sepals *reflexed*, green (Fig. 07.10). Petals bright yellow, glossy. Flowers solitary or few, on
	ungrooved hairy pedicels. Stems erect, pubescent and base *bulbous*. Lower leaves long-
	stalked and *deeply pinnately* divided into *3 lobes*; the latter further divided and coarsely
	toothed; cauline more or less sessile and *pinnately divided*; all pubescent. *A very frequent
	perennial of dry pasture, roadsides, sand-hills, gravelly places, etc.***Ranunculus bulbosus**
	Sepals *spreading* (Fig.07.11). Petals golden yellow, glossy. The flowers solitary or few, axil-
	lary; pedicels grooved, short. Stems shortly pubescent, both *erect* and *prostrate*. Lower
	leaves long-stalked and *deeply pinnately* divided into *3 lobes*; the latter further lobed and
	coarsely toothed; cauline less well divided and short-stalked or nearly sessile; all pubes-
	cent. *An abundant perennial of damp tilled ground, waste places, etc.***Ranunculus repens**

h.	Leaves divided nearly to the *base* into 5-9 palmate lobes (Fig. 07.12); the former orbicular,
	long-stalked, the lobes further divided and toothed; cauline short-stalked or nearly sessile,
	less well divided, and entire or nearly so; all pubescent. Stems erect, pubescent. Flowers
	solitary or few, on ungrooved, hairy pedicels; the sepals 5, hairy; petals 5, bright yellow,
	glossy. *An abundant perennial of meadows, pastures, waste places, etc.***Ranunculus acris**
	Leaves divided to about *half-way* into 3 lobes (Fig. 07.13); lower leaves in a dense rosette,
	long-stalked, with a broad sheaths; the cauline on shorter stalks; all more or less glabrous,
	and with toothed lobes. Flowers small, solitary or few, on glabrous grooved pedicels; sep-
	als 5; petals 5, yellow. *A frequent annual of ditches, streams, etc.* **Ranunculus sceleratus**

19.	Ovary *superior*, or perianth absent, or all flowers staminate .... **23** (page 42)
	Ovary *inferior* **20**

20.	Leaves variously arranged but *not* whorled .... **21** (page 37)
	Leaves *whorled* .... **Rubiaceae**

RUBIACEAE

BEDSTRAW FAMILY: Trees, shrubs or herbs. Leaves *whorled*, with *insignificant stipules.* Stems
quadrangular. Inflorescence *cymes* or *panicles,* rarely in heads. Flowers usually hermaphrodite, *actinomorphic*, with the *epigynous* arrangement. Corolla *gamopetalous*, often *campanulate* with *rotate* lobes,
and of 4-5 petals. Calyx *polysepalous* or *gamosepalous*, of 4-5 sepals; often very small or only represented by an *annular ridge*. Androecium of 4-5 epipetalous stamens. Gynaeceum *syncarpous*; ovary *inferior*.
Fruit a *capsule, berry, drupe*, or dry and *schizocarpic*. This family contains over 400 genera and 5000
species, worldwide. There are 6 common or frequent species which occur as weeds (Figs 07.14-07.19).

a.	Flowers *yellow* or *white* .... b
	Flowers *lilac* (Fig. 07.14); the latter arranged in small head-like clusters with a basal whorl
		of leaf-like bracts; corolla 4-lobed, funnel-shape, with a long slender tube; calyx shortly
		4-lobed, adnate to the ovary, shortly hairy. The stems prostrate or decumbent, pubescent.
		leaves obovate-cuspidate, small, 4-6 per whorl, sessile, mostly bristly adaxially and abaxi-
		ally. *A common annual of cultivated ground, lawns and waste places.* **Sherardia arvensis**

b. Flowers *white* c
Flowers *yellow* (Fig. 07.15); the latter many, in very dense, terminal and axillary compound panicles; corolla tube very short, lobes 4, spreading; calyx minute, annular. Leaves linear or needle-like, rough adaxially, pubescent. Stems slender, covered with minute reflexed hairs. *Common stoloniferous plant of heaths, dry banks and sandy places.* **Galium verum**

c. Stems *smooth* or very nearly so .... d
Stems rough with hooked reflexed *prickles* (Fig. 07.16); the latter scrambling-ascending, diffusely branched. The leaves linear-oblanceolate, mucronate, margins and mid-rib with reflexed prickles. The flowers in axillary clusters; corolla with 4 spreading lobes; calyx minute; fruit prickly. *Abundant annual; cultivated ground, waste places.* **Galium aparine**

d. Fruit or ovary *smooth* or *finely granulated* .... e
Fruit covered with *hook-shaped bristles* (Fig. 07.17). Corolla *funnel-shaped*, bluntly 4-lobed to about half-way. Flowers in long-stalked cymes; pedicels short. Leaves oblanceolate to obovate, cuspidate, with marginal prickles. Stems pubescent beneath the nodes, otherwise glabrous. *A frequent perennial plant of woods and other shaded places.* **Galium odoratum**

e. Leaves relatively *short* and *broad* (Fig. 07.18); the latter obovate-oblanceolate, mucronate, with straight marginal prickles pointing obliquely backwards. Flowers in few-flowered, corymbose cymes or panicles; the corolla with 4 blunt lobes. Fruits with acute tubercles. *A common perennial in heaths, moors, pastures and woods on acid soils.* **Galium saxatile**
Leaves relatively *long* and *narrow* (Fig. 07.19); the latter narrowly oblanceolate, tapered basally, *blunt*, glabrous, but with rough margins, 1-nerved. The stems slender, long, decumbent or ascending, often climbing, long-branched. The flowers in loose, axillary cymes; corolla with 4 acute lobes. *A common perennial of wet marshy places.* **Galium palustre**

21. Sepals *absent* .... **22 (page 38)**
Sepals *present* .... **Onagraceae**

ONAGRACEAE

WILLOWHERB FAMILY: Mostly herbs, rarely shrubs or trees, annuals or more usually perennials, often associated with marshy or wet places. Leaves simple, petiolate, alternate, opposite or very nearly so, rarely spirally arranged or whorled; *exstipulate*. Inflorescence in racemes or of *solitary* flowers. Flowers hermaphrodite, very rarely unisexual, mostly *actinomorphic*, rarely zygomorphic (*Chamerion*), and mostly with the *epigynous* arrangement and an *epigynous* hypanthium; rarely perigynous. Corolla mostly *gamopetalous*, of 2-4 petals. Calyx mostly *gamosepalous*, of 2-4 sepals. Androecium of 2 or 4 short and 4 long stamens. Gynaeceum *syncarpous;* ovary *inferior*, 1-5 celled, with 1-many anatropous ovules in each loculus; style single with a club-shaped or 4-lobed stigma Fruit mostly *capsular*, smooth, rarely club-shaped, indehiscent and prickly. Seeds pubescent, *tuberculate*. A family of about 25 genera and 650 species, it has worldwide distribution. There are 7 common or frequent species (Figs 07.20-07.26).

a. Petals *4* .... b
Petals *2* (Fig. 07.20); the latter rounded basally, deeply emarginate, white or pink; sepals 2, free; stamens 2; fruits with *hooked bristles*; flowers in *racemes*. Leaves broadly ovate to somewhat cordate basally, acuminate apically, opposite and decussate, finely toothed, and slightly pubescent. *A frequent perennial of gardens and shaded places.* **Circaea lutetiana**

b. Flowers *solitary* c
Flowers in large terminal *racemes* (Fig.07.21); the latter frequently irregular; petals *purplish-red*, slightly notched, obovate. Leaves alternate or spirally arranged, narrowly lanceolate, short-stalked, either entire or with small teeth, *glaucous* adaxially, and with *anastomosing* venation. *A frequent perennial of bogs and dry waste places.* **Chamerion angustifolium**

c.   Leaves *tapered* to the stem or *short-stalked* .... d
      Leaves *semi-amplexicaul* (Fig. 07.22); the latter oblong-lanceolate, acute, toothed, strongly
      pubescent, lower opposite, upper sessile and slightly decurrent. Stems densely pubescent.
      Corolla large, *15-20 mm in diameter*; petals deep *purplish-rose*, deeply notched, broadly
      obovate; sepals acute. Flowers solitary, axillary, upper often forming a corymb. *Frequent
      rhizomatous plant; ditches, riversides, streams and other wet places.* **Epilobium hirsutum**

d.   Stems *without* fine ribs or ridges .... e
      Stems *with* 2 or 4 distinctly raised lines (Fig. 07.23); the former erect or decumbent. Leaves
      lanceolate, sessile, scarcely pubescent, entire or obscurely toothed, lower opposite, upper
      alternate. Flowers solitary, axillary; petals *rose-coloured,* shallowly notched; sepals acute,
      glandular-pubescent. *A frequent perennial plant of marshy places.* **Epilobium obscurum**

e.   Stigma *4-lobed* .... f
      Stigma *club-shaped* (Fig. 07.24); flowers often on stem or branch endings; petals *rose-lilac,*
      shallowly notched. Stems without ridges, but often with 2 rows of crisped hairs. Leaves
      linear-lanceolate, cuneate basally, sessile, opposite or upper alternate, entire or obscurely
      toothed, pubescent. *Frequent perennial; bogs and wet marshy ground.***Epilobium palustre**

f.   Leaves apparently *glabrous*, but often with short hairs (Fig.07.25); the latter broadly ovate to
      broadly lanceolate, middle leaves opposite or subopposite, short-stalked, rounded basally,
      toothed, light green, scarcely pubescent. Petals notched, *pink;* sepals frequently reddish.
      *Common perennial; cultivated ground, roadsides, waste areas, etc.***Epilobium montanum**
      Leaves distinctly *pubescent* (Fig. 07.26); the latter finely toothed, lower opposite and stalked,
      the upper alternate and sessile. Stems *densely* covered with soft white hairs. Flowers soli-
      tary, axillary; petals *purplish-rose,* deeply notched; the sepals acute. Stems with soft hairs
      above. *Frequent perennial of ditches and damp waste places, etc.* **Epilobium parviflorum**

22.   Stamens *5*; inflorescence a simple or compound umbel .... **Apiaceae**
      Stamens *1-3* .... **Valerianaceae**

APIACEAE

CARROT FAMILY: Plants mostly biennial or perennial herbs, occasionally shrub-like. Stems often furrowed, pith wide and soft or internodes hollow. Leaves simple or compound, *exstipulate,* but petioles
always with broad *sheath-like* bases; mostly alternate, occasionally subopposite. Inflorescence in *umbels*
or *compound umbels,* rarely of simple umbels or solitary flowers; bracts and bracteoles usually present.
Flowers hermaphrodite or unisexual, mostly *actinomorphic,* or outer flowers somewhat *zygomorphic,* and
with the *'epigynous'* arrangement; often strongly protandrous. Corolla *polypetalous,* of 5 petals. Calyx
mostly *absent* or reduced to 5 small teeth. Androecium of 5 stamens. Gynaeceum *syncarpous;* ovary
inferior, usually bilocular; styles 2, usually with an enlarged base; ovules solitary in each cell, pedulous.
Fruit *schizocarpic,* splitting into halves; each half fruit is termed a mericarp and represents the so-call
seed; the latter generally with *vittae.* This is a very important family (see page 283), containing many economic and poisonous species. Worldwide, there are about 250 genera and 2-3000 species. An alternative
legitimate name for family is Umbelliferae. There are about 10 common or frequent species, most of
which occur as weeds of rough grassland, old pasture and roadsides (Figs 07.27-08.13).

a.   Leaves *pinnately* or *ternately* divided c
      Leaves *simple* or *palmately lobed* .... b

b.   Leaves *simple* (Fig. 07.27); the latter *peltate,* orbicular, crenate, very shallowly lobed, gla-
      brous, veins radiating; petioles longer than peduncles. Umbels simple, rarely compound;
      flowers often greenish. Plant very variable. *A rather common smallish, creeping or some-
      times floating perennial, rooting freely at the nodes; bogs; marshes.* **Hydrocotyle vulgaris**

Fig. 07.00

N.B. DRAWINGS NOT TO SCALE

Leaves deeply *palmately lobed* (Fig. 07.28); the latter mostly all radical, and with 5 obovate lobes, which are further lobed and toothed; glabrous. Flowers small, pink or white; umbels irregular. *Frequent slender perennial; woods and shaded places.* **Sanicula europaea**

c. Fruits or ovaries *not prickly* .... e
    Fruits or ovaries *prickly* .... d

d. Bracts *pinnatifid* (Fig. 08.01). Leaves long-staked, pubescent, tripinnate; leaflets pinnatifid. Umbels compact; rays numerous; the flowers *pinkish-white,* or the central ones often reddish or purplish; bracteoles slender, entire. Stems solid, ridged, covered with bristly hairs. *Common biennial of roadsides, pastures, especially those with light soils.* **Daucus carota**
    Bracts *entire* (Fig. 08.02). The leaves alternate, short-stalked, bipinnately and deeply divided into narrow, pubescent, serrate segments. Umbels with forward projecting hairs; flowers pinkish or purplish; the calyx teeth small; bracteoles present. Stems solid, striate, with reflexed prickles. *Common biennial of hedges, roadsides and waste places.* **Torilis japonica**

e. Upper leaves *not* divided into thread-like segments .... f
    Upper leaves deeply *divided* with thread-like segments (Fig. 08.03); radical leaves few, soon disappearing, ternately divided, glabrous. Umbels terminal; flowers white; calyx teeth absent. Plant tuberous. *A frequent perennial of heaths and old pastures.* **Conopodium majus**

f. Flowers *white*; very rarely pinkish-white .... g
    Flowers *yellow* or *yellowish-green* (Fig. 08.04); petals with inflexed tips; calyx teeth absent; bracts *absent;* bracteoles *very small;* umbels axillary and terminal. Radical leaves *triternate*, the upper cauline opposite, *ternate;* all *glossy-green, with large rhomboid, serrate or lobed leaflets;* glabrous or sheaths hairy. The stems solid, becoming hollow or filled with pith, ridged. *A common biennial; hedges, roadsides, waste places.* **Smyrnium olusatrum**

g. Leaves *bipinnate, tripinnate* or *ternate* .... j
    Leaves *pinnate* or *bipinnately lobed* .... h

h. Umbels clearly *stalked* i
    Umbels *sessile* or nearly so (Fig. 08.05); the latter leaf-opposed, terminal and axillary; bracts few or absent; bracteoles variable, becoming reflexed; petals white; calyx absent. Leaves *pinnate*, bright green, shiny, mostly glabrous; the leaflets elliptic to ovate or lanceolate to ovate, sessile, serrate, often unequally lobed. Stems soft, slender or stout, finely ridged, often rooting basally. *A very common biennial of ditches and streams.* **Apium nodiflorum**

i. Stems *150.0 cm* or less (Fig. 08.06); the latter stout, hollow, ridged, *hispid* with stiff reflexed bristle-like hairs. Leaves variable, mostly *pinnate*, hispid adaxially and abaxially; leaflets broad, irregularly lobed, serrate. The umbel large, mostly terminal, flat-topped; bracts and bracteoles few or absent; petals white, broadly notched, tip inflexed; calyx teeth minute. *Common biennial of old pastures, roadsides, waste places, etc.* **Heracleum sphondylium**
    Stems up to *3.5 m* (Fig. 08.07); the latter hollow, up to 10.0 cm in diameter, ridged, with *stiff, reflexed, bristle-like* hairs. Leaves up to 1 m, pinnately divided, rough. Umbels very large, to 50 cm, flat-topped; bracts and bracteoles present or absent; petals white, up to 12 mm. It contains furocoumarins which cause a blistering of the skin by making it hypersensitive to sunlight. *Occasional, naturalised in many damp places.* **Heracleum mantegazzianum**

j. Plant *not* rhizomatous .... k
    Plant *rhizomatous* (Fig. 08.08). Lower leaves biternate, long-stalked, upper ternate and often opposite; all *glabrous,* with clearly *asymmetrical*, irregularly serrate, lanceolate to ovate leaflets; petioles *bluntly triquetrous.* Umbels terminal; bracts and bracteoles usually *absent*; petals white, apex inflexed; calyx teeth minute. Flowering stems erect, stout, hollow, ridged. *Common perennial of cultivated ground; waste places.* **Aegopodium podagraria**

k.   Calyx teeth *absent* or minute l
Calyx teeth very *conspicuous* (Fig. 08.09). Leaves variously pinnate, glabrous or nearly so, *bluish-green*, with large ovate to suborbicular, serrate segments. Umbels terminal, dome-shaped; bracts present or absent; bracteoles present; petals white, often enlarged, emarginate with a long inflexed point. Stems stout, hollow, ridged. Rootstock with a cluster of *tuberous, slender* roots. *Frequent perennial; ditches, riversides, etc.* **Oenanthe crocata**

l.   Bracteoles *neither* 3 in number *nor* radiating from the compound umbel .... m
Bracteoles *3* in number, at least *twice* as long as the pedicels, closely situated, and *radiating* from the compound umbel (Fig. 08.10). Umbels small, terminal, leaf-opposed; petals unequal, white; calyx minute or absent; bracts *absent*. Leaves few, *glaucous*, glabrous, deltoid, mostly bipinnate; segments deeply pinnatifid and toothed. Stems erect, hollow, finely striate. *Very frequent annual of cultivated ground and waste places.* **Aethusa cynapium**

m.   Petioles *grooved adaxially* .... n
Petioles *ungrooved* (Fig. 08.11); the latter more or less rounded. Leaves bipinnate or tripinnate, glabrous, with deeply pinnately lobed, coarsely serrate segments; plant with *purple spots*. Umbels small, terminal and axillary, short-stalked; bracts and bracteoles present; petals white, tips inflexed; calyx teeth absent. Stems ridged, *purple-spotted*, smooth. *A frequent biennial of farm yards, cultivated ground and waste places.* **Conium maculatum**

n.   Bracteoles *narrow* and *glabrous* or nearly so (Fig. 08.12). The radical leaves bipinnate, with relatively large leaflets in threes; terminal leaflets ovate to obovate, asymmetrical; cauline leaves smaller; leaflets coarsely serrate, and nearly glabrous. Umbels large, terminal and axillary; bracts few or absent; petals white or pinkish, tips inflexed; calyx absent. Stems hollow and ridged. *Common perennial of streamsides, riversides, etc.* **Angelica sylvestris**
Bracteoles *broad*, mostly with a *fringe* of hairs (Fig. 08.13); the latter deflexed and spreading. The leaves bipinnate or tripinnate, with somewhat pubescent, small, ovate, pinnatifid and coarsely serrate segments. Umbels small, terminal; the bracts present or absent; petals emarginate, tips inflexed; calyx absent. Stems hollow, smooth, ridged, mostly glabrous. *An abundant perennial of roadsides, hedgerows, waste places, etc.* **Anthriscus sylvestris**

VALERIANACEAE

VALERIAN FAMILY: Annual or perennial herbs, rarely shrubs or trees; often with strong-smelling rhizomes. Leaves mostly opposite, rarely radical, mostly *entire*, rarely pinnate or pinnatifid; *exstipulate*, petiolate or sessile, the bases often somewhat sheathing. Inflorescence in many-flowered compound paniculate *cymes*, sometimes condensed and capitate. Flowers rather small, hermaphrodite or unisexual, *zygomorphic* or *actinomorphic*, and with the *epigynous* arrangement. Corolla *gamopetalous*, of 5 petals, often spurred basally, sometimes bilobed. Calyx often represented by an *epigynous ring*, or developing into a 'pappus-like' structure in the fruit. Androecium of 1-3 epipetalous stamens which alternate with the corolla lobes. Gynaeceum *syncarpous*; ovary *inferior*, 3-celled, 1 cell fertile, the the other 2 sterile, usually small or almost absent; ovule solitary, pendulous. Fruit an *achene*, the calyx often remaining attached and developing into a winged, or awned 'pappus'. Worldwide, there are about 10 genera and 370 species. Three species are frequent or common (Figs 08.14-08.16).

a.   Leaves *simple* .... b
Leaves *compound* (Fig. 08.14); the latter pinnate, with 9-13 lanceolate, distantly and toothed leaflets. Flowers in cymose clusters; corolla *pale-pink;* calyx teeth minute, but developing into a 'pappus'; stamens 3. *A frequent perennial of wet places.* **Valeriana officinalis**

b.   Corolla *spurred* (Fig. 08.15); the latter *red,* or rarely white; calyx an *annular ring;* stamen *1;* flowers numerous, in panicle-like cymes. The leaves opposite, ovate, entire, short-stalked upper sessile, glabrous, glaucous. *A frequent perennial of dry places.* **Centranthus ruber**

Corolla *not* spurred (Fig. 08.16); the latter *pale lilac*, nearly regular; calyx indistinct; flowers in simple cymes. Leaves small, narrow, sessile, entire, opposite, glabrous or pubescent. Stems rather brittle, well-branched, weakly angled, slightly pubescent below. *A frequent slender, erect annual; sandy areas and light sandy cultivated ground.* **Valerianella locusta**

23. Perianth *differentiated*, or very rarely of *5 white* or *pale blue* segments .... **29** (page 46)

Perianth *not clearly differentiated* into petals and sepals **24**

24. Leaves with *reticulate* or *palmate* venation .... **26** (page 43)

Leaves with 'parallel' venation, or corolla-tube *pubescent* **25**

25. Perianth *petal-like* .... **Liliaceae**

Perianth *sepal-like* .... **Plantaginaceae**

LILIACEAE

ONION FAMILY: Mostly glabrous perennial plants, very rarely small evergreen shrubs; rootstock a rhizome, corm, or tuber; stems erect or climbing, often modified into fleshy subterranean storage organs. Leaves alternate or in several whorls, or all radical, mostly lamellate but sometimes reduced to scales or sheaths, sometimes fleshy with prickly margins; simple, entire, sessile or petiolate, exstipulate and with or without sheathing bases; sometimes reduced to scales and functionally replaced by leaf-like lateral stems (cladodes); venation mostly 'parallel'. Inflorescence variable but mostly in *racemes* or *umbels*. Flowers hermaphrodite, *actinomorphic*, rarely or infrequently zygomorphic, and mostly with the *hypogynous* arrangement, rarely with the epigynous arrangement. Perianth *undifferentiated, petaloid*, often large and showy, of 6 segments, in 2 whorls of 3. Androecium of 6 stamens, in 2 whorls of 3, or inserted on the perianth segments. Gynaeceum *syncarpous*; mostly ovary *superior*, rarely inferior, 3-5 celled with 2-many ovules with axile placentation; styles 3-5 with capitate to linear stigmas. Fruit a *capsule* or *berry*, sometimes a fleshy capsule.

This is a large family of diverse habit and appearance but of uniform floral structure. Some classificatory systems (Stace, 1991) include as many as 8 other families in this taxon (see page 287). Worldwide, there are 250 genera and 2-3000 species. Four species are frequent or common throughout the British Isles (Figs 08.17-08.20).

a. Flowers in a *raceme* c

Flowers in an *umbel* .... b

b. Leaves *cylindrical* (Fig. 08.17); the latter hollow, somewhat grooved, very long and slender, 'parallel-veined'; petioles sheathing the scape. Flowers few, often replaced by bulbils. Perianth segments somewhat campanulate, pink or greenish-white; segments mostly oblong Plant bulbous. *A frequent biennial of roadsides, bushy places and banks.* **Allium vineale**

Leaves blades *flat* (Fig. 08.18); the latter all basal; blades broadly lanceolate, flat, abruptly contracted into a long petiole, 'parallel-veined'; petioles sheathing the scape. Flowers in a loose terminal umbel. Perianth segments lanceolate, acute, spreading, white. Plant bulbous. *A frequent biennial of woods, hedgerows and damp shaded places.* **Allium ursinum**

c. Flowers *yellow* (Fig. 08.19); the latter in a short terminal raceme; perianth segments broadly linear, spreading; stamen filaments densely covered with yellow woolly hairs. Leaves all radical, laterally compressed, 'parallel-veined', alternately arranged, rather stiff. Rhizomes creeping. *A frequent perennial of bogs, wet heaths and moors.* **Narthecium ossifragum**

Flowers *blue* (Fig. 08.20); the latter in a loose drooping, unilateral raceme; each flower subtended by a long narrow bract; the perianth campanulate, with spreading or reflexed, short lobes. Leaves, linear, radical, flat, shiny, with a prominent mid-rib, and 'parallel-veined'. *A very frequent biennial of woods, heaths and grassy places.* **Hyacinthoides non-scripta**

PLANTAGINACEAE

PLANTAIN FAMILY: Herbs, rarely branched undershrubs. Leaves with 'parallel' veins, mostly *basal*; *exstipulate*. Inflorescence in *spikes*. Flowers usually hermaphrodite, *actinomorphic*, with the *hypogynous* arrangement. Perianth *undifferentiated*; segments *4-lobed*. Androecium of 4 stamens. Gynaeceum *syncarpous*; ovary *superior*, 1-4 celled. Fruits *capsular*. A small family, worldwide there are only 3 genera and about 200 species. There are 4 common or frequent species which occur either as weeds of pasture and tillage, or in other habitats of ecological interest (Figs 08.21-08.24).

a. Leaves *linear* or *pinnately* lobed; corolla-tube *pubescent* c
Leaves *broadly ovate* or *lanceolate*; corolla-tube *glabrous* .... b

b. Leaves *broadly ovate* (Fig. 08.21); the latter with *conspicuous* 'parallel' venation, entire or irregularly toothed, *abruptly* narrower into a broad petiole, glabrous or somewhat pubescent. Spike cylindrical, and longer than its *unfurrowed* peduncle; corolla *yellowish-white*. A *common perennial cultivated ground, old pastures, waste places, etc.* **Plantago major**
Leaves *lanceolate* or *broadly-lanceolate* (Fig. 08.22); the latter with *conspicuous* 'parallel' venation, entire or distantly toothed, *gradually* narrowed into a petiole, glabrous or somewhat hairy. Spike short, *ovate*, shorter than its *deeply furrowed* peduncle; corolla *whitish* A *very common perennial of pastures, meadows, waste places, etc.* **Plantago lanceolata**

c. Leaves *pinnately* lobed (Fig. 08.23); the latter variable; mostly pubescent, narrow or linear; rarely entire. Spike cylindrical, shorter than its *unfurrowed*, pubescent peduncle; corolla *pale yellow. Common biennial of sandy soil, especially near the sea.* **Plantago coronopus**
Leaves *not* pinnately divided (Fig. 08.24); the latter *narrow-linear*, fleshy, entire, very rarely toothed, mostly with basal hairs. Spike cylindrical, shorter than its *unfurrowed* peduncle; corolla *whitish. A common perennial of salt-marshes and sea shores.* **Plantago maritima**

26. Perianth *sepal-like* or *absent* .... **28** (page 44)
Perianth *petal-like* **27**

27. Leaves *compound* .... **Fumariaceae**
Leaves *simple* .... **Polygalaceae**

FUMARIACEAE

FUMITORY FAMILY: Herbaceous plants, often lianous. Leaves alternate, compound, usually well divided, *exstipulate*, often *glaucous*. Inflorescence in *racemes* or spikes, rarely of solitary flowers. Flowers hermaphrodite, *zygomorphic*, and with the *hypogynous* arrangement. Perianth *not clearly differentiated*. Calyx of 2 small sepals, *petal-like*, caducous. Corolla of 4 petals, 1 or 2 of the outer whorl *spurred*, often of of a pinkish colour. Androecium of 6 stamens, united in 2 bundles of 3. Gynaeceum *syncarpous*; ovary *superior*, 1-celled. Fruit a capsule or *nutlet*. The family contains over 16 genera and 400 species, worldwide. There are 2 common weed species (Figs 08.25-08.26).

Leaf segments *linear-lanceolate* (Fig. 08.25); leaves bipinnate or tripinnate, *glaucous, glabrous*, often climbing by the petioles. Flowers in *leaf-opposed* racemes; peduncles *shorter* than the latter; corolla forming a *spur*, zygomorphic, *pink*, tips and wings blackish-red; upper petal *dorsally* compressed; lower petal *distinctly* spatulate; sepals 2, toothed, ovate, caducous. A *frequent annual of cultivated ground and waste places.* **Fumaria officinalis**
Leaf segments *oval* or *broadly-triangular* (Fig. 08.26); leaves bipinnate or tripinnate, *glaucous, glabrous*, often climbing by the petioles. The flowers in *leaf-opposed* racemes; peduncles *nearly equalling* or *longer* than the racemes; corolla extended into a *spur*, zygomorphic, pink, tips and wings blackish-red; the upper petal *dorsally* compressed; the lower petal *not* spatulate; sepals 2, irregularly toothed, ovate, caducous; pedicels rarely flexuous. *Very common annual of cultivated ground and waste places, etc.* **Fumaria muralis**

POLYGALACEAE

MILKWORT FAMILY: Herbs, shrubs, climbers, or sometimes small trees. Leaves alternate, rarely opposite or whorled, entire, sometimes scale-like, mostly *exstipulate*, simple. Inflorescence in *racemes*, rarely of solitary flowers. Flowers hermaphrodite, *zygomorphic*, and with the *hypogynous* arrangement. Corolla of 3-5 petals, the 2 outer free or united with the lower to form a tube. Calyx *polysepalous, zygomorphic*, of 5 sepals, 2 of which are very large and often petaloid. Androecium of 8 stamens; the latter mostly *monadelphous*. Gynaeceum *syncarpous;* ovary *superior*. Fruit a *loculicidal capsule*. Worldwide, there are about 10 genera and 700 species. Only 1 genus is represented in the British Isles. There are 2 frequent or common species (Figs 08.27-08.28).

> Leaves, all, *alternate* (Fig. 08.27); lower leaves small, 5-12 mm, lanceolate-linear; upper up to 35 mm; all entire, glabrous. Flowers in racemes; petals blue, pink or white; sepals very unequal. *A perennial, very frequent in pastures, sandy and rocky areas.* **Polygala vulgaris**
> Leaves, at least some, *opposite* (Fig. 08.28); upper somewhat ovate, upper larger; all entire, glabrous. The flowers in racemes; petals mostly gentian blue or slate blue; sepals very unequal. *Very common perennial of heathy ground and hill pastures.* **Polygala serpyllifolia**

28. Perianth *absent*, or *present* but consisting of only *3 sepals* .... **Euphorbiaceae**
Perianth *present*, or *absent* but flower subtended by 2 triangular bracts .... **Chenopodiaceae**

EUPHORBIACEAE

SPURGE FAMILY: Annual to perennial herbs or very rarely woody annuals, often with a milky sap. Mostly *monoecious,* rarely dioecious. Leaves simple or variously compound, stipulate or exstipulate, alternate, sometimes opposite or whorled. Inflorescence variously disposed but usually determinate. Flowers small, mostly *actinomorphic*, unisexual, sometimes much *reduced* by suppression of parts. Corolla mostly *absent*. Calyx *polysepalous*, absent or 3 sepals. Androecium of 1-15 stamens. Gynaeceum *syncarpous*; ovary *superior*, usually 3-celled with 1 ovule per cell; styles 2-3; stigmas strongly papillose or branched; male and female flowers in *Euphorbia* reduced to a *single* stamen and pistil, respectively. Fruit mostly a *capsule,* separating into 3 parts. Worldwide, the family contains up to 290 genera and 7300 species. There are about 5 common or frequent species (Fig 09.01-09.05).

> a. Leaves *alternate* or *spirally* arranged c
> Leaves *opposite* .... b
>
> b. Plant *rhizomatous* (Fig. 09.01). Leaves petiolate, oval to oval-lanceolate, opposite, more or less acute apically, crenate-serrate, dark green, pubescent; stipules *small.* Stems angular. Plant *dioecious.* Flowers small; perianth of *3 sepals*; male in clusters on long, lax, axillary spikes; female axillary, solitary or in pairs, on *long* pedicels; pistil syncarpous; the ovary, tuberculate. *An infrequent perennial in woods and shaded places.* **Mercurialis perennis**
> Plant *not rhizomatous* (Fig. 09.02). Leaves petiolate, broadly lanceolate to oval-lanceolate, opposite, crenate-serrate, somewhat acute, light green, scarcely pubescent; petiole with *2 small glands;* stipules *small.* Stems angular. *Dioecious.* Flowers small; perianth of *3 sepals;* male several, clustered on long-stalked spikes; female mostly solitary, *sessile;* ovary tuberculate. *Frequent annual of cultivated ground and waste places.* **Mercurialis annua**
>
> c. Bracts and leaves *entire* .... d
> Bracts and leaves *toothed* (Fig. 09.03); the latter falling early, spirally arranged, light green, few, *obovate*, tapered, obtuse and *serrulate* apically, more or less glabrous; stipules *absent.* Stems terete, often with *latex.* Umbels *5-rayed*; bracts, *foliaceous.* Flowers small, in a 'glandular cup'; glands *entire;* the perianth *absent;* male flowers many, of 1 stamen, on a *short filament and pedicel;* female *solitary*, on a *long* pedicel; pistil *syncarpous.* Monoecious. *Common annual of cultivated ground and waste places.* **Euphorbia helioscopia**

Fig. 08.00

N.B. DRAWINGS NOT TO SCALE

d.   Leaves *soft* and *thin* (Fig. 09.04); the latter caducous, *obovate,* alternate, light green, tapered, blunt, glabrous; the stipules *absent.* Stems terete, often with latex. Umbels *3-rayed;* bracts *foliaceous.* Flowers in a 'glandular cup'; perianth of *3* sepals; glands *2-horned;* male flowers many, of 1 stamen, on a *short filament and pedicel;* female solitary, on a *long* pedicel. Monoecious. *Common annual of cultivated ground and waste places.* **Euphorbia peplus**

Leaves rather *leathery, fleshy,* dense (Fig.09.05); the latter alternate, sessile, *glaucous,* entire, ovate-oblong. Stems terete. *Monoecious.* Umbel *3-6-rayed;* the bracts ovate, *foliaceous;* The flowers small, in a 'glandular cup'; glands *2-horned;* male flowers many, female solitary. *Frequent perennial of coastal sand-hills and sandy sea shores.* **Euphorbia paralias**

CHENOPODIACEAE

BEET FAMILY (GOOSEFOOT): Mostly *halophytic* annual or perennial herbs, shrubs; rarely trees. Leaves simple, often fleshy, alternate, *exstipulate,* and usually covered with *papillae,* giving a 'mealy' appearance. Inflorescence in small dense *cymes* or *spikes.* Flowers small, green or greenish, hermaphrodite or unisexual, occasionally monoecious, rarely dioecious, mostly *actinomorphic,* and with the *hypogynous* arrangement. Perianth *undifferentiated.* Corolla *absent.* Calyx mostly *gamosepalous,* of 2-5 persistent sepals; rarely absent. Gynaeceum *syncarpous;* ovary *superior* or *semi-superior.* Fruit a small *1-seeded capsule* or *nut,* or a *multigerm glomerule* (see page 255). Worldwide, there are about 100 genera and 1400 species. There are 5 common or frequent species (Figs 09.06-09.10).

a.   Leaves *flat* .... b

Leaves more or less *cylindrical* (Fig. 09.06); the latter small, alternate, glabrous, more or less *glaucous,* fleshy. Flowers sessile, axillary, hermaphrodite; sepals *5, rather fleshy. A frequent prostrate or spreading annual of salt-marshes and sea shores.* **Suaeda maritima**

b.   Leaves *dull green* c

Leaves glossy, *yellowish-green* (Fig. 09.07); the latter variable, frequently oval to rhomboid, petiolate, glabrous, entire or finely and distantly toothed, somewhat fleshy; upper alternate, lower in a rosette. Flowers small, clustered, occurring in loose terminal spikes; sepals fleshy. *Frequent perennial; rocky and gravelly seashores.* **Beta vulgaris** ssp. **maritima**

c.   Flowers *unisexual* .... d

Flowers *hermaphrodite* (Fig. 09.08); the latter in several dense axillary spikes, small; sepals 5, greenish, keeled apically; petals absent. Leaves short-stalked, alternate, lower more or less triangular or almost rhomboid, bluntly and coarsely toothed; upper gradually becoming narrower to almost linear, sessile, entire; all covered with whitish *papillae,* to give a 'mealy' effect. *An abundant annual of cultivated and waste ground.* **Chenopodium album**

d.   Lower leaves *tapered gradually* to the petioles (Fig. 09.09); the latter long-cuneate basally, alternate or opposite, rhomboid to lanceolate and *with 2 spreading basal lobes;* irregularly toothed or nearly entire upwards; the upper becoming narrower; all covered with minute *papillae,* giving a 'mealy' appearance; rarely glabrous. The flowers in slender leafy spikes; male flower with *5 sepals;* female *without* sepals but with *2 toothed* bracts. *A very common, erect or often prostrate annual; cultivated ground and waste places.* **Atriplex patula**

Lower leaves *truncate* or *shortly* cuneate basally (Fig. 09.10); the latter *abruptly* contracted to petiole, alternate or opposite, usually *triangular-hastate,* the margins of at least some making an angle of 90° with the petiole, irregularly toothed or nearly entire; the upper becoming narrower; all with *papillae,* giving a 'mealy' appearance. Flowers in slender leafy spikes; male flower with *5 sepals,* the female *without* sepals but with *2 toothed bracts. A frequent spreading annual of cultivated ground and waste places.* **Atriplex prostrata**

29.   Petals *free,* or only very slightly united but then leaves *grass-like* .... **36** (page 57)

Petals *united,* at least to some extent, or perianth of only *5 white* or *pale blue* segments **30**

30. Ovary *not divided* into 4 lobes .... **32** (page 51)

Ovary deeply *divided* into 4 lobes by 2 vertical clefts **31**

31. Stems *other than* quadrangular; leaves, at least some, *alternate* or *radical* .... **Boraginaceae**

Stems *quadrangular*; leaves, all, *opposite* .... **Lamiaceae**

BORAGINACEAE

BORAGE FAMILY: Annual, biennial or perennial herbs, sometimes shrubs or trees, hispid or scabrid, sometimes glabrous. Leaves simple, alternate, rarely the lower opposite, usually *entire, exstipulate*, occasionally sinuate. Inflorescence usually in *1-sided, coiled cymes*. Flowers mostly hermaphrodite, *actinomorphic*, rarely zygomorphic, and with the *hypogynous* arrangement. Corolla *gamopetalous*, 5-lobed, of 5 petals, rotate, funnel-shaped or *campanulate*, throat often closed by scales or hairs. Calyx *gamosepalous*, of 5 sepals, 5-lobed, sometimes deeply so. Androecium of 5 epipetalous stamens. Gynaeceum *syncarpous*, ovary *superior*, 2-4-celled or deeply divided into 4 lobes; style simple, terminal or from the middle of the 4 lobes. Fruit *schizocarpic*, mostly of *4 nutlets*, rarely a drupe. This is a family of wide distribution and contains about 90 genera and 2000 species, worldwide. There are 8 common or frequent species occurring as weeds in areas of agricultural interest (Figs 09.11-09.18).

a. Ovary or nutlets *not covered* with hooked or barbed prickles .... b

Ovary or nutlets *covered* with hooked or barbed prickles (Fig. 09.11); corolla *dull reddish-purple*, 5-lobed, nearly closed at the mouth with 5 *humps*; calyx 5-lobed, pubescent; flowers in *nodding* axillary cymes. Leaves linear-lanceolate, sessile, entire, and covered with soft hairs. *Frequent, densely hairy biennial; sandy waste places.* **Cynoglossum officinale**

b. Stamens, all, concealed *within* corolla tube c

Stamens, at least some, projecting well *beyond* corolla mouth (Fig. 09.12). Corolla pinkish-purple, later light blue, unequally 5-lobed; the calyx of 5 narrow, hairy sepals; the flowers in axillary spikes. Radical leaves short-stalked, cauline sessile, all lanceolate, entire, pubescent. *A very hispid biennial, frequent in sandy and gravelly places.* **Echium vulgare**

c. Flowers *erect* .... d

Flowers *nodding* (Fig. 09.13); the latter in 1-sided cymes; corolla campanulate, shortly and broadly 5-lobed, and yellowish-white, pink or purple; calyx deeply 5-lobed, hairy. Leaves broadly lanceolate, mostly entire, sparsely pilose or bristly, upper sessile, decurrent, lower short-stalked. *Hispid perennial of damp, shaded, grassy places.* **Symphytum officinale**

d. Leaves *lanceolate, oblanceolate* or *oblong-lanceolate* .... e

Leaves broadly *ovate* (Fig. 09.14); the latter acuminate, entire, alternate, stiffly hairy. Flowers in 1-sided axillary clusters; the corolla *bright blue*, centre white; calyx deeply 5-lobed, hairy. *A very hispid perennial, frequent in dry waste places.* **Pentaglottis sempervirens**

e. Calyx hairs *appressed* .... g

Calyx hairs *spreading* .... f

f. Corolla at first *white*, later becoming blue (Fig. 09.15); the latter with 5 rotate lobes; the tube about *twice* as long as calyx; the latter covered with hooked hairs. Flowers in 1-sided terminal cymes. Radical leaves tapered, cauline sessile, all oblong-lanceolate, entire, pubescent. *Slender annual, frequent in cultivated ground, poor pastures, etc.* **Myosotis discolor**

Corolla *bright blue* (Fig. 09.16); the latter with 5 blunt, rotate, concave lobes, and short tube; calyx deeply divided into 5 narrow lobes covered with *hooked* bristles. Flowers in 1-sided cymes. Radical leaves broadly lanceolate, short-stalked or tapered, cauline narrower, sessile, all entire, covered adaxially and abaxially with *bulbous* hairs, appearing 1-veined. *An annual, very common in most cultivated ground, waste places, etc.* **Myosotis arvensis**

g.   Style *3-4.0 mm* long in fruit (Fig. 09.17). Flowers in forked cymes; corolla *bright blue*, lobes
     flat, emarginate; calyx shortly 5-lobed, with *adnate* hairs. Leaves oblanceolate, sessile,
     with short bulbous hairs. *A frequent perennial of wet muddy places.* **Myosotis scorpioides**
     Style *1-1.5 mm* long in fruit (Fig. 09.18). Corolla bright blue, lobes rounded; calyx deeply 5-
     lobed, hairs adnate. The leaves lanceolate, narrowed basally, hairs adnate adaxially, sub-
     glabrous abaxially. *A common perennial of streams and wet muddy places.* **Myosotis laxa**

LAMIACEAE

LABIATE FAMILY: Predominantly annual or perennial herbs, sometimes shrubs, rarely trees or lianous;
herbage usually with aromatic oils. Stems and branches *quadrangular*. Leaves mostly opposite, rarely
whorled, mostly simple, rarely pinnately dissected and compound; *exstipulate,* toothed; often pubescent
and aromatic. Inflorescence mostly in *whorl-like axillary cymes*, rarely of solitary flowers. Flowers hermaphrodite, mostly *zygomorphic*, rarely actinomorphic, and with the *hypogynous* arrangement. Corolla
gamopetalous, with a well developed basal tube; basically 5-lobed apically, but with the 2 upper lobes
nearly always closely united to form a single lip; the central of the 3 lower lobes usually larger than the 2
laterals. Calyx *gamosepalous*, of 5 sepals; apices free; rarely bilobed. Androecium of 4 *didynamous*,
epipetalous, mostly free, rarely monadelphous (*Coleus*) stamens; rarely reduced to 2; staminodes rarely
present. Gynaeceum *syncarpous*, of 2 carpels, each with 2 ovules; ovary *superior*, apparently equally and
deeply 4-lobed. Fruit *schizocarpic,* separating into 4 *nutlets*. Members of this family form a very natural
taxon and are easily recognised on sight by their quadrangular stems, mostly opposite leaves, flower
arrangement, and distinctive gynaeceum. Worldwide, there are about 200 genera and 3200 species; many
are of great economic importance as a source of volatile aromatic essential oils and garden ornamentals.
Labiatae is an alternative legitimate name for this family. Many species are frequent or common as weed
plants in many habitats (Figs 09.19-10.02).

a.   Stamens *4* .... b
     Stamens *2* (Fig.09.19). The flowers in dense, distant, axillary, sessile clusters; corolla bluish-
          white, with few small purple dots, nearly *equally 4-lobed,* scarcely longer than the calyx;
          the latter deeply divided, and with 5 narrowly triangular, sharp-pointed lobes, pubescent.
          leaves opposite, short-stalked, coarsely toothed or lobulate, *glandular-pitted* abaxially,
          otherwise glabrous. *Frequent perennial; ditches and other wet areas.* **Lycopus europaeus**

b.   Calyx with 5 *more or less equal-sized* teeth .... d
     Calyx clearly *2-lipped* c

c.   Corolla *violet-purple* (Fig. 09.20); the latter 2-lipped, the upper lip very.concave, erect, near-
          ly entire, the lower spreading and unequally 3-lobed; calyx 2-lipped, crimson, the upper
          lip flat, broad, 3-toothed, lower smaller, deeply 2-lobed. Flowers in *dense axillary clus-
          ters.* The leaves petiolate, ovate-lanceolate, entire or finely toothed, sparsely hairy. Stem-
          ribs pubescent. *An abundant perennial; damp pastures, roadsides, etc.* **Prunella vulgaris**
     Corolla *yellow* (Fig. 09.21): the latter of 1 unequally 5-lobed lower lip, the upper lip *absent;*
          calyx pubescent, 2-lipped; upper lip broad and ovate, lower lobed. Flowers in slender ter-
          minal 1-sided racemes. Leaves, cordate, petiolate, crenate, rugose, oblong-lanceolate, pu-
          bescent. *Rhizomatous perennial of woods, heaths and rocky places.* **Teucrium scorodonia**

d.   Corolla *without* 4 equal-sized lobes .... f
     Corolla *with* 4 small, nearly equal-sized lobes .... e

e.   Flowers in *distant* axillary whorls (Fig. 09.22). Corolla *pinkish-mauve*, 4-lobed, lobes nearly
          equal or upper slightly broadest and notched, tube *shorter* than calyx; the latter shallowly
          divided, tubular, with 5 nearly equal lobes, pubescent. Leaves varying from lanceolate to
          nearly round, petiolate, cuneate or rounded basally, finely crenate or serrate, more or less
          pubescent. *A frequent perennial of cultivated ground and waste places.* **Mentha arvensis**

Fig. 09.00

N.B. DRAWINGS NOT TO SCALE

Flowers in *terminal* spikes or heads (Fig. 09.23). Corolla *bluish-mauve*, 4-lobed; lobes nearly equal or the upper slightly the broadest and emarginate; the tube shorter than the calyx; the latter shallowly divided with 5 nearly equal lobes, strongly pubescent. Leaves opposite, petiolate, ovate, *rounded basally,* serrate or crenate, often of a purplish hue, more or less pubescent. Stems more or less erect, simple or branched, often reddish in exposure. *A common perennial, strongly scented, of marshes, lakesides, ditches, etc.* **Mentha aquatica**

f. Corolla obviously *2-lipped* .... g
Corolla *1-lipped* (Fig. 09.24); the latter *blue*, rarely white or pink, its upper lip *very short*, its lower lip *3-lobed*; tube exserted; calyx equally 5-lobed to about the middle, fringed with jointed hairs. The flowers in axillary whorls. Radical leaves in a rosette, rounded apically; the cauline leaves opposite, petiolate to sessile; all leaves entire or obscurely crenate, obovate to oblanceolate, glabrescent. Stems pubescent, often on opposite sides only; stolons short, leafy. *A stoloniferous plant, very frequent in damp shaded places.* **Ajuga reptans**

g. Upper lip of corolla *concave* or *arched* .... h
Upper lip of corolla nearly *flat* (Fig. 09.25); the latter blue, with purple spots on the 4-lobed lower lip, upper lip blunt; the calyx tubular, with 5 nearly equal-sized lobes, hairy. Flowers in axillary whorls. Radical leaves long-stalked, the cauline short-stalked; all opposite, somewhat reniform to ovate-cordate, coarsely crenate, minutely pubescent adaxially, obtuse apically. Flowering stems suberect; stolons occasionally reaching 1.0 m. *A creeping perennial plant, very frequent in woods, hedges and grassy heaths.* **Glechoma hederacea**

h. Subnodal areas *neither* swollen *nor* covered with glandular hairs i
Subnodal areas, at least some, somewhat *swollen* and covered with glandular hairs (Fig. 09. 26). Leaves ovate to ovate-lanceolate, cuneate basally, acuminate, long-stalked, coarsely crenate-serrate, with scattered stiff hairs, at least adaxially. Flowers in dense terminal and axillary whorls. Corolla pale purple or white, 2-lipped; upper *laterally compressed,* lower 3-lobed with 2 *teeth* basally; the calyx divided, with 5 *long narrow, spine-like teeth,*hairy. *A very common coarse annual of cultivated ground and waste places.* **Galeopsis tetrahit**

i. Lateral lobes of lower corolla-lip *well developed* .... k
Lateral lobes of lower corolla-lip *obscure*, short, and with 1 or more small teeth .... j

j. Corolla *white* (Fig. 09.27); the latter up to 2 cm, strongly 2-lipped, laterally compressed; the upper lip hood-like, long, *ciliate*; lateral lobes *well defined,* and 1-toothed; the lower lip 3-toothed; calyx lobes long, narrow. Flowers in axillary clusters. Leaves opposite, petioles shorter than lades, ovate, cordate basally, acuminate apically, coarsely singly or doubly serrate, or crenate-serrate, pubescent, and clearly reticulate and rugose or undulate adaxially. *Perennial, with erect and creeping stems, frequent in grassy places.* **Lamium album**
Corolla *reddish-purple* (Fig. 09.28); the latter up to 15.0 mm, 2-lipped, laterally compressed; the upper lip hood-like, entire, with ring of hairs within base; lateral lobes *small, narrow,* lower lip deeply split; calyx hairy, deeply divided into 5 narrow lobes, often tinged with crimson. Flowers in dense axillary clusters. Leaves opposite, ovate, the lower with petioles longer than the blades; the upper short-stalked; all softly pubescent, clearly reticulate, widely cordate basally, blunt apically, regularly crenate, dull-green. *Very common pubescent, well-branched, annual of cultivated ground and waste places.* **Lamium purpureum**

k. Corolla *clearly* longer than the calyx l
Corolla *scarcely* longer than the calyx (Fig. 09.29); the former *pink or pale purple*, 2-lipped, with a ring of hairs within the tube base; calyx pubescent, tubular, divided into 5 triangular-lanceolate lobes. Flowers in *terminal* clusters. Leaves opposite, petiolate, oval-ovate in general outline, small, truncate or cordate basally, crenate-serrate, blunt apically, pubescent. Stems slender, branched basally, shortly pubescent. *Frequent, small, hairy, annual plant of tilled ground, especially those with non-calcareous soils.* **Stachys arvensis**

1.   Leaves *sessile* or very nearly so (Fig. 10.01); the latter oblong-lanceolate or linear-lanceolate, rounded basally, pointed, crenate-serrate, softly pubescent and clearly reticulate adaxially; the lower very short-stalked, upper sessile. Stems *hollow*, covered with whitish reflexed hairs. Flowers in a terminal spike; corolla 2-lipped, *dull purple*, the upper lip with *gland-tipped* hairs, the lower with 2 small lateral and 1 large central lobe; the calyx deeply divided into 5 triangular lobes, and often *gland-dotted*. *A rhizomatous tuberous plant, common in cultivated ground, ditches, riverbanks and other wet places, etc.* **Stachys palustris**

Leaves clearly *petiolate* (Fig. 10.02); the latter broadly oblong-lanceolate, cordate basally, acuminate, coarsely crenate-serrate, and covered with rather stiff hairs. The stems *solid*, densely hispid; hairs whitish. Flowers in an interrupted terminal spikes; corolla 2-lipped, dark reddish-purple, upper lip somewhat hood-like, pubescent outside; calyx deeply divided into 5 triangular acute lobes, and glandular or eglandular. *A rhizomatous etuberous plant, very frequent hedgerows, woods and waste shaded grassy places.* **Stachys sylvatica**

32. Flowers *not* in leaf-opposite cymes **33**

Flowers *in* leaf-opposite cymes .... **Solanaceae**

SOLANACEAE

POTATO (NIGHTSHADE) FAMILY: Herbs, shrubs, trees, often lianous, the stems with *bicollateral* vascular bundles. Leaves alternate or becoming opposite at or near the inflorescences, *exstipulate*, mostly simple, rarely pinnatisect or lobed. Inflorescence typically in *leaf-opposite cymes* or *axillary cymes*, rarely of solitary flowers. Flowers hermaphrodite, mostly *actinomorphic*, rarely slightly zygomorphic, and with the *hypogynous* arrangement. Corolla usually *gamopetalous*, of 5 petals, rotate or tubular, 5-lobed. Calyx *gamosepalous*, of 5 sepals, rarely more or less, usually persistent. Androecium of 5 epipetalous stamens; anthers *connivent* around the style. Gynaeceum *syncarpous;* ovary *superior*, 2-celled, though often appearing more due to the development of false septae; ovules numerous. Fruit mostly a *berry*, often becoming fleshy; rarely a capsule. This is a very important family with many cultivated plants such as the potato, tomato, peppers and aubergines (see page 271); there are up to 85 genera and 2200 species, worldwide. There are 2 common species which occur as weeds (Figs 10.03-10.04).

Corolla *purple* (Fig. 10.03); the latter with a short tube and 5 spreading lobes; calyx with 5 short, shallow, pubescent rounded lobes; the flowers in branched, stalked, *leaf-opposite cymes*. Leaves alternate, ovate or cordate, softly to thinly pubescent, dark green, pointed, at *least some* with 1 or more basal lobes, margins entire; the petioles short. Stems *woody, often scrambling*, up to 2 m in length. *A pubescent, rarely glabrous, common perennial of hedgerows, ditches, woods, waste places and cultivated ground, etc.* **Solanum dulcamara**

Corolla *white* (Fig. 10.04); the latter with a short tube and 5 *angular* spreading lobes which later become revolute; the calyx with a short tube and 5 obtuse lobes; flowers in scarcely branched *leaf-opposite* cymes. The leaves alternate, dark green, ovate to rhomboid, entire or *sinuate-dentate*, cuneate basally, acute. Stems erect, much branched, glabrous or minutely pubescent. *A frequent annual of waste and cultivated ground.* **Solanum nigrum**

33. Stamens *5-many* .... **34** (page 54)

Stamens *2-4* .... **Scrophulariaceae**

SCROPHULARIACEAE

FIGWORT FAMILY: Mostly herbs or small shrubs, sometimes lianous; some members are parasitic or saprophytic. Leaves variable, mostly simple, pubescent, *exstipulate*, alternate or opposite, rarely whorled. Inflorescence variable, in *panicles, racemes, spikes, cymes* or of *solitary* flowers. Flowers hermaphrodite, mostly *zygomorphic*, rarely actinomorphic, and with the *hypogynous* arrangement. Corolla *gamopeta-*

lous, of 4-5 petals; tube sometimes very short as in *Veronica,* or long as in *Digitalis;* rarely *campanulate* or spurred; sometimes bilobed. Calyx *gamosepalos,* rarely *polysepalous,* of 5 sepals, deeply 4-5 lobed; rarely inflated (*Rhinanthus*). Androecium of 2-4, epipetalous stamens which alternate with the corolla lobes; fifth stamen sometimes represented by a filamentous staminode. Gynaeceum *syncarpous;* ovary *superior,* 2-celled, with numerous ovules. Fruit mostly a *capsule.* There are about 200 genera and 3000 species, worldwide. Twenty-one species are frequent or common (Figs 10.05-10.25).

a. Corolla *not spurred* .... b
Corolla *spurred* (Fig. 10.05); the latter lilac, the tube streaked with crimson, the mouth yellow-spotted; calyx deeply 5-lobed; lobes narrow; flowers on relatively long pedicels, solitary in the axils of the leaves. Leaves long-stalked, often reniform, palmately-veined and broadly lobed, often purplish. *Trailing perennial, common on walls.* **Cymbalaria muralis**

b. Stamens *4-5* .... n
Stamens 2; petals 4, *marginally united* basally; sepals 4, *marginally united* basally c

c. Internodes *without* 2 lines of hairs .... d
Internodes with 2 lines of hairs (Fig.10.06), stems otherwise *hairless.* Leaves opposite, nearly sessile, ovate, coarsely crenate-serrate, dull green, hairy. Flowers in *racemes,* axillary; corolla 4-lobed, deep bright blue, with a *white eye;* the petals unequal; calyx lobes narrow, pubescent. *Common perennial of hedges, woods and old pastures.* **Veronica chamaedrys**

d. Flowers in *racemes* i
Flowers *solitary* .... e

e. Leaves *not reniform;* more than 1.5 cm .... f
Leaves *reniform* (Fig. 10.07); the latter small, up to 10 mm, short-stalked, crenate, pubescent, alternate or opposite. Flowers *solitary,* bright blue or somewhat purplish; petals unequal; pedicels *filiform,* much longer than the petioles; calyx lobes oblong, obtuse. *Free rooting, delicate perennial, abundant; lawns, pastures and damp waste areas.* **Veronica filiformis**

f. Leaves *regularly* toothed or entire .... g
Leaves *not* regularly toothed (Fig. 10.08); the latter alternate, stalked, light green, 3-nerved, with 3-7 well spaced teeth, hairs few, scattered; apical lobe large, entire. Flowers solitary, axillary, pale lilac; calyx lobes ovate-triangular, fringed; pedicels *shorter* than the leaves, hairy. *A common annual of cultivated ground and waste places, etc.* **Veronica hederifolia**

g. Capsule lobes *not* clearly divergent .... h
Capsule lobes *clearly* divergent (Fig. 10.09); the latter ciliate, *nearly twice* as broad as long, its lobes keeled. Flowers solitary, axillary, bright blue; calyx lobes narrow, ciliate; pedicels *longer* than the leaves. Leaves *triangular ovate,* serrate-crenate, light green, pubescent, stalked. *A common annual of cultivated ground and waste places.* **Veronica persica**

h. Leaves, some, *broader* than long (Fig. 10.10); the latter ovate, short-stalked, dull green, and somewhat truncate basally, coarsely and irregularly crenate-serrate, pubescent. Flowers solitary, axillary, bright blue; calyx lobes narrow, pubescent; pedicels *equalling* or *shorter* than the leaves. *A common annual of cultivated and waste ground, etc.* **Veronica polita**
Leaves, all, *longer* than broad (Fig. 10.11); the latter opposite on stem base, becoming alternate upwards, light green, ovate, coarsely and regularly crenate-serrate, tapered abruptly, glabrous or pubescent. Flowers solitary, axillary, pale blue; calyx lobes oblong, obtuse; pedicels *longer* than the leaves; capsule obscurely keeled, *longer than broad,* with *long glandular* hairs. *An infrequent annual of cultivated and waste ground.* **Veronica agrestis**

i. Plant *pubescent* or *puberulent* l
Plant *glabrous* .... j

j.   Leaves *sessile* .... k

Leaves *petiolate* (Fig. 10.12); the latter opposite, somewhat fleshy, *oval-oblong,* abruptly tapered, blunt apically, shallowly crenate-serrate, short-stalked, light green. Flowers in opposite racemes, axillary, slender, blue; calyx lobes ovate. *A succulent perennial with soft rooting stems, common in shallow streams, ponds, wet places, etc.* **Veronica beccabunga**

k.   Racemes *opposite* (Fig.10.13); the latter rather lax; inflorescence branches sometimes glandular; corolla pale blue; calyx lobes ovate-lanceolate, acute; pedicels *ascending* after flowering. The stems shortly creeping and rooting at the base. Leaves lanceolate, acute, nearly entire. *Frequent, fleshy perennial of ditches, muddy places.* **Veronica anagallis-aquatica**

Racemes *alternate* (Fig. 10.14); the latter very lax, slender, few-flowered, pale blue or white; calyx lobes ovate, 3-nerved; the pedicels slender, more than twice as long as the bracts. Leaves linear-lanceolate, opposite, sessile, appearing almost entire, glabrous, yellowish-green. *A very frequent perennial in wet muddy ground, marshes, etc.* **Veronica scutellata**

l.   Leaves *not* rounded apically and basally .... m

Leaves *rounded* apically and basally (Fig. 10.15); the latter sessile or short-stalked, oppopite, *oval-oblong,* glabrous or pubescent, entire or obscurely crenate, light green. Flowers in terminal leafy, loose racemes; corolla pale blue or white with darker lines; calyx lobes oblong. *A common perennial of old pasture, cultivated ground, etc.* **Veronica serpyllifolia**

m.   Racemes *terminal* (Fig. 10.16); the latter occupying most of the stem; corolla blue, *shorter* than the calyx; lobes of the latter lanceolate, fringed with *glandular* and non-glandular hairs; pedicels *short* or absent. Leaves short-stalked or sessile, *triangular-ovate,* toothed, opposite, hairy. *Common annual; cultivated ground and waste places.* **Veronica arvensis**

Racemes *axillary* (Fig. 10.17); the latter long-stalked, spike-like; corolla pale blue; the calyx lobes lanceolate, hairy; the pedicels *very short* or *absent.* Leaves opposite, short-stalked, ovate to obovate, crenate, tapered basally. *A pubescent perennial, very frequent in open woodlands, heaths, dry rocky places, hill pastures, often on dry soils.* **Veronica officinalis**

n.   Leaves *simple* .... p

Leaves *compound* .... o

o.   Upper lip of corolla with 4 teeth (Fig. 10.18); corolla irregular, pinkish-purple, *2-lipped,* the upper lip laterally compressed, toothed, the lower 3-lobed; calyx large, inflated, and with *2 broad, irregularly toothed* lobes, *pubescent outside.* Flowers in terminal spikes or racemes. Leaves *pinnate,* opposite or alternate, slightly pubescent, and with serrate-crenate lobes. *A frequent annual of bogs, wet meadows, marshy areas, etc.* **Pedicularis palustris**

Upper lip of corolla with 2 teeth (Fig. 10.19); corolla irregular, pink, *2-lipped,* upper lip laterally compressed, the lower 3-lobed; calyx 5-angled, with *4 small, leaf-like, 2-3-toothed lobes; the 5th lobe small and linear; pubescent or glabrous.* Flowers in terminal spikes or racemes. Leaves *pinnate,* opposite or alternate, oblong, slightly pubescent; leaflets crenate. *A very frequent perennial of heaths and damp upland pasture.* **Pedicularis sylvatica**

p.   Stems *terete* .... t

Stems *quadrangular* .... q

q.   Calyx *5-toothed* or *5-lobed* .... s

Calyx *4-toothed* or *4-lobed* .... r

r.   Corolla *yellow* (Fig. 10.20); the latter zygomorphic, and 2-lipped; the upper lip laterally compressed, 2-toothed, the lower lip 3-lobed; calyx 4-toothed, inflated and nearly orbicular, covered with a network of veins; the teeth short; flowers in leafy terminal spikes. Leaves opposite, sessile, narrowly lanceolate, coarsely serrate, and covered with very short stiff hairs. *Abundant annual of old pastures, meadows, other grassy places.* **Rhinanthus minor**

Corolla *purplish-pink* (Fig. 10.21); the latter zygomorphic, 2-lipped, the upper lip hood-like, glandular and slightly pubescent, and mostly entire, the lower lip with 3 entire lobes; the calyx campanulate, 4-toothed, covered with bulbous hairs; flowers in 1-sided spike-like racemes. Leaves opposite, sessile, narrowly lanceolate, hispid, remotely toothed. *A small annual plant, common in old pastures, roadsides and dry stony places.*   **Odontites verna**

s.   Stems *winged* (Fig. 10.22). Leaves more or less ovate, crenate, often with 1 or 2 small leaflets at the base, light green, opposite, glabrous, petiolate; the petioles *winged.* Flowers in cymes or panicles; the corolla zygomorphic, wide and inflated, nearly globose, brownish-purple above, greenish on the underside, with 5 small lobes, the upper 2 united basally; calyx lobes rounded with scarious borders. Rhizomes *not* nodular. *A frequent perennial of ditches, streams, shallow waterways and other wet areas, etc.* **Scrophularia auriculata**
Stems *unwinged* (Fig. 10.23). Leaves more or less triangular-acute, opposite, petiolate, *biserrate,* glabrous, truncate basally, usually slightly decurrent on petiole. Flowers in terminal panicles; corolla mostly zygomorphic, 2-lipped; tube greenish, upper lip reddish-brown; calyx *saucer-shaped,* deeply lobed; lobes ovate, with *very narrow* scarious borders. Rhizomes *nodular. Frequent perennial; hedgerows and waste places.* **Scrophularia nodosa**

t.   Leaves *alternate* and *10.0-40.0 cm* (Fig.10.24); the latter ovate-lanceolate, rather abruptly tapered into a stalk, obscurely toothed or nearly entire, covered with *soft white downy hairs.* Flowers in racemes; corolla *campanulate,* very long, pinkish-purple, with crimson spots on the inside, shallowly 5-lobed; calyx deeply 5-lobed, with 4 equal-sized lobes, and the 5th narrow; *much shorter* than the corolla. *Very common biennial or perennial, very tall, up to 150.0 cm, calcifuge plant, on most acid soils; a poisonous plant.* **Digitalis purpurea**
Leaves *opposite* and *3.0 cm or less* (Fig. 10.25); the latter dull green, glabrous, with minute marginal bristles or with a few hairs on margins and veins abaxially, oval-oblong, serrate. Flowers small, axillary, sessile; the corolla irregular, 2-lipped, whitish; upper lip rounded with small ovate notched lobes, lower *reflexed* with notched narrow lobes; calyx small, campanulate; teeth acute. *Frequent annual of heaths and pastures.* **Euphrasia officinalis**

34.   Leaves *other than* hastate or sagittate .... **35** (page 56)
Leaves *hastate* or *sagittate* .... **Convolvulaceae**

CONVOLVULACEAE

BINDWEED FAMILY: Erect or twining herbs, shrubs or small trees, usually with milky sap. Leaves mostly *sagittate* or *hastate,* alternate, simple, entire, mostly *exstipulate.* Inflorescence in *racemes, panicles* or of *solitary* flowers. Flowers hermaphrodite, often large and showy, mostly *actinomorphic,* and with the *hypogynous* arrangement. Corolla *gamopetalous,* funnel- or bell-shaped, of 4-5 petals. Calyx mostly *polysepalous,* usually of 5 sepals. Androecium of 5 stamens, epipetalous at corolla tube and alternate with the lobes. Gynaeceum *syncarpous;* ovary superior, 1-4 celled. Fruit a *capsule,* 2-4-valved, or splitting transversely. The family contains about 40 genera and 1000 species, worldwide. There are 2 common species which occur as weeds of tillage and waste places (Figs 10.26-10.27).

Calyx *enclosed* by 2 large bracts (Fig. 10.26); the latter 5-lobed, *campanulate,* hidden. Flowers *solitary,* axillary, on long pedicels; corolla *campanulate,* white or pink, with *recurved* lobes; *open by day and closed by night.* Leaves up to 15.0 cm, *ovate-triangular, alternate, deeply cordate basally,* acuminate, angular, short-stalked, glabrous; the nerves radiating from base. *Common climbing perennial; hedgerows, bushy places, etc.* **Calystegia sepium**
Calyx *not enclosed* by 2 bracts (Fig. 10.27); the latter deeply divided, with 5 rounded lobes, small. Flowers solitary, axillary, on long pedicels; the bracts small, *about mid-way* on the petiole; corolla *campanulate,* white or pink, shallowly 5-lobed, and apex spreading or recurved. Leaves *hastate* or *sagittate,* alternate, entire, short-stalked, mostly glabrous. *An abundant creeping perennial; cultivated ground, waste places, etc.* **Convolvulus arvensis**

N.B. DRAWINGS NOT TO SCALE

Fig. 10.00

35. Stamens *opposite* corolla lobes, or perianth of only *5 white* or *pale blue* segments .. **Primulaceae**
Stamens *alternating* with corolla lobes .... **Gentianaceae**

PRIMULACEAE

PRIMROSE FAMILY: Perennial or sometimes annual herbs, rarely small undershrubs. Leaves mostly simple, rarely compound, *exstipulate,* mostly opposite though occasionally confined to a basal rosette, very rarely in whorls of 3 or 4, petiolate or very rarely sessile, and either entire or very finely toothed. Inflorescence in *panicles, umbels, whorls* or as *solitary* flowers. The latter mostly hermaphrodite, *actino-morphic*, often heterostylous, bracteate, and with the *hypogynous* arrangement. Corolla *polypetalous*, of 5 petals, rotate, campanulate or funnel-shaped; rarely absent (*Glaux maritima*). Calyx *gamosepalous*, often deeply 5-lobed. Androecium of 5 epipetalous stamens which are situated opposite the corolla lobes, or if the latter are absent, alternating with the petaloid sepals; staminodes sometimes present. Gynaeceum *syncapous*; ovary *superior*, very rarely semi-inferior, 1-celled, with free-central placentation; ovules numerous; style 1; stigma capitate. Fruit a *capsule,* dehiscing by valves. This family contains about 30 genera and 800 species, worldwide. There are 6 frequent or common species (Figs 10.28-11.05).

a. Leaves, at least some, *cauline* c
   Leaves, all, *radical* .... b

b. Flowers *solitary* (Fig. 10.28); the petals creamy-yellow, united basally, rotate apically; calyx tubular, 5-ribbed, 5-lobed, pubescent. Leaves tapered *gradually* into a broad petiole, *ob-ovate to spatulate*; the margins *wavy*; fine-toothed, *glabrous* adaxially except on theveins, hairy abaxially, *wrinkled. Common rhizomatous plant; shaded habitats. **Primula vulgaris***
   Flowers in terminal *clusters* (Fig. 10.29); petals creamy-yellow, united basally, rotate apical-ly; calyxtubular, shortly 5-lobed, pubescent. Leaves *abruptly* contracted basally, finely pubescent *adaxially* and *abaxially*, ovate to oblong, finely toothed, wrinkled. *A frequent rhizomatous plant in old pastures, especially those on calcareous soils, etc.* **Primula veris**

c. Flowers *blue, red* or *pink*, or perianth only of *5* segments .... e
   Flowers *yellow* .... d

d. Stems erect (Fig. 11.01). Leaves in whorls of 3-4, rarely opposite, broadly to narrowly lance-olate, short-stalked, *dotted with orange* or *black glands*, entire or minutely toothed, *some-what* pubescent. Flowers in terminal, leafy, pyramidal panicles; petals yellow, united ba-ally; corolla tube short; sepals lanceolate, fringed with hairs; stamens filaments covered with *glandular* hairs. *Frequent rhizomatous plant on damp habitats.* **Lysimachia vulgaris**
   Stems *procumbent* (Fig. 11.02). Leaves opposite, *ovate*, nearly as long as broad, entire, *gla-brous,* short-stalked, mostly 3-nerved from base. Flowers *solitary*, long-stalked, axillary, yellow; petals united basally, rotate apically; sepals linear-lanceolate. *A frequent peren-nial in woods, shaded hedgerows and other damp shaded habitats.* **Lysimachia nemorum**

e. Flowers clearly *stalked* .... f
   Flowers *sessile* or nearly so (Fig. 11.03); the latter small, axillary, forming a leafy spike-like raceme; corolla *absent*; sepals *white* or *pale pink* and resembling petals, narrowly ob-ovate, united basally; stamens alternating with the sepals. Leaves small, ovate or oblong in general outline, entire, glabrous; lower opposite, upper alternate. *Very common, small, procumbent or suberect, perennial of salt-marshes and muddy shores.*   **Glaux maritima**

f. Stems *quadrangular* and *not* rooting at the nodes (Fig. 11.04). Leaves mostly opposite, rare-ly in whorls of 3, *sessile*, ovate, mostly 3-nerved from the base, glabrous. Flowers *brick-red*, solitary, on slender petioles; the petals shortly united basally, rotate apically, fringed; calyx deeply divided; sepals narrowly lanceolate; filaments *fringed* with jointed hairs. *A common glabrous annual: tillage fields, waste places, sand-hills, etc.* **Anagallis arvensis**

Stems somewhat *rounded* and *rooting* at the nodes (Fig. 11.05). Leaves ovate-orbiicular, opposite, entire, *very short-stalked*, glabrous. Flowers axillary, solitary or in pairs, on slender pedicels, *pale pink*; petals united, *funnel-shaped*; sepals narrowly lanceolate. *Creeping perennial, locally abundant in bogs, damp peaty soils, waste places. Anagallis tenella*

GENTIANACEAE

GENTIAN FAMILY: Mostly glabrous annual herbs, rarely shrubs or undershrubs. Leaves mostly opposite, simple, *exstipulate, entire,* decussate, sessile or very nearly so. Vascular bundles *bicollateral*. Inflorescence in terminal dichasial or monochasial *cymes*, rarely solitary. Flowers hermaphrodite, *actinomorphic*, and with the *hypogynous* arrangement. Corolla *gamopetalous*, of 8 petals, with a cylindrical tube and spreading lobes, variously coloured, persistent in fruit. Calyx *polysepalous* or *gamosepalous*, of 8 sepals. Androecium with epipetalous stamens, equalling in number, and alternating with, the corolla lobes. Gynaeceum *syncarpous*; ovary *superior*, usually unilocular, with 2 parietal placentas, with numerous anatropous ovules, or rarely bilocular by the intrusion of the placentas; style simple with a single or bilobed stigma or 2 stigmas. Fruit a *septicidal capsule*. Worldwide, there are about 70 genera and 800 species. The opposite leaves and unilocular ovary with numerous ovules with parietal placentation, and actinomorphic flowers, are characteristic. Two are frequent or common species throughout the British Isles (Figs 11.06-11.07).

Corolla *yellow* (Fig. 11.06); the latter rotate with 8 spreading lobes; calyx deeply divided into 8 linear-subulate segments. The flowers in terminal dichasial cymes. Leaves oval, glaucous, united in pairs basally, the stem apparently passing through the centre of the 'composite' leaf. *A frequent annual; dry calcareous grasslands, dunes. Blackstonia perfoliata*
Corolla *pink* (Fig. 11.07); the latter funnel-shaped, contracted apically and with 5 spreading lobes; calyx deeply 5-lobed; lobes narrow, keeled. Flowers numerous, crowded in dense corymb-like cymes. Radical leaves obovate or somewhat spatulate; cauline shorter, sessile. *Frequent annual of dry grasslands, banks and sand-hills, etc.Centaurium erythraea*

36.   Leaves *alternate, opposite, whorled,* or *spirally* arranged .... **38** (page 58)
        Leaves *all radical* **37**

37.   Leaves *compound* .... **Oxalidaceae**
        Leaves *simple* .... **Plumbaginaceae**

OXALIDACEAE

WOOD-SORREL FAMILY: Perennials, rarely annuals, often slightly succulent herbs, occasionally with bulbs or rhizomes. Leaves mostly radical, rarely alternate, stipulate or *exstipulate*, petiolate, pinnate or palmate, or simple by the suppression of leaflets; entire; often showing sleep movement. Inflorescence in axillary *cymes* or *racemes,* or of *solitary* flowers. Flowers hermaphrodite, *actinomorphic*, often heterostylous, heterochlamydeous, and with the *hypogynous* arrangement. Corolla *polypetalous*, of 5 petals, the latter *contorted*. Calyx *polysepalous*, of 5 usually imbricate sepals. Androecium of 10, rarely 15, basally *connate*, weakly obdiplostemonous stamens. Gynaeceum *syncarpous*; ovary *superior*, 3-5-celled; with 1-many anatropous ovules with axile placentation; styles 3-5; stigmas minute, capitate. Fruit a *loculicidal capsule*, rarely a berry. Worldwide, there are 7 genera and about 1000 species. The ternate, or less often palmate, leaves and the conspicuous actinomorphic flowers, with contorted petals, are characteristic. There is 1 common species throughout the British Isles (Fig. 11.08).

Leaves all *radical* (Fig. 11.08); the latter trifoliate; petioles long; leaflets ciliate and with appressed hairs, cuneiform-obcordate. Flowers *solitary*; pedicels at least equalling petioles. Petals *white*, obovate, emarginate apically. Sepals obovate, slightly pubescent. Rhizomes slender, creeping, scaly. *Common perennial; woods and shaded places.Oxalis acetosella*

PLUMBAGINACEAE

SEA-LAVENDER FAMILY: Mostly perennial herbs, rarely annuals or shrubs. Leaves mostly *radical*, rarely spirally arranged, entire, *exstipulate*. Inflorescence racemose or *cymose*, the cymes sometimes closely aggregated into heads. Flowers small, numerous, hermaphrodite, *actinomorphic*, and with the *hypogynous* arrangement. Corolla *gamopetalous* or *polypetalous*, of 5 petals. Calyx *polysepalous*, of 5 sepals, persistent, scarious and often angled or winged. Androecium of 5 free or epipetalous stamens. Gynaeceum *syncarpous*; ovary *superior* but remaining enclosed in the calyx tube, unilocular, with 1 basal anatropous ovule; styles 5 or 1 with a 5-lobed stigma. Fruit mostly a minute *1-seeded capsule*. World-wide, there are about 10 genera and 300 species. Two species are frequent or common (Figs 11.09-11.10).

Flowering stem *branched* (Fig.11.09); flowers small, short-stalked, in clusters, several; each cluster with 3 scale-like bracts. Corolla *bluish-mauve*. Calyx tubular, shortly 5-lobed. The leaves radical, oblong-oblanceolate, narrowed basally into long petioles, glabrous; lateral veins obscure. *Frequent perennial of salt-marshes; muddy sea shores.* **Limonium humile**
Flowering stem *unbranched* (Fig. 11.10); flowers in more or less *globular heads*; peduncles mostly *shortly pubescent*. Corolla *pink*, rarely white. Calyx tube with hairy ribs; the calyx teeth with long bristles. Leaves all radical, rather fleshy, pubescent, acute or blunt, *grass-like*, linear. *Common perennial of salt-marshes and dry coastal areas.* **Armeria maritima**

38.   Flowers *actinomorphic* **39**
           Flowers *zygomorphic* .... **Resedaceae**

RESEDACEAE

MIGNONETTE FAMILY: Annual or perennial herbs, rarely woody plants. Leaves alternate or spirally arranged, simple or pinnately divided, rarely *glandular stipulate*. Inflorescence in *spike-like racemes* or *spikes*. Flowers mostly hermaphrodite, *zygomorphic*, and with the *hypogynous* or *perigynous* arrangement. Corolla *polypetalous*, of 4-6 divided petals. Calyx *zygomorphic, polysepalous*, of 4-6 sepals. Androecium of 3-40 stamens, usually borne on a nectar-secreting disc. Gynaeceum *syncarpous*, of 2-6 carpels; ovary *superior*, unilocular, usually *open* apically; stigmas absent; ovules numerous. Fruit a *capsule*. Worldwide, there are 6 genera and 70 species. Two species are frequent (Figs 11.11-12).

Leaves *entire* (Fig. 11.11); the latter both radical and cauline, linear or narrowly lanceolate, glabrous, with rather *wavy margins*; cauline narrower, sessile. Flowers in long, terminal, spike-like racemes; pedicels slender; petals 4-5, *yellowish-green*, some deeply 3-lobed; sepals 4, small; stamens 20-25. *Frequent biennial of dry gravelly places.* **Reseda luteola**
Leaves *pinnately lobed* (Fig. 11.12); the latter both radical and cauline, simple, and glabrous; lobes narrowly oblong, blunt, entire; upper leaves sessile. Flowers in long, stiff, terminal, spike-like racemes; petals usually 6, *yellowish-green*, some deeply 3-lobed; sepals usually 6, linear, unequal; stamens 12-20. *Frequent biennial; dry gravelly places.* **Reseda lutea**

39.   Stamens *10* or *less* .... **41** (page 60)
           Stamens *12* or *more* **40**

40.   Leaves *entire* .... **Hypericaceae**
           Leaves *lobed* or *toothed* .... **Papaveraceae**

HYPERICACEAE

ST JOHN'S-WORT FAMILY: Trees, shrubs or herbs with resinous juice. Leaves *entire*, mostly *opposite*, rarely whorled, *sessile, exstipulate, punctate*. Inflorescence mostly in terminal, solitary or in branched *cymes.* Flowers showy, hermaphrodite, *actinomorphic*, and with the *hypogynous* arrangement. Corolla *polypetalous*, of 5 petals, *yellow*. Calyx *polysepalous*, of 5 sepals. Androecium of *numerous* stamens; the latter often *fascicled*; anthers versatile. Gynaeceum *syncarpous*, ovary *superior*, 1-5-celled; ovules numerous, anatropous, with axile placentation; styles 3 or 5. Fruit a *capsule*, rarely a drupe or berry. This is a rather small family with about 8 genera and 375 species, worldwide. Economically, the family is of little importance. Some species contain chemicals which have *photosensitizing* properties. Others are useful garden ornamentals. There are 5 common or frequent species (Figs 11.13-11.17).

a. Stems *erect, firm* .... b
Stems *prostrate, creeping* (Fig. 11.13); the latter softly pubescent, soft, rooting at the nodes. Leaves opposite, sessile, rounded or suborbicular, softly pubescent and *semi-amplexicaul*. Flowers in axillary cymes; the petals much longer than sepals; the latter fringed with short stalked glands. *A common stoloniferous plant of wet marshy places.* **Hypericum elodes**

b. Stems *without* 2 raised lines .... d
Stems *with* 2 raised lines c

c. Leaves *mostly 3-nerved* from the base (Fig. 11.14); the latter up to 3.0 cm, opposite, entire, ovate-oblong, abundantly furnished with *translucent glandular dots*. Flowers in terminal axillary cymes; petals twice as long as the sepals, with marginal glands; stamens united into *3 bundles*. *A common perennial of dry banks, roadsides, etc.* **Hypericum perforatum**
Leaves with 1 central vein and *many* laterals (Fig. 11.15); the latter large, up to 8 cm, broadly ovate, rounded basally, blunt, with *few* translucent dots. Flowers few in terminal cymes sepals *very unequal*, the larger equalling petals; the stamens united into *5 bundles*. *A frequent perennial of damp woods and other damp rocky places.* **Hypericum androsaemum**

d. Stems *quadrangular* (Fig. 11.16); the latter acutely 4-angled or winged, glabrous, stiff, erect, with slender stolons at base, and branched. Leaves up to 3.0 cm, ovate, sessile, opposite, entire, obtuse, semi-amplexicaul, furnished with numerous small *translucent dots*. Flowers in many-flowered, flat-topped cymes; petals *not* gland-tipped; sepals entire, lanceolate, acute. *Frequent perennial; ditches and damp marshy places.* **Hypericum tetrapterum**
Stems *terete* (Fig. 11.17), erect or ascending, and slender. The leaves opposite, sessile, glabrous, entire, broadly ovate, cordate, rounded apically, and semi-amplexicaul; *translucent dots* towards the margins. Flowers rather few, in a loose cyme; the petals with marginal sessile black glands; sepals densely *fringed* with sessile *black glands*. *A frequent perennial; dry woods and rough grassy places on non-calcareous soils.* **Hypericum pulchrum**

PAPAVERACEAE

POPPY FAMILY: Herbaceous annual or perennial plants, rarely shrubs or trees; milky or coloured sap often present. Leaves alternate, rarely whorled or spirally arranged, entire to *pinnately* or palmately divided, *exstipulate*. Inflorescence mostly of *solitary* flowers. The latter hermaphrodite, *actinomorphic*, often showy, and with the *hypogynous* arrangement; rarely perigynous. Corolla *polypetalous*, mostly of 4 petals, often crumpled at first. Calyx *polysepalous*, mostly of 2 sepals; the latter frequently *caducous*; rarely absent. Androecium of *numerous*, rarely 4, stamens, in several whorls. Gynaeceum *syncarpous*; ovary *superior*, unilocular; ovules numerous, with parietal placentation. Fruit usually a *capsule*, opening by valves or apical pores; seeds *numerous*. Worldwide, the family contains about 30 genera and 430 species. Four species are frequent or common throughout the British Isles (Figs 11.18-11.21).

a. Petals *scarlet, pink* or *white* .... b
Petals *bright yellow* (Fig. 11.18); the latter broadly obovate, caducous; sepals 2, small, pubescent; fruit *long* and *narrow*, up to 5 cm, cylindrical. Leaves *irregularly pinnate*, lobes coarsely toothed, pubescent, slightly glaucous. The stems brittle, slightly glaucous, often with yellow latex. *A perennial, frequent on walls and in hedgerows.* **Chelidonium majus**

b.   Leaves *pinnatisect* c
Leaves *toothed* or *shallowly* lobed (Fig.11.19); the latter ovate-oblong, glabrous; lobes irregular, coarsely toothed; the cauline clasping the stems; the latter mostly pubescent. Corolla
bluish-white; sepals 2, caducous. *A frequent annual of waste places.* **Papaver somniferum**

c.   Flowering stems with *appressed* hairs (Fig. 11.20). Leaves bristly pubescent, divided almost
to the mid-rib, the lobes further divided into oblong segments which are rather *broad* and
abruptly acute; terminal segments *relatively* small. Petals *orange-pink*; sepals hairy; fruit
long and *narrow. Common annual; cultivated ground and waste places.* **Papaver dubium**
Flowering stems with *exserted* hairs (Fig. 11.21). Leaves bristly pubescent, divided almost
to the mid-rib, the lobes further divided into segments which are rather *narrow* and *acute*
terminal lobes equalling laterals. Petals large, *scarlet*, rarely white; sepals pubescent; fruit
short and *broad. Common annual of cultivated ground and waste places.* **Papaver rhoeas**

41.   Flowers in *other than* in racemes .... **42** (page 63)
Flowers in *racemes* .... **Brassicaceae**

BRASSICACEAE

TURNIP FAMILY: Annual, biennial or perennial herbs, rarely woody plants; forked or *stellate unicellular hairs* often present. Leaves alternate, rarely subopposite, simple or compound, *exstipulate*. Inflorescence in *racemes*. Flowers hermaphrodite, *actinomorphic*, and with the *hypogynous* arrangement. Corolla
polypetalous, of 4 petals; the latter rarely absent. Calyx *polysepalous*, of 4 sepals. Androecium of *2, 4* or
6 free stamens. Gynaeceum *syncarpous*; ovary *superior*. Fruit a *capsule,* mostly a *siliqua* or a *silicula*;
rarely *schizocarpic*. An important family (see page 233), it has many economic species.   An alternative
legitimate name for this taxon is the *Cruciferae*. Worldwide, it contains about 350 genera and 3000 species.   There are about 17 common or frequent species (Figs 11.22-11.10).

a.   Leaves, *none*, amplexicaul .... e
Leaves, at least *some,* amplexicaul .... b

b.   Corolla *yellow* .... d
Corolla *white* or *mauve* c

c.   Fruit *cordate* apically (Fig. 11.22); the latter tapered basally; flowers very small, in a slender
terminal raceme; petals *white*, spatulate. Radical leaves in a dense rosette, mostly deeply
pinnatifid, rarely entire, oblanceolate, tapered basally into a petiole; cauline small, entire
or somewhat toothed, always with acute auricles; all leaves dull green, with *stellate* hairs.
Very common annual of cultivated ground and waste places, etc. **Capsella bursa-pastoris**
Fruit *not* cordate apically (Fig.11.23); the latter *broadly ellipsoidal*; the flowers in *corymbose*
racemes; the petals *white* or *mauve*. Radical leaves in a loose rosette, long-stalked, ovate;
the base widely *cordate*; entire or toothed; cauline sessile, auriculate, somewhat 5-sided;
all glabrous. *A common fleshy plant; rocky and muddy sea-shores.* **Cochlearia officinalis**

d.   Stems *terete* (Fig. 11.24); the latter mostly smooth. The radical leaves petiolate, usually with
stiff bristly hairs, lyrate-pinnatifid; cauline sessile, auriculate, oblong-lanceolate, mostly
entire, *glaucous*, more or less glabrous. Flowers in a terminal raceme; petals *bright yellow*, longer than sepals; the latter spreading; the fruit *long and narrow*, ending with a long
beak. *A common annual or biennial of cultivated ground and waste places.* **Brassica rapa**
Stems *angular* (Fig.11.25); the latter branched, glabrous. Radical leaves *lyrate-pinatisect*,
with a rounded, often cordate, terminal lobe; stalked; lower cauline leaves deeply divided,
with a large terminal and small lateral lobes; the upper cauline sessile, auriculate, mostly
toothed; all deep green, shiny, glabrous. Flowers in axillary racemes; petals bright yellow. *A frequent biennial or perennial of roadsides and waste places.* **Barbarea vulgaris**

N.B. DRAWINGS NOT TO SCALE

Fig. 11.00

e.   Leaves *not* pinnatisect; petals *mostly present* .... g
       Leaves *pinnatisect*; petals *mostly absent* .... f

f.   Fruit *mucronate* (Fig. 11.26); the latter very small, bilobed, shorter than its stalk, constricted vertically. Leaves *finely divided*; lobes narrow, entire or toothed; mostly glabrous; *foetid*; the basal and lower stem leaves stalked; upper leaves sessile. Stems prostrate or ascending, branched. Flowers minute, in *leaf-opposed* racemes; the petals *absent*; sepals minute. Stamens 2. *A frequent annual of cultivated ground and waste places.* **Coronopus didymus**
       Fruit *emarginate* (Fig. 11.27); the latter very small, bilobed, longer than its stalk, constricted vertically. Leaves *finely divided*; lobes narrow, entire or toothed; mostly glabrous; *foetid*; the basal and lower stem leaves stalked; upper leaves sessile. Stems prostrate or ascending, branched. Flowers minute, in *leaf-opposed* racemes; the petals *absent*; sepals minute. Stamens 2. *Frequent annual; cultivated ground and waste places.* **Coronopus squamatus**

g.   Leaves *entire, toothed* or *lobed* l
       Leaves *pinnate* .... h

h.   Stems *erect* .... j
       Stems *prostrate* i

i.   Fruit with 2 rows of seeds in each loculus (Fig. 11.28). Stems *weak, stout*, angular, glabrous, *floating* or *creeping*. Leaves alternate, and with rounded or oval, toothed or entire leaflets; terminal leaflet the largest. Racemes corymbose; flowers small; petals *white*, nearly twice as long as the sepals. *A common perennial of ditches, streams, etc.* **Nasturtium officinale**
       Fruit with *1* row of seeds in each loculus (Fig. 11.29). Stems *weak, stout*, angular, glabrous, *floating* or *creeping*. Leaves alternate, and with rounded or oval, toothed or entire leaflets; terminal leaflet the largest. Racemes corymbose; flowers small; petals *white*, nearly twice as long as sepals. *A common perennial of ditches, streams, etc.* **Nasturtium microphyllum**

j.   Corolla *white* .... k
       Corolla *lilac* (Fig.12.01); the latter about thrice as long as sepals; the anthers *yellow;* racemes corymbose. The stems erect or ascending, usually simple, terete. Leaves pinnate; leaflets of the radical leaves larger than those of the cauline leaves; the former ovate or rounded; those of the latter lanceolate; the terminal leaflets largest; all more or less glabrous. *Common, often stoloniferous perennial; marshes, wet grassy areas, etc.* **Cardamine pratensis**

k.   Stamens *4* (Fig. 12.02); flowers small, at first in corymbose racemes; petals narrow, twice as long as the sepals; the latter with white margins. Stems *straight.* Radical leaves in a *compact* rosette, numerous, stalked, with broadly ovate or rounded leaflets; the cauline leaves fewer, almost sessile, with narrower leaflets; leaves either pubescent or glabrous. *A slender, erect, common annual of cultivated ground and waste places, etc.* **Cardamine hirsuta**
       Stamens *6* (Fig. 12.03); flowers small, at first in corymbose racemes; petals narrow, twice as long as the sepals; the latter with white margins. Stems *flexuous.* The radical leaves in a *loose* rosette, with rounded or broadly ovate leaflets; cauline leaflets narrower; glabrous or pubescent. *A common annual; mostly damp or wet waste places.* **Cardamine flexuosa**

l.   Corolla *yellow* .... o
       Corolla *white* or *lilac* .... m

m.   Leaves *toothed* .... n
       Leaves, at least some, deeply *pinnatifid* (Fig.12.04); the lower leaves *lyrate-pinnatifid,* with a large rounded terminal lobe, and usually with much smaller laterals; upper oblong, pinnately lobed or toothed; all deep green, more or less *bristly*. Petals *white* or *lilac*, usually dark veined; racemes large, terminal, showy; fruit with *few* seeds, *long-beaked*, and *constricted* between seeds. *Frequent annual; sandy and peaty soils.* **Raphanus raphanistrum**

n. Leaves, at least *some*, cordate basally (Fig. 12.05); radical leaves in a rosette, long-stalked, *cordate* or *reniform*; cauline short-stalked, *cordate*; all pointed, toothed, glabrous or nearly so. Flowers in terminal and lateral racemes; fruits very long, narrow. Plant with garlic smell when crushed. *A frequent biennial of roadsides and waste places.* **Alliaria petiolata**
Leaves, *none*, cordate (Fig. 12.06); the latter mostly basal, oblanceolate, petiolate, dentate, mostly with branched hairs; cauline few, sessile. Flowers small, in a long, loose raceme; fruits *very narrow*. *Annual of cultivated ground and waste places.* **Arabidopsis thaliana**

o. Beak of fruit more or less *cylindrical* .... p
Beak of fruit laterally *compressed, sword-shaped* (Fig. 12.07); the latter long, often *curved*; fruit few-seeded, often *bristly*; the flowers in long terminal racemes. Leaves *deeply pinnatifid*, terminal lobe largest, covered with *bristly hairs* or rarely glabrous; lobes toothed; venation distinct. *A frequent annual of cultivated ground and waste places.* **Sinapsis alba**

p. Beak relatively long, *distinct* .... q
Beak *minute* (Fig. 12.08); the fruits *stiffly erect* and more or less *appressed* to the stems, glabrous or pubescent. Stems rather *woody*. The leaves deeply and *unevenly* pinnately lobed; all lobes irregularly toothed, bristly hairy. Flowers very small, in short, terminal racemes. *A common annual of cultivated ground, roadsides, waste places.* **Sisymbrium officinale**

q. Fruit *deeply constricted* between seeds (Fig. 12.09); the latter with *few* seeds and a *very long slender beak*, glabrous. Lower leaves *deeply lyrate-pinnatifid*, with a large rounded terminal lobe, and usually with much smaller laterals; the upper oblong, pinnately lobed or toothed; all deep green, more or less *bristly*. Petals *white*, usually dark-veined; racemes large, terminal, showy. *Frequent annual; sandy and peaty soils.* **Raphanus raphanistrum**
Fruit *not* deeply constricted between the seeds (Fig. 12.10); the latter, at most, *slightly* constricted, with many seeds, and a *stout short* beak; glabrous or stiffly hairy. Stems mostly covered with *short, bristly* hairs. Flowers large, in loose terminal racemes. Leaves variable; lower short-stalked and irregularly lobed; middle cauline more toothed than lobed, sessile, broadly ovate; the upper cauline often entire; all deep green, usually covered with short, stiff hairs. *Abundant annual; cultivated ground and waste places.* **Sinapis arvensis**

42. Stamens *free* .... **Caryophyllaceae**
Stamens *connate basally* .... **Linaceae**

CARYOPHYLLACEAE

CHICKWEED FAMILY: Annual or perennial herbs, sometimes suffrutescent shrubs. Leaves mostly *opposite, decussate,* rarely alternate or apparently whorled, *entire*, mostly *exstipulate*, often connected basally by a transverse line, mostly glabrous or with a few hairs axially. Inflorescence mostly in *cymes*, rarely of solitary flowers. Flowers hermaphrodite, *actinomorphic*, and mostly with the *hypogynous* arrangement, rarely perigynous. Corolla *polypetalous*, of 4-5 petals, sometimes *notched* or deeply *lobed;* rarely absent. Calyx *polysepalous* or *gamosepalous*, of 5, rarely 4, sepals. Androecium of 3-10 free stamens. Gynaeceum *syncarpous*; ovary *superior*, 1-celled; styles 2-5; ovules 1-many, campylotropous; placentation free-central; seeds numerous. Fruit mostly a *capsule*. Worldwide, there are about 80 genera and 2000 species. Fifteen species are common or frequent in the British Isles (Figs 12.11-12.26).

a. Sepals *free* to the base, or very nearly so .... e
Sepals *united* into tubular calyx, free only at the apices .... b

b. Styles *3-5* c
Styles *2* (Fig. 12.11); flowers in cymes; petals *entire* or *nearly* so, large, *pale-pink* or nearly white; calyx 5-toothed. Leaves broadly ovate, acute, connected by a small rim, mostly 3-5-nerved, usually glabrous. *A frequent perennial of grassy places.* **Saponaria officinalis**

c. Petals *2-lobed* .... d
Petals *4-lobed* (Fig. 12.12); the latter *reddish,* rarely white; lobes long and narrow; the calyx shortly 5-lobed; flowers in terminal and lateral, loose, cymes. Lower leaves oblanceolate, tapered basally into a petiole-like base, acute; upper narrower and sessile; all glabrous or fringed with hairs basally. *Very frequent perennial of marshy ground.* **Lychnis flos-cuculi**

d. Styles *5,* or flowers *staminate* only (Fig. 12.13). Plant *dioecious,* the stems pubescent. Lower leaves oblanceolate, narrowed into stalks; upper lanceolate, sessile, acuminate, connected basally by a narrow hairy rim; all opposite and pubescent. Flowers *unisexual,* in a few-flowered loose cyme; petals large, *white,* with a narrow claw; the calyx shortly 5-toothed, mostly hairy. *A frequent perennial of dry cultivated ground and waste places.* **Silene alba**
Styles *3* (Fig. 12.14). Lower leaves opposite, broadly lanceolate, pointed, abruptly rounded basally and short-stalked; upper narrower, sessile; all glabrous or margins minutely hairy, often somewhat glaucous. Flowers in cymes; petals *white;* calyx *inflated* in fruit, *clearly veined,* mostly downy, 5-toothed. *A frequent perennial; dry grassy places.* **Silene vulgaris**

e. Petals *notched* .... j
Petals *entire* .... f

f. Corolla *white* .... g
Corolla *pink* or *mauve* (Fig. 12.15); petals exceeding sepals, ovate, blunt; sepals glabrous or hairy; flowers solitary or in cymes. Leaves linear, c. 2.5 cm, *fleshy,* blunt to acute, horny-tipped, often shortly mucronate, flat adaxially, rounded abaxially, grey-green, glabrous. *A frequent perennial of salt-marshes of muddy and gravelly seashores.* **Spergularia media**

g. Cauline leaves *opposite* .... h
Cauline leaves apparently *whorled* (Fig. 12.16); the latter *linear,* almost *needle-like,* up to 5.0 cm, grey-green, blunt, somewhat pubescent or glabrous; small *insignificant* stipules often present. Flowers numerous in terminal cymes; pedicels often drooping, minutely rough or hairy; petals *white,* broadly oblong-elliptic, blunt; sepals hairy. *A very frequent annual of cultivated ground and waste place places, especially on acid soils.* **Spergula arvensis**

h. Leaves *ovate* i
Leaves *linear* or *needle-like* (Fig. 12.17); the latter radical and cauline; the former in a dense rosette; abruptly pointed, small, smooth or rough. Flowers small, few or solitary; petals small. *Tufted perennial, common in poor pastures, paths, lawns, etc.* **Sagina procumbens**

i. Flowers in *cymes* (Fig. 12.18); the former small; petals white, obovate, *shorter* than the sepals. The main stem never elongating or flowering, prostrate and rooting at the nodes, then ascending; plant pubescent. Leaves up to 6 mm, mid green, *joined at the base,* blunt apically, 3-5 nerved. *Frequent annual or biennial of dry sandy places.* **Arenaria serpyllifolia**
Flowers *solitary* (Fig. 12.19); the latter in the forks of the stems and upper leaf-axils; petals greenish-white. The plant succulent, stoloniferous, with leafy and non-flowering shoots, decumbent basally, then erect. Leaves *fleshy,* deep green, 6-20.0 mm long, slightly *frilled,* sessile. *Frequent prostrate perennial; sandy and gravelly seashores.* **Honkenya peploides**

j. Styles *3;* leaves mostly glabrous .... m
Styles *4* or *5;* leaves mostly pubescent .... k

k. Flowers in a *spreading* cyme l
Flowers in a relatively *compact* cyme (Fig. 12.20); flowering branches very short, *pubescent,* and *glandular,* at least apically. Leaves small, about 2.5 cm, *pale yellowish-green, sticky,* covered with long white hairs, apiculate; lower oblanceolate or ovate, narrowed basally; upper broadly ovate or elliptic-ovate. Petals *white;* sepals lanceolate, acute; stamens 10. *A frequent annual of cultivated ground, wall-tops, waste places, etc.* **Cerastium glomeratum**

l.   Petals and sepals *about equal* (Fig. 12.21). Shoots diffusely branched, decumbent or ascending, dull dark green, densely covered with *viscous* and *glandular* hairs. Leaves about 12.0 mm, covered with *non-glandular* and *short glandular* hairs; lower oblanceolate; the upper ovate or oblong-ovate, acute. Flowers in spreading cymes; branches *glandular*; petals *white*; stamens 4-5. *Frequent annual; light sandy soils near the sea.* **Cerastium diffusum**

Petals *longer* than sepals (Fig. 12.22). Flowering shoots erect and decumbent, simple; non-flowering short, decumbent, leafy; all slender, more or less hairy but *not* viscous. Leaves on non-flowering shoots oblanceolate, blunt, tapered into a stalk-like base; those of flowering shoots ovate-oblong, subacute, sessile, and in distant pairs; all *densely* covered with long white hairs, and *dark greyish-green*. Flowers in loose cymes; petals *white*; stamens 10, rarely 5. *Abundant perennial of old pastures and waste places.* **Cerastium fontanum**

m.   Stem-internodes *glabrous* .... n

Stem-internodes with 1 alternating *line of hairs* (Fig. 12.23); the former otherwise *glabrous*, diffusely branched, decumbent or ascending. Leaves very variable; all light green; lower mostly broadly ovate, acute, long-stalked; upper ovate or broadly elliptic, acute or shortly acuminate, more or less sessile, *glabrous* or *ciliate* at the base; some or all often with undulate margins. Flowers numerous, usually in branched cymes, rarely solitary; petals *white, very deeply 2-lobed*, not exceeding the sepals; the latter lanceolate, pubescent, with thin margins. *An abundant annual of cultivated ground and waste places.* **Stellaria media**

n.   Sepals *shorter* than petals .... o

Sepals *longer* than petals (Fig. 12.24); the latter divided *nearly to the base* into narrow, widely divergent lobes, *white*; the lobes lanceolate, acute; the flowers small, in few-flowered cymes. Leaves sessile, oval or broadly lanceolate, often *slightly glaucous*, glabrous or ciliate basally. Stems 4-angled. *Common perennial; bogs, marshy areas, etc.* **Stellaria alsine**

o.   Petals divided *almost* to the base (Fig. 12.25); the latter *white*, and with 2 divergent, narrow lobes; sepals slightly united basally, *prominently* 3-veined, nearly equalling petals, often with marginal hairs; flowers numerous, in a loose cyme. Stems slender, clearly *4-angled*. Leaves *linear-lanceolate*, acute, up to 4.0 cm, margins mostly smooth, rarely ciliate basally. *A very frequent perennial of grasslands and waste grassy places.* **Stellaria graminea**

Petals divided to *about half-way* (Fig. 12.26); the latter *white*, about twice as long as the sepals, cuneate basally; lobes divergent; sepals ovate-lanceolate, acuminate, *inconspicuously* 3-nerved, with thin margins. Stems *4-angled*. Leaves *grass-like*, lanceolate, firm, tapered from base to apex, up to 7.0 cm, *slightly* glaucous, rough on margins and keel, sessile. *A very frequent perennial of hedgerows, roadsides, waste grassy places.* **Stellaria holostea**

LINACEAE

FLAX FAMILY: Herbs, rarely shrubs or trees. Leaves alternate or opposite, rarely whorled, mostly entire, *exstipulate* or *stipulate*. Inflorescence in *cymes* or *racemes*, rarely of solitary flowers. Flowers hermaphrodite, often *heterostylous, actinomorphic*, with the *hypogynous* arrangement. Corolla *polypetalous*, of 5, rarely 4, petals. Calyx *polysepalous*, of 5, rarely 4, sepals. Androecium mostly of 5 stamens; the latter *connate* basally; staminodes often present. Gynaeceum *syncarpous*; ovary *superior*, 2-5-celled, with mostly 2 ovules in each cell; styles free, rarely more or less united. There are about 9 genera and 200 species. Two species are frequent species throughout the British Isles (Figs 12.27-12.28).

Corolla *blue* (Fig. 12.27); petals obovate, clawed basally; calyx of 5 acuminate or apiculate sepals. Leaves numerous, alternate, narrowly lanceolate, acute, glabrous, entire, about 2.5 cm. Flowers in a cyme. *Frequent annual; dry pastures and gravelly banks.* **Linum bienne**

Corolla *white* (Fig. 12.28); petals narrowly obovate; calyx of 5 ovate-lanceolate, acuminate sepals. Flowers in a loose cyme. Leaves opposite, up to 16.0 mm, glabrous, oblong to obovate, sessile. *Common annual; dry pastures, heaths, dry banks, etc.* **Linum catharticum**

Vegetative Phases

Weed plants remain in flower, when recognition is easiest, for a comparatively short period of time. Identification, however, is often necessary long before the flowers appear. Weed control, in any situation, is more effective when the plants are at an early stage of growth. One of the most important factors in the selection of an appropriate herbicide is the identity of the 'weed plants' present. To this end, the following 'key' has been developed and is based entirely on vegetative characteristics. It embraces a total of 178 species which are likely to occur in areas of immediate agricultural interest.

KEY TO THE VEGETATIVE PHASES

1. Branches present or absent, variously arranged but *not* whorled 3
   Branches *whorled,* grooved; true leaves absent; scale-leaves present, *whorled* 2

2. Branches *hollow* (Fig. 13.01). Sterile and fertile stems similar; both erect or nearly prostrate, usually with suberect branches, rarely simple, slightly rough, with 4-6 grooves and ridges; rhizomes glabrous. *Common in streams, marshes, ditches and in shallow waters, etc. **Equisetum palustre***
   Branches *solid* (Fig.13.02). Sterile stems erect or decumbent, rather rough, with 6-18 deep grooves; the branches mostly spreading, numerous, usually simple, 4-angled; the fertile stems simple; rhizomes pubescent. *Common in cultivated ground, roadsides, hedgerows, etc.**Equisetum arvense***

3. Leaves *neither* grass-like *nor* bristle-like 30
   Leaves somewhat *grass-like* or *bristle-like*, or sheaths present but blades absent 4

4. Leaves *other than* opposite 7
   Leaves *opposite* 5

5. Stems *quadrangular* 6
   Stems *terete* (Fig. 13.03). Plant small. Leaves acute, minute, *linear* or *needle-like*, and both radical and cauline; the former in a dense rosette, latter opposite; abruptly and sharply pointed, smooth or minutely hairy. *Common perennial; poor pastures, lawns, pathways, etc.**Sagina procumbens***

6. Leaves *with strong comb-like* teeth on the margins (Fig. 13.04); the former up to 7.0 cm, lanceolate, firm, sessile, narrowed from base to apex, slightly *glaucous*, margins and keels rough. The stems quadrangular. *Frequent perennial of hedgerows, roadsides and waste places. **Stellaria holostea***
   Leaves *without* strong comb-like teeth (Fig. 13.05); the former up to 4 cm, margins mostly smooth but often softly ciliate towards the base, narrowly lanceolate, acute, opposite. Stems quadrangular. *Very frequent perennial of dry pastures, woods, heaths and waste places.**Stellaria graminea***

7. Cauline leaves *not* whorled 8
   Cauline leaves *apparently* whorled (Fig. 13.06); the latter *linear* and almost needle-like, up to 5 cm long, grey-green, blunt, slightly pubescent or glabrous, *very minute* stipules often present. Stems erect, branched. *Very frequent annual of cultivated ground and waste places. **Spergula arvensis***

8. Blades, if present, *not* keeled over the entire abaxial surface 22
   Blades *keeled* over the entire abaxial surface 9

9. Stems *terete* 10
   Stems *trigonous* or *trigonous-terete* 12

10. Plant *pubescent* 11
    Plant *glabrous* (Fig. 13.07). Leaves rather narrow, 5-25 cm x 0.5-1.0 mm, more or less stiff, keeled, keeled, dark green, blunt; the ligule *minute*; becoming brown. Rhizomes shortly creeping; shoots often densely tufted. *Frequent perennial in bogs, heaths and mountain pasture. **Carex pulicaris***

N.B. DRAWINGS NOT TO SCALE

Fig. 12.00

11. Plant *shortly stoloniferous* (Fig. 13.08). Leaves *linear, grass-like,* usually 2.0-4.0 mm wide, with a small truncate swelling at the apex, sparsely covered with whitish or colourless hairs. *Common, loosely tufted perennial, with short stolons, of heaths, open moorlands, etc.* **Luzula campestris**
Plant *cespitose* (Fig.13.09). Leaves *linear, grass-like,* 3.0-6.0 mm wide, sparsely covered with long, whitish or colourless hairs, and bright green. *A densely tufted perennial, with few or no stolons; taller than the above species; very frequent in upland heaths, moorlands, etc.* **Luzula multiflora**

12. Plant *glabrous* 13
Plant, at least basally, *pubescent* (Fig. 13.10). Leaves 10-50 cm x 2-5 mm, pubescent adaxially and abaxially, flat or somewhat keeled, gradually tapered apically; sheaths pubescent; ligule 1-2 mm and fringed with hairs. *Frequent rhizomatous plant of wet grasslands, marshes, etc.* **Carex hirta**

13. Plant *yellowish-green, grey-green, mid-green* or *dark green* 14
Plant *glaucous* all over or on the adaxial surface 20

14. Colour *mid-green, grey-green* or *dark green* 15
Colour *yellowish-green* (Fig. 13.11). The leaves 5-60 cm x 3-6 mm, soft, slightly keeled or plicate, abruptly narrowed to fine sharp points; sheaths hyaline becoming brown; ligule obtuse, about 2 mm long. *A frequent plant; heavy wet soils in woods and other shaded places.* **Carex sylvatica**

15. Rhizomes *short* 16
Rhizomes *very extensive* (Fig. 13.12). The leaves 20-60 cm x 1.5-3.5 mm, more or less flat, rough, *rigid,* thick, often recurved, tapered gradually into a *trigonous* point, *dark green,* usually shiny; sheaths grey-brown; ligule 3-5 mm, blunt. *Frequent; sand-hills and sea shores.* **Carex arenaria**

16. Leaf-tips more or less *flat* 18
Leaf-tips *trigonous* 17

17. Blades up to *20.0 cm x 1.5-2.5 mm* (Fig. 13.13); the latter often *recurved,* and dark green; sheaths brown; ligule 1-2 mm, obtuse. Rhizomes shortly creeping; shoots loosely tufted; roots purple-brown or dark brown in colour. *Frequent plant in dry pastures and heaths.* **Carex caryophyllea**
Blades up to *50.0 cm x 1.0-3.0 mm* (Fig. 13.14); the latter more or less *soft,* pointed, and dark green, thin, more or less flat, gradually tapered into a *trigonous* point; sheaths becoming pink-brown or grey-brown; ligule about 1.0 mm, obtuse. *A very common in wet upland pastures.* **Carex ovalis**

18. Blades *without* pendulous apices 19
Blades *with* pendulous apices (Fig. 13.15); the former 20-60 cm x 1.5-2.0 mm, *channelled,* gradually tapered into long pendulous tips, mid-green; the sheaths yellowish-brown; ligule 1.0-2 mm, rounded or obtuse. *Rhizomatous, very frequent in woods and damp shaded places.* **Carex remota**

19. Blades *gradually* narrowed into acute apices (Fig. 13.16); the latter 10-15 cm x 2-6 mm, *rigid,* more or less erect, flat, grey-green but becoming brown; sheaths dark to mid-brown; ligule from 2-3.0 mm, obtuse. *Frequent rhizomatous plant in salt-marshes and other coastal areas.* **Carex distans**
Blades *abruptly* narrowed into acute apices (Fig. 13.17); the former 7-30 cm x 2-6 mm, dark brown, *rigid,* often arcuate in dwarf plants, keeled or more or less flat; sheaths dull, red-brown. Stems *trigonous-terete,* often with a single groove. *Common in moorlands and heaths.* **Carex binervis**

20. Blades glaucous adaxially *or* abaxially but *not* both 21
Blades glaucous *both* adaxially and abaxially (Fig.13.18); the former up to 90 cm x 1.5-3 mm, thin, more or less flat, gradually tapered into fine apices; the sheaths brown, black, rarely red; ligule 1-3 mm, rounded. Rhizomes extensive. *Common in marshes and other wet places.* **Carex nigra**

21. Blades glaucous *adaxially* and dark green and shiny *abaxially* (Fig. 13.19); the latter 30.0-120 cm x 2-7.0 mm, rough, rigid, keeled, plicate or inrolled, tapered to a long point; sheaths thick, more or less spongy, dark brown; ligule 2.0-3.0 mm, rounded. *Common in marshy places.* **Carex rostrata**

Blades glaucous *abaxially* and dark, dull green *adaxially* (Fig. 13.20); the former up to 50 cm x 1.5-4.0 mm, rigid, often arcuate, flat, and gradually tapered into a fine point; sheaths becoming dark brown; ligule 2.0-3 mm, rounded. *Common in pastures, heaths and grassy places.* **Carex flacca**

22. Blades *absent* or *minute* 23
    Blades *present*, distinct 25

23. Blades *absent* but sheaths *present* 24
    Blades *minute* (Fig. 13.21). Sheaths *numerous* and *crowded* at the base of the plant, constricted and short-pointed apically, overlapping, brown or green, *shiny*. Plant *densely* tufted; roots thick, numerous. *A common perennial of bogs, moors, heaths, mountains, etc.* **Trichophorum cespitosum**

24. Stems glaucous (Fig. 13.22); the latter wiry, stiffly erect, ribbed, grooved; internal pith *interrupted*; sheaths *glossy*, dark or blackish. *A common perennial on wet calcareous soils.* **Juncus inflexus**
    Stems *yellowish-green* or *bright green* (Fig. 13.23); the latter *soft, smooth, glossy*; the internal pith *continuous*; sheaths reddish to dark brown. *Common perennial on wet acid soils.* **Juncus effusus**

25. Blades *with* septa 26
    Blades *without* septa 27

26. Leaves strongly *laterally compressed*, deep green, hollow, and with up to 25 *inconspicuous septa* (Fig. 13.24). Stems *prostrate basally*. *Common perennial of marshy places.* **Juncus articulatus**
    Leaves slightly *compressed* (Fig. 13.25); the latter deep green and with up to 25 *conspicuous* septa. Stems *stiffly* erect; rhizomes *extensive*. *Common perennial of marshy places.* **Juncus acutiflorus**

27. Blades *stiff, shiny* 28
    Blades *soft, not* shiny 29

28. Sheaths *tightly arranged* around base of plant (Fig. 13.26); the latter dark reddish-brown or almost black, shiny. Leaf-blades very *narrow*, short, involute or terete, rigid. Stems smooth, wiry. Plant *densely* tufted. *A common perennial of fens, bogs, heaths and marshes, etc.* **Schoenus nigricans**
    Sheaths *loosely arranged*, rather broad (Fig. 13.27). Leaves usually all *radical*; blades *stiff*, numerous, spreading, rounded abaxially, deeply channelled adaxially, glossy, very narrow. The stems rather *stiff*, erect and slender. *A common perennial; moors, heaths, bogs, etc.* **Juncus squarrosus**

29. Blades *3.0-6.0 mm wide* (Fig.13.28); the latter *channelled*, tapered into a long *triquetrous* point, numerous; cauline leaves few, mostly reduced to loose sheaths, or sometimes with short leafy tips. Rhizomes very extensive. *A common perennial of moors and bogs.* **Eriophorum angustifolium**
    Blades *1.0-2.0 mm wide, setaceous* (Fig.13. 29); the latter *triquetrous*; upper blades very short, with *inflated* sheaths. *Common tufted-rhizomatous plant of moors, bogs, etc.* **Eriophorum vaginatum**

30. Leaves arranged on *overground* stems 67
    Leaves, all, *radical* or appearing to arise from *underground* stems 31

31. Blades *longer than* broad 34
    Blades more or less *rounded, reniform, cordate* basally or *broadly-triangular* 32

32. Blades up to *16.0 cm, pubescent*, especially abaxially 33
    Blades *1.0-4.0 cm* (Fig. 13.30); the latter *cordate, reniform* or *somewhat rounded* in general outline mostly glabrous, long-stalked, entire or finely toothed, rather fleshy, yellowish-green; rootstock of small tubers. *A frequent perennial of damp, shaded places and gardens.* **Ranunculus ficaria**

33. Blades somewhat *angled* (Fig. 13.31); the latter *cordate* basally, long-stalked, densely covered with white cottony hairs abaxially, and more or less rounded but shallowly lobed; petioles grooved and lobes toothed. *Common perennial of cultivated ground, waste places, etc.* **Tussilago farfara**

70

Blades *not* angled (Fig. 13.32); the latter somewhat rounded, cordate basally, long-stalked, serrate, 10-15 cm, rarely up to 20, across; mostly green adaxially and abaxially; somewhat downy abaxially. *A very frequent rhizomatous perennial of shaded banks, roadsides, etc.* **Petasites fragrans**

34. Leaves *simple* 41
    Leaves *compound* 35

35. Leaves *exstipulate* 37
    Leaves *stipulate* 36

36. Leaves *silvery* adaxially and abaxially (Fig. 13.33); the latter pinnate; leaflets alternating large and small, oblanceolate, deeply serrate; stipules entire or finely toothed, adnate to petioles. Runners short, slender. *Abundant perennial of cultivated ground, waste places, etc.* **Potentilla anserina**
    Leaves *dark green* adaxially (Fig.13.34); the latter pinnate; the leaflets alternating large and small, ovate, crenate, *glabrous* adaxially, *white-tomentose* abaxially, rarely glabrous; terminal leaflet largest, 3-lobed; stipules leafy. *Abundant perennial of damp grassy places.* **Filipendula ulmaria**

37. Leaves *pinnate*, relatively small, up to 20 cm 39
    Leaves *bipinnate* or *tripinnate,* long, up to 200 cm; plant *fern-like* 38

38. Lowest pinnae slightly *shorter, equal to* or *larger,* than those in the middle (Fig. 13.35). Leaves up to 200.0 cm, *bi-* or *tripinnate*, petioles erect, pubescent when young and arising from rhizomes; pinnules narrow-oblong, numerous and entire; the sori *continuous* on the margins. *An abundant perennial in open woodlands, neglected pastures, shaded hedgerows, etc.* **Pteridium aquilinum**
    Lowest pinnae clearly *decreasing* in size basally (Fig. 13.36). Leaves 40-100 cm, *bipinnate* or *tripinnate;* the petioles and mid-ribs moderately covered with brownish scales; pinnules undivided, oblong, rounded apically and somewhat toothed laterally and apically; sori 3-6 *in a row* on each margin. *An abundant perennial of shaded hedgerows and woodlands areas.* **Dryopteris filix-mas**

39. Plant with *some* hairs 40
    Plant *glabrous* (Fig.13.37). Leaves pinnate; leaflets ovate or rounded; terminal largest, *reniform*; all mostly entire. *Common perennial of marshes areas, wet grassy places, etc.* **Cardamine pratensis**

40. Radical leaves in a *loose* rosette (Fig. 13.38); leaflets rounded, reniform or ovate; terminal largest; all toothed and weakly hairy. *Common annual; mostly damp or wet places.* **Cardamine flexuosa**
    Radical leaves in a *compact* rosette (Fig. 13.39); leaflets rounded or broadly ovate; terminal largest; all lobed or toothed. *Common annual of cultivated ground and waste places.* **Cardamine hirsuta**

41. Blades *prickly* 42
    Blades *not* prickly 44

42. Plant *not* rhizomatous 43
    Plant *rhizomatous* (Fig. 13.40). Leaves in a *compact* rosette, oblong-lanceolate, tapered basally into a short stalk, and usually pinnatifid and with undulating lobes; the latter toothed, spiny-ciliate, and ending in strong sharp spines. *A common perennial in several habitats.* **Cirsium arvense**

43. Blades, adaxial surfaces, *rough* or *prickly* (Fig. 13.41); the latter, at least some, obovate-lanceolate, narrowed into a stalk-like base, often deeply pinnatifid, usually with 2-lobed segments, the upper lobe toothed towards the base, the lower entire; terminal lobe long, pointed and *spear-like*; all ending in long stout, firm spines. *Biennial, very common in several habitats.* **Cirsium vulgare**
    Blades, adaxial surfaces, *smooth* (Fig. 13.42); some, at least, narrowly oblanceolate, petiolate, shallowly pinnatifid, spinous. *An abundant biennial; marshes and wet grasslands.* **Cirsium palustre**

44. Blades with *1* longitudinal vein 46
    Blades with *several* longitudinal veins 45

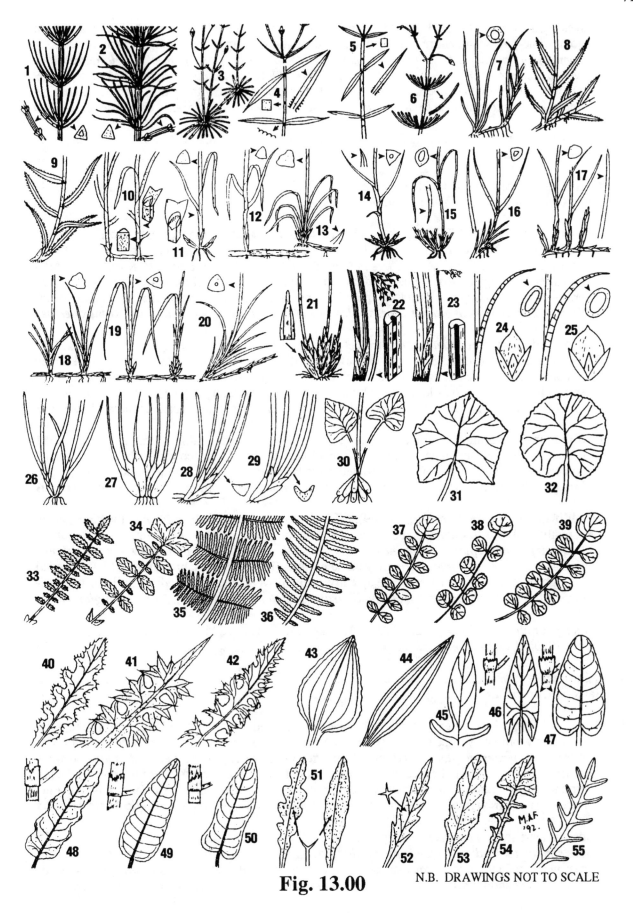

Fig. 13.00

N.B. DRAWINGS NOT TO SCALE

45. Leaves *broadly ovate* (Fig. 13.43); the latter *abruptly* constricted into a broad petiole, glabrous or
     somewhat pubescent, entire or irregularly toothed and with up to 9 prominent 'parallel' veins. *A
     common perennial plant; cultivated ground, trodden pastures, waste places, etc.***Plantago major**
     Leaves *lanceolate* or *broadly-lanceolate* (Fig. 13.44); the latter *gradually* narrowed into a petiole,
     glabrous or somewhat pubescent, entire or distantly toothed and with up to 7 prominent 'parallel'
     veins. *A very common perennial of pastures, meadows, waste places, etc.* **Plantago lanceolata**

46. Leaves *stipulate* 47
     Leaves *exstipulate* 52

47. Leaves *neither* hastate *nor* sagittate 49
     Leaves, at least some, *hastate* or *sagittate* 48

48. Leaves *hastate* (Fig. 13.45); the latter stalked, narrowly lanceolate to oblanceolate exceptfor the ba-
     sal lobes which spread at right angles to the mid-rib, covered with minute tubercular protuber-
     ances, up to 6 cm, entire. *A calcifuge perennial, common in heaths, bogs, etc.***Rumex acetosella**
     Leaves *sagittate* (Fig. 13.46); the latter with prominent pointed basal lobes, somewhat thick, mostly
     entire or nearly so, up to 18 cm long; petioles long, and with a bitter taste. *A very common calci-
     fuge perennial plant of old pastures, shaded places, especially on wet acid soils.* **Rumex acetosa**

49. Base of blade *square, rounded* or *subcordate* 50
     Base of blade *cordate* (Fig.13.47); the lower broad, elliptical to oblong-ovate in outline, dark green
     in colour, up to 25 cm long, gradually and then abruptly pointed, glabrous throughout, entire or
     somewhat toothed; nerves looped and branched; all with sheathing stipules. *A very common per-
     ennial plant of roadsides, grasslands, cultivated ground, waste places, etc.* **Rumex obtusifolius**

50. Leaf-margins *not* strongly undulate 51
     Leaf-margins *strongly* undulate and crisped (Fig.13.48); the blades lanceolate to oblong-lanceolate,
     somewhat rounded or abruptly constricted basally, tapered from about the middle, obtuse, gla-
     brous, entire or finely toothed. *Common perennial; roadsides, waste places, etc.***Rumex crispus**

51. Lower blades *ovate-lanceolate* (Fig. 13.49); the latter rounded or slightly cordate at the bases, acute
     apically, glabrous throughout and entire or nearly so, veins *rust-red*; all dull green in colour. *A
     frequent perennial plant; woods, hedgerows, grassy and damp waste places.***Rumex sanguineus**
     Lower blades *oblong* or *panduriform* in general outline (Fig. 13.50), glabrous throughout, rounded
     or slightly cordate at the bases, acute apically, entire or very finely toothed; all dull green in col-
     our. *A frequent perennial plant of woods, grassy and damp waste place.* **Rumex conglomeratus**

52. Hairs *soft* or *absent* entirely 57
     Hairs *stiff, bifid,* or *stellate* 53

53. Hairs *simple* 55
     Hairs *bifid* or *stellate* 54

54. Hairs *bifid* (Fig. 13.51). Leaves mostly oblanceolate, tapered basally into a long petiole-like base,
     irregularly pinnately lobed or nearly entire, mostly with bifid hairs, very rarely nearly glabrous.
     A frequent perennial plant of old gravelly pastures and sandy places. **Leontodon taraxacoides**
     Hairs *stellate* (Fig. 13.52). Leaves in a dense rosette, tapered basally into a petiole, mostly deeply
     pinnatifid, very rarely entire, up to 10.0 cm, rarely more, and lanceolate or oblanceolate in out-
     line. *Very common annual plant of cultivated ground and waste places.***Capsella bursa-pastoris**

55. Blades, at least some, *very deeply* lobed 56
     Blades *sinuate-toothed* to *shallowly pinnatifid* (Fig. 13.53); the latter broadlyoblong-lanceolate, and
     narrowed gradually into a broad stalk-like base, dull green adaxially and somewhat glaucous ab-
     axially, hairs short, stiff, plentiful. *Abundant perennial; most grasslands.* **Hypochoeris radicata**

56. Terminal-lobe *clearly larger* than the laterals (Fig. 13.54); blades frequently *lyrate-pinnatifid*, with lobes variable in length and width, usually oblanceolate, obtuse or somewhat acute, frequently covered with *dark, short, stiff* hairs. *Common biennial of dry grassy situations.* **Crepis vesicaria**
    Terminal-lobe *not* clearly larger than laterals (Fig. 13.55); lobes very narrow; leaves all in a rosette, and like a small dandelion in general appearance. The blades deeply pinnatifid, mostly oblanceolate; hairs few, rarely absent. *A common perennial of many grasslands.* **Leontodon autumnalis**

57. Blades *unwrinkled, or* wrinkled *but* deeply and *irregularly lobed* 59
    Blades *clearly wrinkled* and either entire or somewhat toothed 58

58. Blades tapered *gradually* into a broad petiole-like bases (Fig. 14.01); the former obovate to spatulate, obtuse, glabrous adaxially except for the veins, woolly abaxially, coarsely reticulate, margins wavy, and finely toothed. *Common rhizomatous plant in shaded habitats.* **Primula vulgaris**
    Blades tapered *abruptly* into a winged petioles (Fig. 14.02); the former finely pubescent adaxially and abaxially, ovate to oblong in outline, obtuse, wrinkled, crenulate or more or less serrate. *A frequent rhizomatous plant of old pastures, especially those on calcareous soils.* **Primula veris**

59. Blades *lobed* 63
    Blades *entire* or *toothed* 60

60. Blades *toothed* 61
    Blades *entire* (Fig. 14.03); the lower broadly lanceolate in general outline, forming a dense rosette, blunt apically, dark green, short-stalked or tapered; mid-rib distinct; all abundantly covered with *bulbous* hairs. *A very common annual of cultivated ground, waste places, etc.* **Myosotis arvensis**

61. Leaves *lanceolate* to *oblanceolate* in general outline 62
    Leaves *spatulate* to *obovate* (Fig. 14.04); the latter up to 8.0 cm long, broad and somewhat rounded apically, tapered basally, more or less finely toothed, rarely entire, sparsely pubescent, and light green in colour. *An ubiquitous perennial; old pastures, lawns, waste places, etc.* **Bellis perennis**

62. Leaves ending in *long, narrow stalks* (Fig. 14.05); the former *oblanceolate* or lanceolate, frequently long and narrow, up to 20.0 cm long, merely toothed, pubescent or glabrous, rather thin, and dull green (see couplet no. 66). *A common annual or biennial of dry grassy places.* **Crepis capillaris**
    Leaves tapered *to the base* (Fig. 14.06); the former *oblanceolate* to lanceolate, mostly long and narrow, up to 20 cm long, merely toothed, rather thin, light green, glabrous; taproot thick (see also couplet no. 66). *Ubiquitous perennial plant of several habitats.* **Taraxacum officinale sensu lato**

63. Blades *unwrinkled*, thin, not fleshy 65
    Blades *wrinkled* or rather fleshy 64

64. Blades *deeply* and *irregularly lobed, puckered* or *wrinkled* (Fig. 14.07); the latter in a dense rosette, lyrate-pinnatifid, dull green, petiolate; lobes lobulate or toothed, minutely scurfy pubescent adaxially and abaxially, especially on the veins, and at least some *crowded and overlapping. An ubiquitous biennial; pastures, meadows, grassy hedgerows, waste places, etc.* **Senecio jacobaea**
    Blade lobes *well spaced* but *decreasing* in size towards the base (Fig. 14.08); blades long-stalked, mostly lyrate to pinnatifid, coarsely lobulate and toothed; terminal lobe largest, laterals smaller and *directed forward. Common biennial plant of wet grasslands, ditches, etc.* **Senecio aquaticus**

65. Sini *rounded* or *acute* 66
    Sini *broad* (Fig. 14.09); all lobes narrow, directed apically; terminal lobe *not* clearly larger than laterals; leaves all in a rosette and like a small dandelion in general appearance; blades deeply pinnatifid, mostly oblanceolate. *A common perennial of many grasslands.* **Leontodon autumnalis**

66. Lobes *triangular, directed basally* (Fig. 14.10); the blades variable, mostly *lyrate*, tapered, glabrous, soft, light green. *Ubiquitous perennial plant; several habitats.* **Taraxacum officinale sensu lato**

Lobes *spreading* (Fig. 14.11); blades long-stalked, irregularly pinnately and distantly lobed; pubescent or glabrous, rather thin. *A common annual or biennial; dry grassy places.* **Crepis capillaris**

67. Leaves variously arranged *but not* whorled 73
    Leaves *whorled* 68

68. Stems *quadrangular* 69
    Stems *terete* (Fig. 14.12); the latter mostly erect. The leaves *linear*, almost needle-like, up to 5.0 cm
    long grey green, blunt, slightly pubescent or glabrous, *insignificant* stipules often present. Stems
    mostly glabrous. *Very frequent annual; cultivated ground and waste places.* **Spergula arvensis**

69. Blades *cuspidate* or *mucronate* 71
    Blades *blunt* or *acute* but *never* mucronate 70

70. Leaves *acute* (Fig. 14.13); the latter obovate-cuspidate, small, sessile, mostly glabrous but margins
    and underside of the mid-rib scabrid with prickles. Stems rough or hairy, and erect, prostrate or
    decumbent. *A common annual of cultivated ground, lawns, waste places, etc.***Sherardia arvensis**
    Blades *blunt* (Fig. 14.14); the latter relatively long and narrow, oblanceolate, tapered basally, glabrous except for rough prickly margins. Stems up to 120.0 cm long, weak, decumbent, glabrous,
    smooth or more or less scabrid. *A very common perennial; wet marshy ground.* **Galium palustre**

71. Stems-angles *smooth* or only *slightly rough* 72
    Stems-angles *markedly rough* with *recurved* prickles (Fig. 14.15); the former up to 2 m long, prostrate or, more often, scrambling-ascending and diffusely branched. Leaves linear-oblanceolate,
    mucronate, 1-veined, glabrous or bristly adaxially, and the margins with backwardly projecting
    prickles. *Abundant annual of cultivated ground, hedgerows, waste places, etc.* **Galium aparine**

72. Blades *linear* (Fig. 14.16); the latter often needle-like, dark green, rough adaxially, pale and pubescent abaxially; margins often rolled downwards. Stems slender, covered with minute reflexed
    hairs. *A very common stoloniferous plant of heaths, dry banks and sandy places.* **Galium verum**
    Blades *obovate* or obovate-oblanceolate (Fig. 14.17), rather short and broad, mucronate, and with
    straight marginal prickles pointing obliquely forward. The stems erect or prostrate, smooth, glabrous, branched. *Common perennial of heaths, moors, pastures and acid soils.* **Galium saxatile**

73. Leaves *alternate* or in *clusters* on stolons 119
    Leaves mostly *opposite* or *subopposite* 74

74. Leaves, at least some, *ternate* 169
    Leaves, all, *other than* ternate 75

75. Blades *15.0 mm*, or *more*, in length; or rarely less and then at least the upper *sessile* 76
    Blades *12.0 mm*, or *less, in length* (Fig. 14.18); the latter either reniform or bluntly cordate, crenate,
    pubescent; petioles about equal to or shorter than the blades; stems rooting at the nodes, slender,
    numerous, short. *A common perennial of old pastures, lawns, gardens, etc.* **Veronica filiformis**

76. Leaves *exstipulate* 82
    Leaves *stipulate* 77

77. Blades *crenately toothed* 80
    Blades *pinnately* or *palmately lobed* 78

78. Blades *palmately* lobed 79
    Blades *divided into 3 large lobes* which are again deeply divided (Fig. 14.19); final lobes divided
    or toothed; pubescent. The stems decumbent or ascending, often pinkish, and with an unpleasant
    smell when crushed; pubescent. *An abundant annual of shaded places.* **Geranium robertianum**

79. Sini *extending to the leaf base* or very nearly so (Fig.14.20). Leaves pubescent,long-stalked, orbicular or reniform; the lobes cuneiform and further divided into narrow segments. Stems branched, straggling. *A common annual of dry pastures, roadsides, waste places, etc.* **Geranium dissectum**
Sini extending to *about half-way* or slightly beyond (Fig.14.21). Leaves divided into 5-9 lobes, pubescent, long-stalked; lobes variously divided or toothed; petioles long. Stems somewhat prostrate. *A common annual plant of dry grasslands, roadsides, waste places, etc.* **Geranium molle**

80. Stem-internodes *lacking* a thick rib 81
Stem-internodes *showing* a thick rib (Fig. 14.22). Stems angular, erect, firm, and either glabrous or sparsely hairy. Leaves lanceolate to oval-lanceolate, crenate-serrate, somewhat acute, scarcely ly pubescent, light green; petiole short and with a pair of minute glands on junction with blade; stipules minute. *A frequent annual of cultivated ground and waste places.* **Mercurialis annua**

81. Blades, at least some, *cordate* or nearly so (Fig. 14.23); the latter coarsely crenate-serrate, acuminate, petiolate, dull green, with *stinging*; stipules small. Stems pubescent, 4-angled. *An abundant rhizomatous plant; pastures, cultivated ground, roadsides, waste places, etc.* **Urtica dioica**
Blades *ovate to elliptic* (Fig. 14.24); the latter acuminate to obtuse, coarsely crenate-serrate, petiolate, mostly *3-nerved*, dark or grey-green, with *stinging hairs*; the stipules small, narrow. Stems erect, pubescent, 4-angled. *A common annual of tilled ground, waste places, etc.* **Urtica urens**

82. Stems *terete*, or *ridged* but *not* quadrangular 97
Stems *quadrangular*, or *quadrangular* apart from *raised ribs* on opposing sides 83

83. Leaf-margins *clearly toothed* 85
Leaf-margins *entire* or very *nearly* so 84

84. Stem-angles *sharply winged* (Fig. 14.25); stems stiff, glabrous; plant stoloniferous. Leaves up to 3 cm, ovate, sessile, entire, semi-amplexicaul, *translucent* dots present, yellowish-green, mostly 3-nerved, glabrous. *Frequent perennial; ditches and damp marshy places.* **Hypericum tetrapterum**
Stem-angles *unwinged* (Fig. 14.26); the latter prostrate or ascending, weak. Blades mostly opposite, rarely in whorls of 3, sessile, ovate, mostly 3-nerved, glabrous, and dotted with black glands abaxially. *A common annual of cultivated ground, waste places, sand-hills, etc.* **Anagallis arvensis**

85. Blades *clearly petiolate* 88
Blades *sessile* or very *nearly* so 86

86. Stems *distinctly hairy* 87
Stems *glabrous* or with *few hairs* (Fig. 14.27); the latter mostly erect. Leaves narrowly lanceolate, coarsely serrate, sessile, light green, with short stiff hairs, and with pinnate venation. *An abundant semi-parasitic annual; old pastures, meadows and other grassy places.* **Rhinanthus minor**

87. Blades *tapered* basally (Fig. 14.28); the latter mostly 5.0 x 1.0 cm, remotely toothed, lanceolate to linear-lanceolate, tapered, hispid with short *bulbous-based* hairs. The stems *solid*, hispid with reflexed hairs. *Small common annual; pastures, roadsides and dry stony places.* **Odontites verna**
Blades *rounded* basally (Fig. 14.29); the latter oblong-lanceolate or linear-lanceolate, tapered apically, crenate-serrate, pubescent. Stems *hollow*, stout, with reflexed hairs.*Rhizomatous, tuberous plant, common in cultivated ground, ditches, riverbanks and other wet places.* **Stachys palustris**

88. Blades, all, *tapered* or *rounded* basally 93
Blades, at least some, *cordate* or *cut off square* basally 89

89. Blades *not reniform* in outline 90
Blades, at least some, *reniform* or very nearly so (Fig. 14.30); radical leaves long-stalked, cauline short-stalked; slightly pubescent, crenate. The stems creeping, unbranched, internodes relatively long. *Frequent creeping perennial; woods, hedgerows and grassy places.* **Glechoma hederacea**

90. Plant *lacking* rhizomes; blades small, 1-3, rarely 5, cm; margins mostly crenate 92
    Plant *rhizomatous*; blades large, up to 10 cm, margins mostly serrate 91

91. Stems *solid* (Fig.14.31; the latter rather densely hispid with whitish stiff hairs throughout, and up to
    100 cm in height, often branched; foetid when squashed; long creeping, non-tuberous rhizomes
    often present. Leaves cordate basally, with a broad triangular apex, broadly oblong-lanceolate in
    general outline, crenate-serrate, dull green, up to 10 cm long, and with rather stiff hairs; petioles
    long. *A perennial plant, often frequent in hedgerows, woods and waste places.* **Stachys sylvatica**
    Stems *hollow* (Fig. 14.32); the former slightly pubescent with spreading hairs, particularly in the
    upper half; up to 50.0 cm in height; short stolons often present. Leaves broadly ovate, cordate to
    slightly rounded basally, acutely acuminate, up to 5.0 x 4.0 cm; colour pale green; often doubly
    serrate or crenate-serrate, pubescent, clearly reticulate and rugose or undulate; petioles shorter
    than the blades. *An erect perennial plant, often frequent in many grassy places.* **Lamium album**

92. Plant *dull green* or *purple tinted* (Fig.14.33). The stems branched from near the base, spreading and
    ascending, with short reflexed hairs. The leaves ovate, widely cordate or truncate basally, blunt,
    upper with petioles longer than the blades, lower petioles shorter; all softly pubescent, and regu-
    larly crenate. *A common annual of cultivated ground and waste places.*   **Lamium purpureum**
    Plant *green* or *bright green* (Fig. 14.34). Stems slender, usually branched from near the base, pu-
    bescent, and ranging from 10-30 cm in height. Leaves oval-ovate in general outline, small; cor-
    date or truncate basally, blunt apically; crenate-serrate, pubescent. *A frequent annual plant of
    cultivated ground and waste places, especially those with non-calcareous soils.* **Stachys arvensis**

93. Blades *gradually* or *abruptly* tapered into a *distinct petiole* 94
    Blades *gradually tapered* to the *stem* (Fig. 14.35); the latter radical and cauline; those of the latter
    petiolate to sessile; all entire or obscurely crenate, obovate to oblanceolate, glabrescent, blunt,
    margins often wavy. Stems erect, some creeping, unbranched with white weak hairs below and
    between the leaf bases. *A perennial plant, very frequent in damp shaded places.* **Ajuga reptans**

94. Ridges *absent* from stems 95
    Ridges, more or less *covered with dark hairs, present* on stems (Fig. 14.36); the latter often creep-
    ing, and crimson in colour. Leaves short-stalked, ovate-lanceolate, or broadly wedge-shaped,
    with about 3 pairs of lateral nerves, entire or obscurely toothed, thinly covered on both surfaces
    with multi-celled hairs. *Abundant perennial of damp pastures, roadsides, etc.***Prunella vulgaris**

95. Nodes *neither* swollen *nor* with glandular hairs 96
    Nodes, at least some, *swollen* and covered *with glandular* hairs (Fig. 14.37). Stems erect, branched,
    and with bristly hairs. Leaves fairly long-stalked, dark green, ovate to ovate-lanceolate, rounded
    to wedge-shaped basally, pointed apically, coarsely toothed, with scattered stiff hairs, at least
    adaxially. *Very common coarse annual of cultivated ground and waste places.***Galeopsis tetrahit**

96. Plant *purplish* or *reddish,* and strongly *pungent* when squashed (Fig. 14.38). The rootstock slender,
    often creeping. Stems simple or branched, up to 40 cm. Leaves often ovate, short-stalked, often
    rounded basally, triangular apically, crenate-serrate, more or less pubescent adaxially and abaxi-
    ally. *A common perennial plant of marshes, lakesides, riversides, ditches, etc.***Mentha aquatica**
    Plant *light green* and *not* pungent (Fig. 14.39). Stems erect or ascending, and simple or branched.
    Leaves varying from lanceolate to nearly round, petiolate, cuneate or rounded at the base, usual-
    ly obtuse, shallowly crenately or serrately toothed, more or less pubescent adaxially and abaxi-
    ally, petiolate. *A frequent perennial of cultivated ground and waste places.*   **Mentha arvensis**

97. Leaves *simple* 98
    Leaves *pinnate* (Fig. 14.40); the former sessile or short-stalked, mostly opposite, lanceolate, always
    deeply cut into cristate segments; the latter serrate-crenate toothed or pinnatifid and slightly pu-
    bescent. Stems single or much-branched from near the base up to the middle, leafy all over, and
    slightly pubescent. *A frequent annual plant; bogs, marshes and wet places.***Pedicularis palustris**

98. Blades *neither* round *nor* reniform 99
    Blades, at least some, *cordate* or *reniform* in outline (Fig. 14.41); the latter small, 1.0-4.0 cm long,
    entire or distantly and finely toothed, long-stalked, mostly glabrous or nearly so; lower in a ro-
    sette; upper *opposite* and in *unequal-sized* pairs; all rather fleshy; yellowish-green. Rootstock of
    a few small tubers. *Frequent perennial of gardens and damp shaded places.* **Ranunculus ficaria**

99. Leaves *not* 'mealy', *i.e.* papillae *absent* 101
    Leaves 'mealy', *i.e.* papillae *present* 100

100. Lower leaves *tapered gradually* to the petiole (Fig. 14.42); the latter long-cuneate at the base, often
    opposite, mostly rhomboid to lanceolate, with *2 spreading basal lobes,* toothed or nearly entire
    upwards; upper narrower; all with minute *papillae,* giving a 'mealy' appearance; rarely glabrous.
    Very common, erect or prostrate, annual of cultivated ground and waste places. **Atriplex patula**
    Lower leaves *truncate* or *shortly cuneate* basally (Fig. 14.43); the latter *abruptly* contracted to the
    petioles, often opposite, usually *triangular-hastate,* some margins making an angle of 90° with
    the petiole, toothed or nearly entire; upper narrower; all with *papillae,* giving a 'mealy' appear-
    ance. *Very frequent, spreading annual of cultivated ground and waste places.* **Atriplex prostrata**

101. Leaves *entire* 110
    Leaves *toothed* 102

102. Plant with at least *some* hairs 103
    Plant *glabrous* through (Fig. 14.44). Stems creeping, then ascending, hollow, rooting at the nodes,
    thick, soft or fleshy. The leaves rather fleshy, generally opposite, *oval or oblong* in outline, blunt
    apically, abruptly tapered basally, shallowly crenate-serrate, short-stalked, light green in colour.
    A perennial, common in shallow streams, muddy ponds, wet places, etc. **Veronica beccabunga**

103. Internodes *lacking* lines of hairs 104
    Internodes *showing 2 lines of long white hairs* on opposite sides (Fig. 14.45). Stems prostrate, then
    erect or ascending, and rooting towards the base. Leaves ovate in outline, nearly sessile, mostly
    opposite, coarsely toothed, dull green, pubescent, subcordate to rounded basally. *A common per-
    ennial of hedgerows, woodlands, waste grassy places and old pastures.* **Veronica chamaedrys**

104. Lower leaves *sessile* 105
    Lower leaves *petiolate* 108

105. Blades *relatively large,* at least *4 cm* long 106
    Blades *very small,* 2 *cm* or *less* (Fig. 14.46); the latter triangular-ovate in general outline, upper ses-
    sile, becoming bract-like, lower short-stalked and opposite; all pubescent adaxially, dull green
    and coarsely crenate-serrate abaxially. Stems simple or with many ascending branches from near
    the base. *An annual plant, common in cultivated ground and waste places.* **Veronica arvensis**

106. Blades *other than* amplexicaul or semi-amplexicaul 107
    Blades, at least some, *semi-amplexicaul* (Fig. 14.47); the latter oblong-lanceolate, acute, pubescent,
    dull green, upper sessile, lower short-stalked, slightly decurrent, the margins ciliate and with un-
    equal, slender, and incurved branches. Stems glandular-pubescent, erect, almost terete. *A fre-
    quent rhizomatous plant; ditches, streams, riversides and other wet places.* **Epilobium hirsutum**

107. Hairs *few, scattered* and *minute* (Fig. 14.48). Leaves broadly ovate to broadly lanceolate, opposite
    or alternate, rarely in threes, mostly rounded basally, tapering to the apex, *light green* in colour,
    sharply toothed, sparsely pubescent. Stems sparsely pubescent or glabrous, often pinkish. *Very
    common perennial of cultivated ground, roadsides, waste places, etc.* **Epilobium montanum**
    Hairs *plentiful, easily seen* (Fig. 14.49). Leaves lanceolate, mostly opposite, often pointed, rounded
    basally, very finely toothed, and often reddish. The stems erect, slender, and *covered* with short
    spreading hairs. *A frequent perennial of ditches and damp waste places.* **Epilobium parviflorum**

108.   Blades mostly *ovate*; pedicels, if present, *equal to* or *shorter than* the blades 109
        Blades mostly *triangular-ovate* (Fig. 14.50); the latter 1.0-3 cm long, coarsely crenate-serrate, light green, short-stalked, mostly pubescent on the veins adaxially, and minutely ciliate. Pedicels, if present, *longer* than the blades. Stems mostly weak, branched at the base, and with decumbent branches 10-40 cm. *A common annual of cultivated ground and waste places.* ***Veronica persica***

109.   Lower blades *broader than long* (Fig. 14.51), latter opposite or nearly so, triangular-ovate in general outline, short-stalked, small, 1.0-3 cm, light green in colour, usually coarsely crenate-serrate, pubescent on the veins, abaxially, and minutely ciliate. Stems weak, pubescent, branched, with decumbent branches. *A common annual of cultivated ground and waste places.* ***Veronica polita***
        Lower blades *longer than broad* (Fig. 14.52); the latter broadly triangular-ovate in general outline, opposite, or nearly so, short-stalked, small, 1-3 cm, light green, regularly crenate-serrate, pubescent on the veins abaxially; minutely ciliate. Stems weak, pubescent, branched from the base; branches decumbent. *Common annual of cultivated ground, waste places, etc.* ***Veronica agrestis***

110.   Blades *not* punctate 112
        Blades *punctate,* mostly *3-veined* from base 111

111.   Stems *terete* and *showing 2 raised* lines (Fig. 14.53); the latter mostly woody basally. Leaves *mostly 3-nerved* from the base, up to 3 cm long, all opposite, sessile, entire, ovate-oblong, abundantly furnished with *translucent* dots, obtuse, sometimes mucronate, and glabrous. Plant rhizomatous; poisonous. *A very common perennial plant of dry banks, roadsides, etc.* ***Hypericum perforatum***
        Stems *quadrangular* (Fig. 14.54); the latter erect, mostly glabrous, with sharply winged angles, and often with slender stolons at the base. Leaves up to 3 cm in length, mostly ovate in outline, sessile, opposite, entire, obtuse, 3-nerved from base, semi-amplexicaul, furnished with many *translucent* dots. *A frequent perennial of ditches and damp marshy places.* ***Hypericum tetrapterum***

112.   Internodes *lacking* lines of hairs 114
        Internodes *showing* 1 or 2 lines of hairs 113

113.   Leaves, at least some, *petiolate* (Fig. 14.55); the latter rather variable, often small with undulating margins, entire; the lower mostly broadly ovate in general outline, tapered apically, long-stalked, glabrous or ciliate at the base; upper sessile or nearly so; all light green in colour. Stems slender, soft, diffusely branched, decumbent or ascending, and showing *1, alternating, line of hairs* on each internode. *Abundant annual plant of cultivated ground and waste places.* ***Stellaria media***
        Leaves, all, *sessile* (Fig. 15.01); the latter lanceolate to linear-lanceolate in outline, cuneate basally, mostly opposite, rarely alternate, entire or obscurely toothed, subglabrous or with crisped hairs on the margins and nerves, dull green in colour, and gradually tapered into a blunt apex. Stems erect or ascending, bearing subterranean stolons, simple or branched, and with *2 lines of crisped hairs* on each internode. *Frequent perennial; bogs and wet marshy ground.* ***Epilobium palustre***

114.   Blades *not* sharply pointed 116
        Blades *acutely* and *sharply* pointed 115

115.   Leaves *up to 7.0 cm* long (Fig.15.02); the latter narrow and *grass-like*, lanceolate-acuminate in general outline, tapered from base to a long fine point, entire, rather firm, slightly *glaucous*, sessile, glabrous except for basal *comb-like* teeth. Stems *roughly 4-angled*, weak, slender, erect, straggling. *A frequent perennial of hedgerows, roadsides and waste grassy places.* ***Stellaria holostea***
        Leaves *up to 4.0 cm* long (Fig.15.03); the latter narrow and *grass-like*, linear-lanceolate or narrowly elliptical in general outline, pointed apically, entire, with *smooth* margins, glabrous or often ciliate basally, and mid-green in colour. Stems *sharply 4-angled*, slender, numerous, up to 40.0 cm long. *A very frequent perennial plant of grasslands and waste grassy places.* ***Stellaria graminea***

116.   Plant *glabrous* or *nearly* so 118
        Plant distinctly *pubescent* 117

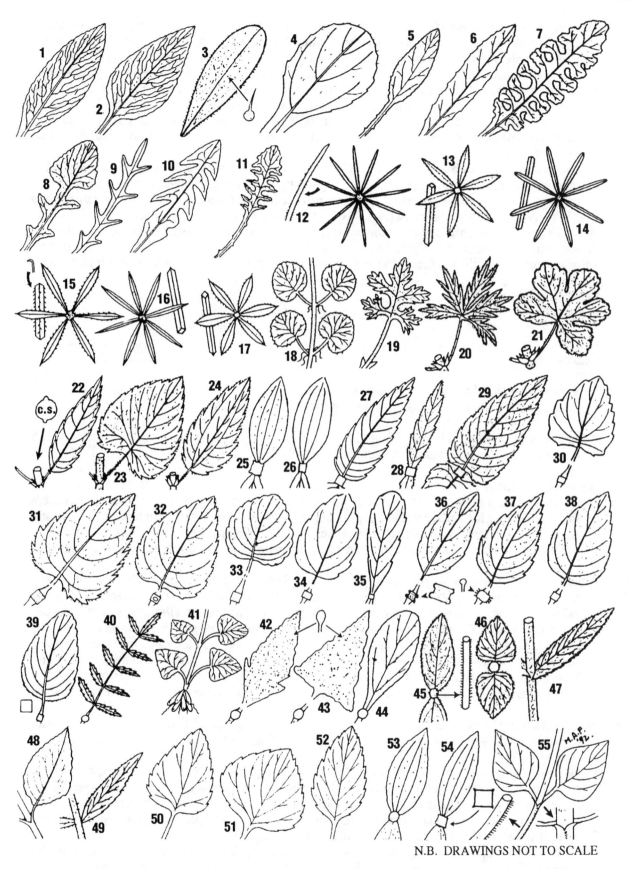

N.B. DRAWINGS NOT TO SCALE

Fig. 14.00

117. Plant very hairy and *sticky* (Fig. 15.04). The stems erect, weak, pubescent; glandular, at least above. Leaves sessile, rather small, about 2.5 cm, entire; lower oblanceolate, narrowed basally; upper elliptic-ovate or broadly ovate; all *pale yellowish-green* in colour, covered with many long white hairs. *A frequent annual of cultivated ground, wall tops, waste places.* **Cerastium glomeratum**
Plant very hairy but *not* sticky (Fig. 15.05). Stems erect, weak, pubescent; eglandular; 5.0-50.0 cm long. Leaves sessile, small, about 2.5 cm, entire; lower oblanceolate narrowed basally; the upper elliptic-ovate or broadly ovate; all opposite, *pale yellowish-green* in colour, covered with long white hairs. *An abundant perennial plant of old pastures and waste places.***Cerastium fontanum**

118. Blades *rounded* apically and basally (Fig. 15.06); the latter 1-3 cm, sessile or short-stalked, oval or oblong, entire or obscurely crenate, light green; hairs few or none. Stems erect, or creeping and rooting. *A very common perennial of old pastures, cultivated ground, etc.***Veronica serpyllifolia**
Blades *tapered* apically and basally (Fig. 15.07); the lower oblanceolate in general outline, tapered into a petiole-like base, acute, somewhat rough; upper narrower, sessile, oblong-lanceolate; all glabrous or pubescent basally, entire, exstipulate, up to 15 cm. The stems erect, weak, branched, often with a few reflexed hairs. *A very frequent perennial of marshy ground.***Lychnis flos-cuculi**

119. Leaves *exstipulate* 153
Leaves *stipulate* 120

120. Leaves *compound* 137
Leaves *simple* 121

121. Stipules *neither* sheathing *nor* encircling the stem 130
Stipules *either* sheathing *or* partially encircling the stem 122

122. Blades *without* hastate or sagittate lobes 125
Blades, at least some, *with* hastate or sagittate basal lobes 123

123. Basal lobes *sagittate* 124
Basal lobes, at least some, *hastate* (Fig. 15.08); leaves narrowly lanceolate to oblanceolate, except for basal lobes which are spreading at right angles to the mid-rib, up to 6 cm long, entire, mostly covered with minute wart-like protuberances; stipules hyaline. The stems erect or decumbent, 7-30 cm in height. *A small calcifuge perennial, common in heaths, bogs, etc.* **Rumex acetosella**

124. Apices of blades *abruptly long-pointed* (Fig. 15.09); blades longer than the petioles, outline mostly triangular, bases deeply *cordate-sagittate*, entire, glabrous, scurfy abaxially, nearly smooth adaxially, 2-8 cm; stipules obliquely truncate. Stems weak, twining or climbing, slender, angular or ribbed, 'mealy'. *Common annual of cultivated ground and waste places.***Fallopia convolvulus**
Apices of blades *blunt* (Fig. 15.10); the blades sagittate with long, pointed basal lobes; upper short-stalked or nearly sessile, clasping the stems; the lower on longer stalks; all entire, mostly glabrous, light green, and with a bitter taste. Stems up to 1.0 m, though often less; stipules fringed. *A very common perennial of pastures, shaded places, particularly on acid soils.* **Rumex acetosa**

125. Petioles *clearly* evident 127
Petioles *absent*, or *very short* and *not* clearly evident 126

126. Stems *mostly prostrate, rarely* erect (Fig. 15.11). Blades 1.5-5.0 cm long, rarely more, and 0.3-2.0 cm wide, dull green in colour, outline lanceolate to ovate-lanceolate or linear, narrowed basally, sessile or very short-stalked, entire, glabrous, and acute or subobtuse; stipules *whitish*, nearly entire or split. *Abundant annual of cultivated ground and waste places.* **Polygonum aviculare**
Stems *mostly erect* (Fig. 15.12); the latter often of a reddish colour and somewhat swollen at the nodes. Blades up to 12 x 5 cm long, lanceolate, tapered, glabrous, ciliate, or sometimes woolly abaxially, entire or nearly so, often *black-purple blotched*, adaxially; stipules usually fringed, truncate. *An abundant annual of cultivated ground and waste places, etc.* **Polygonum persicaria**

127. Blade bases and petioles forming an approximate 90° angle; or base *rounded* or *subcordate* ... 127
   Blade bases, at least some, *cordate* (Fig. 15.13); lower blades broadly elliptic, dark green in colour, up to 25.0 x 12.0 cm, gradually and abruptly pointed, margins often slightly wavy, mostly entire, rarely finely toothed; the upper blades ovate-lanceolate to lanceolate; all with numerous looped and branched nerves, mostly glabrous, rarely pubescent on the under surface; stipules membranous. *A very common perennial of roadsides, grasslands, waste places, etc.* **Rumex obtusifolius**

128. Blade-margins *not* very wavy 129
   Blade-margins *very* wavy (Fig.15.14); the latter lanceolate or oblong-lanceolate, somewhat rounded or constricted at the base, tapered from middle to an obtuse point, dull green, mostly glabrous, up to 30 cm, entire or finely toothed; upper blades narrower, often linear-lanceolate; the stipules membranous. *A common perennial of roadsides, grasslands, waste places, etc.* **Rumex crispus**

129. Lower blades *ovate-lanceolate* (Fig. 15.15); the latter ovate-lanceolate in general outline, rounded or slightly cordate basally, acute, glabrous throughout, entire or nearly so; veins frequently *rust-red*; upper narrower, narrowly lanceolate; all dull green in colour; stipules membranous, sheathing. *A frequent perennial plant of woods, grassy and damp waste places.* **Rumex sanguineus**
   Lower blades *oblong* or frequently *panduriform* in general outline (Fig. 15.16); the latter rounded to somewhat cordate basally, acute apically, entire or very finely toothed, glabrous throughout, nerves looped and branched; upper narrowly lanceolate; dull green; the stipules membranous, sheathing. *A frequent perennial of woods, grassy and damp waste places.***Rumex conglomeratus**

130. Stipules *deeply* lobed 136
   Stipules *entire* or *toothed* 131

131. Blades *cordate* or *sagittate* 132
   Blades or leaflets broadly *oval* in general outline (Fig.15.17); the latter tapered or rounded and nearly entire basally, toothed apically; dull green in colour; the stipules *large,* adnate to the petiole, toothed, with *sticky hairs.* Stems prostrate, branched from near the base, densely leafy, covered with long, whitish, many-celled hairs. *A frequent perennial of dry sandy places.* **Ononis repens**

132. Blades *not acutely pointed* apically 133
   Blades *acutely pointed* apically (Fig. 15.18); the latter deeply cordate and *angled* basally, otherwise nearly entire; large, up to 15 cm, glabrous or slightly pubescent; petioles usually shorter than the blades; light green in colour; stipules *minute.* Stems up to 3.0 m, twining in a counter clockwise direction or trailing. *A frequent perennial of hedgerows, bushy places, etc.* **Calystegia sepium**

133. Stipules *fringed* with long, narrow, comb-like teeth 134
   Stipules *without* comb-like teeth 135

134. Blades *gland-dotted* (Fig. 15.19); the latter orbicular-ovate in outline, cordate basally, bluntly acuminate, crenate or crenulate, scarcely pubescent; the stipules broader than the glabrous petioles, tapered, fringed. Plant *not* stoloniferous. The rootstock covered with persistent bases of previous year's leaves. *A common perennial of upland pastures, banks and sandy places.* **Viola riviniana**
   Blades *not* gland-dotted (Fig. 15.20); the latter orbicular-ovate in outline, bluntly acuminate, cordate basally, scarcely pubescent or glabrous, crenate; stipules broader than the petioles, tapered apically, fringed on the margins. Plant with short, slender stolons, and persistent bases of previous year's leaves. *Frequent perennial of woods and shaded places, etc.***Viola reichenbachiana**

135. Blades *up to 2.5 cm* in diameter (Fig. 15.21); the former more or less ovate, cordate basally, blunt apically, poorly crenate or crenate-serrate, scarcely pubescent; stipules entire or toothed. Rootstock with few persistent leaf-bases. *A frequent perennial, mostly of dry habitats.* **Viola canina**
   Blades *up to 5.0 cm* in diameter (Fig. 15.22); the latter orbicular or reniform, cordate basally, blunt apically, toothed, glabrous; the stipules toothed or fringed with glands. Stolons covered with the remains of hard scale-like leaf-bases. *Very frequent perennial of marshy ground.***Viola palustris**

136. Terminal stipule-lobe *lanceolate* and *entire* (Fig. 15.23). Blades oval to oblanceolate, tapered from apex to base, toothed, glabrous or scarcely pubescent. Stems mostly erect, rarely some creeping *Very frequent annual of dry sandy cultivated ground and other and sandy places*. **Viola tricolor**
Terminal stipule-lobe *somewhat ovate* and *toothed* (Fig.15.24). The blades oval to oblanceolate, tapered from apex, toothed, glabrous or barely pubescent. Stems mostly erect, rarely some creeping. *Very frequent annual of dry sandy cultivated ground and dry gravelly places*.**Viola arvensis**

137. Leaflets *entire* 146
Leaflets *toothed* 138

138. Blades *trifoliate* or *palmate* 141
Blades *pinnate* 139

139. Stipules *small, not* leaflet-like 140
Stipules *large* and *leaflet-like* (Fig. 15.25). Radical leaves long-stalked, with 2-3 pairs of unequal-sized leaflets, terminal leaflet much larger than the laterals, and 3-lobed or tooted; the cauline, at least some, ternate or 3-lobed; all leaflets toothed and pubescent. Stems erect and creeping; rhizomes short, thick. *A common perennial of hedges, woods and shaded places*. **Geum urbanum**

140. Blades *silvery*, densely covered with long silky hairs adaxially and abaxially (Fig. 15.26); the former 8-15 cm; leaflets crowded, mostly alternating large and small, deeply and regularly serrate, oblanceolate; stipules entire or finely toothed, adnate to the petioles. Stems slender and creeping, short. *An abundant perennial of cultivated ground, waste places, etc.* **Potentilla anserina**
Blades *dark green* (Fig. 15.27); the latter glabrous adaxially, and usually white-tomentose abaxially, or green and glabrous or pubescent; up to 60 cm long; leaflets ovate, sharply and doubly crenate, often alternating large and small; terminal leaflet largest, usually lobed. Stems both erect, up to 120 cm, and creeping. *An abundant perennial of damp grassy places*. **Filipendula ulmaria**

141. Leaflets *3* 142
Leaflets *5*, rarely more or less (Fig. 15.28); leaves palmate, on long slender petioles, and mostly of 5 obovate or oblanceolate, sparingly pubescent, coarsely serrate leaflets; stipules small, lanceolate, entire or fine-toothed; radical leaves on long stalks; cauline stalks shorter. Runners extensive, rooting. *Abundant perennial of poor pastures, roadsides, waste places*. **Potentilla reptans**

142. Stipules *toothed* 144
Stipules *entire* 143

143. Petioles *equalling the leaflets* in length, or nearly so (Fig. 15.29). Leaflets cuneiform, toothed and emarginate apically, pubescent or glabrous; stipules small, acuminate. Stems slender, short, procumbent or ascending. *A small common annual of pastures and waste places*. **Trifolium dubium**
Petioles *several times longer* than the leaflets (Fig. 15.30). Stolons long, wiry, spreading and rooting at the nodes. Leaflet variable in size and shape, often obovate or nearly round, emarginate apically, toothed, *glabrous*, and mostly with white crescent-shaped markings; stipules large, entire, acuminate. *Ubiquitous perennial of pastures, meadows, waste places, etc.* **Trifolium repens**

144. Leaflets *not* mucronate 145
Leaflets *mucronate* (Fig. 15.31); the latter small, obovate or nearly round in general outline, central vein exserted as a small projection, finely serrate on upper half, glabrous or pubescent. Leaves short-stalked; the stipules small, lanceolate, and toothed. Stems straggling or prostrate, usually pubescent. *A common, small, annual plant of old pastures and waste places*. **Medicago lupulina**

145. Blades or leaflets broadly *oval* in general outline (Fig.15.32); the latter tapered or rounded and nearly entire basally, toothed apically; dull green in colour; stipules *large,* and adnate to the petiole, toothed, with *sticky hairs*. Stems prostrate, branched from near the base, densely leafy, covered with long, whitish, many-celled hairs. *A frequent perennial of dry sandy places*. **Ononis repens**

Leaflets *mainly cuneiform* (Fig. 15.33); the latter coarsely serrate apically, entire basally, pubescent or glabrous, *without* sticky glandular hairs, small; stipules *not adnate* to petiole, deeply toothed Stems mostly erect or rarely trailing. *A common perennial of bogs, heaths, etc.* **Potentilla erecta**

     Leaflets *2* (Fig. 15.34); the latter lanceolate in outline, sagittate basally, tapered from the base to an acute apex, entire, scarcely pubescent, 'parallel-veined'; stipules lanceolate, leaflet-like, sagittate basally, tapered apically; tendrils simple or branched. Stems sharply angular, erect or scrambling. *Common perennial plant; damp meadows, hedgerows and ditches, etc.* **Lathyrus pratensis**

     Leaflets *tapered to a short point* (Fig. 15.35); the latter numerous, linear-lanceolate in general outline, small, up to 25 mm, mostly acute, entire, pubescent or nearly glabrous, the lowermost nearly at base of the very short petiole; stipules inconspicuous; tendrils branched.The stems erect, up to 140 cm. *A very common perennial plant of hedgerows and waste grassy places.* **Vicia cracca**

149. Leaflets *narrow* (Fig. 15.36); the latter linear-obovate, acute, or emarginate and mucronate, scarcely pubescent, 15-35 mm; stipules semi-sagittate, entire or toothed, often with a dark central spot. Stems erect, climbing or trailing. *Common annual of dry gravelly places.* **Vicia sativa** ssp. **nigra**
     Leaflets *broad* (Fig. 15.37); the latter ovate, rounded basally, blunt and mucronate, 10-25 mm, and glabrous or scarcely pubescent; stipules semi-sagittate, often toothed; tendrils branched. Stems climbing or straggling. *Common annual; hedgerows, roadsides, waste places, etc.* **Vicia sepium**

     Terminal leaflets *larger* than the laterals (Fig. 15.38); lateral leaflets ovate-elliptic in general outline; terminal largest and broadly lanceolate in outline; all acute, entire, mostly pubescent; hairs appressed; petioles very short; stipules inconspicuous. Stems spreading or ascending, up to 40.0 cm in height. *A common perennial plant of dry banks and sandy pastures.* **Anthyllis vulneraria**

     Leaves *clearly trifoliate* (Fig. 15.39). The leaflets very variable in size and outline, obovate, nearly round, or obcordate, emarginate, pubescent, entire; white crescent-shaped markings invariably present; stipules large, tapered; lower petioles long, upper short. The stems erect or straggling, soft or firm, or wiry. *A very common perennial of grassy and waste places.* **Trifolium pratense**

152. Stems *solid* or nearly so (Fig. 15.40); the latter decumbent or ascending, up to 30.0 cm, usually glabrous. Leaves trifoliate, short-stalked; leaflets obovate to ovate, entire, obtuse or apiculate, 5-12 mm, slightly pubescent or *almost glabrous*; petioles very short; stipules large and *leaflet-like*. *A very common perennial plant of dry grasslands, roadsides, sand-hills, etc.* **Lotus corniculatus**
     Stems *hollow* (Fig. 15.41); the latter erect or ascending, up to 60.0 cm in height, glabrous or slightly pubescent. Leaves trifoliate, relatively large, 15-20 mm; petioles short; leaflets obovate in outline, often obliquely so, blunt apically, *mostly pubescent*, rarely glabrous, 10-20 mm; the stipules large, *leaflet-like*. *A common perennial of damp meadows, marshy ground, etc.* **Lotus uliginosus**

156. Leaf-segments *linear-lanceolate* (Fig. 15.42); the latter small; leaves bipinnate or tripinnate, *clearly glaucous*, glabrous, and often climbing by the petioles. Stems erect or straggling. *A frequent annual of cultivated ground; and waste places; often forming compact tufts.* **Fumaria officinalis**
Leaf-segments *oval* or *broadly-triangular* (Fig. 15.43); the latter small; leaves bipinnate or tripinnate, *distinctly glaucous*, glabrous, often climbing by the petioles. Stems erect or straggling. *A common annual of cultivated ground, waste places;* often forming dense tufts; **Fumaria muralis**

157. Leaves *not very* finely divided, segments *neither* linear *nor* hair-like 160
Leaves *very finely* divided, segments linear or *hair-like* 158

158. Plant *not* emitting a pineapple odour though often scented 159
Plant, when crushed, *emitting a pineapple odour* (Fig. 15.44). Leaves bipinnate or tripinnate, segments *linear*, bristle-pointed, yellowish-green. Stems erect, glabrous, green, often purplish basally. *An aromatic annual, common in farm-yards, waste places, etc.* **Matricaria matricarioides**

159. Stems *furrowed,* more or less *woolly* (Fig. 15.45); the latter both erect and creeping. Basal leaves petiolate; cauline sessile; all bipinnate or tripinnate, more or less woolly, grey-green; ultimate segments linear and pointed, numerous, crowded. Plant often strongly aromatic. *A very common stoloniferous, aromatic plant, of roadsides, old pastures, waste places, etc.* **Achillea millefolium**
Stems *neither* furrowed *nor* woolly (Fig. 15.46); the latter erect, decumbent or prostrate. Leaves bipinnate or tripinnate, glabrous; segments *linear*, mostly sessile; ultimate segments *not* crowded. *A common biennial of cultivated ground and waste places, etc.*   **Tripleurospermum inodorum**

160. Stems *solid, firm, erect* 161
Stems *hollow, weak,* rather *stout* (Fig. 15.47); the latter creeping, angular. Leaves pinnate with oval leaflets; terminal largest; all entire or toothed, glabrous. Two *common annual* species, of *ditches and streams*; vegetatively, they cannot be separated. **Nasturtium officinale** or **N. microphyllum**

161. Stem *straight* 162
Stem *flexuous, particularly toward the base* (Fig. 15.48); the latter, slender, up to 30 cm in height. Basal leaves few, in a *loose* rosette, petiolate; leaflets rounded, ovate or reniform; upper leaves with shorter petioles; leaflets becoming narrower up the stem; all leaflets mostly entire or faintly toothed and sparsely ciliate. *Common annual; mostly damp or wet places.* **Cardamine flexuosa**

162. Plant *with some* hairs (Fig. 15.49). Radical leaflets *mostly rounded* or broadly *ovate*, compact; terminal leaflets largest, mostly reniform; the cauline narrower; all lobed or toothed, sparsely hairy adaxially and on margins; leaves petiolate. Stem erect, slender, never distinctly flexuous, more or less glabrous. *A common annual of cultivated ground and waste places.* **Cardamine hirsuta**
Plant *glabrous* (Fig. 15.50). Lower leaves in a loose rosette, relatively long-stalked; leaflets ovate or rounded in outline; terminal leaflets largest and more or less reniform; upper leaves with lanceolate leaflets; rather short-stalked; all leaflets mostly entire or very finely toothed. Stems erect, straight, up to 70 cm. *Very common perennial of marshes, wet places, etc.* **Cardamine pratensis**

163. Lower leaves divided into *relatively large non fern-like* segments 168
Lower leaves *finely divided* into somewhat *fern-like* 164

164. Upper leaves *not* finely divided into bristle-like segments 165
Upper leaves *deeply divided* into *thread-like* or *bristle-like* segments (Fig. 15.51); the radical leaves few, deltoid, soon disappearing, ternate, glabrous; petioles slender, flexuous and mostly arising from subterranean tubers. *Frequent perennial of heaths and old pastures.* **Conopodium majus**

165. Plant with *at least some hairs* 166
Plant *glabrous* (Fig. 15.52). The leaves relatively small, few, dull bluish-green or glaucous, deltoid, and mostly bipinnate; segments ovate, pinnatifid, toothed. Stems erect, up to 120.0 cm, hollow, finely striate. *Very frequent perennial of cultivated ground and waste places.* **Aethusa cynapium**

N.B. DRAWINGS NOT TO SCALE

Fig. 15.00

166. Hairs *short, stiff, bristle-like*, plentiful 167
Hairs *soft, few*, or *absent* (Fig. 15.53). Lower leaves in a rosette, rather large, up to 35 cm long, bipinnate or tripinnate, with long petioles; upper leaves gradually becoming smaller and sessile; petiole grooved adaxially, with widely expanded, sheath-like bases which are often fringed; all segments scarcely pubescent, small, ovate, pinnatifid, serrate, yellowish-green. Stems erect, up to 124.0 cm in height, hollow, grooved, becoming glabrous higher up except at the nodes which are fringed with hairs. *Abundant perennial of roadsides, waste places, etc.* **Anthriscus sylvestris**

167. Stems *hollow* (Fig. 15.54); the latter erect or slightly ascending, grooved, and covered with short backwardly directed hairs; hairs, at least some, *bulbous* basally, and plentiful. Leaves alternate, short-stalked, bipinnate or tripinnate, deeply divided into narrow pubescent segments; sheaths margins membranous. *Common biennial of hedges, roadsides and waste places.***Torilis japonica**
Stems *solid* (Fig. 16.01); the latter erect, grooved, and covered with reflexed hairs; hairs short, stiff, *not* bulbous. The leaves mostly tripinnate, dull green; sheath with narrow membranous margins; upper smaller than lower; leaflets pinnatifid; segments deeply cut into acute lanceolate lobes. *Common biennial of roadsides, pastures, especially those with light sandy soils.* **Daucus carota**

168. Leaves *pinnate, bipinnate* or *tripinnate* 170
Leaves *ternate, biternate* or *triternate* 169

169. Plant *deep glossy-green* (Fig. 16.02). Lower leaves often triternate; upper opposite, ternate; leaflets rhomboid, serrately toothed or lobed, glabrous; bases more or less asymmetrical; sheaths wide, glabrous or pubescent. Stems at first solid, later becoming hollow or filled with pith, grooved, glabrous. *A common biennial of roadsides, hedgerows and waste places.* **Smyrnium olusatrum**
Plant *dull green* (Fig. 16.03). Lower leaves biternate, long-stalked, upper ternate and often opposite; all glabrous; leaflets broadly lanceolate to ovate, up to 6.0 cm, unevenly serrate, and at least some clearly *asymmetrical* basally; petioles bluntly *triquetrous*. Stems erect, glabrous, grooved. *A common rhizomatous plant of cultivated ground, waste places, etc.* **Aegopodium podagraria**

170. Leaves simply *pinnate* 171
Leaves *bipinnate* or *tripinnate* (Fig.16.04). Lower leaves mostly bipinnate, with rather large asymmetrical leaflets in threes; the cauline smaller; terminal leaflets ovate to obovate; all deeply and regularly toothed, mostly glabrous; petiole grooved. Stems hollow, grooved, often purplish and pruinose, mostly erect. *A common perennial of streamsides, riversides, etc.* **Angelica sylvestris**

171. Leaves *bright green,* often shiny (Fig. 16.05); the latter mostly glabrous, simply pinnate; leaflets of 2-5 pairs, elliptic or lanceolate-ovate in general outline, serrate or crenate; petioles with widely expanded sheath-like bases. Stems soft, slender or stout, mostly procumbent or ascending, finely grooved, often rooting. *A very common biennial of ditches and streams.* **Apium nodiflorum**
Leaves *dull* or *grey-green* (Fig. 16.06); the latter variable, mostly pinnate, hispid adaxially and abaxially; leaflets variously lobed or toothed, often serrate, upper often largest. Stems stout, erect, somewhat angular, grooved, hollow, and hispid with an abundance of reflexed, bristle-like, stiff hairs. *A common biennial of old pastures, roadsides, waste places, etc.***Heracleum sphondylium**

172. Leaves *neither* amplexicaul *nor* semi-amplexicaul 188
Leaves, at least some, *either* amplexicaul *or* semi-amplexicaul 173

173. Stems *mostly terete,* grooved or ribbed, and *with* or *without* auricles 176
Stems *mostly 5-sided, often purplish* basally, *distinctly* auriculate, and frequently with latex ... 174

174. Auricles *rounded* and *appressed* 175
Auricles, at least some, *acute* and *spreading* (Fig. 16.07). Leaves variable; the basal petiolate, *not* prickly, with a large terminal lobe wider than the laterals; upper similar but with *spreading* auricles; all glaucous, mostly glabrous, toothed; petioles winged. Stems erect, often 5-faced, angular, hollow, soft. *Ubiquitous annual of cultivated ground, waste places, etc.* **Sonchus oleraceus**

175. Plant *rhizomatous* (Fig.16.08). The stems angular, often 5-faced, soft, hollow, glabrous or cottony towards the base, and usually glandular-hairy above. The lower leaves variously divided, mostly pinnatifid, tapered basally; lobes triangular, spinous-ciliate; cauline leaves similar but sessile, less well divided, and with *rounded appressed* auricles; all light green or rather glaucous. *Common perennial of cultivated ground, waste places, hedgerows, roadsides, etc.* **Sonchus arvensis**
Plant *tufted* (Fig. 16.09). The stems angular, often 5-faced, soft, hollow, and mostly glabrous. Lower leaves, pinnatifid, often with terminal lobe narrower than the upper laterals; the upper similar but sessile; the *auricles rounded, appressed*; all dark glossy-green, and with *crisped* and *sharply spinous-ciliate* margins. *Abundant annual of cultivated ground, waste places, etc.* **Sonchus asper**

176. Blades *not* prickly 179
Blades *very* prickly 177

177. Plant *not* rhizomatous 178
Plant *rhizomatous* (Fig. 16.10). Lower leaves in a *compact* rosette, oblong-lanceolate, tapered into a short stalk, and usually pinnatifid, and with undulating lobes; the latter toothed, spiny-ciliate, and ending in strong sharp spines; cauline similar but sessile, and semi-amplexicaul. Stems *not continuously spiny-winged. A very common perennial plant in several habitats.* **Cirsium arvense**

178. Blades, adaxial surfaces, *rough* (Fig.16.11); the latter, at least some, obovate-lanceolate, often deeply pinnatifid, segments 2-lobed, upper lobe toothed at the base, lower entire; terminal lobe long, pointed. Stems *continuously* winged. *Very common plant in several situations.* **Cirsium vulgare**
Blades, adaxial surfaces, *smooth* (Fig. 16.12); some, at least, narrowly oblanceolate, petiolate, shallowly pinnatifid, spinous; cauline leaves sessile, decurrent, and deeply pinnatifid. Stems cottony, *continuously spiny-winged. Abundant biennial; marshes and wet grasslands.* **Cirsium palustre**

179. Leaves *not* glaucous 182
Leaves, at least some, *glaucous* or somewhat *bluish-green* 180

180. Auricles *short, rarely* extending beyond the stem 181
Auricles *broad, long, very prominent* (Fig.16.13); the latter extending well beyond the stem. Stems erect. The lower leaves petiolate, usually with *stiff bristly hairs*, lyrate-pinnatifid; cauline sessile, *auriculate*, oblong-lanceolate, mostly entire, or short-lobed or somewhat toothed, *glaucous*, mostly glabrous. *A common annual or biennial; tilled ground and waste places.* **Brassica rapa**

181. Blades *fleshy, up to 25.0 x 2-6 cm* (Fig. 16.14); basal leaves coarsely toothed or pinnatifid, oblong-lanceolate in outline, tapered into a winged petiole; the upper sessile, oblong, toothed or nearly entire, *semi-amplexicaul*; all *glaucous.* Stems erect, glabrous. *Annual plant, frequent in cultivated ground and waste places, but particularly those with acid soils.* **Chrysanthemum segetum**
Blades *not* fleshy, up to 12 x 1-3 cm (Fig. 16.15); lower leaves stalked, broadly lanceolate, regularly or irregularly lobed to about mid-way, toothed, glaucous, glabrous or white cottony; the upper *semi-amplexicaul. Ubiquitous annual of cultivated ground, waste places, etc.* **Senecio vulgaris**

182. Blades, at least some, *lobed* or *clearly toothed* 183
Blades, all, *entire* or distantly toothed (Fig.16.16); radical leaves tapered basally; cauline *clearly* amplexicaul, and oblong-lanceolate to lanceolate; all densely covered with *white woolly hairs.* Stems erect, hairy. *A frequent perennial of damp meadows, marshes, etc.* **Pulicaria dysenterica**

183. Stem-ribs *similar* 184
Stem showing some *contrasting* ribs (Fig. 16.17); some ribs rather broad and dark, others narrow and light coloured. Basal leaves obovate to spatulate, tapered into long petioles, mostly covered with whitish hairs; upper small, few, blunt, distantly toothed, dark green, *semi-amplexicaul*, and sessile. *An abundant perennial, mostly of dry grassy and waste places.* **Leucanthemum vulgare**

184. Hairs, all, *simple* 185
Hairs, at least some, *stellate* (Fig. 16.18). The lower leaves in a dense rosette, lanceolate or oblan-
ceolate in general outline, varying from deeply pinnatifid to nearly entire, narrowed basally, up
to 10 cm long, rarely more; cauline leaves also very variable in shape, becoming smaller, entire
or finely toothed, *auriculate*, always clasping the stems; all dull green. Stems erect, up to 50 cm
slender. *A very common annual of cultivated ground and waste places.* **Capsella bursa-pastoris**

185. Hairs, all, *whitish, soft; some matted* 187
Hairs, at least some, *short stiff* and *blackish* 186

186. Terminal lobe of basal leaves *several times larger* than the laterals (Fig. 16.19); the former oblanceolate, acute or blunt; the latter variable, mostly lyrate-pinnatifid or rarely merely toothed; the
cauline leaves sessile and amplexicaul or short-stalked, somewhat entire or pinnatifid; all finely
pubescent adaxially and abaxially; the stems grooved, erect, ribbed, scabrid with *short dark, stiff
hairs* and often with patches of purple. *A common biennial of dry grassy places.Crepis vesicaria*
Terminal lobe of basal leaves *only slightly larger* than the laterals (Fig. 16.20); the former lanceolate or oblanceolate; the latter mostly narrowed into a stalk-like base, lyrate-pinnatifid, with well
spaced, often backwardly directed, toothed lobes; cauline leaves lanceolate, amplexicaul with
distinct basal auricles; all sparsely hairy, or very rarely glabrous. Stems erect or ascending, with
at least *some short hairs. A common annual or biennial of dry grassy places.* **Crepis capillaris**

187. Blades *deeply* and *irregularly lobed, puckered or wrinkled* (Fig. 16.21); the latter in a dense rosette,
lyrate-pinnatifid, dull green, petiolate; lobes lobulate or toothed, minutely scurfy pubescent adaxially and abaxially, especially on the veins, at least some crowded and overlapping; terminal lobes largest; the cauline leaves, at least some, semi-amplexicaul. Stems cottony, rarely glabrous. *An ubiquitous biennial plant of pastures, meadows, waste places, etc.* **Senecio jacobaea**
Blade lobes *well spaced* but *decreasing* in size from apex to base (Fig. 16.22); blades long-stalked,
mostly lyrate to pinnatifid, coarsely lobulate and toothed; terminal lobe largest, laterals smaller
and *directed forwards*; cauline leaves becoming smaller, at least some, semi-amplexicaul; all
crenate to coarsely serrate, and more or less glabrous. Stems mostly erect, cottony or glabrous
above, often reddish. *Common biennial plant of wet grasslands, ditches, etc.* **Senecio aquaticus**

188. Blades *not* 'mealy', *i.e.* papillae *absent* 191
Blades 'mealy', *i.e.* papillae *present* 189

189. Blades, all, *clearly tapered* to the petiole 190
Blades, lower, *truncate* or *shortly cuneate* basally (Fig. 16.23); the latter *abruptly* contracted to the
petioles, often alternate, usually *triangular-hastate,* some margins making an angle of 90° with
the petiole, toothed or nearly entire; upper narrower; all with *papillae*, giving a 'mealy' appearance. *A frequent spreading annual of cultivated ground and waste places, etc.Atriplex prostrata*

190. Leaves *with 2 spreading lower lobes* (Fig. 16.24); the former long-cuneate basally, often alternate,
mostly rhomboid to lanceolate, *tapered gradually to the petioles,* toothed or nearly entire upwards; upper narrower; covered with minute *papillae* giving a 'mealy' appearance; rarely glabrous. *Common, erect or prostrate annual; cultivated ground and waste places. Atriplex patula*
Leaves *without* 2 spreading basal lobes (Fig. 16.25); lower more or less triangular or almost *rhomboid,* short-stalked, bluntly and coarsely toothed; upper often narrower to almost linear, sessile
and entire; all covered with *papillae* giving a 'mealy' appearance. The stem erect, grooved, often
reddish, 'mealy'. *Abundant annual of cultivated ground and waste places.* **Chenopodium album**

191. Blade bases *neither* cordate, hastate, sagittate *nor* rounded 199
Blade bases. at least some, *either* cordate, hastate, sagittate *or* blades more or less orbicular ... 192

192. Blades *large, 2.5-40 cm* 194
Blades *1.0 cm or less* 193

193. Blades *entire* (Fig. 16.26); the latter very small, at most 0.5 x 0.5 mm, suborbicular, short-stalked, mid-green, scarcely pubescent, alternate. Stems very delicate, slender, weak, translucent, scarcely pubescent, mostly decumbent, often rooting at the nodes. *A very common perennial of poor lawns, gardens and rockeries where it often forms large evergreen parches.* **Soleirolia soleirolii**
   Blades *toothed* (Fig. 16.27); the latter either *widely reniform* or *bluntly cordate*, crenate, pubescent all over, up to 10.0 x 10.0 mm, opposite or alternate, and the petioles about equal to or shorter than blades. Stems small, numerous, slender, pubescent, creeping, and rooting at the nodes. *A creeping perennial, often forming large patches, in old pastures, lawns, etc.* **Veronica filiformis**

194. Blades, all, *lacking* deep lateral-basal lobes 195
   Blades, at least some, *with 1 or 2 lateral-basal lobes* (Fig. 16.28); the latter ovate or cordate, mostly scarcely pubescent, somewhat pointed, at least *some* with 1 or more basal lobes; the margins entire. Stems rather woody, scrambling or climbing and from 15.0-225 cm long. *A common perennial of hedgerows, ditches, woods, waste places and cultivated ground.* **Solanum dulcamara**

195. Plant *not* emitting a garlic odour when crushed 196
   Plant *emitting a strong garlic odour* when crushed (Fig. 16.29). Leaves, at least some, *cordate* basally or *reniform*; radical in a rosette, long-stalked; cauline short-stalked; all toothed, mostly glabrous. Stems erect. *A frequent biennial of cultivated ground and waste places.* **Alliaria petiolata**

196. Stems *neither* twining *nor* climbing, *i.e.* generally erect 198
   Stems *twining* and *climbing*, or trailing on the ground 197

197. Blades *acutely pointed* apically (Fig. 16.30); the latter deeply cordate and angled basally, otherwise nearly entire; large, up to 15 cm, glabrous, or slightly pubescent; petioles usually shorter than the blades; light green in colour. Stems up to 3.0 m in length, usually twining in a counter clockwise direction or trailing. *A frequent perennial of hedgerows, bushy places, etc.* **Calystegia sepium**
   Blades *sagittate* or *hastate* (Fig. 16.31); the latter entire, acuminate, short-stalked, and mostly glabrous. Stems climbing by twisting counter clockwise onto neighbouring plants or objects. Plant rhizomatous. *Abundant perennial of cultivated ground, waste places, etc.* **Convolvulus arvensis**

198. Blades, at least some, *ovate-oblong* (Fig.16.32); basal leaves entire or distantly toothed, nearly glabrous adaxially, somewhat cottony but green abaxially, often blunt; the upper leaves becoming much smaller and entire or almost entire. Stems grooved, stout, somewhat cottony, frequently reddish, erect or spreading. *A common biennial of roadsides, waste places, etc.* **Arctium minus**
   Blades *reniform* or *rounded* (Fig. 16.33); the latter large, up to 12 x 12 cm, blunt apically, regularly crenate, glabrous; lower long-stalked; upper leaves becoming sessile. Stems stout, erect or prostrate, some creeping underground. *A common perennial of wet marshy ground.* **Caltha palustris**

199. Stems and blades *lacking* long spines 201
   Stems and blades *prickly*, with long sharp spines 200

200. Blades, adaxial surfaces, *rough* or *prickly* (Fig. 16.34); the latter, at least some, obovate-lanceolate, deeply pinnatifid, usually with 2-lobed segments, upper lobe toothed towards the base, the lower entire; terminal lobe *spear-like*; cauline leaves similar but sessile and decurrent; all long-spined. Stems *continuously* winged. *Very common biennial plant in several situations.* **Cirsium vulgare**
   Blades, adaxial surfaces, *smooth* (Fig. 16.35); some, at least, narrowly oblanceolate in outline, petiolate, shallowly pinnatifid; the lobes shallow with spinous margins; cauline leaves sessile, long-decurrent, and deeply pinnatifid and undulate. Stems cottony, up to 150.0 cm in height, *continuously spiny-winged.* *Abundant biennial plant of marshes and wet grasslands.* **Cirsium palustre**

201. Blades, at least some, *very deeply lobed* 207
   Blades *entire* or *toothed* 202

202. Leaves *ovate, lanceolate* or *broadly lanceolate* 204
Leaves *obovate, glabrous*, and either *yellowish-green* or *bluish-green* 203

203. Blades *toothed* (Fig. 16.36); the latter 1.0-4.0 cm long, short-stalked, soft and thin, alternate, falling early, yellowish-green. Stems simple or branched, usually *5-rayed* apically, usually *containing latex*, smooth. *A very common annual; cultivated ground; waste places.* **Euphorbia helioscopia**
Blades *entire* (Fig. 16.37); the latter 0.5-3.5 cm long, short-stalked, soft and thin, alternate, falling early, yellowish-green. Stems simple or branched, usually *3-rayed* apically, usually *containing latex*, smooth. *A very common annual of cultivated ground, waste places, etc.* **Euphorbia peplus**

204. Blades *toothed* 205
Blades *entire* (Fig. 16.38); radical eaves *mostly oblanceolate*, somewhat pointed, short-stalked or tapered, often in a dense rosette; upper leaves narrower, sessile; all *densely* covered adaxially and abaxially with *bulbous hairs*, and dull green. The stems erect, terete, ascending, and covered with *appressed* hairs. *Common annual; cultivated ground, waste places, etc.* **Myosotis arvensis**

205. Leaves, at least some, *large, 5-35 cm* 206
Leaves small, 1-3 cm (Fig.16.39); the latter with 3-7 well-spaced teeth, petiolate, alternate, *ivy-like;* more or less truncate basally, rather thick, the terminal lobe the largest and entire; ciliate with a few scattered short hairs; the upper leaves becoming smaller. Stems weak; branches decumbent from the base. *Common annual of cultivated ground and waste places, etc.* **Veronica hederifolia**

206. Hairs *soft*, plentiful (Fig.16.40). The lower leaves *ovate-lanceolate* or *broadly lanceolate*, obscurely toothed, rarely entire, abruptly narrowed into stalks, softly pubescent adaxially, grey-tomentose abaxially; upper becoming smaller, mostly lanceolate. Stems usually simple, greyish-tomentose or glabrous basally. *A very common biennial or perennial calcifuge plant.* **Digitalis purpurea**
Hairs, at least some, *short, stiff* and *bristly*, few (Fig. 16.41). Leaves variable; lower short-stalked, coarsely toothed or shallowly lobed; middle cauline leaves sessile, broadly ovate; upper usually entire; all deep or yellowish-green. Stems frequently stiffly hairy, at least at the base, but sometimes glabrous. *Abundant annual plant of cultivated ground and waste places.* **Sinapis arvensis**

207. Plant some colour *other than* glaucous 208
Plant *glaucous* (Fig. 16.42). The lower leaves petiolate, broadly lanceolate, regularly or irregularly lobed to about mid-way; lobes sinuate, pinnate, toothed, glabrous or white cottony; upper leaves *sessile*, otherwise similar. The stems erect, succulent, glabrous or covered with whitish cottony hairs, branched. *An ubiquitous annual of cultivated ground, waste places, etc.* **Senecio vulgaris**

208. Blades *pinnatifid* or with a *few* deep lateral lobes 211
Blades, at least some, *palmatifid* or *deeply 3-lobed*; lobes further *divided* or *toothed* 209

209. Basal and middle leaves *deeply trilobed close to the mid-rib* 210
Basal and middle leaves stalked, divided nearly *to the base* into 5-9 *palmatifid* lobes (Fig. 16.43); the latter divided and toothed; the former alternate, more or less orbicular in outline; upper sessile or short-stalked, entire or nearly so; all mostly pubescent. Stems erect, mostly thinly pubescent, branched. *Abundant perennial of meadows, pastures, waste places, etc.* **Ranunculus acris**

210. Plant *stoloniferous* (Fig. 16.44). Lower leaves with long membranous sheathing stalks, and *deeply pinnately* divided into *3 lobes*; middle lobe long-stalked and projecting beyond the other 2; all lobes further 3-lobed once or twice; the middle upper blades less well divided; the uppermost blades sessile, narrow, and entire or nearly so; all pubescent. The stems both erect and prostrate, pubescent. *Abundant perennial; damp cultivated ground, waste places, etc.* **Ranunculus repens**
Plant *not* stoloniferous (Fig. 16.45). Lower leaves *deeply pinnately* divided into *3 lobes*, *relatively compact;* the middle lobe long-stalked and projecting beyond the other 2; lobes further 3-lobed once or twice; middle upper less well all pubescent. The stems *erect, bulbous* basally, pubescent. *A very frequent perennial of dry pastures, roadsides, gravelly places, etc.* **Ranunculus bulbosus**

N.B. DRAWINGS NOT TO SCALE

Fig. 16.00

211. Main stems *erect*; plant *not* foetid 213
      Main stems *prostrate*; plant *foetid* 212

212. Fruits, which appear early, *mucronate* (Fig. 16.46); the latter mostly suborbicular, laterally com-
     pressed, constricted in the centre, cordate basally. The leaves *finely* and *deeply pinnatisect*, deep
     green; segments of lower leaves toothed; those of the upper leaves narrower, mostly entire; all
     usually minutely pubescent. *A frequent annual of cultivated ground, etc.* **Coronopus squamatus**
     Fruits, which appear early, *emarginate* (Fig. 16.47); the latter mostly suborbicular, laterally com-
     pressed, constricted in the centre, cordate basally. The leaves *finely* and *deeply pinnatisect*, deep
     green; segments of lower leaves toothed; those of the upper leaves narrower, mostly entire; all
     usually minutely pubescent. *Frequent annual plant; cultivated ground, etc.* **Coronopus didymus**

213. Sini extending *very close* to the mid-rib 214
     Sini stopping *well short* of the mid-rib (Fig. 16.48). Leaves variable; lower short-stalked, coarsely
     toothed or very shallowly lobed; middle-cauline leaves mostly sessile, broadly ovate; upper cau-
     line often entire, small; all deep or yellowish-green, with at least some *short, stiff, bristly* hairs.
     Stems erect, simple or branched, up to 80 cm, usually stiffly hairy at least at the base, but some-
     times glabrous. *Abundant annual plant of cultivated ground and waste places.* **Sinapis arvensis**

214. Terminal lobe of the lower leaves *divided* into at least 3 secondary lobes 216
     Terminal lobe of the lower leaves *entire* or *toothed* 215

215. Colour *deep yellowish-green* (Fig. 16.49). Basal leaves long-stalked, lyrate-pinnatifid, with a very
     large, ovate-cordate, somewhat toothed terminal lobe; lateral lobes very few and small; middle
     upper short-stalked or sessile, toothed or entire, ovate-rhomboid to lanceolate, acute; all sparse-
     ly and softly pubescent. Stems mostly erect with spreading hairs towards the base, and usually
     glabrous apically. *Common annual of cultivated ground, waste places, etc.* **Lapsanna communis**
     Colour *dull grey-green* (Fig. 16.50). Basal leaves deeply and irregularly pinnately lobed, the latter
     usually unequally toothed or lobulate; middle leaves mostly oblong, pinnately lobed or toothed;
     all thinly and very shortly pubescent adaxially and abaxially, frequently with the terminal lobe
     the largest and somewhat hastate. Stems erect, woody basally and covered with reflexed bristly
     hairs. *Common annual of cultivated ground, roadsides, waste places, etc.* **Sisymbrium officinale**

216. Hairs *few, scattered* or *absent* 218
     Hairs *dense, abundant* 217

217. Ultimate segments of the lower blades *narrow, acute* (Fig. 16.51); the latter once or twice *deeply
     pinnately* divided into *coarsely* toothed or short bristle-pointed segments; upper becoming ses-
     sile, and often 3-lobed; the central lobe *elongate lanceolate;* all leaves bristly. Stems erect, and
     with *spreading* hairs. *A common annual of cultivated ground and waste places.* **Papaver rhoeas**
     Ultimate segments of the lower blades short, *rather broad* and *abruptly acute* (Fig. 16.52); the lat-
     ter once or twice *deeply pinnately* divided into *coarsely* toothed or short bristle-pointed seg-
     ments; upper sessile, often 3-lobed; the central lobe *relatively small*; leaves bristly. Stems erect,
     with *appressed* hairs. *A common annual of cultivated ground and waste places.* **Papaver dubium**

218. Veins *clearly anastomosing* (Fig.16.53). Leaves deeply lyrate-pinnatifid or pinnate with the termin-
     al lobe larger than the laterals, bright green, glabrous except for short scattered bristle hairs; the
     terminal further 3-lobed to about the middle; lobes often increasing in size upwards; the edges
     coarsely and irregularly toothed. Stems erect, simple or branched, ribbed, and mostly with a few
     short reflexed bristles. *A frequent annual of cultivated ground and waste places.* **Sinapis alba**
     Veins *not anastomosing* (Fig. 16.54). Basal leaves deeply and irregularly pinnately lobed, the latter
     usually unequally toothed or lobulate; middle leaves mostly oblong, pinnately lobed or toothed;
     all thinly and very shortly pubescent adaxially and abaxially, frequently with the terminal lobe
     the largest and somewhat hastate. Stems erect, woody basally, and covered with reflexed bristly
     hairs. *A common annual; cultivated ground, roadsides, waste places, etc.* **Sisymbrium officinale**

3 POISONOUS PLANTS

General: From the earliest of times some plants were venerated because they were known to have healing properties. Indeed, knowledge of the existence of plants containing narcotics and poisonous substances is almost as old as society itself. Man, and even animals, exploited these plants without actually knowing the nature of the properties involved or their mode of action. Some Indian tribes, in primitive jungle areas, have long understood the curative and toxic properties of an extensive range of wild plants and have used them to their advantage. They used a wide range of plant extracts to kill or stun fish, to enhance lethal arrowheads, or as a curative treatment for a variety of ailments. While they lacked detailed knowledge of the chemical properties of the plants involved, they were aware that certain species had specific effects. Fortunately, modern man has a better, though not a complete, understanding of the therapeutic effects of a wide range of plants. Countless drugs have been extracted, refined and evaluated for their medicinal properties. The search for new drugs from plant sources is ongoing. With the help of the chemical and pharmaceutical industries, man no longer depends on natural plant sources as these are no longer sufficient to meet the needs of commerce. Many drugs which originated in plants are now synthesised. Nevertheless, plants remain an active source of chemical substances which cause serious illness or even death in man and other animals.

Poisoning, particularly of livestock, resulting from ingestion of a wide range of plants with poisonous properties is a fairly frequent occurrence. Furthermore, there have been several incidences in Ireland, in recent years, of human poisoning as a result of the ingestion of plants mistaken for edible species. What is a poisonous plant? There are many definitions. One of the most widely used in given by Forsyth (1954): 'a poisonous plant is one which gives rise to a serious departure from normal health when a small quantity of its seed, root or vegetation is eaten by a creature which is susceptible to its effects'. Adopting this very broad definition there are many plants, either common or rare, throughout the British Isles which can be termed poisonous.

Biotoxicology is the science concerned with the study of poisons produced by plants and animals. Biotoxins, or biological poisons, are poisons produced by biological organisms. The term phytoxin is more specific in that it refers to any toxin originating in a plant. This term is also commonly used by plant pathologists to refer to any substance that is toxic to plants. Poison is a term used to refer to any substance which if injected, ingested, absorbed, or applied to the body of an animal in relatively small quantities, may cause by its chemical action, damage to a structure, a disturbance of function or, if the quantity is sufficient, even death.

Classification

A study of the wide range of plants involved suggests that they may be classified under 4 headings: (i) plants which are poisonous to eat; (ii) plants that are poisonous to touch; (iii) plants that produce photosensitisation, usually in light-coloured animals; and (iv) plants responsible for airborne allergies. Other possible classificatory systems involve grouping on the basis of their morphology, on the basis of the poisonous substances they contain, and on the basis of the symptoms they produced in the affected animal. A consideration of all the plants under the morphology heading will show that they constitute a very artificial group, ranging from trees 30 m or more in height to microscopic fungi or bacteria. Wild herbaceous plants are the principal offenders, though many food plants have proved poisonous to man and livestock in the past. Practically all members of the potato family, the Solanaceae, have some poisonous properties. Green tubers of *Solanum tuberosum* and the stalks of *Lycopersicon esculentum*, which contain alkaloids, have proved fatal to livestock in other countries. The fruits of both plants are wholesome. In addition, some food storage organs such as *Beta vulgaris* may also prove poisonous if fed to livestock in a fresh state. The seeds of some cultivated *Brassica* species may have a similar effect under certain circumstances.

In the present discussion, microscopic fungi and bacteria may be considered to be the least important. Microbial toxins are produced by bacteria, blue-green and golden-brown algae, dinoflagellates, fungi and others. From the bacterial point of view, species of *Corynebacterium, Staphylococcus, Streptococcus, Clostridium,* and *Escherichia* are the most important. Their ability to cause severe infections and septicaemia is well known. Microfungal organisms worthy of mention include species of *Claviceps, Stachybotrys, Fusarium,* and *Cladosporium.*

Macrofungi, represented by mushrooms, or toadstools as they are commonly called, are widely distributed throughout the British Isles. Most are poison-free but a few, particularly some species of *Amanita, Agaricus, Cortinarius* and *Inocybe* are very poisonous. Poisonous toadstools are rarely eaten by livestock because they occur in Autumn when forage is plentiful, or because they occur in areas which are not accessible to animals. The most important species is undoubtedly *Amanita phalloides*. It is responsible for more deaths and severe cases of poisoning in human beings than any other fungus. Animals are rarely affected. It is relatively large, 4-12 cm in height, with a very distinctive appearance. The cap is flat or convex, up to 12 cm in diameter, smooth with radiating streaks, and yellowish olive-green in colour externally; some caps may be almost totally white. Cap-flesh is white except beneath the skin, where it is tinged with yellow or green. The stem, which is up to 12 cm x 20 mm, is smooth, white, and thickens towards a large bulbous base enclosed in a large white, rarely yellow, sheath or *saccate volva*. Remnants of a white membrane, which joined the stem to the cap when young, are evident at the top. *Amanita phalloides* is very poisonous and occurs mostly in, or at the edges of, oak and beech woodland.

The study of poisonous algae is known as phycotoxicology. Algae compromise some of the smallest forms of chlorophyll bearing plants. They live in all parts of the aquatic environment that receive light. Members responsible for illness and death in animals include species of *Anacystis, Anabaena, Aphanizomenon* and *Lyngbya*. Poisoning by this source is of no great significance.

Phytotoxins

A study based on the chemical properties responsible for the poisoning effect is possibly the most useful. These chemicals differ widely in their formulation and in their effect on the body. Some, for example, alter heart beat. Others effect the respiratory processes, while more have the effect of deadening pain. In addition, they may be stable or unstable. Stable substances are effective in both fresh and dried material, while unstable chemicals are only toxic in fresh growing plants. Furthermore, poisonous substances are not necessarily distributed uniformly throughout the plant. Some parts, for example the seeds, may be more poisonous than any other part of the plant. In some food plants the position is reversed. Other features which influence the potency of poisonous plants include the climatic and soil conditions under which they are grown, and the plant type or strain.

Plant toxins constitute a vast range of biologically active chemical substances, such as glycosides, alkaloids, amines, oxalates, resins, various oils, acrid substances, and other miscellaneous substances whose chemical structure is not fully understood. Some of these are very poisonous in small amounts but, in reality, the poison is in the dose. Others have to be ingested in large quantities to effect a deleterious consequence. Most of the more poisonous plants, *e.g. Taxus baccata*, contain a combination of these substances. Animal susceptibility is also an important factor, as they can react differently to a given substance. Some of the more important poisonous substances are considered further hereunder.

Glycosides: These are complex substances which break down on fermentation or by interaction with dilute mineral acids to give a sugar and another substance, the aglycone, which is extremely poisonous. Plants containing glycosides usually have the enzymes capable of breaking them down. When the enzyme and glycoside come together, as during mastication of plant material containing them, the poison is released. The most usual poison involved is prussic acid or HCN. This is very potent and invariably brings about sudden death. Cyanogenetic glycosides are found, *inter alia*, in *Prunus laurocerasus* and in *Taxus baccata*. Their presence in the latter provides an immediate explanation as to why death, resulting from the ingestion of its twigs, is so rapid (R.A.F.).

Alkaloids: Alkaloids, most of which are found in plants, are compounds of carbon, hydrogen, oxygen and nitrogen with a chemically basic reaction, are noted chiefly for their physiological activity. It has been estimated that about 10% of all plant species contain some type of alkaloid. They are usually bitter in taste. Many have histories as poisons, narcotics and medicinal agents. Only a few of the approximate 5000 alkaloids have no known biological activity. They are found in many poisonous plants and are not confined to any one plant group. Most alkaloids are isolated from natural sources and so bear coined names. Such nomenclature may reflect the name of the discoverer, the botanical source (strychnine from *Strychnos nuxvomica)* or the biological effect. For most, the name is obtained by dropping the last syllable of the generic or specific name of the plant which is its main source and adding the suffix 'ine'. Thus, the principle alkaloid in *Atropa belladonna* is atropine.

Many alkaloids are of medicinal importance because they are the cause of occasional poisoning of livestock or man, an example being the group of ergot alkaloids, produced by the fungus ergot. The alkaloids of the *Equisetum* species have come under suspicion in instances of livestock poisoning. *Conium maculatum* contains lethal amounts of coniine and other alkaloids which are potent only in the growing plant; they volatalise when the plant is cut and left to dry. Others remain active in fresh and dried material. Some well known alkaloids include morphine from *Papaver somniferum*; opium, or the congealed latex derived from the fruits of this plant, contains up to 30 alkaloids. Other well known alkaloids include Nicotine from *Nicotiana tabacum* and Cocaine from *Erythroxylon coca* (D.A.F.).

Volatile Oils and Acrid Substances: These may be termed the 'safest' of the plant poisons as they rarely prove fatal. They can, however, give rise to great discomforts. They are present in many plants and exert their effect by irritation and, in some cases, by coagulating the proteins of the stomach and bowel walls. If absorbed into the blood stream they may cause damage to any part of the body. They may be excreted in sweat, saliva, milk or urine. Most plants which contain them are harmful only in the fresh or growing stage because these poisons either volatalise or are precipitated into non-poisonous components when the material is cut and allowed to dry. *Ranunculus* species are good examples.

For convenience, plants which produce airborne allergies, may be mentioned here. The pollen of a few plants may cause a respiratory allergy, as anyone who suffers from hay fever in the summer can testify. Some plants may bring about an eczematous dermatitis of the exposed parts of the body. Grasses are the principal offenders. Other plants in this category include *Acer negundo, Fraxinus excelsior* and species of *Populus* and *Ulmus*. Dermatogenic species represent another group of poisonous plants of very minor significance. These are capable of inflicting pain by means of parenteral contact. Few such species occur in the British Isles. The most common examples are species of *Euphorbia* and *Urtica*.

Photodynamic Substances: A few plants contain chemicals, photodynamic substances, which, on being absorbed into the blood stream, render the skin of white-skinned animals very sensitive to light. Animals with heavily pigmented skins or with skins covered with pigmented hair are not harmed. The affected part of the skin develops scars which are often slow to heal. *Narthecium ossifragum, Hypericum perforatum, Heracleum mantegazzianum* and *Fagopyrum esculentum* are 4 plants which contain chemical substances with photosensitising properties. The first 2 are fairly common and are harmful to animals. In addition, some forms of *Trifolium repens* are also suspected of having similar properties.

The third plant, *H. mantegazzianum* produces dermatitis in man. It contains furocoumarins which render the skin hypersensitive to sunlight. After contact with the plant and exposure to sunlight, the skin becomes red and develops blisters within 48 hours. This plant is relatively rare in the British Isles, but there are indications, in recent years, that it is becoming more widespread. It occurs mostly in damp places, particularly in old estates. The general appearance is very similar to that of the common *Heracleum sphondylium*, but it is much larger in all its parts; it often attains a height of 3 m. The fourth plant, *F. esculentum*, is cultivated to a very limited extent in Britain as a fodder crop, particularly in fens. It occurs as a casual weed in some parts of Ireland. In other European countries it is grown for its flour which is used in a particular type of pancake. It is a member of the *Polygonaceae* and is rather similar to *Polygonum persicaria* in many respects. It has the characteristic stipules of the family but the leaves are cordate-sagittate, pointed and about as broad as long. It is a weak, little-branched annual. The flowers are in pink or white cymose panicles. The black pyramidal fruits are wholesome but the vegetative parts, particularly when fresh, cause photosensitisation in white-skinned animal.

Circumstances Under Which Poisoning Occurs

Fortunately, most of the plants which contain poisonous properties are rarely eaten by livestock under normal circumstances. This is due, in most instances, to the nature of the poisonous properties involved and to the ecological preference of these plants. Many poisonous plants, for example, *Digitalis purpurea*, have a very bitter taste and are immediately rejected by livestock when taken into the mouth. Others, such as species of *Allium*, have an unpleasant odour. These plants are usually avoided by livestock. *Ranunculus* species are poisonous only in the fresh state; their poisonous properties are precipitated into non-poisonous components on drying, as happens when these plants occur in hay fields. As already stated, animals differ in their susceptibility to poisonous plants. Apparently *Senecio jacobaea* can be eaten by sheep with impunity. Poisoning by this plant generally occurs in other animals and arises from the ingestion of dried material, such as hay. *Conium maculatum*, arguably the most poisonous herbaceous plant in the British Isles, is rarely attractive in the mature stage. In spring, however, when the young leaves are intermingled with grasses the plant is occasionally eaten with fatal consequences.

Some poisonous plant grow only in wet ditches, streams, lakesides and similar ecological situations, areas which are not normally frequented by animals. In addition, some dangerous shrubs, such as the *Taxus baccata* and *Prunus laurocerasus*, are invariably found in areas which are not accessible to animals. All poisonous herbaceous ornamentals occur in similar situations. In times of scarcity, *i.e.* lack of drinking water in summer or fodder in winter due to snow-covered pastures, however, animals are often attracted to the lush vegetation in wet areas or the green plants in hedgerows or enclosed areas. It is under these circumstances that plants, such as *Oenathe crocata*, *Cicuta virosa*, *Prunus laurocerasus* and *Taxus baccata* and some herbaceous ornamentals cause poisoning. Also, it is very advisable when ditches and streams, which contain poisonous plants, are cleaned that the debris be removed to a safe area. Animals, out of curiosity, are often attracted to the debris. These plants, because of the stable nature of their poisonous substances, remain a source of danger. The removal of poisonous shrubs or trees requires careful attention. The author is aware of two instances where the felling of yew trees caused the death of a number of animals! Apparently, in the felling and removal operations, small pieces of twigs and grass-coloured leaves were left on the ground, and these were eaten with fatal consequences. It is very advisable, therefore, that all material, however small, should be removed to a safe disposal area.

IDENTIFICATION

The poisonous plants given in the accompanying 'key' form a very artificial group, and range from small prostrate herbaceous plants to trees 30 m or more in height. All are reasonably plentiful throughout the British Isles and pose a threat to animal and human health. A person trying to identify a poisonous plant, or wanting to know if a given plant has worthwhile poisonous properties, should use the 'key' given hereafter. All the important poisonous species are included. If the plant cannot be located in this section, then, in all probability, it has no significant poisonous substances. Information on the general characteristics, poisonous properties, and the areas where it is likely to occur, is given for each plant as it occurs in the 'key'.

Several plants are omitted from the 'key' because they contain small insignificant amounts of known poisonous properties, or because there is no definitive information on their poisonous properties, or because they occur in areas which are not readily accessible to livestock. Some are ornamental and are unlikely to be eaten except under unusual circumstances. Some of the more well known plants in this category include:

Atriplex patula. This is a herbaceous annual of the beet family, it occurs commonly in tilled fields and waste ground. All parts of the plant are covered with 'mealy' hairs or *papillae*. Ingestion is said to have caused severe photosensitisation in white-skinned animals.

Beta vulgaris. Included under this name are the sugar beet, fodder beet and mangolds or mangels. Nitrates accumulate in these plants and are present in freshly harvested 'roots' and tops. Nitrates act as severe irritants. In addition, these plants contain oxalic acid and soluble oxalates.

Brassica species and related plants. Many Brassica seeds contain *glucosinolates* which give the poisonous substances thiocyanates, isothiocyanates and nitriles. Plants of importance include *Brassica rapa* (turnips and turnip rapes), *B. oleracea* (cabbage type plants), *B. napus* (swede and swede rapes).

Capsella bursa-pastoris. This is a well-known annual of cultivated ground and waste places. It may be identified by its deeply lobed leaves, very small white flowers and heart-shaped fruits. This plant, shepherd's-purse, contains very small amounts of glycosides which are said to taint milk.

Chelidonium majus. A member of the poppy family, it is an occasional plant of hedgerows, banks, walls, particularly near habitations. It has the general poppy-type flower, but the colour is greenish-yellow; the fruits are long and cylindrical. Leaves are glaucous and almost pinnate; it contains a bright orange latex.

Crataegus monogyna. A deciduous shrub with spiny branches, its leaves are small, dull green, lobed, short-stalked and glabrous except for tufts of hairs in the axils of the lower veins. Flowers are white. Rarely eaten by livestock, but its reddish drupes are said to contain small amounts of a toxic glycoside.

Fraxinus excelsior A deciduous tree, it contains poisonous substances which may cause an illness if the leaves and fruits are eaten in quantity.

Galeopsis tetrahit. This is rather similar to *Lamium purpureum*, described elsewhere, but is bigger in all its parts and has glandular nodes. It is alleged to have poisoned animals in other European countries.

Grasses. *Lolium temulentum* is a poisonous plant, though the nature of its toxins is not fully understood. *Anthoxanthum odoratum* contains chemicals which may interfere with blood clotting. *Secale cereale* is the only cereal suspected of having poisonous properties. The ergot fungus is often associated with many grass species. It contains many alkaloids, although all are not poisonous. Fodder containing ergot has a musty smell and a disagreeable odour.

Juncus species. There are 6 common species throughout the British Isles. These are *acutiflorus, articulatus, bufonius, effusus, inflexus* and *squarrosus*. They occur in wet areas or heaths. Most are said to contain small amounts of cyanogenetic glycosides.

Leucanthemum vulgare. A common plant of dry grassy places, it contains a substance which taints milk.

Lotus corniculatus. A common legume of roadsides, old pastures, sand-hills and waste grassy places, it is reputed to contain cyanogenetic glycosides.

Mentha species. There are several species; all contain a chemical which can prevent the clotting of blood. *Mentha aquatica* is the most important species as it occurs frequently in wet grassland.

Oenanthe aquatica. Closely related to the more accessible *O. crocata*, it occurs in slow-moving rivers, canals, lakesides, etc. It contains similar poisons, but in lesser amounts.

KEY TO POISONOUS PLANTS

1. Plant *herbaceous* 25
   Plant *woody,* either a *tree, shrub* or *shrublet* 2

2. Branches *not* thread-like 3
   Branches *thread-like* (Fig.17.01). Tall tree, reaching up to 25 m, either pyramidal or becoming flat-topped; bark red-brown in colour, ridged; the final branches thread-like and covered with minute overlapping leaves; the latter are less than 5 mm, deltoid, obtuse, dark green in colour, in opposite, alternating, appressed pairs; often furrowed. Poisoning, particularly of abortion in cows, is said to have occurred in some countries. *Frequently planted shelter tree.* **Cupressus macrocarpa**

3. Leaves *other than* trifoliate 5
   Leaves *trifoliate* 4

4. Twigs *terete* (Fig. 17.02). Small tree, up to an average of 7 m in height. Leaves trifoliate; each leaflet mostly oval-lanceolate in outline, and borne on short stalks. Flowers in very long, pendulous, yellow-coloured racemes. Fruits or pods are brown, and frequently persist over winter. Children are often attracted to the pods. All parts are very toxic. Nearing maturity, the poisonous alkaloid, *cytisine,* concentrates in the seed. *Very common ornamental plant.* **Laburnum anagyroides**
   Twigs *5-angled* (Fig. 17.03). Shrub up to an average of 2 m in height; branches freely; the latter often occurring in bundles. Leaves small, usually trifoliate. Stems and branches ridged or angled. The flowers, which are yellow-coloured, occur 1-2-3 in leaf axils. Contains very small amounts of *cytisine, sparteine* and *scoparine.* Animals need to eat large quantities before serious poisoning occurs. *Very frequent calcifuge plant; both wild and garden ornamental.* **Cytisus scoparius**

5. Leaves *pinnate, pinnately lobed, toothed* or *entire* 6
   Leaves *palmately lobed* (Fig. 17.04); the latter variable, often 3-5 lobed, dark green above. Plant climbing, reaching several metres in height or creeping along the ground. Stems densely clothed in adhesive roots. Flowers in yellowish-green umbels; fruits berry-like, small, black; said to be the most poisonous part of the plant. Poisoning, which is not fatal, can occur in winter when all other plants are scarce. Plant *occurs commonly on walls, trees, old buildings, etc.* **Hedera helix**

6. Leaves *simple* 7
   Leaves *pinnate* (Fig. 17.05); the latter with about 5-7 narrowly ovate, acute or long-pointed leaflets which are subcordate or rounded basally, coarsely toothed or entire, hairless or slightly pubescent; opposite. Flowers in lax axillary, terminal, fragrant panicles. Perianth-segments greenish-white and densely pubescent on the outside; petals absent; stamens and carpels several; fruits or achenes with long thread-like styles, and covered in a mass of long silky hairs. Plant contains *protoanemonin* which is only active in the fresh stage; causes very severe abdominal pain. *Slender perennial, woody climber, it is frequent on trees, shrubs and in hedgerows.* **Clematis vitalba**

7. Leaves *toothed* or *entire* 9
   Leaves *pinnately lobed* 8

8. Blades *tapered* basally (Fig. 17.06); the latter pinnately lobed, short-stalked. Tall tree, up to 30.0 m in height. Fruits or acorns are attached directly to the main stem; the former long, up to 30 mm, and subtended by short leathery cupules. Oak contains *tannic acid* which acts as an astringent. Poisoning usually occurs in spring on the ingestion of young leaves or, later in the year, when the acorns are formed. *Frequent, especially in woods in mountainous districts.* **Quercus petraea**
   Blades *rounded* basally (Fig. 17.07); the latter deeply pinnately lobed, short-stalked. Tall tree, up to 30 m. Fruits or acorns are attached to the main stem by short pedicels; the former long, up to 30 mm, and subtended by short leathery cupules. Oak contains *tannic acid* which acts as an astringent. Poisoning usually occurs on the ingestion of the young leaves or later when mature acorns fall to the ground. *Frequent in woods and rocky places, mainly in lowland areas.* **Quercus robur**

9. Leaves *entire* 14
    Leaves *toothed* 10

10. Leaves *alternate* or *spirally* arranged 12
    Leaves *opposite* or *subopposite* 11

11. Plant *not* spiny (Fig. 17.08). A deciduous shrub or small tree, from 2-6 m in height; twigs with 3-4 indistinct ridges. Leaves opposite, oval-lanceolate, 3-14 cm long, finely toothed. Flowers in leaf-axil clusters, greenish-white; petals 4. Fruits most characteristic, 4-lobed, bright coral-pink in colour. They are particularly attractive to children. Ingestion of twigs, leaves and fruits causes severe purgation. *Frequent plant; thickets, rocky places and lakeshores. Euonymus europaeus*
    Plant *spiny* (Fig. 17.09). Shrub or small tree up to 6 m in height; usually with some thorns. Leaves opposite or subopposite, ovate, 3-6 cm long, toothed. Venation characteristic; side-veins curving towards the tip of the blade. Plant dioecious; flowers in small green clusters, on short spurs. Petals 4, green. Fruits black, about the same size as black currants. They are particularly dangerous to children. Ingestion of leaves, fruit and bark, which contain *glycosides,* causes super purgation and even collapse. *Occasional plant; rocky places and lakeshores. Rhamnus catharticus*

12. Leaves *clearly* toothed 13
    Leaves minutely toothed (Fig. 17.10); the latter large, leathery, yellowish-green, glossy, up to 18 x 6 cm, mostly obovate in outline, denticulate, short-stalked, and usually with 2 conspicuous glandular pits towards the base. Flowers in erect whitish racemes. Shrub contains the cyanogenetic glycoside, *amygdalin,* which produces *hydrocyanic acid*; others, including *prunasin,* also present in very small amounts; very rarely eaten under normal circumstances; poisoning occurs during times of scarcity. *A common evergreen hedgerow or ornamental plant. Prunus laurocerasus*

13. Young leaves *glabrous* (Fig.17.11); all leaves leathery, glossy, strongly toothed, spotted yellow, up to 20.0 cm. A small dense evergreen laurel-like shrub, up to 3 m in height. Some forms produce scarlet-coloured berries late in the season. The poisonous property is a cyanogenetic *glycoside.* Poisoning is a real possibility if sufficient quantities of leaf material are ingested, though, under normal circumstances this is unlikely. *Very common garden ornamental plant. Aucuba japonica*
    Young leaves *pubescent* (Fig. 17.12); all leaves leathery, untoothed basally, but characteristically and distinctly toothed apically, 3-9 cm long; downy abaxially and adaxially; the abaxial surfaces becoming glabrous and dark glossy-green; young leaves light green. Flowers, as in all oaks, are unisexual. Plant contains *tannic acid* which acts as an astringent. Poisoning may occur on the ingestion of the leaves or acorns. *Frequent tree, up to 28 m, in gardens and parks. Quercus ilex*

14. Leaves *neither* sharp-pointed *nor* less than 30 mm 16
    Leaves somewhat *sharp-pointed*, short, *30 mm or less* 15

15. Leaves in *whorls of 3* (Fig. 17.13); the latter sharp-pointed, needle-like, about 12.0 mm long, with a broad grey, adaxial band; rarely scale-like, appressed and opposite. Shrub or tree up to 4.0 m in height, habit varying from procumbent to erect and narrow; bark scaly, rough, reddish-brown. Severe digestive upsets may occur on the ingestion of some leaves and bark. It *occurs in rocky places, mountain heaths and lakeshores*, particularly in western counties. *Juniperus communis*
    Leaves spreading *in 2 ranks* (Fig. 17.14); the latter attached all around the stem, about 25 mm long, needle-like, short-stalked, glossy-green, short-pointed; central vein usually raised; abaxial surface light green in colour. Evergreen poisonous shrub, it contains a number of poisonous properties such as *taxine, ephedrine,* traces of *hydrocyanic acid* and *volatile oils.* Death is sudden, and usually results from the ingestion of twigs and leaves on the ground when the tree is felled. *A common garden ornamental;* also *in graveyards and other enclosed locations. Taxus baccata*

16. Leaves *clearly opposite* 22
    Leaves *alternate, spirally arranged, clustered* or *subopposite* 17

17. Leaves *narrow* 18
    Leaves *relatively broad* 19

18. Stems *terete* (Fig. 17.15); the latter bluish-green. Leaves, when present, simple, linear-lanceolate, small, falling early. Flowers large, yellow, zygomorphic, and occur 1-3 in the leaf axils. Small shrub, up to 3 m in height. Plant contains the alkaloids *sparteine* and *isosparteine*; more in lesser amounts include *cytisine, genistein, lupinidine* and *sarothamnine*. Poisoning by this plant only occurs under unusual circumstances. *Very frequent garden ornamental plant.***Spartium junceum**
    Stems *5-angled* (Fig. 17.16); the latter strict, green. Leaves simple or trifoliate; leaflets narrowly elliptical to obovate, acute, with appressed hairs. Flowers zygomorphic, yellow, large, and occur 1-3 in the leaf axils. A small shrub, up to 3.0 m. Plant contains *sparteine* and *isosparteine*; other alkaloids include *cytisine, genistein, lupinidine* and *sarothamnine*. Poisoning by this plant only occurs under unusual circumstances. *Very frequent calcifuge plant of heaths.* **Cytisus scoparius**

19. Leaves, at least some, *spirally arranged* or *clustered*; *leathery* 20
    Leaves *clearly alternate* (Fig. 17.17); the latter soft, broad, entire, widest in their upper half. Young twigs greyish-brown, minutely pubescent. Flowers mostly in small clusters on young wood and with 5 greenish petals. Berries green, then red, and finally black. An unarmed deciduous shrub or small tree, up to 5.0 m in height. Ingestion of leaves, fruit or bark, which contain *glycosides,* can cause severe purgation and even collapse. *A frequent plant of rocky places.* **Frangula alnus**

20. Adaxial and abaxial surfaces of a *uniform* colour 21
    Adaxial and abaxial surfaces *clearly different* in colour (Fig. 17.18); the latter dark green adaxially, paler green abaxially, 6-15 cm long, coriaceous, somewhat oblanceolate; young leaves brighter green and less leathery. Flowers large, showy, in terminal racemes or clusters; corolla campanulate, deeply divided and dull-purple. An evergreen shrub up to 3.0 m in height. Plant contains several toxic *diterpenoids* but is not eaten under normal circumstances. In times of scarcity, it is dangerous if ingested. *A naturalised plant throughout the British Isles.***Rhododendron ponticum**

21. Blades *light green* (Fig. 17.19); the latter 3-10 mm, oblanceolate, acute, glabrous, mostly clustered, thin, not entire, coriaceous. Flowers, which occur before the leaves, purple or rarely white, often pubescent externally, fragrant, and arise in the axils of the previous year's growth. Small shrublet, up to 1.0 m in height. The leaves and berries contain an *active irritant glycoside*. Plant *local* and rare throughout the British Isles; often planted *as a garden ornamental.* **Daphne mezereum**
    Blades *dark glossy-green* (Fig. 17.20); the latter obovate-lanceolate or oblanceolate, acute, entire, 5-12 cm, glabrous, coriaceous. Flowers in short axillary racemes, green and glabrous externally. Erect glabrous shrub, sparingly branched; thick stems showing scars of fallen leaves. It contains an *acrid juice,* and small amounts of *glycosides* and *resins;* not normally eaten by animals because of its unpalatability but it may prove poisonous under unusual circumstances. *A frequent plant throughout the British Isles; calcareous soils; also a garden ornamental.***Daphne laureola**

22. Blades *tapered* apically 23
    Blades *blunt* apically (Fig. 17.21); the latter opposite, small, oval, obtuse, short-stalked, entire, and green or dark brown in colour. Flowers small, clustered, and whitish-green. A small evergreen shrub; twigs angled, downy. All parts contain the alkaloid *buxine,* some *resins* and *oils*. Poisoning, in the form of digestive upset, accompanied by severe pain and salivation, can result from the ingestion of the bark or leaves. *A common hedgerows or garden plant.* **Buxus sempervirens**

23. Stems and branches *without* 2 raised lines 24
    Stems and branches with *2 raised lines* (Fig. 17.22); the latter otherwise glabrous. Leaves very distinct, broadly ovate, blunt apically, 5-10 cm, entire, sessile or very nearly so, often with minute translucent glands. Flowers in panicles, showy, yellow; petals and sepals 5; stamens numerous. A shrub, varying in height from 10.0-100 cm. Contains a red-coloured pigment, *hypericin* which causes photosensitisation in white-skinned animals. *A frequent plant in hedges, thickets, rocky places and in and around old buildings; often an ornamental plant.* **Hypericum androsaemum**

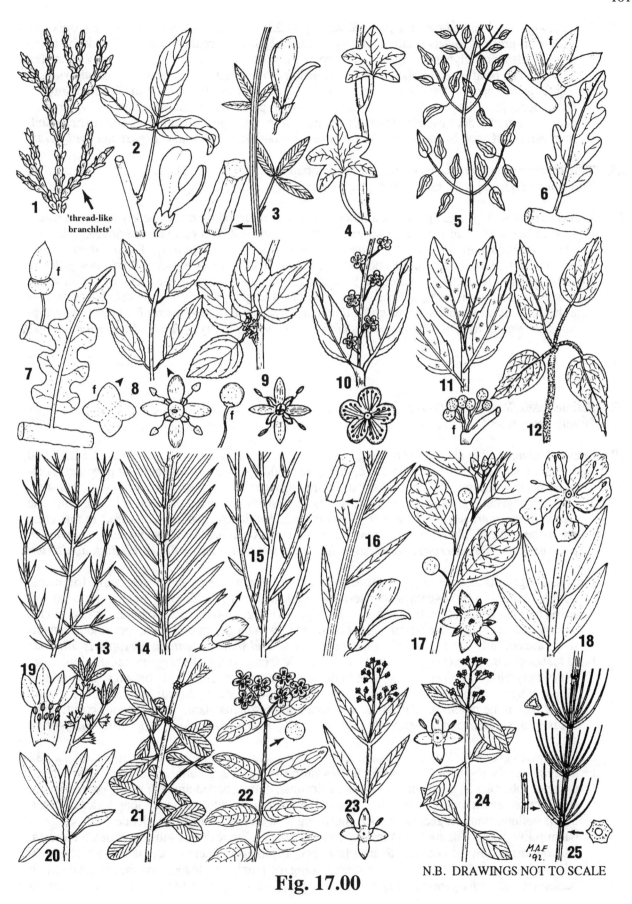

Fig. 17.00

24. Young twigs and panicle branches *puberulent* (Fig.17.23). The leaves lanceolate, 3-6 cm, opposite, yellowish-green. Flowers small, in clusters, white; petals and sepals 4; stamens 2. Shrub, it contains a poisonous glycoside, *ligustrin*. Poisoning results on the ingestion of leaves and berries. Children can be affected by the berries. *A common garden hedgerow plant.* **Ligustrum vulgare**
Young twigs and panicle branches *glabrous* (Fig. 17.24). Leaves elliptic-oval to elliptic-oblong, 3-6 cm, opposite, yellowish. Flowers in clusters, white; petals and sepals 4; stamens 2. A shrub, it contains a glycoside, *ligustrin*. Poisoning results from the ingestion of leaves and berries. Children can be affected by the latter. *Very common garden hedgerow plant.* **Ligustrum ovalifolium**

25. True leaves *present;* branches, if present, *not* whorled 27
True leaves *absent;* branches *whorled* 26

26. Branches *hollow* (Fig. 17.25). Sterile and fertile stems similar; both erect or nearly prostrate, usually with suberect branches, rarely simple, slightly rough, with 4-8 grooves and ridges; rhizomes glabrous. It contain the enzyme, *thiaminase,* which destroys *vitamin B_1.* Poisoning symptoms are similar to those of *vitamin B_1* deficiency. *Common plant in wet places.* **Equisetum palustre**
Branches *solid* (Fig. 18.01). Sterile stems erect or decumbent, slightly rough, with 6-18 grooves; branches usually spreading, several, usually simple, 4-angled; the fertile stems simple; rhizomes pubescent. Plant contains the enzyme, *thiaminase,* which destroys *vitamin B_1.* Poisoning symptoms are similar to those of *vitamin B_1* deficiency. *Common in dry ground.* **Equisetum arvense**

27. Leaves *simple* 34
Leaves *compound* 28

28. Petiole-bases *flat* and *sheath-like* 31
Petiole-bases *round* 29

29. Leaves *bipinnate;* rhizomes *not* spreading 30
Leaves *tripinnate* (Fig.18.02), up to 200 cm long, light glossy-green; young petioles pubescent; pinnules narrow-oblong, and 'parallel-sided', entire. Rhizomatous. Contains several toxic constituents; cyanogenetic glycoside, *prunasin,* is present in small amounts; other chemicals include the carcinogens, *quercetin* and *kaempferol*; an enzyme, *thiaminase,* has a similar effect to that present in the horsetail plants; other haemorrhage inducing constituents are also present. *Abundant in poor grasslands, hedgerows, woods, especially those on light sandy soils.* **Pteridium aquilinum**

30. Final leaf-segments *toothed* (Fig. 18.03); leaves 40.0 to 100 cm, mostly bipinnate, rarely tripinnate; petioles and mid-ribs covered with pale-brown scales; pinnules undivided, oblong, rounded apically, more or less toothed laterally and apically; sori large, 3-6 on each pinnule; lowest pinnae clearly decreasing in size basally. This is not an important poisonous plant; poisonous constituents are not fully understood. *Common in woodland and shaded locations.* **Dryopteris filix-mas**
Final leaf-segments *lobed* (Fig.18.04); leaves 20 to 100.0 cm, flaccid, light green, mostly bipinnate, very rarely tripinnate; petioles, at least in the lower halves, covered with brown scales; middle pinnae longest; pinnules oblong-lanceolate, pinnatifid or pinnately lobed; sori forming a row on either side of the pinnules. This is not an important poisonous plant; its poisonous constituents are not fully understood. *Frequent in woodlands and other shaded places.* **Athyrium filix-femina**

31. Leaf-segments *green* 32
Leaf-segments *bluish-green* (Fig. 18.05); leaves variously pinnate, glabrous or nearly so, with large ovate or suborbicular segments. Flowers in terminal, dome-shaped umbels; bracts present or absent; bracteoles present; petals white, often enlarged, emarginate, each with long reflexed point. Stems hollow, stout, ridged. Rootstock consisting of a cluster of tuberous, slender, carrot-like or parsnip-like 'root'. The roots and leaves contain a very toxic resinous substance, *oenanthetoxin,* which turns yellow on exposure to air. Plant very poisonous to man and animals; poisoning usually occurs when ditches and streams, containing the plant, are cleaned and the material left on banksides, etc. *A frequent plant of ditches, streams, lakesides, riversides, etc.* **Oenanthe crocata**

32. Petioles *grooved* 33

Petioles *ungrooved*, more or less rounded (Fig. 18.06). Leaves bipinnate or tripinnate, with deeply lobed, toothed segments; all parts with purple blotches. Flowers in small, terminal and axillary, short-stalked umbels; bracts and bracteoles present; petals white; calyx absent. Stems smooth, ridged. Extremely poisonous, contains 4 unstable alkaloids: *coniine, methylconiine, coniceine* and *conhydrine;* the plant is poisonous only in the fresh state. Most instances of poisoning occur when animals eat the leaves, intermingled with grass leaves, in early spring. *A frequent plant in and around old buildings, farmyards, hedgerows, and waste grassy places.* **Conium maculatum**

33. Leaflets somewhat *ovate*; deeply pinnatifid, toothed (Fig. 1.07). Leaves deltoid, usually few, somewhat glaucous, often bipinnate. Flowers in leaf-opposed umbels; petals white; sepals and bracts absent; bracteoles 3, at least twice as long as pedicels, radiating from the umbels. Stems hollow, finely striate. It contains the alkaloids, *coniine* and *cynapine*; large quantities need to be eaten before poisoning occurs. *Frequent annual of tilled ground and waste places.* **Aethusa cynapium**

Leaflets *linear-lanceolate* (Fig. 18.08), opposite, 2.0-10 cm, glabrous, acutely serrate. Leaves bipinnate or tripinnate; light green. Stems hollow, thick, somewhat ridged; rootstock with lacunae. Flowers in terminal umbels; petals white; sepals reduced to sharp teeth; bracts absent; bracteoles narrow. It contains a stable substance, *cicutoxin*, a potent, convulsant poison; poisoning occurs when streams, ditches, lakesides, containing the plant, are cleaned and the debris is left accessible to animals. *Infrequent plant of wet places such as riversides, lakesides, etc.* **Cicuta virosa**

34. Perianth-segments some number *other than* 6 38

Perianth-segments 6 in number 35

35. Leaves *neither* compressed *nor* sword-like 36

Leaves *strongly compressed, sword-like* (Fig. 18.09); the latter all radical, 'parallel-veined', rigid and often curved, in 2 opposite rows like those of the Iris plant to which it bears a strong resemblance in miniature. Flowers small; racemes terminal, dense or lax; perianth-segments 6, bright yellow; stamens 6, filaments densely covered with yellow woolly hairs. Plant rhizomatous. It contains 2 saponins, *narthecin* and *xylosin* which effect the liver, and cause photosensitisation in some white-skinned animals. *Very frequent in bogs, wet moors, etc.* **Narthecium ossifragum**

36. Blades *not* cylindrical 37

Blades *more or less cylindrical* (Fig. 18.10); the latter hollow, somewhat grooved, very long and terete, 'parallel-veined'; the petioles sheathing the scape. Flowers in umbels, few, often replaced by bulbils; pedicels much longer than petals; perianth-segments somewhat campanulate, pink or greenish-white. The plant contains *volatile compounds* which taint milk and flesh; whole plant smells of garlic. *A frequent biennial plant of roadsides, bushy places and banks.* **Allium vineale**

37. Petioles *distinct* (Fig. 18.11); the latter sheathing the scape; leaf blades 'parallel-veined', flat, and all radical. Flowers in loose terminal umbels; perianth-segments lanceolate, acute, spreading, white. Plant bulbous. Plant contains *volatile compounds* which taints milk and flesh; the whole plant smells of garlic. *Frequent biennial of woods, hedgerows, damp shaded places.* **Allium ursinum**

Petioles *not well defined* (Fig. 18.12). Leaves linear-lanceolate, dark green, broad, 'parallel-veined', and appearing in spring when poisoning most likely to occur. Flowers large, pinkish-purple, solitary, and arising from onion-like corms, in the autumn, when leaves have died-back; perianth-segments 6. Very poisonous to man and animals, all parts of the plant contain alkaloids - *colchicine* and *colchiceine*; both able to withstand drying, storage and boiling. They effect the nervous system. *An infrequent plant of damp meadows and woods on basic soils.* **Colchicum autumnale**

38. Stipules *absent* 45

Stipules *present* 39

39. Leaves *alternate* or *basal* 41

Leaves *opposite* 40

40. Plant *rhizomatous* (Fig. 18.14). Leaves petiolate, broadly lanceolate to oval-lanceolate, opposite, crenate-serrate, somewhat acute, light green, scarcely pubescent; petioles with 2 small glands; stipules small. Stems angular. Dioecious plant; flowers small; perianth of 3 sepals; male flowers many, clustered on long-stalked spikes; female mostly solitary, sessile; ovary tuberculate. Plant contains the poisonous alkaloid, *mercurialine,* and some *saponins* and *volatile oils*; not a very poisonous plant. *An Infrequent perennial of woodlands and shaded places.***Mercurialis perennis**
Plant *lacking* rhizomes (Fig. 18.13). Leaves petiolate, broadly lanceolate to oval-lanceolate, opposite, crenate-serrate, somewhat acute, light green, scarcely pubescent; the petioles with 2 small glands; stipules small. Stems angular. Dioecious; flowers small; perianth of 3 sepals; male flowers several, clustered on long-stalked spikes; female mostly solitary, sessile; ovary tuberculate. The plant contains a poisonous alkaloid, *mercurialine,* and some *saponins* and *volatile oils*; not very poisonous. *A frequent annual of cultivated ground and waste places.* **Mercurialis annua**

41. Leaves *clearly lobed* basally 43
Leaves *tapered* basally 42

42. Stems *mostly prostrate*, rarely erect (Fig. 18.15). Blades 1.5-5.0 cm long, rarely more, and 0.3-2.0 cm wide, dull green, lanceolate to ovate-lanceolate or linear in outline, narrowed basally, sessile or very short-stalked, entire, glabrous, and acute or subobtuse; stipules whitish, nearly entire or split. Flowers in inconspicuous axillary, almost sessile clusters; sepals united basally, petaloid, colour green with pink or white margins. Plant contains an *acrid juice* which can cause gastro-intestinal irritation. *Abundant annual; cultivated ground and waste places.***Polygonum aviculare**
Stems *mostly erect* (Fig. 18.16); the latter usually of a reddish colour and somewhat swollen at the nodes. Blades up to 12 x 5 cm, lanceolate, tapered, glabrous, ciliate, or woolly abaxially, entire or nearly so, frequently black-purple blotched adaxially; stipules usually fringed. Flowers compact, in erect, terminal, axillary, spike-like panicles; sepals 5 in number, usually pinkish in colour, rarely whitish; petals absent. Plant contains an *acrid juice* which can cause gastrointestinal irritation. *Abundant annual plant of cultivated ground and waste places.***Polygonum persicaria**

43. Basal lobes *sagittate* 44
Basal lobes, at least some, *hastate* (Fig. 18.17); leaves narrowly lanceolate to oblanceolate, except for the basal lobes which are spreading at right angles to the mid-rib, up to 6 cm, entire, mostly covered with minute wart-like protuberances; stipules hyaline. The stems erect or decumbent, 7-30 cm in height. Flowers small, brown, clustered in leafless, loose, slender, terminal panicles; usually unisexual and then dioecious; sepals small, glabrous. Plant contains *oxalic acid* which combines with the blood calcium to form calcium oxalate. In time the animal could suffer from calcium deficiency. *A small calcifuge perennial, common in heaths, bogs, etc.***Rumex acetosella**

44. Apices of blades *blunt* (Fig. 18.18); blades sagittate with long, pointed basal lobes; the upper short-stalked or nearly sessile, clasping the stems; the lower on longer stalks; all entire, mostly glabrous, light green, and with a bitter taste. Stems up to 1.0 m, though often less; stipules fringed. Flowers clustered in long, loose, terminal panicles; dioecious; perianth-segments 6, dark brown; inner segments broadly oval and concealing the fruit; outer reflexed and appressed to the pedicels after flowering. Stems erect. Plant contains *oxalic acid* which combines with the blood calcium to form calcium oxalate. In time the animal could suffer from calcium deficiency. *A very common perennial of old pastures, shaded places, especially those on acid soils.***Rumex acetosa**
Apices of blades *abruptly acuminate* (Fig. 18.19); blades longer than the petioles, outline mostly triangular, bases deeply cordate-sagittate, entire, glabrous 'mealy' abaxially, nearly smooth adaxially, 2-8 cm; stipules obliquely truncate.Stems weak, twining or climbing, slender, angular or ribbed. Flowers small, clustered in axillary panicles; perianth-segments 5, glabrous, green, with white margins; the outer 3 large and keeled. The plant contains *irritants* which cause itching and scratching. *A very common annual of cultivated ground and waste places.***Fallopia convolvulus**

45. Leaf base *without 2* prominent entire lobes 47
Leaf base *with 2* prominent entire lobes 46

105

Fig. 18.00

N.B. DRAWINGS NOT TO SCALE

46. Basal lobes *pointed* (Fig. 18.20); leaves sagittate, entire, all radical, long-stalked, frequently spotted with purple, shiny, dark. Flowers unisexual, sessile, carried on a fleshy axis, upper part of which is barren and bright purple in colour; perianth absent; male flowers massed above the female; fruit of red berries. Plant contains an *acute irritant*. All parts, including the underground tubers, are poisonous. *A frequent perennial of hedgerows and other shaded places.* **Arum maculatum**
    Basal lobes *rounded* (Fig. 18.21); leaves alternate, triangular-ovate, with rounded, deeply cordate, basal lobes, mostly tapered, with several veins from near the base. Stems twining. Plant dioecious; flowers in racemes, yellowish-green; male flowers stalked, female subsessile; perianth of 6 narrow, somewhat recurved lobes; fruits globose, pale red. The poisonous substance is thought to be a *glycoside. An infrequent perennial plant of woodlands, hedgerows, etc.* **Tamus communis**

47. Stems *terete* 49
    Stems *quadrangular* 48

48. Stem *winged* (Fig. 18.22); the latter up to 100.0 cm, smooth. The leaves somewhat ovate in outline, obtuse, often with 1-2 small pinnae at the base, crenate, often with 1 or 2 small basal lobes, petiolate; petioles winged. Flowers in cymes or panicles or cymes; corolla brownish-purple above, globular, greenish on the underside, and with 5 small lobes; the upper 2 united basally; the calyx rounded. Rhizomes not nodular. Plant contains *glycosides*; not very poisonous. *A very frequent perennial of ditches, streams, shallow waterways and other wet places.* **Scrophularia auriculata**
    Stem *wingless* (Fig. 18.23); the latter sharply angular but not winged, up to 80 cm. Leaves more or less triangular-acute in outline, opposite, petiolate, coarsely and unequally biserrate, glabrous, more or less truncate basally, usually slightly decurrent on petioles. Flowers in terminal panicles or cymes; corolla 2-lipped; tube greenish; upper lip reddish-brown; calyx deeply lobed; the lobes ovate, with narrow scarious borders. Rhizomes swollen and nodular. Plant contains *glycosides*; not very poisonous. *A frequent perennial of hedgerows and waste places.* **Scrophularia nodosa**

49. Blades *neither* obovate *nor* 'mealy', *i.e.* papillae *absent* 52
    Blades *either* obovate *or* 'mealy', *i.e.* papillae *present* 50

50. Blades *not* angular 51
    Blades *angular* (Fig. 18.24); the latter more or less triangular or almost rhomboid in general outline, short-stalked, bluntly and coarsely toothed; upper leaves frequently narrower to almost linear, sessile and entire; all covered with papillae giving a 'mealy' appearance. Flowers in several dense axillary spikes, small; sepals 5, greenish, keeled apically; petals absent. Plant accumulates *nitrates*, and also contains *oxalates* which are thought to be responsible for most cases of livestock poisoning. *Abundant annual of cultivated ground and waste places.* **Chenopodium album**

51. Blades *toothed* (Fig. 18.25); the latter 1.0-4.0 cm long, short-stalked, soft and thin, alternate, falling early, yellowish-green. Stem simple or branched, mostly 5-rayed apically, usually containing latex, smooth; monoecious plant; umbel 5-rayed; bracts many; flowers small, in a 'glandular cup'; glands entire; perianth absent; male flowers many, of 1 stamen on a short filament and pedicel; female solitary, on a long pedicel; pistil syncarpous. Plant is said to contain a *resin,* an *alkaloid,* a *glycoside* and a complex substance called *euphorbiosteroid;* nevertheless, not an important poisonous plant. *Common annual of cultivated ground and waste places.* **Euphorbia helioscopia**
    Blades *entire* (Fig. 19.01); the latter 1.0-4.0 cm long, short-stalked, soft and thin, alternate, falling early, yellowish-green. Stem simple or branched, mostly 3-rayed apically, usually containing latex, smooth; monoecious plant; umbel 3-rayed; bracts many; flowers small, in a 'glandular cup'; glands horned; perianth absent; male flowers many, of 1 stamen on a short filament and pedicel; female solitary, on a long pedicel; pistil syncarpous. Plant said to contain a *resin,* an *alkaloid,* a *glycoside* and a very complex substance called *euphorbiosteroid;* nevertheless, not an important poisonous plant. *Common annual of cultivated ground and waste places.* **Euphorbia peplus**

52. Leaves *alternate* or *basal* 55
    Leaves *opposite* 53

53. Stem-internodes *without* a line of hairs 54

Stem-internodes *showing 1 line of hairs* (Fig. 19.02). The leaves petiolate, glabrous or ciliate, often with undulate margins, light green, entire, opposite, broadly ovate, acute. The flowers usually in branched cymes, rarely solitary; petals white, deeply 2-lobed. It contains *saponins* which effect the red blood corpuscles. *Abundant annual; cultivated ground and waste places.* **Stellaria media**

54. Stems *terete* and *showing 2 raised* lines (Fig. 19.03). Leaves mostly 3-nerved, up to 3 cm, opposite, sessile, entire, ovate-oblong, punctate, obtuse, sometimes mucronate, glabrous. Plant rhizomatous. Flowers in terminal axillary cymes; petals yellow, with marginal glands; stamens united in bundles of 3. Plant contains the pigment *hypericin*, a photosensitising agent, which can effect white-skinned animals. *Common perennial of dry banks, roadsides, etc.* **Hypericum perforatum**

Stems *lacking* 2 raised lines (Fig. 19.04); the latter stout, more or less glabrous, and arising from creeping rhizomes. Leaves broadly ovate-elliptical, 5-10 cm, 3-5 veined, more or less glabrous, opposite, entire, connected basally by a narrow rim. Flowers in compact corymbs; the calyx-tube often reddish; corolla pale pink or nearly white. Plant contains a *colloidal glycoside* which can prove poisonous. *A frequent perennial of hedges and waste grassy places.* **Saponaria officinalis**

55. Leaves *toothed* or *entire* 66

Leaves *lobed* 56

56. Blades *neither* wrinkled *nor* puckered; flowers *not* in composite heads 58

Blades *wrinkled* or *puckered*; flowers *in* composite heads 57

57. Blade *deeply* and *irregularly lobed, puckered, or wrinkled* (Fig. 19.05); the latter in a dense rosette, lyrate-pinnatifid, dull green, petiolate; lobes lobulate or toothed, minutely scurfy pubescent adaxially and abaxially, especially on the veins, at least some crowded and overlapping; terminal lobe largest; the cauline leaves, at least some, semi-amplexicaul. Stems cottony, rarely glabrous. The flowers in compact heads, yellow. Plant contains at least 4 stable alkaloids, *jacobine, jaconine, Senecionine* and *seneciphylline*; all are cumulative poisons, and effect the nervous and digestive systems. *An ubiquitous perennial; old grasslands, waste places, etc.* **Senecio jacobaea**

Blade lobes *well spaced* but *decreasing* in size from apex to base (Fig. 19.06); blades long-stalked, mostly lyrate to pinnatifid, coarsely lobulate and toothed; terminal lobe largest, laterals smaller and directed forwards; the cauline leaves becoming smaller, at least some, semi-amplexicaul; all crenate to coarsely serrate, and more or less glabrous. Stems mostly erect, cottony or glabrous above, usually reddish. Flowers in composite loose heads, very conspicuous, yellow. Plant contains small amounts of the above named *alkaloids*; they act in a similar manner; not as poisonous as the above species. *A common biennial of wet grasslands, ditches, etc.* **Senecio aquaticus**

58. Flowers *not* campanulate 59

Flowers *campanulate* (Fig. 19.07), in 2-rowed, subsessile cymes; corolla yellowish, purple-veined, 5-lobed; calyx subglobose, with triangular teeth; anthers purple. Leaves with a few large teeth or lobes or rarely nearly entire; lower stalked, upper smaller, amplexicaul. Contains the poisonous alkaloid *hyoscyamine*; *hyoscine* and *atropine* are also present. *An infrequent viscid, pubescent, annual plant of sandy places, particularly those near old buildings, etc.* **Hyoscyamus niger**

59. Blades *palmately* divided 62

Blades *pinnately* divided 60

60. Blades divided into *3 large lobes*, each of which is *further* lobed and toothed 61

Blades, at least some, with *1 or 2 lateral-basal lobes* (Fig. 19.08); the latter ovate or cordate, mostly scarcely pubescent, somewhat pointed, at least some with 1 or more basal lobes, the margins entire. The stems rather woody, scrambling or climbing and from 15-225 cm. Flowers in stalked, branched, leaf-opposite cymes; corolla purple; tube short, 5-lobed; the fruit of crimson-coloured berries. Plant contains *solanine* and its related alkaloids; berries are very poisonous. *A common perennial; hedgerows, ditches, woods, waste places and cultivated ground.* **Solanum dulcamara**

61. Plant *not* creeping (Fig. 19.09). Lower leaves deeply pinnately divided into 3 lobes, the latter further lobed and coarsely toothed; cauline more or less sessile, pinnately divided; all pubescent. Flowers yellow; sepals reflexed; stamens and carpels numerous. Plant contains *volatile oils* and *acrid substances* which cause irritation and burning; dried material is harmless. *A very frequent perennial of old dry pasture, roadsides, sand-hills, gravelly places, etc.* **Ranunculus bulbosus**
    Plant *stoloniferous* (Fig. 19.10). Leaves deeply pinnately divided into 3 lobes; lobes further 3-lobed once or twice; upper blades sessile, narrow, entire or nearly so; pubescent. Stems both erect and prostrate. Flowers yellow; sepals spreading; stamens and carpels several. Contains *volatile oils* and *acrid substances* which cause irritation and burning; dried material is harmless. *An abundant perennial plant; damp cultivated ground, roadsides and waste places.* **Ranunculus repens**

62. Flowers *yellow, mauve* or *bluish-mauve* 64
    Flowers *greenish* 63

63. Leaves, all, *cauline* (Fig. 19.11); lower stem leaves pedate, on long petioles which gradually widen basally; leaf segments 3-9, narrowly lanceolate, acute, serrate except towards the base; middle leaves with enlarged sheaths and reduced blades. Stems stout, branched, overwintering. Flowers several, regular, globose, in corymb-like branched cymes, green tipped with purple. The plant contains the glycosides, *helleborin, helleborein* and *helleborigenin*; all are stable compounds. *A frequent foetid plant of woods, stony places, especially on calcareous soils.* **Helleborus foetidus**
    Leaves, some, *radical* (Fig. 19.12); the latter usually 2, on long petioles, digitate-pedate; segments 7-11, sessile, the central free, laterals connected at the base; all elliptical, shortly acuminate, serrate, with prominent veins; the cauline leaves smaller, sessile; segments narrower. Flowers in cymes, half-drooping, often 2 to 4 together; perianth yellowish-green, nearly flat, actinomorphic. Plant contains the glycosides, *helleborin, helleborein* and *helleborigenin*; all are stable chemicals. *Infrequent plant of woods, stony places, especially on calcareous soils.* **Helleborus viridis**

64. Flowers *yellow* 65
    Flowers *mauve* to *bluish-mauve, helmet-shaped;* lower sepals reflexed. Blades *deeply divided* into 3-5 segments (Fig. 19.13); lower stem-leaves pentagonal, palmatifid, short-stalked, middle segment wedge-shaped at the base; the upper smaller, sessile; all light green, soft, pubescent or glabrous. Tuberous. The plant contains several alkaloids, particularly the very potent *aconite*; rarely eaten by livestock but, because of its very poisonous properties, is a very important poisonous plant. *Infrequent perennial plant of wet shaded areas; also an ornamental.* **Aconitum napellus**

65. Lower blades divided *close* to the base (Fig. 19.14); the latter and middle leaves stalked, divided into 5-9 palmatifid lobes; lobes further divided and toothed; upper short-stalked or sessile; all alternate, more or less orbicular, mostly pubescent. Petals yellow; stamens and carpels numerous. Plant contains *volatile oils* and *acrid substances* which cause irritation and burning; dried material is harmless. *An abundant perennial of grassy and waste places, etc.* **Ranunculus acris**
    Lower blades divided *approximately half-way* to the base (Fig. 19.15); the former mostly 3-lobed, long-stalked; lobes shallowly lobed or toothed; the middle blades deeply divided, short-stalked. Flowers yellow, small; stamens and carpels numerous; fruiting heads cylindrical. The plant contains *volatile oils* and *acrid substances* which cause irritation and burning; dried material harmless; very poisonous. *An annual, frequent in wet peaty soils, ditches, etc.* **Ranunculus sceleratus**

66. Corolla *not* campanulate 69
    Corolla *campanulate* 67

67. Leaves *not* in unequal-size pairs 68
    Leaves in *unequal-sized, 1-sided pairs* (Fig. 19.16); the latter often ovate, tapered basally, acuminate, entire, often with 5-8 pairs of widely spreading arched veins, glabrous or pubescent. Stems stout, much branched. Flowers drooping, solitary; corolla campanulate, 5-lobed, somewhat irregular, violet or greenish; fruits are black berries. Very poisonous; plant contains the alkaloids *hyoscyamine, atropine* and *hyoscine*. *Infrequent perennial of rocky places.* **Atropa belladonna**

Fig. 19.00

N.B. DRAWINGS NOT TO SCALE

68. Blades *truncate* or *abruptly contracted* at the base (Fig. 19.17); the former with large teeth or lobes; lower stalked, oblong-ovate; the cauline smaller, and amplexicaul. Flowers, large, with 5 obtuse lobes, in 2-rowed cymes, and dull yellow but purple-veined. All parts are poisonous; the active alkaloid is *hyoscyamine*, but others such as *hyoscine* and *atropine* also present; very poisonous. *A pubescent annual of sandy places, especially those near old buildings, etc.* **Hyoscyamus niger**
   Blades tapered *gradually* or *abruptly* into long broad petioles (Fig. 19.18); the former mostly ovate to broadly lanceolate, finely toothed or entire, and covered with soft downy hairs; stem-leaves becoming smaller. Flowers large, campanulate, usually drooping, in long racemes; corolla campanulate, shallowly 5-lobed, light purple and usually spotted internally with crimson. Very poisonous but mostly avoided by livestock except in times of scarcity. It contains a series of glycosides, the most important being *digitoxin*. *A frequent perennial of acid soils.* **Digitalis purpurea**

69. Blades *gradually* or *abruptly* tapered basally 72
    Blades *rounded* or *cordate* basally 70

70. Blades *clearly toothed* 71
    Blades *entire* or *nearly so* (Fig. 19.19); the latter small, mostly glabrous, unequal-sized, usually basal. Rootstock of a few small tubers. Flowers often solitary; pedicels long; petals yellow; sepals often 3. The plant contains *volatile oils* and *acrid substances* which cause irritation and burning; the dried material is harmless. *A common perennial of damp shaded places.* **Ranunculus ficaria**

71. Stems, when crushed, *smelling of garlic* (Fig. 19.20). Leaves, at least some, cordate basally; radical leaves in a rosette, long-stalked, often cordate or reniform in outline; cauline short-stalked; all pointed, glabrous or nearly so, and light green in colour. Flowers in terminal and lateral racemes; petals white, 4 in number; sepals 4; fruits long, narrow. Plant contains *volatile oils* which taint milk and animal flesh. *A frequent biennial of roadsides and waste places.* **Alliaria petiolata**
    Stems, when crushed, *not* smelling of garlic (Fig. 19.21). Leaves large, cordate or reniform, blunt, regularly crenate, and glabrous; lower long-stalked, the upper sessile. Flowers large and showy, on long pedicels, solitary, or 2-3 together; the petals absent; sepals yellow; carpels and stamens numerous. The plant contains *volatile oils* and *acrid substances* which cause irritation and burning; the dried material is harmless. *A common perennial of wet marshy ground.* **Caltha palustris**

72. Leaves *lanceolate* 74
    Leaves *abruptly* tapered basally 73

73. Blades *entire* or *nearly so* (Fig. 19.22); the latter small, mostly glabrous, unequal-sized, usually basal. Rootstock of a few small tubers. Flowers often solitary; pedicels long; petals yellow; sepals often 3. The plant contains *volatile oils* and *acrid substances* which cause irritation and burning; the dried material is harmless. *A common perennial of damp shaded places.* **Ranunculus ficaria**
    Blades, at least some, *ovate* or *rhomboid* (Fig. 19.23); the latter alternate, sinuate-dentate or entire, dark green, cuneate basally, acute. Flowers in scarcely branched, leaf-opposite cymes; corolla white, short-tubed, and with 5 spreading lobes which become revolute; calyx 5-lobed; fruits are black berries. Contains small amounts of *alkaloids* and the glycoalkaloid *solanine*; berries are very poisonous. *A bushy annual, frequent in tilled fields and waste ground.* **Solanum nigrum**

74. Blades with clear *reticulate venation* (Fig. 19.24); the lower leaves long-stalked, mostly lanceolate; sheaths broad; cauline becoming sessile; all entire or distantly and finely toothed, more or less glabrous. Flowers few, in loose corymbs; sepals 5, pubescent; petals 5, pale yellow. Plant contains small amounts of *volatile oils* and *acrid substances*, causing irritation and burning; dried material is harmless. *Common perennial; ditches and wet marshy areas.* **Ranunculus flammula**
    Blades with 'parallel' venation (Fig. 19.25); lower leaves long-stalked, mostly lanceolate; sheaths broad; cauline becoming sessile; all entire or distantly and finely toothed, more or less glabrous. Flowers few, in loose corymbs; sepals 5, pubescent; petals 5, pale yellow. Plant contains small amounts of *volatile oils* and *acrid substances* which cause irritation and burning; dried material is harmless. *A very common perennial of ditches and wet marshy places.* **Ranunculus lingua**

4   SEEDS - FRUITS - SEEDLINGS

'SEEDS'

General: The 'seeds' of the various crop and weed plants are a mixture of true seeds, whole, partial or multiple fruits, ripened florets, and whole or multiple spikelets. Identification is a simple problem once the distinction between the various types, and the terminology used to describe variation within same, are understood. A seed may be defined as a ripened ovule. Starting with the process of pollination the development of the ovule to the ripened seed may be described briefly as follows. A more comprehensive account of its development is given in Lowson's (1977) *Textbook of Botany*.

The essential organs necessary for its development are illustrated in Figure 20.00. Following pollination, the pollen grain germinates and the pollen-tube penetrates the stigma, passes through the style and enters the ovule, usually through the micropyle. During the elongation phase of the pollen-tube its nucleus divides to form 2 nuclei. When the tip of the pollen-tube enters the embryo sac 1 of the nuclei fuses (fertilises) with the egg-cell nucleus while the other fuses with the 2 polar nuclei. The zygote resulting from the first union eventually gives rise to the embryo, while the second union gives rise to the endosperm, a food tissue of varying degrees of importance in different species. After fertilization, the synergid and antipodal cells usually degenerate. The nucellus food reserve goes to the developing embryo so that by the time the seed reaches maturity little or none remains. The integuments increase in size, keeping pace with the developing embryo, undergo changes and eventually form the seed-coats.

NON-GRASS 'SEEDS'

True Seeds: Many plant families of agricultural and horticultural importance produce true seeds in abundance. Good examples would be most species of the Brassicaceae, Plantaginaceae, Scrophulariaceae, Juncaceae, Onagraceae, Fabaceae and Papaveraceae. These families produce capsular fruits which, on maturing, rupture in various ways to shed their seeds. True seeds are characterised by the presence of a hilum, or funicular scar, which is usually a neatly sculptured feature. Its features are generally consistent for seeds of any given species. Seeds of the same species usually show many other consistent features.

False Seeds: Often, what is called a seed is not really a seed in a botanical sense, but a simple fruit, a partial fruit, a multiple fruit, or modifications of 1 or more of the foregoing. There are many examples. Thus, the 'seeds' of such families as Asteraceae, Polygonaceae, Urticaceae and most members of the Ranunculaceae are simple fruits. They develop from 1-ovule ovaries, and the mature ovary wall persists around each seed. Some other families, *e.g.* the Apiaceae, Lamiaceae, Boraginaceae and the Lamiaceae, have fruits which split into segments, and each segment is the so-called seed. These 'seed' types do not have a hilum. Instead, an *attachment-scar*, which results from the splitting of the fruit, is shown. An attachment-scar is not a neatly sculpture feature.

Characters of Use in Identification of Non-grass 'Seeds'

True seeds and false seeds show a wide range of morphological features which are of great importance in seed identification. Unfortunately some features, such as shape and length, are influenced by environmental conditions, and great care must be exercised when basing a final decision on these 2 criteria. However, most of the following characteristics are reasonably consistent. For convenience, the various 'seed' types are referred to, hereafter, as the seed, and the outer coverings as the seed coats.

1. Colour: The colour of the individual seed is a reasonably dependable feature, and as such is important in seed identification. However, the colour of bulk seed is a much more useful characteristic. Many colours are represented in the seeds of crop and weed plants.

2. Shape: Shape of the base, apex, or overall, is obviously an important consideration. While it may show some variation for seeds of any given species, it is, nevertheless, reasonably consistent for the species as a whole. The shapes, as indicated by various technical names, are illustrated in Figure 20.00. All are represented in the seeds of crop and weed plants. Shape, as exemplified by the cross-section, also shows important differences. Thus, most members of the Lamiaceae show 1 convex and 2 flat faces. Others, for example *Pastinaca sativa* and *Heracleum sphondylium* are so strongly laterally compressed as to be disc-like.

3. Pubescence: Seeds of several plant species are consistently hairy. Hair type - whether *papillate, stellate, crisped*, or *simple* or *branched* - and their area of attachment, are important. Papillae are microscopic hairs, but their presence, particularly in the Chenopodiaceae, usually gives a 'mealy' or 'glistening' affect to the seed surface. *Spergula arvensis* is characterised by the presence of whitish papillae. Many species of the Asteraceae retain their pappus which is used as a dispersal aid; pappus hairs may be simple or branched.

4. Smooth: Most leguminous seeds, and others, are smooth, or at least appear so under low magnification. Others with this characteristic include 'seeds' of all *Rumex* and *Cirsium* species.

5. Rugose: Seeds of *Fumaria* species, and many more, have irregularly undulating or wrinkled surfaces; technically, they are described as being *rugose*.

6. Ribs or Ridges: Most seeds of the Apiaceae and Asteraceae show longitudinal ribs. The former have mostly 5, rarely 9, which are not uniformly arranged around the circumference; the latter have up to 20 such ribs, uniformly arranged. The ribs may be sharp, blunt, straight or wavy. Intercostal-zones may also vary; they may be shallow, deep, and contrast in colour with the ribs. Many seeds, for example *Cardamine* and *Plantago* species, have irregularly inter-connected ribs.

7. Vittae: These are resinous canals, present in the pericarps of some seeds, particularly those of the family Apiaceae. Some, *e.g. Carum carvi*, contain aromatic compounds which give a distinctive flavour to the seed. In a few other species, particularly *Pastinaca sativa* and *Heracleum sphondylium*, they are clearly visible as broad, dark longitudinal bands. The degree to which they extend over the surface of the seed can be an important diagnostic feature.

8. Scales: Species of the family Asteraceae show a wide range of variation in relation to the presence or absence the pappus. In a few, such as *Cichorium intybus* and *Centaurea nigra*, it is represented by a dense covering of small scales.

9. Tuberculation: This feature refers to the presence of wart-like structures or *tubercles*. They are found on a wide range of seeds, and vary in size from those clearly visible to the naked eye on some *Rumex* species, to microscopic types found on *Epilobium* seeds. Other families where they occur in abundance include the Caryophyllaceae and the Convolvulaceae.

10. Teeth and Prickles: Small, short, sharp projections are known as teeth. They occur on many seeds. Prominent, sharp teeth, are known as *prickles*. The latter occur on the seeds of *Galium aparine, Torilis japonica, Daucus carota, Sherardia arvensis*, and others. Teeth can be uniformly distributed of confined to a particular part of the seed. Some seeds of Asteraceae plants lack a pappus. In a few instances it is represented by 4-5 short teeth.

11. Reticulation: Seeds of *Papaver* species and most cruciferous plants display raised, irregular, network or reticulation. The latter are *finely reticulate*; those of *Mercurialis annua* may be described as being *broadly reticulate*. Other seeds, for example some species of *Geranium*, show partial reticulation.

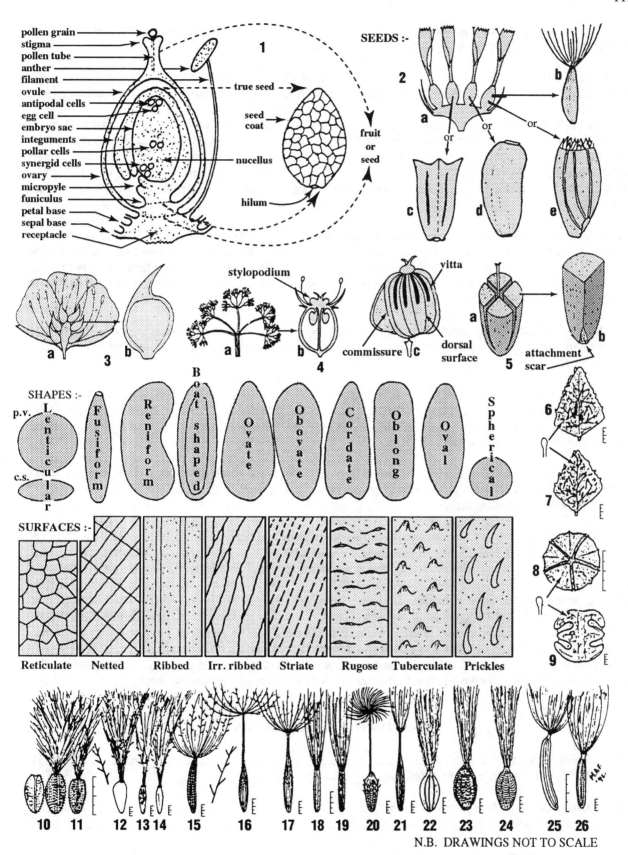

pollen grain
stigma
pollen tube
anther
filament
ovule
antipodal cells
egg cell
embryo sac
integuments
pollar cells
synergid cells
ovary
micropyle
funiculus
petal base
sepal base
receptacle

1

true seed

seed coat

nucellus

hilum

fruit or seed

SEEDS :-

2
a   b   c   d   e
or   or   or

3
a   b

stylopodium
4
a   b

vitta
commissure   dorsal surface
c

5
a   b
attachment scar

6   7   8   9

SHAPES :-

p.v.
Lenticular
c.s.

Fusiform   Reniform   Boat shaped   Ovate   Obovate   Cordate   Oblong   Oval   Spherical

SURFACES :-

Reticulate   Netted   Ribbed   Irr. ribbed   Striate   Rugose   Tuberculate   Prickles

10   11   12   13   14   15   16   17   18   19   20   21   22   23   24   25   26

N.B. DRAWINGS NOT TO SCALE

Fig. 20.00

12. Striate: *Allium cepa, A. porrum* and *Polygonum aviculare* seeds, and many more, are striated or covered with *striate*. These, essentially, are fine, raised, interrupted, more or less 'parallel' lines.

13. Netted: The best example of a *netted* surface is shown by seeds of *Atriplex patula* and *A. prostrata*; these are characterised by the presence of a regular arrangement of intersecting lines and interstices.

14. Hilum: This feature is, of course, only visible in a true seed. Its location on the seed, shape, size relative to that of the seed, and colour are important considerations. Thus, in *Glycine max* it is long and narrow and laterally placed; in *trifolium* species, it is rounded and basally situated, whereas in *Plantago major* it is located about mid-point on the ventral surface. In the latter species, also, it is small and circular in outline. Species of *Vicia* can be separated by the degree to which it extends around the circumference of the seed.

15. Attachment-scar: Reference has been made already to the attachment-scar of true seeds, *viz*. the hilum. The term *attachment-scar*, however, really applies to the fracture of achenial and schizocarpic type 'seeds'. It, like the hilum, may vary in outline, shape, location and size. Unlike the latter, it is not neatly sculptured and generally has a ragged appearance. It may be laterally, basally or obliquely placed, and may be wide or narrow.

16. Length: This criterion, *per se*, is not completely reliable. A wide range of variation, due to environmental factors, may be found for the seeds of any 1 species. However, the average length of several seeds is a much better criterion.

17. Border: This feature is present in a few species. Thus, the more or less spherical seeds of *Spergula arvensis* always show a narrow peripheral or circumferential area distinct from the body of the seed. Distinct peripheral areas or borders are also displayed by *Angelica sylvestris, Rhinanthus minor* and some members of the Asteraceae family.

18. Perianth-segments: These segments often persist, at least for some time, around the seed of some species. Thus, seed of *medicago lupulina, Polygonum aviculare, P. persicaria* and all *Rumex* species, may show total or partial segments.

The following 'keys' are intended as an aid to the identification of the seeds of most cultivated crops and all common weed plants. Where possible, the characteristics of several seeds of the same species should be studied before making a final decision on identification. This is desirable because of the wide range of genetical and environmental variation found within species. All seeds show a number of distinguishing features and it is on these the 'keys' are based. The reader is advised to be particularly careful when basing a final diagnostic decision on such features as colour and length as these can be altered considerably by environmental factors. Furthermore, it is necessary to be aware of the possibility of immature, shrunken, shrivelled or even diseased seeds. These can only be identified with difficulty and no provision is made for their identification in the present 'keys'. Finally, the line drawings accompanying the 'keys' are only intended to give a general guide to the outline and markings on the seeds.

KEY TO NON-GRASS 'SEEDS'

1. Hairs *totally* absent 31
   Hairs, either visible to the naked eye or microscopic, *present* 2

2. Bracts or perianth-segments *present* 3
   Bracts or perianth-segments *absent* 6

3. Bracts or segments 2 4
   Bracts or segments *5-many* 5

4. Bracts united *at extreme* base (Fig. 20.06); the latter triangular-ovate, acute or acuminate, subcuneate to subcordate at the base, irregularly toothed, 3-12 mm long, mostly covered with 'mealy' papillae; a small lenticular-shaped fruit enclosed; rarely tubercled dorsally. ***Atriplex prostrata***
Bracts united some distance *away* from the base (Fig. 20.07); the latter triangular-ovate, acute or acuminate, subcuneate to subcordate at the base, irregularly toothed, 4-8 mm long, mostly covered with 'mealy' papillae, and enclose a small fruit; rarely tubercled dorsally. ***Atriplex patula***

5. Segments *5, regularly arranged* (Fig. 20.08); the latter more or less triangular, or at least angular, keeled basally, tapered apically, covered with minute papillae giving a 'mealy' or 'glistening' affect; seed horizontally compressed; diameter c.1.5 mm; colour greyish.***Chenopodium album***
Segments *many, irregularly arranged* (Fig. 20.09); the latter rather woody, firm, of a brownish colour, covered with papillae giving a 'mealy' or 'glistening' effect; surface very deeply puckered; seed relatively large; diameter 4-8 mm; attachment-scar not clearly visible. ***Beta vulgaris***

6. Hairs *scattered* on body of seed 23
Hairs present in the form of *an apical tuft* 7

7. Seed *atuberculate* 9
Seed *tuberculate* on dorsal surface 8

8. Length about 1.0 mm, width about 0.4 mm (Fig. 20.10). Seed mostly obovate in general outline; ventral surface grooved; tubercles minute; hair silky, and easily detached.***Epilobium hirsutum***
Length about 0.8 mm, width about 0.3 mm (Fig. 20.11). Seed mostly obovate in general outline; ventral surface grooved; tubercles minute; hairs silky, easily detached. ***Epilobium montanum***

9. Hairs *simple* 15
Hairs, at least some, *branched* or *feathery* 10

10. Seed *prominently* ribbed 13
Seed *smooth* or at most *indistinctly* ribbed 11

11. Length *3.0 mm or less* 12
Length *4.0 mm or more* (Figs 20.12 and 21.32). Seed obliquely obovate in general outline, with 1 broad and 1 narrow shoulder; creamy-white or grey in colour; the terminal and basal scars distinct; cusp frequently present; hairs long, whitish, branched, easily detached. ***Cirsium vulgare***

12. Seed *pale brown* in colour (Figs 20.13 and 21.33); the former narrowly obovate in general outline, smooth; terminal and basal scars distinct; hairs long, whitish, easily detached.***Cirsium palustre***
Seed *dark brown* in colour (Figs 20.14 and 21.34);the former narrowly obovate in general outline, smooth; terminal and basal scars distinct; hairs long, whitish, easily detached.***Cirsium arvense***

13. Seed *beaked* 14
Seed *unbeaked,* or at most *somewhat tapered* apically (Figs. 20.15 and 21.26); the latter somewhat oblanceolate in general outline, marked with prominent cross-striations, and mostly reddish-brown in colour; length 4-6 mm; hairs dull white, spreading, branched. ***Leontodon autumnalis***

14. Beak *greater than* one-half of overall length (Fig. 20.16); the latter narrowly oblanceolate in general outline, length 4-7 mm, mostly of a somewhat orange colour, strongly muricate; hairs dull white, inner row branched, outer row shorter, simple; all easily detached.***Hypochoeris radicata***
Beak *less than* one-half overall length (Fig.20.17); the latter narrowly oblanceolate in general outline, length 3-5 mm, mostly brownish in colour, the longitudinal ribs prominent strongly muricate; hairs brownish-white, branched, spreading, easily detached. ***Leontodon taraxacoides***

15. Hairs *absent* from seed body 17
Hairs *present* on seed body 16

16. Body-hairs *few* and *scattered* (Figs 20.18 and 21.01); seed mostly fusiform, 2.2 x 0.5 mm, mostly with 8 broad longitudinal ribs; apical hairs about twice as long as the seed. ***Senecio jacobaea***
Body-hairs *abundant* and *closely set* (Figs 20.19 and 21.02); seed mostly fusiform, 2.2 x 0.3 mm; longitudinal ribs narrow, about 8; apical hairs about twice as long as the seed. ***Senecio vulgaris***

17. Seed *unbeaked* 19
Seed narrowed into a *beak* 18

18. Width *at least 1.0 mm* (Figs 20.20 and 21.22); length 3.5-5.5 mm; seed mostly obovate, brownish, greenish or grey, toothed on its upper half, often transversely wrinkled, and with a beak 2-4 times as long as the seed; apical hairs spreading, whitish. ***Taraxacum officinale, sensu lato***
Width *0.8 mm* or *less* (Figs 20.21 and 21.23); length about 4.5 mm; the seed narrowly cylindrical in outline, gradually narrowed into a beak about as long as the seed, colour pale brown, with about 8 more or less rough longitudinal ribs; apical tuft of hairs white, simple. ***Crepis vesicaria***

19. Seed more or less *cylindrical* 22
Seed strongly *compressed*, somewhat 2-faced 20

20. Surface *transversely* wrinkled 21
Surface smooth (Figs 20.22 and 21.19); seed elliptical to obovate in outline, strongly compressed, with 3 well-spaced longitudinal ribs on each face, usually yellowish-brown though variable in colour, about 2.5 mm in length; apical hairs simple, in 2 rows, easily detached. ***Sonchus asper***

21. Colour *dark chocolate-brown* (Figs 20.23 and 21.20); seed mostly narrowly ellipsoidal in outline, beak absent, with about 5 prominent, strongly wrinkled, longitudinal ribs on each face, length about 3.50 mm; apical tuft of hairs white, in 2 rows, simple, easily detached. ***Sonchus arvensis***
Colour *yellowish-brown* (Figs 20.24 and 21.21); seed mostly oblanceolate in outline, beak absent, with about 3 poorly developed and moderately wrinkled longitudinal ribs on each face; length about 2.8 mm; apical tuft of hairs white, in 2 rows, simple, easily detached. ***Sonchus oleraceus***

22. Cross-walls, between longitudinal ribs, *not evident* (Figs 20.26 and 21.31); seed cylindrical, often curved, short-beaked, with about 10, more or less smooth longitudinal, moderately developed ribs, length 4-6 mm; colour pale brown; apical hairs in many rows, simple. ***Crepis capillaris***
Cross-walls *clearly evident* (Figs 20.25 and 21.27); seed narrowly elliptical to cylindrical, curved, short-beaked, with about 12 prominent, smooth or minutely toothed, longitudinal ribs; colour pale brown, length 5-10 mm; apical tuft of hairs in many rows, pure white. ***Tussilago farfara***

23. Hairs scattered on *body* of seed; apical tuft present or absent 25
Hairs confined to the *intercostal-zones* 24

24. Body-hairs *few* and *scattered* (Figs 20.18 and 21.01); seed mostly fusiform, 2.2 x 0.5 mm, mostly with 8 broad longitudinal ribs; the apical and basal attachment-scars evident. ***Senecio jacobaea***
Body-hairs *abundant* and *closely set* (Figs 21.02 and 20.19); seed mostly fusiform, 2.2 x 0.3 mm; longitudinal ribs narrow, about 8; apical and basal attachment-scars evident. ***Senecio vulgaris***

25. Seed *not* so strongly dorso-ventrally compressed as to be *disc-like* 26
Seed *very strongly dorso-ventrally* compressed (Fig. 21.03); general outline mostly suborbicular, disc-like, often with a thin border, length 7-12.0 mm, colour yellowish or whitish; longitudinal ribs 5, confined to the dorsal surface; 4 dark-coloured oil glands extend over two-thirds of dorsal surface; hairs few, scattered; short reflexed style often persisting. ***Heracleum sphondylium***

26. General outline *other than* spherical 27
General outline *spherical* (Fig.21.04); seed very small, approximately 1.25 mm in diameter, of a bluish-black colour, frequently sparsely covered with small whitish papillae, frequently tuberculate, and mostly with a thin narrow, wavy, whitish circumferential border. ***Spergula arvensis***

27.   Seed *sparsely* pubescent 28
        Seed *densely* covered with white frost-like hairs (Fig. 21.05); seed mostly irregularly oblanceolate
        in outline; laterally compressed; length 3-5 mm; colour light brown. ***Lycopersicon esculentum***

28.   Surface *not* strongly reticulate 29
        Surface *strongly* reticulate (Fig. 21.06); seed reniform; length 2.0-3.0 mm; of a blackish colour,
        hairs entirely confined to the persistent, light-coloured, sepals and pedicel. ***Medicago lupulina***

29.   Seed *not* abruptly constricted apically 30
        Seed *abruptly constricted* and *angled* apically (Fig. 21.07); general outline mostly flask-shaped,
        'neck' portion approximately half overall length, lower half more or less spherical; basal scar
        indistinct; length about 3 mm; of a yellowish colour; hairs few, scattered. ***Ranunculus ficaria***

30.   Length *1.5 mm* or *less* (Fig. 21.08); seed often ellipsoidal in general outline, with a thick circum-
        ferential border; basal scar obscure; terminal scar distinct, circular; yellowish in colour; longi-
        tudinal ribs present or absent; dorso-ventrally compressed; hairs few, scattered. ***Bellis perennis***
        Length *2.5 mm* or *more* (Fig. 21.09); the seed mostly obovate, but truncate apically and frequently
        crowned with small scales; longitudinal ribs present or absent; the lower scar clearly defined,
        obliquely placed; colour mostly pale brown; surface more or less pubescent. ***Centaurea nigra***

31.   Longitudinal ribs *absent;* or present but *not uniformly* arranged and attachment-scar *lateral* ... 55
        Longitudinal ribs *present* and *uniformly arranged* on circumference; seed *not* 3-sided 32

32.   Attachment-scar *basal* 33
        Attachment-scar *obliquely basal* (Fig. 21.09, without hairs); the latter clearly defined; length 2.5
        mm or more; the seed mostly obovate, truncate apically, frequently crowned with small scales;
        longitudinal ribs present or absent; the colour pale brown see couplet no.30). ***Centaurea nigra***

33.   Seed *lacking* a border 35
        Seed *bordered* 34

34.   Longitudinal ribs *few* or *absent* (Fig.21.10). Seed obovate; terminal scar circular; basal scar indis-
        tinct; small cusp often present; centre of seed ribless or with 1-2 short, mostly basal ribs; sur-
        face grey, covered with fine striate; border yellowish; length 1.5-2.0 mm. ***Achillea millefolium***
        Longitudinal ribs *many* and *very well* defined (Fig. 21.11); sini deep, acute; outline variable but
        mostly cylindrical-oblong; terminal and basal scars well defined; longitudinal border present
        or absent; length c. 2 mm; colour yellowish-brown or creamy-white. ***Chrysanthemum segetum***

35.   Seed *not* crowned with scales or teeth 40
        Seed *crowned* with teeth or scales 36

36.   Surface *not* blackish and mottled brown 37
        Surface *blackish* and *mottled* brown (Fig. 21.12); seed obovate, but blunt apically, tapered basal-
        ly; terminal scar wide, with or without a cusp; the basal scar small; longitudinal ribs narrow,
        sharp, well spaced; sini often longitudinally wrinkled; teeth obscure; 5-7 mm. ***Arctium minus***

37.   Seed base *other than* cordate 38
        Seed base cordate (Fig. 21.13); the seed about as broad as it is long, 2-3 mm; somewhat 2-lobed;
        apex crowned with the persistent sepals which form 6 very prominent teeth; longitudinal ribs
        very obscure; surface covered with sharp prickles; colour mostly brownish. ***Sherardia arvensis***

38.   Apex crowned with *a few* prominent or obscure, well-spaced teeth 39
        Apex *crowned with scales* (Fig. 21.14); the latter rarely absent but terminal scar very broad and
        clearly flat-topped; seed somewhat oblong but blunt apically and basally, tapered from apex to
        base; irregularly angular; 2-3 mm long; colour pale brown or creamy-white. ***Cichorium intybus***

39. Longitudinal ribs *3-4* (Fig. 21.15); apical scar very wide and containing a small central cusp; basal scar much smaller; terminal teeth about 4; longitudinal ribs very prominent, 3 somewhat acute, the fourth blunt, glandular; 2 dark-coloured longitudinal dark oil glands present in the sini; seed gradually tapered from apex to base; about 2 mm long.***Tripleurospermum inodorum***
    Longitudinal ribs *several* and *well* defined (Fig. 21.16); the latter blunt and uniformly arranged around the circumference; the sini deep, acute; outline variable but mostly cylindrical-oblong; basal attachment-scar more or less circular and well defined; apex very blunt; longitudinal border present or absent; c. 2.0 mm; yellowish-brown or creamy-white. ***Chrysanthemum segetum***

40. Ribs *10* or *less* 42
    Ribs *12* or *more* 41

41. Ribs *unequally* developed (Fig. 21.17); the latter about 20, closely set, and some very well developed; ellipsoidal or oblanceolate, dorso-ventrally compressed; often curved; seed short-necked apically, basal scar small, circular; colour pale-brown; 3.0-5.5 mm long. ***Lapsanna communis***
    Ribs *equally* developed (Fig. 21.18); the latter closely set and about 16; seed oblanceolate, strongly dorso-ventrally compressed, often somewhat curved; seed tapered apically to a short acute point and rather abruptly basally; the colour grey or dark brown; 3-4 mm long. ***Lactuca sativa***

42. Seed more or less *cylindrical* 45
    Seed strongly *compressed*, somewhat 2-faced 43

43. Surface *transversely* wrinkled 44
    Surface smooth (Figs 20.22 and 21.19); seed elliptical to obovate in outline, strongly compressed, with 3 well-spaced longitudinal ribs on each face, usually yellowish-brown though variable in colour, about 2.5 mm in length; (similar to couplet no.20, apart from the hairs). ***Sonchus asper***

44. Colour *dark chocolate-brown* (Figs 20.23 and 21.20); seed mostly narrowly ellipsoidal in outline, beak absent, with about 5 prominent, strongly wrinkled, longitudinal ribs on each face, length about 3.50 mm; (similar to couplet no. 21, apart from the absence of hairs). ***Sonchus arvensis***
    Colour *yellowish-brown* (Figs 20.24 and 21.21); seed mostly oblanceolate in outline, beak absent, with about 3 poorly developed and moderately wrinkled, longitudinal ribs on each face, length about 2.8 mm; (similar to couplet no. 21, apart from the absence of hairs). ***Sonchus oleraceus***

45. Seed *not* ending in a beak 49
    Seed *ending* in a beak 46

46. Cusp *absent* 47
    Cusp *present* (Figs 20.20 and 21.22); width at least 1.0 mm; 3.5-5.5 mm long; the seed obovate, brownish, greenish or grey, toothed on its upper half, often transversely wrinkled, with beak 2-4 times as long as seed; (similar to no.18, apart from hairs). ***Taraxacum officinale, sensu lato***

47. Ribs *more strongly* toothed apically than basally 48
    Ribs *equally toothed* throughout (Figs 20.21 and 21.23); about 4.5 mm long; narrowly cylindrical in outline, gradually narrowed into a beak about as long as the seed; colour pale brown; with c. 8 more or less rough longitudinal ribs; (similar to couplet no.18; no hairs). ***Crepis vesicaria***

48. Beak *at least as long* as the seed (Figs 20.16 and 21.24); the latter narrowly oblanceolate in general outline; 4-7 mm long; mostly of an orange colour; strongly muricate. ***Hypochoeris radicata***
    Beak *shorter* than the seed (Figs 20.17 and 21.25); the latter narrowly oblanceolate in general outline; length 3-5 mm; mostly brownish in colour; the longitudinal ribs prominent strongly muricate; very similar to couplet no.14, apart from the absence of hairs. ***Leontodon taraxacoides***

49. Seed *not* clearly ribbed or angled 55
    Seed *clearly* ribbed or angled 50

50. Seed *not* transversely toothed 51
Seed *transversely* toothed, and somewhat tapered apically (Figs 20.15 and 21.26); the latter somewhat oblanceolate in general outline; marked with prominent cross-striations; mostly reddish-
brown in colour; length 4-6 mm; (similar to couplet no. 13; no hairs). ***Leontodon autumnalis***

51. Cross-walls *not* evident 52
Cross-walls *clearly evident* (Figs 20.25 and 21.27); seed narrowly elliptical to cylindrical, curved,
short-beaked, with about 12 prominent, smooth or minutely toothed, longitudinal ribs; colour
pale brown, length 5-10 mm; (similar to couplet no. 22, apart from hairs). ***Tussilago farfara***

52. Sini *not* contrasting in colour with the ribs 53
Sini *black* and *contrasting* with the light-coloured ribs (Fig. 21.28); the latter very prominent and
raised, yellowish-brown; seed cylindrical-oblong, tapered from apex to base; apical scar distinct; basal scar smaller; border present or absent; length 2.0-2.5 mm. ***Leucanthemum vulgare***

53. Longitudinal ribs *10* or *less* 54
Longitudinal ribs *many* and *very well* defined (Fig. 21.29); sini deep, acute; outline variable but
mostly cylindrical-oblong; terminal and basal scars well defined; longitudinal border present
or absent; length c. 2 mm; colour yellowish-brown or creamy-white.***Chrysanthemum segetum***

54. Longitudinal ribs *about 5* (Fig. 21.30); the latter well-spaced, rather inconspicuous; terminal scar
obliquely placed, bordered with an obscure rim; the basal scar small; terminal cusp present or
absent; outline somewhat obovate; colour pale brown; 1.0-1.5 mm. ***Matricaria matricarioides***
Longitudinal ribs *about* 10 (Figs 20.26 and 21.31); the latter more or less smooth, moderately developed, closely set; the seed cylindrical, often curved, short-beaked; terminal and basal scars
small; 4-6 mm; the colour pale brown; (similar to couplet no.22; no hairs). ***Crepis capillaris***

55. Seed *without* a distinct terminal collar and terminal cusp 58
Seed *with* a distinct terminal collar; terminal cusp often *present* 56

56. Length *3 mm* or *less* 57
Length *4 mm* or *more* (Figs 20.12 and 21.32). Seed obliquely obovate in general outline, with 1
broad and 1 narrow shoulder, and creamy-white or grey in colour; terminal and basal scars distinct; cusp frequently present; (similar to couplet no.11 but the hairs absent). ***Cirsium vulgare***

57. Colour *pale brown* (Figs 20.13 and 21.33). The seed narrowly obovate or somewhat cylindrical
in outline, smooth; the attachment-scar small, distinct and somewhat circular; apex blunt, usually with a cusp (similar to couplet no. 12, apart from the absence of hairs). ***Cirsium palustre***
Colour *dark brown* (Figs 20.14 and 21.34). The seed narrowly obovate or somewhat cylindrical
in outline, smooth; the attachment-scar small, distinct and somewhat circular; apex blunt, usually with a cusp (similar to couplet no. 12, apart from the absence of hairs). ***Cirsium arvense***

58. Longitudinal ribs *absent* or, if present, *indistinct* and attachment-scar *basal* 75
Longitudinal ribs *3-5*, rarely 9; rarely on neck region only; attachment-scar *lateral* 59

59. Seed *laterally compressed;* attachment-scar *narrow* 67
Seed *dorso-ventrally* compressed; attachment-scar *broad* 60

60. Seed *flat* and *disc-like* 61
Seed *not* disc-like 62

61. Vittae or oil glands 4, reaching *to the base* of seed (Fig.21.35); the former dark-coloured and confined to the dorsal surface of the seed; 4-7 mm long; general outline mostly suborbicular; seed
often with a thin border; colour yellowish or whitish; the longitudinal ribs 5, not well defined,
confined to dorsal surface; attachment-scar broad, lateral; style often present.***Pastinaca sativa***

Vittae or oil glands 4, extending over *two-thirds* of the seed (Fig.21.36); the former dark-coloured, confined to the dorsal surface; the attachment-scar broad, lateral; general outline mostly suborbicular; seed often with a thin border; length 7-12 mm; colour yellowish or whitish; longitudinal ribs 5, and confined to the dorsal surface; style often present. **Heracleum sphondylium**

62. Teeth *absent* 64
Teeth *present* on the ribs or on seed *apex* 63

63. Ribs *toothed* (Fig. 21.37); teeth of various sizes; mostly with apices removed in commercial seed; secondary ribs larger than the primary; the length 2.5-4.0 mm; general outline ovoid-oblong to broadly ovoid; the attachment-scar lateral; reflexed style often present; brown. **Daucus carota**
Ribs *smooth* (Fig. 21.38); teeth about 5, confined to the apex; the outline somewhat spherical; primary ribs low, slender, secondary broader; vittae obscure, solitary beneath secondary ribs; the length 3.0-4.5 mm; reddish-brown; a small reflexed style often present. **Coriandrum sativum**

64. Seed *not* winged 65
Seed *winged* (Fig. 21.39); the wing very broad, extending laterally, very conspicuous; general outline mostly orbicular; dorso-ventrally compressed; longitudinal ribs 5, smooth, very well defined; the intercostal-zones deep, v-shaped; 4.5-5.5 mm long; pale brown or yellowish; attachment scar broad and laterally placed; a small reflexed style often present. **Angelica sylvestris**

65. Longitudinal ribs *very well developed* 66
Longitudinal ribs rather *inconspicuous* though extending throughout the entire length of the seed (Fig. 21.40); the latter rather narrow, often rather wavy on the ridges, well spaced; intercostal-zones v-shaped; outline ovoid to oblanceolate, dorso-ventrally compressed; attachment-scar very broad, laterally placed; colour brown or yellowish; about 3 mm long. **Pimpinella anisum**

66. Intercostal-zones rather *deep* and v-*shaped* (Fig. 21.41); general outline broadly ovoid to obovate, dorso-ventrally compressed; longitudinal ribs very prominent, acute on the ridges, extending throughout the entire surface, smooth; sini deep; the attachment-scar very broad, lateral placed; mostly creamy-white; 3.0-4.0 mm long; small reflexed style often present. **Aethusa cynapium**
Intercostal-zones *flat* (Fig.21.42); outline ovoid to suborbicular; seed dorso-ventrally compressed; the longitudinal ribs 5, very prominent, extending throughout the entire surface; ridges rather narrow, minutely rough, often wavy; the attachment-scar very broad, laterally placed; brown or yellowish; length about 3.0 mm; small reflexed style frequently present. **Conium maculatum**

67. Teeth or prickles *absent* 69
Teeth or prickles *present* 68

68. Seed apex ending in *3 prominent teeth* (Fig. 21.43); the latter well spaced, and represent the persistent sepals; general outline obovoid; attachment-scar lateral; 2-3 mm long; the longitudinal ribs obscure; surface covered with sharp prickles; colour mostly brownish. **Sherardia arvensis**
Seed apex *not* ending in 3 prominent teeth (Fig. 21.44); the former blunt or with a small reflexed style; seed narrowly ovoid, attachment-scar narrow, lateral; longitudinal ribs not well defined, densely covered with prominent hooked prickles; the latter with swollen bases. **Torilis japonica**

69. Ribs extending over the *entire* length of the seed 71
Ribs *short* and *confined* to the neck of the seed 70

70. Width *0.5 mm* or *less* (Fig.21.45); length c. 9 mm; general outline narrowly oblanceolate, tapered into a long beak; often curved; the longitudinal ribs inconspicuous and confined to the upper region; attachment-scar very narrow, laterally placed; colour dark brown. **Foeniculum vulgare**
Width *1.0 mm* or *more* (Fig. 21.46); length about 8 mm; outline narrowly oblanceolate, tapered into a long beak; often somewhat curved; the longitudinal ribs inconspicuous and confined to the upper region; attachment-scar very narrow, lateral; colour dark brown. **Anthriscus sylvestris**

71. Length *2.0 mm* or *more* 72
     Length *1.2 mm* or *less* (Fig. 21.47); general outline broadly elliptic; longitudinal ribs rather prominent, acute, light brown, and contrasting with the intercostal-zones; the latter dark brown; attachment-scar narrow, laterally placed; small reflexed style often present. ***Apium graveolens***

72. Width *2 mm* or *less* 73
     Width *3-5 mm* (Fig. 21.48); length up to 8.0 mm, rarely more; general outline broadly ovoid; seed laterally compressed; teeth absent; attachment-scar very narrow, laterally placed; longitudinal ribs sharply defined and extending over surface of the seed; intercostal-zones very wide, wrinkled; colour black or blackish; a small reflexed style frequently present.***Smyrninum olusatrum***

73. Ribs *contrasting* in colour with the intercostal-zones 74
     Ribs *not* contrasting in colour with the intercostal-zones (Fig. 21.49); the latter wide; seed mostly ovoid; laterally compressed, attachment-scar narrow, lateral; longitudinal ribs reasonablydefined; colour dark brown; c. 4 mm long; reflexed style often present. ***Aegopodium podagraria***

74. Outline *elliptical* (Fig. 21.50); seed often curved; laterally compressed, attachment-scar narrow, laterally placed; longitudinal ribs well defined, yellowish, and contrasting with the intercostalzones; the latter brown, wide; c. 4.5 mm long; small reflexed style often present. ***Carum carvi***
     Outline *ovoid* (Fig. 21.51); seed often curved; laterally compressed, attachment-scar narrow, lateral; 2.0-2.7 mm; longitudinal ribs well defined, creamy-white, often wavy, contrasting with the intercostal-zones; the latter brown, wide; reflexed style often present. ***Petroselinum crispum***

75. Seed *neither* prickly *nor* toothed, and *without* complete or incomplete perianth-segments 91
     Seed *either* prickly *or* toothed, or *with* complete or incomplete perianth-segments *attached* ... 76

76. Prickles *absent* 78
     Prickles *present* 77

77. Bristles *hook-shaped* apically (Fig. 21.52); the latter with tuberculate bases and extending over the entire surface; the general outline more or less spherical or orbicular; diameter 2.5-5.0 mm; attachment-scar round or elliptical on basal surface, depressed; colour greyish.***Galium aparine***
     Bristles *straight* (Fig. 21.53); seed with a cordate base, about as broad as long, 2-3 mm, somewhat 2-lobed, apex crowned with the persistent sepals which form 6 very prominent teeth; the longitudinal ribs obscure; surface covered with sharp prickles; colour brownish. ***Sherardia arvensis***

78. Perianth-segments *atuberculate* 82
     Perianth-segments, at least some, *tuberculate* 79

79. Large perianth-segments *not* clearly toothed 80
     Large perianth-segments *clearly toothed* (Fig. 21.54); the latter dull brown or dark brown in colour; the 1 large segment with a large tubercle, other large segments with or without tubercles; small segments atuberculate; c.4 mm; segments enclose a triangular nut. ***Rumex obtusifolius***

80. Tubercles, large, *1 in number,* small tubercles present or absent 81
     Tubercles, large, *3 in number* (Fig. 21.55); large segment entire; the inner segment narrow, entire, without tubercles; all dull brown; length 1.5-2.5 mm; inner segments enclose a small, triangular, brown nut; the latter sharp-angled, and pointed apically and basally.***Rumex conglomeratus***

81. Inner segments *narrow* (Fig. 21.56); the latter entire; large segments entire, c. 3 mm, dull brown, with 1 large tubercle; the smaller tubercles present or absent; the inner segments enclose small, triangular, brown nut; the latter sharp-angled, pointed apically and basally. ***Rumex sanguineus***
     Inner segments *broad* (Fig. 21.57); the latter entire; large segments entire, about 4.5 mm long, dull brown, with 1 large tubercle; other 2 with smaller tubercles; inner segments enclose a small, triangular, brown nut; the latter sharp-angled, and pointed apically and basally. ***Rumex crispus***

82. Segments *2, 5, 10* or *many,* and *either* complete *or* partial 84
    Segments *6,* 3 large inner and 3 small outer 83

83. Outer segments *reflexed* (Fig. 21.58); the latter usually appressed to the persistent pedicel, 3.0-4.5 mm long, entire, orbicular-cordate in outline; inner segments loosely attached, usually broadly winged; all light brown; inner enclose a small broadly triangular, brown nut. ***Rumex acetosa***
    Outer segments *closely adhering* to the inner (Fig. 21.59); the former rather small, about 1.5 mm long, entire, tapered apically; inner much larger, clearly reticulate, without tubercles, and enclose a small, bluntly angular, brown nut; the nut blunt apically and basally. ***Rumex acetosella***

84. Segments *partial* 89
    Segments *complete* 85

85. Segments *equal-sized* 86
    Segments *unequal,* 3 large, 2 small (Fig. 21.60); large keeled or winged, and adhering to fruit; all glabrous, blunt, greenish, and margins white; the enclosed fruit triangular, with 3 equal-sized concave faces; angles blunt; the fruit black, tapered apically and basally. ***Fallopia convolvulus***

86. Segments extending over the *entire* seed, or very nearly so 88
    Segments *not* extending over entire seed 87

87. Outline more or less *ovate* (Fig. 22.01); the segments 5 in number, united close to the base, more or less triangular in outline, blunt apically, and extending over approximately one-half of the fruit; glabrous or minutely pubescent; the length 3.5-5.5 mm; greenish. ***Convolvulus arvensis***
    Outline, general, *reniform* or nearly so (Fig. 22.02); persistent segments 5 in number, united basally, gradually tapered apically, pubescent throughout; short stalk often present; colour black; reticulate throughout; length from 2 to 3 mm; smooth (see couplet no.28). ***Medicago lupulina***

88. Seed somewhat laterally *compressed* (Fig. 22.03); perianth-segments 5 in number, frequently of a pinkish colour, closely adhering to the fruit, blunt, united for a short distance basally; the veins well developed; segments enclose a black-coloured fruit; 2-3 mm long. ***Polygonum persicaria***
    Seed somewhat *angular* (Fig. 22.04); perianth-segments 5 in number, keeled, of a pinkish colour, closely adhering to the fruit, blunt apically, united for a short distance basally; veins not well developed; segments enclose a black-coloured fruit; length 2.5-3.0 mm. ***Polygonum aviculare***

89. Seed *not* showing 3 equal-sized concave faces 90
    Seed showing *3 equal-sized* concave faces and usually triangular in cross-section (Fig. 22.05); abruptly tapered to a blunt apex, gradually tapered basally; angles very prominent and extending from base to apex, blunt, smooth; segments partial, confined to the base; 4-5 mm long; black or blackish, often shiny; covered with irregular striate (see couplet no.85). ***Fallopia convolvulus***

90. Seed somewhat laterally *compressed* (Fig. 22.06); the latter variable in shape; abruptly tapered into a short point; colour either black, blackish, or dark brown; glossy; length 2-3 mm; surface frequently pitted; perianth-segments partial, basal; (see couplet no.88). ***Polygonum persicaria***
    Seed somewhat *angular* (Fig. 22.07); the latter variable in shape; sides usually unequal; gradually tapered into a blunt point; the colour black or blackish; dull; 2.5-3.0 mm; surface marked with irregular striate; perianth-segments partial, basal; (see couplet no.88). ***Polygonum aviculare***

91. Base *not* attached to a stalk-like structure 98
    Base abruptly constricted into a *stalk-like* structure or attached to an *actual stalk* 92

92. Seed attached to a *stalk* 93
    Seed abruptly constricted into a *stalk-like structure* (Fig. 22.08); the latter very short; general outline very variable, mostly bluntly triangular, somewhat laterally compressed; surface faintly wrinkled, widely pitted; length about 3 mm; colour mostly greyish-brown. ***Spinacia oleracea***

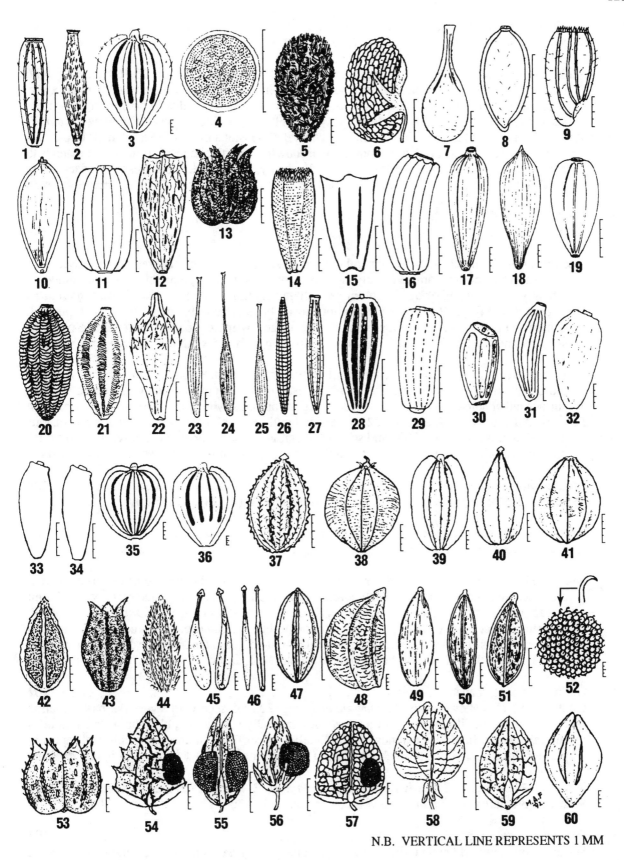

N.B. VERTICAL LINE REPRESENTS 1 MM

Fig. 21.00

93. Vertical constriction *present* 95
    Vertical constriction *absent* 94

94. Width mostly *greater than* length (Fig. 22.09); general dimensions mostly 2.5-3.0 x 2.0-2.5 mm; outline orbicular or suborbicular; seed somewhat 2-lobed, laterally compressed, showing a vertical constriction which extends to both faces, very blunt and shallowly notched apically; basal stalk short; surface puckered; keeled, at least basally; colour greyish. *Fumaria officinalis*
    Width *about equal* to the length (Fig. 22.10); general dimensions mostly 2.5 x 2.5 mm; outline orbicular; seed 2-lobed, laterally compressed, showing a vertical constriction which extends to both faces, very blunt, and shallowly notched apically; basal stalk short, fleshy; surface finely wrinkled; keeled, at least basally; greyish in colour (very similar to above). *Fumaria muralis*

95. Seed *3-lobed* 97
    Seed *2-lobed* 96

96. Apex *mucronate* (Fig. 22.11); the general outline mostly suborbicular; laterally compressed; constricted in the centre; cordate basally; surface strongly wrinkled, puckered, reticulate-pitted, or strongly and irregularly ridged; colour greyish; mostly 2.0 x 3.0 mm. *Coronopus squamatus*
    Apex *emarginate* (Fig. 22.12); general outline mostly broadly suborbicular, laterally compressed; constricted in the centre; cordate basally; the surface wrinkled, puckered, reticulate-pitted, or strongly and irregularly ridged; the colour greyish; mostly 2.5 x 1.5 mm. *Coronopus didymus*

97. Faces *not* keeled (Fig.22.13); the general outline subglobose, with 3 convex protrusions as seen in cross-section, widest and somewhat cordate basally, tapered apically; diameter about 2.5 mm; surface either smooth or with minute tubercles; colour mostly greyish. *Euphorbia helioscopia*
    Faces with distinct *wavy keels* (Fig. 22.14); general outline subglobose, with 3 convex protrusions as seen in cross-section, widest and somewhat cordate basally, tapered apically; diameter c.1.5 mm; surface either smooth or with minute tubercles; colour mostly greyish. *Euphorbia peplus*

98. Seed *neither* 3-sided *nor* irregularly angular 129
    Seed *regularly* or *irregularly* 3-sided, *or* irregularly angular (see Figs 22.15-20) 99

99. All sides or faces *not* convex 101
    All sides or faces *convex* (*e.g.* Fig. 22.15) 100

100. Faces *not* keeled (Fig.22.13); the general outline subglobose, with 3 convex protrusions as seen in cross-section, widest and somewhat cordate basally, tapered apically; diameter about 2.5 mm; surface either smooth or with minute tubercles; colour mostly greyish. *Euphorbia helioscopia*
    Faces with distinct *wavy keels* (Fig. 22.14); general outline subglobose, with 3 convex protrusions as seen in cross-section, widest and somewhat cordate basally, tapered apically; diameter c.1.5 mm; surface either smooth or with minute tubercles; colour mostly greyish. *Euphorbia peplus*

101. Outline *not* cone-shaped 102
    Outline, general, somewhat bluntly *cone-shaped* (Fig. 22.21); seed with a very flat base which is either circular or elliptical in outline; the sides gradually tapered to a blunt apex; a continuous central angle or fine ridge extends longitudinally over surface; hilum small, elliptical, apically placed; surface reticulate throughout; dark in colour; length about 1.5 mm. *Anagallis arvensis*

102. Seed with 1 *convex* and 2 *mostly* flat sides, or *irregularly* 3-sided, or *irregularly* angular 111
    Seed with *3 equal-sized flat* or *concave* sides 103

103. Surface *atuberculate* 104
    Surface prominently *tubercled* (Fig. 22.22); the outline very variable, often triangular to obovate; seed with 2 flat and 1 convex sides; longitudinal angle obscure; apex very blunt; hilum clearly visible, basal; the length 2.5-5.5 mm; colour blackish or dark brown. *Convolvulus arvensis*

104. Surface *not* clearly striate; length 3 mm or *less* 105
Surface *clearly striate* (Fig. 22.23); seed bluntly triangular in cross-section, with equal-sized concave faces; angles very well developed but blunt or rounded; tapered gradually to a blunt apical point, and more abruptly basally; 3-5 mm long; colour mostly black. ***Fallopia convolvulus***

105. Apex *not* abruptly tapered to an acute point 106
Apex *abruptly tapered* to a somewhat short acute point (Fig. 22.24); seed more gradually tapered
or rounded on the base; outline variable, often ovate or ovoid; cross-section nearly triangular,
particularly towards the lower half; sides mostly flat; the attachment-scar small, basally placed;
surface often pitted; black, blackish or dark brown; shiny; 2-3 mm long. ***Polygonum persicaria***

106. Length *1.5 mm* or *more* 107
Length *1 mm* or *less* (Fig.22.25); general outline of seed bluntly triangular in cross-section, with
more or less flat sides; usually uniformly tapered basally and apically; the apex and base blunt;
attachment-scar obscure, basal; surface rough or rugose; colour brownish. ***Rumex acetosella***

107. Angles *blunt* 108
Angles *sharp* (Fig. 22.26); seed sharply triangular in cross-section, widest in the middle, thereaf-
ter uniformly tapered apically and basally; apex and base blunt; attachment-scar very obscure,
basal; sides more or less flat; colour mostly glossy-brown; length about 2 mm. ***Rumex acetosa***

108. Length *1.5 mm* or *less* 110
Length *2.0 mm* or *more* 109

109. Seed *gradually tapered* or *rounded* basally (Fig. 22.27); seed sharply triangular in cross-section;
widest in the lower half; tapered gradually towards the apex; the latter and base blunt; attachment-scar obscure, basal; sides flat; colour glossy-brown; length c. 2 mm. ***Rumex obtusifolius***
Seed *very abruptly* tapered basally (Fig. 22.28); outline sharply triangular in cross-section; widest
in the lower half; tapered rather abruptly towards the apex; latter and base blunt; attachment-
scar very obscure, basal; sides flat; colour glossy-brown; length about 2.0 mm. ***Rumex crispus***

110. Seed tapered apically for *about one-half* its length (Fig. 22.29); cross-section sharply triangular in
outline, widest in lower half, tapered or rounded basally; apex and base blunt; attachment-scar
obscure, basal; sides flat; colour glossy-brown; length about 1.2 mm. ***Rumex conglomeratus***
Seed tapered apically for *about two-thirds* of its length (Fig. 22.30); cross-section sharply triangular in outline; widest in lower half, and tapered or rounded basally; apex and base blunt; attachment scar obscure, basal; sides flat; colour glossy-brown; c. 1.5 mm long. ***Rumex sanguineus***

111. Seed irregularly *3-sided* (*e.g.* Fig. 22.19) or *angular* (*e.g.* Fig. 22.20) 123
Seed with 1 *convex* and 2 more or less *flat* sides (*e.g.* Fig. 22.18) 112

112. Caruncle *absent* 113
Caruncle *present* (Fig. 22.31); the latter white in colour; outline oblanceolate or oblong but rather
rounded apically and tapered basally; surface smooth but showing 1 narrow, longitudinal ventral groove; colour yellowish-brown, with black streaks; length 2.0-2.5 mm. ***Prunella vulgaris***

113. Apex *gradually* tapered or *blunt* 114
Apex *abruptly* tapered (Fig.22.32); general outline variable, somewhat lanceolate; the seed with 1
convex and 2 more or less flat faces, particularly towards the upper half; dull; black or blackish; 2.5-3.0 mm long; surface with irregular striate; (see couplet no.90). ***Polygonum aviculare***

114. Seed base *not* cordate 115
Seed base *somewhat cordate* (Fig. 22.33); general outline more or less ovate; rounded and cordate
basally, narrowed gradually apically, somewhat dorso-ventrally compressed; the surface either
smooth or minutely tuberculate; colour light brown or yellowish; c.1.5 mm long. ***Urtica urens***

115.   Seed *not* bordered 116
       Seed showing a *thick border* (Fig.22.34); the latter distinct; outline mostly, narrowed apically to a
       blunt point, rounded basally, somewhat compressed; attachment-scar basal, clearly visible on
       the ventral surface; colour glossy-black; surface smooth; length c. 1.5 mm. *Myosotis arvensis*

116.   Length *1.25 mm* or *more* 117
       Length *1 mm* or *less* (Fig.22.35; general outline somewhat oblong, rounded or very blunt apical-
       ly, somewhat tapered basally; the attachment-scar basal distinct; longitudinal ridge indistinct;
       the surface minutely reticulate; colour either yellowish-brown or pale brown. *Mentha arvensis*

117.   Surface *not* reticulate 118
       Surface *distinctly reticulate* (Fig. 22.36); oblong, blunt; the ventral ridge obscure; scar prominent,
       about one-half of the ventral surface; colour greyish or pale brown; c. 2.0 mm. *Ajuga reptans*

118.   Surface *atuberculate* 120
       Surface *tuberculate* 119

119.   Length *2.5-5.5 mm* (Fig. 22.37); outline variable; often triangular to obovate; surface prominently
       tubercled; seed with 2 flat and 1 convex sides; longitudinal angle obscure; apex blunt; hilum
       clearly visible, basal; length 2.5-5.5 mm; colour blackish or dark brown. *Convolvulus arvensis*
       Length *c. 1.5 mm* (Fig. 22.38); general outline mostly obovate; rounded or blunt apically, some-
       what tapered basally; longitudinal ridge not well defined; the attachment-scar basal, indistinct;
       surface mostly brownish, but frequently covered with minute dark tubercles. *Stachys arvensis*

120.   Ventral ridge *blunt* 122
       Ventral ridge *sharply defined* 121

121.   Ridge *not extending* to the apex (Fig. 22.39); the seed with 1 convex and 2 flat sides, oblanceolate
       but blunt or truncate apically; ridge well defined, sharp; attachment-scar basal, distinct; length
       c. 2 mm; brownish-grey; sparingly covered with fine frost-like papillae. *Lamium purpureum*
       Ridge *extending* to the apex of the seed (Fig. 22.40); outline obovate, gradually rounded or blunt
       apically, tapered to the base; attachment-scar basal, indistinct; length 1.5-2.0 mm; surface fine-
       ly pitted; colour glossy-black; sparingly covered with grey frost-like papillae.*Stachys sylvatica*

122.   Outline *broadly obovate* (Fig. 22.41); the seed rounded or blunt apically, gradually tapered and
       rounded basally; the attachment-scar basal; longitudinal ridge not well defined; length 2.5 mm
       or more; brownish or greyish; surface often covered with frost-like papillae. *Galeopsis tetrahit*
       Outline mostly *broadly elliptical* (Fig. 22.42); seed usually uniformly rounded or tapered apical-
       ly and basally; the attachment-scar distinct, basal; longitudinal ridge not well defined;the latter
       rounded or blunt; surface smooth or pitted; colour brownish; length c. 2 mm. *Stachys palustris*

123.   Surface *atuberculate* 124
       Surface *tuberculate* (Fig. 22.43); outline variable; sometimes irregularly 3-sided in cross-section;
       apex very blunt; hilum clearly visible, basal; surface prominently tubercled; longitudinal ridge
       rounded or blunt; the length 2.5-5.5 mm; colour blackish or dark brown. *Convolvulus arvensis*

124.   Seed irregularly *angular* (*e.g.* Fig. 22.20) 128
       Seed irregularly *3-sided* (*e.g.* Fig. 22.19) 125

125.   Angles, at least 1, *wavy* 127
       Angles, all, *straight* 126

126.   Apex *abruptly tapered* to a short acute point (Fig. 22.44); seed abruptly tapered or rounded basal-
       ly; outline very variable; sometimes irregularly 3-sided; the attachment-scar basal;the surface
       often pitted; colour black, blackish or dark brown; shiny; 2-3 mm long. *Polygonum persicaria*

Apex *gradually* tapered apically (Fig. 22.45); outline of the seed very variable; sides sometimes unequal; occasionally irregularly 3-sided; gradually tapered to a blunt apex; colour black or blackish; dull; length 2.5-3.0 mm; usually marked with irregular striate. ***Polygonum aviculare***

127. Length *2.5-3.5 mm*, width *2.0-2.5 mm* (Fig. 22.46); general outline variable; frequently somewhat elliptic; rather blunt basally; often constricted near the apex; somewhat puckered; often irregularly angular; dorsal ridge wavy; colour dull-black; surface covered with striate. ***Allium cepa***
Length *2.0-2.5 mm*, width *1.5-2.0 mm* (Fig. 22.47); general outline variable; frequently somewhat elliptic; rather blunt basally; often constricted near the apex; somewhat puckered; often irregularly angular; dorsal ridge wavy; colour dull-black; striate present on surface. ***Allium porrum***

128. Hilum *basal* (Fig. 22.48); the latter basal; outline variable; mostly oblong and angular; blunt basally, often abruptly tapered apically; laterally compressed; radicle position evident; mostly of a brownish colour; covered with fine irregular ribs; 1.25-1.75 mm long. ***Sisymbrium officinale***
Hilum *ventral* (Fig. 22.49); seed irregular in outline, but frequently rhomboid; sharply angular; always dorso-ventrally compressed; hilum centrally located on ventral surface; both surfaces covered with very fine irregular ribs; colour mostly light brown; c. 1.5 mm. ***Plantago major***

129. Seed *not* ending in a short projection (caruncle *not* considered) 134
Seed *ending* in a short projection (caruncle *not* considered) 130

130. Surface *smooth* or *nearly so* 131
Surface *obliquely ribbed* (Fig.22.50); ribs minute, and often anastomosing; general outline mostly oblong, but the seed with a short, obliquely placed, terminal projection; attachment-scar basal; rather elliptical in cross-section; length 1.5-2.5 mm; colour mostly brownish. ***Geranium molle***

131. Seed *laterally* compressed 132
Seed, in lower half, *nearly circular* in cross-section (Fig. 22.51); the outline mostly flask-shaped; 'neck' portion approximately one-half overall length, often angled; shoulders nearly equal in size; attachment-scar basal, obscure; length about 3 mm; colour yellowish.***Ranunculus ficaria***

132. Terminal projection clearly *longer* than the width of the seed-border 133
Terminal projection or beak *minute* (Fig. 22.52); the latter equal to or only slightly longer than seed-border; often obovate but seed with an obliquely placed terminal projection; seed clearly bordered; central portion slightly raised; length 3-4 mm; colour yellowish.***Ranunculus acris***

133. Large shoulder *gently rounded* (Fig. 22.53); terminal beak or projection usually longer than the seed-border; mostly obovate but seed with an obliquely placed terminal projection; seed bordered; central portion slightly raised; 2.5-3 mm long; colour yellowish. ***Ranunculus bulbosus***
Large shoulder *flat* (Fig. 22.54); terminal beak or projection clearly longer than seed-border; general outline mostly obovate but seed with an obliquely placed terminal projection; seed clearly bordered; central portion slightly raised; 3-4 mm long; colour yellowish.***Ranunculus repens***

134. Seed *not* visibly puckered or indented 140
Seed *visibly* puckered or indented 135

135. Caruncle *absent* 136
Caruncle, white, *present* (Fig. 22.55); seed irregularly oblong; dorsal surface marked with broad dark grey indentations; ventral surface showing 2 oblong indentations, and a linear groove; the base blunt; attachment-scar basal; overall colour light grey; c. 1.25-1.5 mm. ***Euphorbia peplus***

136. Apex *not* mucronate 137
Apex *mucronate* (Fig. 22.56); the general outline mostly suborbicular; laterally compressed; constricted in the centre; cordate basally; surface strongly wrinkled, puckered, reticulate-pitted, or strongly and irregularly ridged; colour greyish; mostly 2.0 x 3.0 mm.***Coronopus squamatus***

137. Hilum *clearly evident* 138
Hilum or attachment-scar *not clearly* evident (Fig. 22.57); entire seed enveloped with numerous, more or less triangular or angular, short, firm, woody, persistent perianth-segments; surface very irregular, puckered and deeply pitted throughout, generally glistening; the general outline more or less spherical or globular; diameter from 4-8 mm; colour light brown. ***Beta vulgaris***

138. Seed *not wider* at hilum end than elsewhere 139
Seed *wider at hilum end* than elsewhere (Fig. 22.58); the general outline irregularly and widely oblong, clearly widest basally, narrowed upwards, and laterally compressed; hilum long, narrow, basal, distinct; surface smooth, at least to the naked eye; greenish; 15-22 mm. ***Vicia faba***

139. Colour *green* or *greenish* (Fig. 22.59); the general outline mostly spherical or globular; surface apparently smooth, at least so to the naked eye, though frequently somewhat puckered or indented; hilum clearly visible, longer than broad, basally placed; c. 10 mm long. ***Pisum sativum***
Colour *dark chocolate-brown* (Fig. 22.60); general outline irregularly oval-oblong; surface apparently smooth, at least so to the naked eye, though frequently somewhat puckered or indented; hilum clearly visible, much longer than broad, and basally placed; length c. 10 mm. ***Vicia faba***

140. Surface *reticulate, grooved, cup-shaped, boat-shaped, finely ribbed* or *wrinkled* 164
Surface *smooth*, at least so to the naked eye 141

141. Hilum or attachment-scar *round* or *nearly* so 152
Hilum or attachment-scar *longer* than broad 142

142. Length *2 mm* or *more* 143
Length *1.5 mm* or *less* (Fig. 22.61); general outline oblong, though rounded apically and somewhat blunt basally; surface smooth, at least so to the naked eye; elliptical in cross-section; colour mostly yellowish or pale brown; hilum circular in outline, basally placed. ***Geranium molle***

143. Outline *not* reniform 145
Outline *reniform* 144

144. Colour *black* and *mottled red* (Fig. 22.62); seed outline reniform; somewhat laterally compressed; often bluntly angled dorsally; the surface smooth, at least to the naked eye; hilum frequently white and contrasting with seed coat, laterally placed; up to 22 mm long. ***Phaseolus coccineus***
Colour *dark glossy-brown* (Fig. 22.63); outline reniform; often laterally compressed; often bluntly angled dorsally; surface smooth, at least to the naked eye; hilum often white, contrasting with seed coat; lateral, 2-3 times longer than broad; the length upto 22 mm. ***Phaseolus vulgaris***

145. Hilum *1 mm* or *more* wide 149
Hilum *linear*, about 0.5 mm wide 146

146. Length or diameter *3.5 mm* or *less* 147
Length or diameter *about 5 mm* (Fig. 23.01); seed spherical to orbicular, sometimes slightly compressed; surface smooth, at least so to the naked eye; hilum very narrow, several times longer than wide, not contrasting in colour with seed coat; dull bluish-black. ***Vicia sativa*** ssp. ***sativa***

147. Hilum *not* extending half-way around the seed 148
Hilum extending *more than half-way* around the circumference (Fig. 23.02); outline often subglobose, often slightly compressed; surface smooth, at least so to the naked eye; the hilum very narrow, elongate; the colour dark brown, mottled; length or diameter c. 4.0 mm. ***Vicia sepium***

148. Hilum occupying *approximately one-third* of the circumference (Fig. 23.03); the former many times longer than broad; seed subglobose; slightly compressed; the surface appearing smooth; length or diameter c. 2.5 mm; brown or reddish-brown, often glossy. ***Vicia sativa*** ssp. ***nigra***

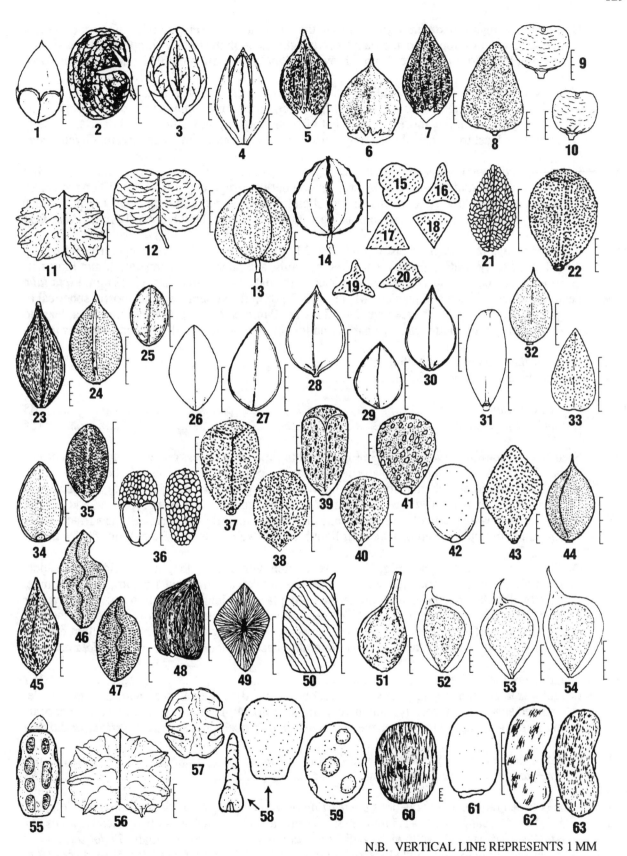

129

N.B. VERTICAL LINE REPRESENTS 1 MM

Fig. 22.00

Hilum occupying *approximately one-fifth* of the circumference (Fig. 23.04); the former narrow and several times longer than broad; seed subglobose; often slightly compressed; the surface smooth; length or diameter c. 3 mm; brown or light brown, often mottled. ***Lathyrus pratensis***

149. Hilum situated *on short* side of seed 150
Hilum situated *on long side* of the seed (Fig. 23.05); the former relatively wide but longer than broad; general outline mostly oblong, but often slightly laterally compressed; surface usually smooth, at least to the naked eye; length up to 9 mm; colour often creamy white.***Glycine max***

150. Colour *light brown* or *greenish* 151
Colour *dark chocolate-brown* (Fig. 23.06); general outline irregularly oval-oblong; surface apparently smooth, at least so to the naked eye; slightly compressed; the hilum clearly visible, much longer than broad, basal, rarely contrasting with the seed coat; length about 10 mm. ***Vicia faba***

151. Seed *wider at hilum end* than elsewhere (Fig. 23.07); the general outline irregularly and widely oblong, clearly widest basally, narrowed upwards, and laterally compressed; hilum long, narrow, basal, distinct; surface smooth, at least to the naked eye; greenish; 15-22 mm.***Vicia faba***
Seed *not* wider at hilum end than elsewhere (Fig. 23.08); the general outline mostly spherical or globular; colour green or greenish; length about 10 mm; surface apparently smooth, at least to the naked eye; the hilum clearly visible, and longer than broad; basally placed. ***Pisum sativum***

152. Outline *other than* ovate in outline 156
Outline *ovate* in outline 153

153. Caruncle *present* 155
Caruncle *absent* 154

154. Seed base *somewhat cordate* (Fig. 23.09); general outline more or less ovate; rounded and cordate basally, narrowed gradually apically, somewhat dorso-ventrally compressed; the surface either smooth or minutely tuberculate; colour light brown or yellowish; c. 1.5 mm long. ***Urtica urens***
Seed base *abruptly narrowed* (Fig.23.10); general outline more or less ovate; rounded and cordate basally, narrowed gradually apically, somewhat dorso-ventrally compressed; the surface either smooth or minutely tuberculate; colour light brown or yellowish; c.1.5 mm long. ***Urtica dioica***

155. Outline of seed *broadly ovate* (Fig. 23.11); seed often with a blunt longitudinal ridge on the dorsal surface; surface smooth, at least so to the naked eye; length about 1.5 mm; hilum minute, basally placed; caruncle small, terminal-laterally placed; colour often yellowish. ***Viola tricolor***
Outline of seed *narrowly ovate* (Fig. 23.12); seed often with a blunt longitudinal ridge on the dorsal surface; surface smooth, at least so to the naked eye; length about 1.5 mm; hilum minute, basally placed; caruncle small, terminal-laterally placed; colour often yellowish.***Viola arvensis***

156. Length *more* than 1.0 mm 157
Length *less* than 1.0 mm (Fig. 23.13); general outline mostly elliptical; seed widest in the middle, rounded apically and basally; hilum very small, obliquely-lateral in position; surface appearing smooth, at least to the naked eye; yellowish;radicle position discernable.***Trifolium dubium***

157. Outline *other than* bluntly triangular 159
Outline *more or less* bluntly triangular 158

158. Zone between radicle and cotyledons *contrasting in colour* with remainder of seed coat (Fig. 23. 14); bluntly triangular; 1.0-1.5 mm long; emarginate and somewhat 2-lobed apically; lobes unequal; surface smooth; yellowish; radicle position discernable; hilum small. ***Trifolium repens***
Zone between radicle and cotyledons *not* contrasting in colour with remainder of seed coat (Fig. 23.15); bluntly triangular; 1.0-1.5 mm; emarginate and rather 2-lobed apically; lobes unequal; surface smooth; colour yellowish; radicle position evident; hilum small. ***Trifolium hybridum***

159.   Seed more or less *spherical* or *reniform* 160
Seed clearly *wider in one-half* than in the other, gradually rounded on the base, more abruptly narrowed apically; somewhat 2-lobed apically; lobes very unequal; irregularly obovate (Fig. 23. 16); surface apparently smooth; length 1.5-2.5 mm; hilum very small, located in a lateral deep concavity; yellowish-green or purplish-green; radicle position discernable. ***Trifolium pratense***

160.   Outline more or less *reniform* 162
Outline more or less *spherical* 161

161.   Seed coat *somewhat glossy* (Fig.23.17); the general outline somewhat irregularly spherical; length from 1.2-1.5 mm; surface smooth, at least so to the naked eye; hilum small, circular in outline, not contrasting in colour with the seed coat; colour often chocolate-brown. ***Lotus corniculatus***
Seed coat somewhat *dull* (Fig. 23.18); the general outline somewhat irregularly spherical; length from 1.2-1.5 mm; surface smooth, at least so to the naked eye; hilum small, circular in outline, not contrasting in colour with the seed coat; colour mostly chocolate-brown. ***Lotus uliginosus***

162.   Seed *without* a small lateral projection 163
Seed with a *small lateral* projection (Fig. 23.19); outline more or less reniform, slightly laterally compressed, rounded apically and basally ; surface smooth, at least so to the naked eye; radicle position marked by a lateral projection; length 2-3 mm; colour yellowish. ***Medicago lupulina***

163.   Seed showing 2 *unequal-sized* lobes (Fig. 23.20); outline more or less reniform, slightly laterally compressed, rounded apically and basally; surface smooth, at least so to the naked eye; length 2-3 mm; yellowish; hilum small, circular, located in the lateral indentation. ***Medicago sativa***
Seed showing 2 *equal-sized* or *nearly* equal-sized lobes (Fig. 23.21); length 1.8-2.2 mm; smooth; colour yellowish; the hilum small, circular, located in the lateral indentation. Outline more or less reniform, slightly laterally compressed, rounded apically and basally. ***Anthyllis vulneraria***

164.   Seed coat *not* broadly reticulate 170
Seed coat *broadly* reticulate 165

165.   Attachment-scar or hilum *narrow* 166
Attachment-scar *broad* and occupying approximately *one-half* of the ventral surface (Fig. 23.22); the latter white; surface distinctly reticulate; outline mostly oblong, very blunt apically and basally; longitudinal ridge not well defined; greyish or pale brown; length c.2 mm.***Ajuga reptans***

166.   Attachment-scar *not* obliquely placed 167
Attachment-scar *obliquely* placed (Fig. 23.23); the latter somewhat circular in outline, large, very distinct, whitish; seed rounded basally, more gradually narrowed apically; mostly obovate; surface broadly reticulate throughout; ventral surface showing a long linear shallow groove; dorsal surface with a shorter groove; length about 2.0 mm; colour greyish. ***Euphorbia helioscopia***

167.   Hilum *circular* 169
Hilum *elongate* 168

168.   Seed mostly *oblong* or somewhat *cylindrical* in outline (Fig. 23.24); seed more or less circular in cross-section; rounded basally and apically; the attachment-scar elongated, narrow and basally placed; surface sharply reticulate; 1.25-1.75 mm; greyish or brownish. ***Geranium dissectum***
Seed *laterally* and strongly *compressed* (Fig. 23.25); general outline somewhat obovate, but blunt basally and rounded apically; seed surrounded by a very narrow membranous border; radicle position clearly evident, lateral; length about 1.25 mm; yellowish-brown. ***Cardamine flexuosa***

169.   Microscopic cross-striations *evident* (Fig.23.26); outline somewhat reniform; seed mostly elliptical in cross-section; the surface, relative to seed size, broadly reticulate; colour mostly black or very dark brown; hilum minute, circular, laterally placed; length c. 0.7 mm. ***Papaver dubium***

Microscopic cross-striations *absent* (Fig. 23.27); outline somewhat reniform; seed mostly elliptical in cross-section; surface, relative to seed size, broadly reticulate; colour mostly steel-blue;
the hilum minute, usually circular and laterally placed; length about 0.8 mm. **Papaver rhoeas**

171. Reticulation *complete* 172
Reticulation *incomplete* (Fig. 23.28); general outline oblong, though rounded apically and somewhat blunt basally; surface smooth, except for small area; often elliptical in cross-section; colour mostly yellowish or pale brown; hilum circular in outline, basally placed. **Geranium molle**

172. Hilum *basal* or *ventral* 173
Hilum *obliquely placed* (Fig.23.29); the latter small, circular or narrowly elliptical in outline; general outline somewhat oblanceolate, asymmetrical in side-view, cross-section elliptical; blunt
basally, more gradually tapered and blunt apically; often somewhat curved; surface very finely
reticulate throughout; length about 1.5 mm; brown or brownish in colour. **Potentilla anserina**

173. Seed *not* cylindrical and parallel-sided 175
Seed *cylindrical* and *parallel-sided* 174

174. Length about *1.0 mm* (Fig. 23.30); general outline cylindrical, clearly truncate apically and basally, often with a minute terminal mucro; bluntly quadrangular in cross-section; surface finely
reticulate; hilum minute, circular, basal; colour usually dark brown. **Hypericum perforatum**
Length about *0.6 mm* (Fig. 23.31); general outline cylindrical, clearly truncate apically and basally, often with a minute terminal mucro; bluntly quadrangular in cross-section; surface finely
reticulate; hilum minute, circular, basal; colour usually dark brown. **Hypericum tetrapterum**

175. Radicle position *not* evident on broad side 177
Radicle position *evident* on broad side 176

176. Peripheral border *absent* (Fig. 23.32); outline mostly oblong; seed rounded apically and basally;
narrowly elliptical or bluntly triangular in cross-section; radicle position evident; length about
1.0 mm; yellowish-brown; hilum visible, small, circular, and basal. **Capsella bursa-pastoris**
Peripheral border *present* (Fig. 23.33); the latter very narrow; outline oblong, very blunt apically
and basally; very strongly laterally compressed; radicle position clearly visible; surface finely
reticulate; yellow or yellowish-brown; length 1.0 mm or less; hilum basal. **Cardamine hirsuta**

177. Seed nearly *spherical, ovate, obovate* or *globose* 178
Seed very strongly *dorso-ventrally compressed* (Fig. 23.34); general outline mostly oblanceolate;
base often somewhat constricted; margins rather acute; length 5-7 mm; surface very finely reticulate; colour mostly dark brown, glossy; hilum small, basally placed. **Linum usitatissimum**

178. Outline more or less *spherical* or *globose* 179
Outline more or less *ovate* or *obovate* (Fig. 23.35); seed somewhat elliptical in cross-section; radicle position and cotyledon margins clearly evident laterally; the colour mostly pink or pinkish; length from 2-3 mm; surface very finely reticulate; hilum small, basal. **Raphanus sativus**

179. Colour *other than* yellow or yellowish 180
Colour yellow or yellowish (Fig. 23.36); outline spherical or globose; seed often somewhat elliptical in cross-section; the radicle and cotyledon positions mostly evident, laterally; surface very
finely reticulate; length or diameter 1.5-2.5 mm; hilum located on blunt the end. **Sinapis alba**

180. Cotyledon and radicle positions *not* evident on narrow side 182
Cotyledon and radicle positions *evident* on narrow side 181

181.   Colour *grey-brown* (Fig. 23.37); general outline spherical or globose; seed often somewhat elliptical in cross-section; radicle and cotyledon positions frequently evident laterally; surface finely reticulate; length or diameter c. 2 mm; hilum small, located on blunt end. ***Brassica oleracea***
         Colour *dark brown* (Fig. 23.38); outline spherical or globose; radicle and cotyledon positions frequently evident laterally; seed often somewhat elliptical in cross-section; surface very finely reticulate; length or diameter 1.6-2 mm; hilum small and located on blunt end. ***Brassica napus***

182.   Colour *dark brown,* often *glossy* (Fig. 23.39); outline spherical or globose; seed often somewhat elliptical in cross-section; radicle and cotyledon positions not evident laterally; surface very finely reticulate; length or diameter 1-2.5 mm; hilum located on the blunt end. ***Brassica rapa***
         Colour *bluish-brown* (Fig. 23.40); outline spherical or globose; radicle and cotyledon positions not evident laterally; seed often somewhat elliptical in cross-section; surface very finely reticulate; the length or diameter from 1.5-2 mm; hilum small, located on blunt end.*Sinapis arvensis*

183.   Seed *neither* boat-shaped *nor* cup-shaped 190
         Seed *boat-shaped* or *cup-shaped,* or narrowed into a beak 184

184.   Seed *not* boat-shaped 185
         Seed *boat-shaped* (Fig. 23.41); the general outline narrowly oblong; usually dorso-ventrally compressed; ventral surface with a long trough; the hilum small, ventrally placed; dorsal surface with light-coloured longitudinal band; c. 2.0 mm in length; glossy-brown. ***Plantago lanceolata***

185.   Seed *not* beaked 187
         Seed *beaked* 186

186.   Width *0.5 mm* or *less* (Fig. 23.42); length c. 9 mm; the general outline narrowly oblanceolate, tapered into a long beak; often curved; the longitudinal ribs indistinct and confined to the upper region; attachment-scar very narrow, laterally placed; colour dark brown. ***Foeniculum vulgare***
         Width *1.0 mm* or *more* (Fig. 23.43); length about 8.0 mm; outline narrowly oblanceolate, tapered into a long beak; often somewhat curved; the longitudinal ribs inconspicuous and confined to the upper region; attachment-scar very narrow, lateral; colour darkbrown.*Anthriscus sylvestris*

187.   Length *1.7 mm* or *less* 188
         Length *2.0 mm* or *more* (Fig. 23.44); general outline somewhat cup-shaped and globose; dorsal surface irregularly and roughly ribbed; radiating ribs on ventral surface; the latter with a deep depression; the colour light or dark brown; hilum small, ventrally placed. ***Veronica hederifolia***

188.   Radiating ridges *atuberculate* 189
         Radiating ridges *coarsely tubercled* (Fig. 23.45); outline mostly obovate; seed rounded apically, tapered basally, usually with a deep, wide ventral depression; dorsal surface rounded, irregularly ribbed or ridged; colour pale brown; length 1.5 x 1 mm; hilum ventral. ***Veronica persica***

189.   General outline *obovate* (Fig. 23.46); seed rounded apically, tapered basally, usually with a deep, wide ventral depression; the dorsal surface rounded, irregularly ribbed or ridged; colour either pale brown or yellowish; length 1-1.5 mm; hilum small, ventrally placed. ***Veronica agrestis***
         General outline *broadly elliptical* (Fig. 23.47); the seed rounded apically, tapered basally, usually with a deep, wide ventral depression; dorsal surface rounded, irregularly ribbed or ridged; colour mostly pale or dull brown; 1.6 x 1.2 mm; hilum small, ventrally placed. ***Veronica polita***

190.   Seed *lacking* a border 192
         Seed *with* a complete or crescentic-shaped *border*; outline *spherical* or *disc-shaped* 191

191.   Length or diameter *3-5 mm* (Fig.23.48); general outline irregularly orbicular but seed very strongly compressed and disc-like; surrounded by a wide, very conspicuous yellowish, incomplete border; centre raised, dark brown in colour; hilum small, laterally placed. ***Rhinanthus minor***

Length or diameter *approximately 1.25 mm* (Fig. 23.49); the general outline spherical; seed surrounded by a narrow, indistinct, whitish, wavy, thin border; surface frequently minutely tuberculate; colour mostly bluish-black; papillae absent or present; hilum minute. *Spergula arvensis*

194. Apex *very blunt* (Fig. 23.50); general outline very variable; sometimes very irregularly 3-sided in cross-section; hilum clearly visible, basal; surface clearly tubercled; ventral surface often with rounded or blunt ridge; the length 2.5-5.5 mm; blackish or dark brown. *Convolvulus arvensis*
   Apex *somewhat beaked* (Fig. 23.51); the latter broad, flat membranous; outline mostly obovate; seed broadly elliptical in cross-section; surface covered with an abundance of prominent tubercles; grey or silvery-grey; length or diameter 2.5-4.0 mm; hilum distinct. *Mercurialis annua*

196. Length *approximately 1.0 mm* (Fig. 23.52); outline obovoid; seed very blunt or truncate apically, somewhat rounded basally; more or less circular in cross-section; surface with an abundance of minute tubercles; ventral surface with a narrow longitudinal groove. *Epilobium montanum*
   Length *approximately 0.8 mm* (Fig. 23.53); outline obovoid; seed very blunt or truncate apically, somewhat rounded basally; more or less circular in cross-section; surface with an abundance of minute tubercles; ventral surface with a narrow longitudinal groove. *Epilobium hirsutum*

198. Hilum *circular* (Fig. 23.54); the latter very narrow and located on narrow end of seed; the outline mostly orbicular; seed mostly broadly elliptical in cross-section; surface with an abundance of small, evenly distributed, tubercles; colour reddish-brown; length 1.25-2.0 mm. *Ononis repens*
   Hilum *elongate* (Fig. 23.55); the latter clearly visible on flat surface; outline irregularly circular; laterally compressed and narrowly elliptical in cross-section; surface covered with small, unevenly scattered, tubercles; yellowish-brown; length or diameter .75-1.25 mm. *Stellaria media*

199. Seed base *somewhat cordate* (Fig. 23.56); general outline more or less ovate; rounded and cordate basally, narrowed gradually apically, somewhat dorso-ventrally compressed; the surface either smooth or minutely tuberculate; colour light brown or yellowish; c. 1.5 mm long. *Urtica urens*
   Seed base *abruptly narrowed* (Fig. 23.57); general outline more or less ovate; rounded and cordate basally, narrowed gradually apically, somewhat dorso-ventrally compressed; the surface either smooth or minutely tuberculate; colour light brown or yellowish; c. 1.5 mm long. *Urtica dioica*

   Surface *obliquely ribbed* (Fig.23.58); ribs minute, and often anastomosing; mostly oblong; attachment-scar basal; elliptical in cross-section; 1.5-2.5 mm long; mostly brownish. *Geranium molle*

202. Hilum *ventral* (Fig. 23.59); seed irregular in outline, but frequently rhomboid; sharply angular; always dorso-ventrally compressed; hilum centrally located on ventral surface; surfaces covered with very fine irregular ribs; colour mostly light brown; c. 1.5 mm long. *Plantago major*

135

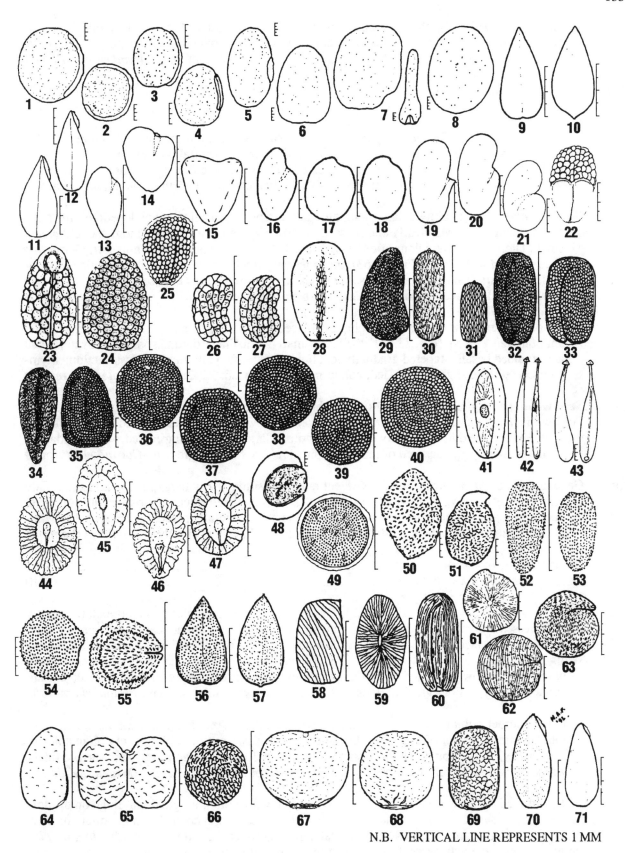

N.B. VERTICAL LINE REPRESENTS 1 MM

Fig. 23.00

136

Hilum *basal* (Fig. 23.60); the latter basal; outline variable; mostly oblong and angular; blunt basally, often abruptly tapered apically; laterally compressed; radicle position evident; mostly of a brownish colour; covered with fine irregular ribs; 1.25-1.75 mm long. *Sisymbrium officinale*

203. Seed *other than* lenticular 206
Seed *lenticular* 204

204. Surface covered with a *fine network*, or *striate* which *do not* radiate from the seed-centre 205
Surface covered with fine *delicate striate*, which radiate from a central position (Fig. 23.61); outline mostly lenticular; seed clearly compressed and narrowly elliptical in cross-section; hilum very small, laterally placed; mostly glossy-black; diameter about 1.5 mm. *Chenopodium album*

205. Surface *netted* (Fig. 23.62); the latter longitudinally arranged, very finely developed; outline more or less lenticular; seed clearly compressed, narrowly elliptical in cross-section; the hilum very small, laterally placed; colour often greyish-black; diameter about 1.5-3.5 mm. *Atriplex patula*
Surface *covered with striate* (Fig. 23.63); the latter very finely developed; general outline more or less lenticular; seed clearly compressed and narrowly elliptical in cross-section; the hilum very small, laterally placed; colour often greyish-black; diameter 1.5 to 3.5 mm. *Atriplex prostrata*

206. Hilum or attachment-scar *other than* obliquely placed 207
Hilum *obliquely placed* (Fig. 23.64); the latter small but distinct; outline somewhat ovate, broad and rounded basally, tapered gradually to a blunt apex; often elliptical in cross-section; asymmetrical; surface slightly wrinkled; colour mostly yellowish; 1.2-2 mm long. *Potentilla reptans*

207. Seed *without* a longitudinal constriction; attachment-scar or hilum *not* long and narrow 208
Seed with a *longitudinal constriction*, or with a *long, narrow* and *laterally* placed attachment-scar (Fig. 23.65); the outline broadly orbicular or orbicular; laterally compressed; base cordate or rounded; wrinkled, puckered or reticulate-pitted; c. 2 x 1.5 mm; greyish. *Coronopus didymus*

208. General outline *oblong, ovate,* or *globose* but radicle position *not* evident 209
General outline *somewhat globose* (Fig. 23.66); seed broadly elliptical in cross-section; radicle. position clearly defined; the hilum very small, indistinct, laterally placed; surface clearly wrinkled; colour mostly pale or reddish-brown; length or diameter c.1.0 mm. *Stellaria graminea*

209. Outline *oblong* or *somewhat ovate* 211
Outline *spherical* or *nearly so* 210

210. Width mostly *greater than* length (Fig. 23.67); general outline orbicular or suborbicular, laterally compressed, very blunt, and shallowly notched apically; attachment-scar basal; surface finely wrinkled; keeled, at least basally; greyish; mostly 2.5-3.0 x 2.0-2.5 mm. *Fumaria officinalis*
Width *about equal* to the length (Fig. 23.68); general outline orbicular, laterally compressed, very blunt, and shallowly notched apically; attachment-scar basal; surface finely wrinkled; keeled, at least basally; colour greyish; mostly 2.5 x 2.5 mm. (v. similar to above). *Fumaria muralis*

211. Peripheral border *absent* 212
Peripheral border *present* (Fig. 23.69); the latter very narrow; outline oblong, very blunt apically and basally; very strongly laterally compressed; radicle position learly visible; surface finely wrinkled; yellow or yellowish-brown; length 1.0 mm or less; hilum basal. *Cardamine hirsuta*

212. Outline of seed *broadly ovate* (Fig. 23.70); seed often with a blunt longitudinal ridge on the dorsal surface; surface smooth, at least so to the naked eye; length about 1.5 mm; hilum minute, basally placed; caruncle small, terminal-laterally placed; colour often yellowish. *Viola tricolor*
Outline of seed *narrowly ovate* (Fig. 23.71); seed often with a blunt longitudinal ridge on the dorsal surface; surface smooth, at least so to the naked eye; length about 1.5 mm; hilum minute, basally placed; caruncle small, terminal-laterally placed; colour often yellowish. *Viola arvensis*

GRASS 'SEEDS'

A knowledge of the structure of the grass floret and spikelet, as given on pages 186 and 188, is a necessary prerequisite to an understanding of the grass 'seed'. Technically there is no such entity as a grass seed. The 'seed' of the vast majority of Poaceae members is a dry, 1-seeded, indehiscent fruit, termed a *caryopsis*; berries or nuts are produced by some bamboo species. Other dry, 1-seeded, indehiscent fruits have been mentioned already in relation to families Asteraceae, Polygonaceae and Urticaceae, but in these families the fruits, because the seed-coat and fruit-wall do not fuse, are called *achenes*.

The ovary wall of the grass pistil encloses a single orthotropous ovule. The latter has 2 integuments, each of which is usually composed of 2 layers of cells. After fertilisation the stigma shrivels, though it often persists as a 'tuft of short hairs' on the caryopsis. The outer integument is absorbed while the the inner persists and eventually gives rise to the seed coat. During maturation the inner portion of the fruit wall also disintegrates allowing the remainder of the wall to fuse with the seed coat. The resulting fruit is termed a caryopsis and is characterised by the lateral-oblique position of its embryo. When the caryopsis is not enclosed within other attachments the position of the embryo is clearly visible (Figs 24.1-6). Where grasses of the British Isles are concerned the caryopsis, when ripe, is shed from its protective flowering pales or *lemma* and *palea* in a comparatively small number of species (Figs 24.1-6). The free caryopsis, as it occurs in *Triticum aestivum*, *Secale cereale* and *Zea mays*, is commonly known as a grain. Some *Agrostis* species and *Phleum pratense* are examples where the grain may be easily shed or retained within the flowering pales. In most species the grain is enclosed within the flowering pales, or within the latter and the the empty pales or *glumes*. In 1 species, *Hordeum murinum* (Fig. 24.07), it is enclosed within a cluster of spikelets. In this plant the spikelets fall off in clusters of 3, the central spikelet contains the grain, whereas the laterals are staminate and are incapable of forming grains.

The 'seed' of about 6 species is represented by the entire spikelet. At maturity the spikelets of *Briza media* may be shed intact or may disarticulate at each rachilla node to give a caryopsis enclosed within the lemma and palea. The spikelet of *Elymus repens* is shed in its entirety even though it may contain more than 1 caryopsis. A similar situation may be noted for the *Alopecurus* (Figs 24.10-12) and *Holcus* (Figs 24.13 and 24.14) species where the entire spikelet also represents the 'seed'.

There are several examples of enclosed caryopses remaining attached to additional floret-like structures on reaching maturity (Figs 24.15-23). In *Anthoxanthum odoratum* (Fig. 24.20) and *Phalaris canariensis* and *P. arundinacea* (Figs 24.15 and Fig. 24.16) additional asexual lemmas persist. *Dactylis glomerata* often sheds its enclosed caryopses in clusters (Fig. 24.17). Spikelets of *Arrhenatherum elatius* are usually 2-flowered, the lower being hermaphrodite and the upper staminate. On reaching maturity the lower, which contains the caryopsis, remains attached to the upper floret (Fig. 24.21). Only rarely is it detached from the staminate floret. *Deschampsia flexuosa* and *D. cespitosa* normally contain 2 hermaphrodite florets per spikelet. Each floret is normally shed separately (Fig. 24.38-39), though occasionally they remain attached (Fig. 24.22 and 24.23, respectively). On rare occasions the caryopses of *Holcus lanatus* and *H. mollis* become detached from their glumes (Figs 24.18-19).

As already stated, the majority of caryopses are shed enclosed within the flowering pales (Figs 24.24-25.41). Also, 1 rachilla internode often remains attached, and this is commonly known as the *seed-stalk*. Variation in these pales and seed-stalks provides dependable criteria for identification. Thus, such features of the lower pale - or lemma - as the number of nerves or vascular bundles, texture, length, shape, pubescence, the presence or absence of an awn or bristle, are dependable criteria for use in identification. In addition, the smaller pale - or *palea* - can provide a few diagnostic features. Here, its length relative to that of the bigger pale, the prominence of its 2 keels, the presence or absence of teeth, and the location of the latter, are important. Finally, the presence or absence of a seed-stalk, and variations in its shape, length and pubescence, provide additional information.

Some difficulty may be experienced in determining the number of nerves per outer flowering pale, but the difficulty can be minimised by presoaking the 'seed' in 10% alcohol. Presoaking makes the nerves

138

more translucent and thus easier to count. Alternatively, a simpler method involves placing the separated lemma on the stage of a microscope. Using transmitted rather than incident light, the nerves stand out as dark outlines when viewed through the eyepiece. Again, some difficulty may arise in relation to keeled seeds. The latter tend to lay on their side. The total number of nerves can be determined by counting those on the side exposed to the eyepiece, doubling that number, and adding 1 for the keel. For identification purposes, the important numbers are 3, 5 and 7. Fortunately, determination is easy where the number is 3 or 7 and the difficulty tends to arise only where the number is 5. About 90% of 'seeds' have the latter number.

The presence of an awn and its point of insertion are useful clues (Figs 24.17-41). However, caution should be exercised when trying to identify commercial 'seed' which normally have awns. The awns of such 'seeds' as *Dactylis glomerata* (Fig. 25.27), *Festuca rubra* (Fig. 25.39) and *Lolium multiflorum* (Fig. 25.40) are easily dislodged during cleaning processes. Provision is made in the following 'key' for this possibility. Another difficulty often arises with the 'seeds' of the *Poa* species (Figs 25.28-32). Most of these tend to have a delicate *web* of silky hairs attached to the base of each flowering pale. The web is often dislodged during the cleaning of commercial 'seed' but invariably present when the 'seed' is hand-picked from wild material in the field.

The following 'key' is intended as an aid to the identification of the more common grass 'seeds'. For the reasons stated earlier (see page 114), the characteristics of several 'seeds' of the same species should be studied before making a final decision on identification. Furthermore, it is important to note that provision is not made for the identification of immature or undeveloped grass seeds. Such seeds tend to occur, *inter alia*, in species of *Dactylis* and *Poa*.

KEY TO GRASS 'SEEDS'

1. Germ position *hidden;* grain *covered* with lemma and pale, or glumes 7
   Germ position *clearly visible;* grain *not* covered 2

2. Grain clearly *longer than broad* 3
   Grain nearly as *broad as long* (Fig. 24.01); the latter obovate in outline; flat, rounded or indented apically, tapered to a blunt base, dorso-ventrally compressed, flat-sided; position of embryonal tissues visible basally; 9-10 mm long; yellow or yellowish in colour. *A cereal grain.* **Zea mays**

3. Length *2.5 mm* or *less* 4
   Length *4.0 mm* or *more* 6

4. Grain *dorso-ventrally* compressed and showing a narrow longitudinal groove 5
   Grain *laterally* compressed, resulting in the embryonal tissues occupying an obliquely lateral position (Fig.24.02); the grain rather obovate; apex abruptly tapered into a very small projection; no longitudinal groove evident; length 1.5-2.2 mm. *A common pasture plant.* **Phleum pratense**

5. Grain tapered more gradually *basally* than apically (Fig. 24.03); the latter very small, about 1 mm, oblanceolate in outline; ventral longitudinal groove narrow; brownish in colour; often minutely wrinkled and very finely ribbed. *A frequent grass on heaths.* **Agrostis canina** or **A. vinealis**
   Grain tapered more gradually *apically* than basally (Fig. 24.04); the latter very small, about 1 mm, somewhat lanceolate in outline; ventral longitudinal groove narrow; colour brownish; surface often minutely wrinkled and very finely ribbed. *Very common wild species.* **Agrostis capillaris**

6. Grain *5-7 mm* in length (Fig. 24.05); the latter somewhat oblong or ovate in outline, blunt apically and basally, rounded dorsally; ventral surface with a deep longitudinal groove; colour yellow or yellowish; minutely hairy apically or glabrous. *Common cereal grain.* **Triticum aestivum**
   Grain *7-9 x 2-3 mm* long (Fig. 24.06); the former narrowly elliptical in outline, somewhat keeled dorsally; the apex blunt; tapered and rather pointed basally; rounded dorsally; either minutely hairy or glabrous; ventral groove narrow; surface wrinkled. *A cereal grain.* **Secale cereale**

7. Seed represented by a floret or a cluster of florets; glumes *absent* 15
    Seed represented by a spikelet or a cluster of spikelets; glumes *present* 8

8. Seed represented by a *single* spikelet 9
    Seed represented by a *cluster* of 3 spikelets (Fig.24.07); glumes 6, linear in outline, each with very
    long bristles up to 30.0 mm; central pair fringed with hairs; the laterals toothed; fertile lemma
    lanceolate, others very narrow; all with long awns. *A frequent weed grass.* **Hordeum murinum**

9. Lemmas *concealed* within glumes 11
    Lemmas clearly *visible* 10

10. Lemmas deeply *concave* (Fig.24.08); the latter rounded apically and dorsally, and usually cordate
    basally, about 4 mm in length and 7-9-nerved. Spikelet mostly ovate in outline, laterally com-
    pressed, shiny, and containing 5-12 florets. *A common plant on calcareous soils.* **Briza media**
    Lemmas *neither* concave *nor* rounded apically (Fig. 24.09); outline lanceolate; awned or pointed,
    keeled apically, firm, and often minutely rough; 7.0-14.0 mm in length; both glumes very sim-
    ilar. *A very common troublesome grass weed of cultivated and waste ground.* **Elymus repens**

11. Lower glume *1-nerved* 14
    Lower glume *3-nerved* 12

12. Glumes *united* basally 13
    Glumes *free* to the base (Fig. 24.10); the latter frequently dark brown in colour with traces of pur-
    ple, membranous, 3-nerved, lateral nerves close to the margins; fringed; 2.5-3.5 mm; lemma
    shorter than glumes, blunt, with an awn. *Very frequent; wet pastures.* **Alopecurus geniculatus**

13. Glumes united for *one-quarter,* or less, of their length (Fig. 24.11); the latter lanceolate, pointed,
    4.0-8.0 mm long, both 3-nerved; nerves, keels and margins fringed with hairs; lemma with a
    long awn; palea absent; *common plant in old pastures, grassy areas, etc.* **Alopecurus pratensis**
    Glumes united for about *one-third,* or more, of their length (Fig. 24.12); the latter pointed, 4.0-7.0
    mm long, winged on the minutely hairy keels. *Common tillage weed.* **Alopecurus myosuroides**

14. Awn *hook-shaped,* short and not clearly visible (Fig. 24.13); glumes yellowish or pinkish-white,
    4-6 mm long, obscurely short-awned and fringed with short silky hairs; upper 3-nerved, lower
    1-nerved; lemmas very short, shiny, bearded, 1 awnless. *Very common plant.* **Holcus lanatus**
    Awn *straight* (Fig. 24.14); glumes thin, 4-7.0 mm long, slightly unequal, pointed, lower 1-nerved,
    upper 3-nerved, and both with short stiff hairs on the veins. *Frequent plant from open wood-
    lands, heaths, etc.; also occurs, though rarely, as a tillage weed in sandy soils.* **Holcus mollis**

15. Seed represented by the structures of a *single* floret 24
    Seed represented by structures *other than* those of a single floret 16

16. Awn *present* 18
    Awn *absent* 17

17. Additional asexual lemmas *similar* to the normal lemmas but shorter and narrower (Fig. 24.15);
    seed broad, 4-6 x 2.5-3.5 mm; *a very frequent impurity in seed-mixtures.* **Phalaris canariensis**
    Additional asexual lemmas very small, laterally placed, *densely hairy* (Fig. 24.16); seed glossy,
    3-4 x 1.0 mm; *common from wet habitats; cultivated in some countries.* **Phalaris arundinacea**

18. Awn *subterminal,* dorsal or basal 19
    Awn *terminal,* short (Fig. 24.17); lemma strongly keeled, at least on its upper half, fringed with
    stiff hairs or distinctly toothed; 4-7 mm; somewhat asymmetrical, 2-5 floret-like structures in
    each cluster; the upper florets generally undeveloped; glumes, if present, firm, unequal, similar
    to the lemmas. *Very common cultivated plant; also occurs in grassy areas.* **Dactylis glomerata**

140

19. Lemmas, all, *with* awns 22
    Lemmas of the floret containing the caryopsis *without* awns 20

20. Awn *not* hook-shaped 21
    Awn *hook-shaped* (Fig. 24.18); usually 2 florets held together by a short seed-stalk; both small, 1.5-2.0 mm, glossy, smooth, membranous, blunt apically, bearded basally, 1 with an awn, the other awnless; bearded basally; a rare form of Figure 24.13. *Ubiquitous plant.* **Holcus lanatus**

21. Lemmas *bearded* basally (Fig. 24.19); usually 2 florets held together by a short seed-stalk; small, 1.5-2.0 mm, glossy, smooth, membranous, blunt apically, bearded basally, and 1 with an awn, the other awnless; bearded basally; rare form of Figure 24.14. *A frequent plant.* **Holcus mollis**
    Lemmas, asexual, hairy *throughout* (Fig.24.20); the latter dark brown; awns attached at different levels; fertile lemma c. 2 mm, smooth, shiny-brown. *Common plant.***Anthoxanthum odoratum**

22. Awns, *all*, basal or nearly so 23
    Awns, *1*, of the lower structure attached dorsally, or very nearly so, and twisted and bent (Fig. 24.21); the upper structure awnless or with a short straight awn, and enclosing the caryopsis; both structures minutely bearded basally. Lemmas 7-nerved, 7.0-11.0 mm, rounded dorsally, somewhat pointed, firm except for membranous apices; upper lemma or both loosely hairy or hairless. *Common from rough grasslands, hedgerows, grassy places, etc.* **Arrhenatherum elatius**

23. Awn extending *far* beyond the apex of the lemma (Fig. 24.22); lemma glossy, 4-6 mm, membranous, entire or minutely toothed apically, bearded; palea blunt; usually 2 florets held together by a short hairy seed-stalk. *A frequent plant, mostly from heaths, etc.* **Deschampsia flexuosa**
    Awn *not* extending far beyond the apex of the lemma (Fig. 24.23); lemma 3-4.0 mm long, glossy, membranous, very ragged or torn apically, bearded basally; 2 florets held together by a short hairy seed-stalk; palea *bifid. Frequent in rough, damp grassy places.* **Deschampsia cespitosa**

24. Lemma *without* a distinct awn 51
    Lemma *with* a distinct awn 25

25. Lemma *5-nerved* or *less* 36
    Lemma *7-nerved* or more 26

26. Awn *terminal* or *subterminal* 30
    Awn *dorsal* or *basal* 27

27. Lemma *blunt*, pointed or somewhat toothed 28
    Lemma ending in 2 distinct awn-like *projections* (Fig. 24.24); the former either hairless or minutely hairy basally, 10-22 mm long, 7-nerved; awn long, stout, bent, and basal. *Cereal plant, formerly cultivated in areas of poor soils, but now rather frequent as a weed.* **Avena strigosa**

28. Seed-base *lacking* a circular depression 29
    Seed-base *with* a distinct, circular, long-bearded, shallow depression (Fig.24.25); lemma very variable in colour and pubescence but often very dark brown; 10-23.0 mm, 7-9-nerved, rough and somewhat toothed or ragged apically; seed-stalk long, slender, very hairy; awn stout, geniculate, bent and arising from a dorsal position. *A weed of tillage, waste places, etc.* **Avena fatua**

29. Seed-base *not* bearded (Fig. 24.26); lemma yellowish in colour, often glossy, smooth, tapered or ragged apically, mostly rounded dorsally, but veins distinct apically; 10-22 mm. *An infrequent form of cereal grains of limited economic importance which usually lack awns.* **Avena sativa**
    Seed-base *bearded* (Fig. 24.27); awn attached near the apex, rarely absent, short, straight. Lemma 7-nerved, 7.0-11.0 mm, rounded dorsally, somewhat pointed, firm except for membranous apex, greenish or greyish-green in colour, not glossy, loosely hairy or hairless above. *Common from rough grasslands and grassy places*; see also couplet no. 22.**Arrhenatherum elatius**

30.   Lemma *rounded* or *flat* dorsally 33
      Lemma keeled *throughout* or *apically* 31

31.   Lemma keeled *throughout* 32
      Lemma keeled *apically* (Fig. 24.28); the latter minutely hairy, particularly towards the base; veins
      frequently minutely toothed; the apex with 2 long, narrow, awn-like, membranous projections;
      awn equal to, or longer than, its lemma. *Frequent as a weed of dry sandy soils.* **Bromus sterilis**

32.   Awn *terminal* (Fig.24.29); the latter straight or flexuous, 3-8.0 mm in length, and arising from the
      lemma apex; the latter narrowly-lanceolate, shortly bifid or pointed, firm but with membra-
      nous margins, 7-nerved. *Plant often locally abundant on dry calcareous soils.* **Bromus erectus**
      Awn *subterminal* (Fig. 24.30). Lemma narrowly lanceolate-oblong in outline, firm, 10.0-16.0 mm
      long, 7-nerved, and with short hairs near the margins and back; awn straight, subterminal, and
      clearly longer than its lemma. *Plant frequent on moist soils in shaded places.* **Bromus ramosus**

33.   Awn *subterminal* 34
      Awn *terminal* (Fig. 24.31). Seed-stalk narrow, cylindrical, tapered basally. Seed flat, 8-12.0 mm
      long, mostly stiffly hairy, rarely smooth or rough, parallel-sided; the lemmas 7-9-nerved; awn
      straight, at least equal to its lemma. *Very frequent; shaded places.* **Brachypodium sylvaticum**

34.   Lemma *bifid* 35
      Lemma *blunt* (Fig.24.32); seed-stalk conspicuous, straight in side-view, 1-3 mm, smooth, tapered
      basally, elliptical in cross-section; seed 5.0-8.5 x 2.2-3.5 mm, rounded dorsally; the minutely
      rough keels of the palea clearly evident; the latter often separated from the lemma by the pro-
      truding dark caryopsis. *Weed, or valuable forage plant in some countries.* **Lolium temulentum**

35.   Seed length *3.5-6.0 mm* (Fig. 24.33); caryopsis mostly protruding from between the lemma and
      palea; lemma obovate in outline, 7-nerved, clearly 2-toothed apically, and rounded dorsally but
      somewhat angular towards the apex, mostly hairless, rarely hairy; the awn straight and arising
      behind apex. *Frequent annual or biennial in pasture and other grassy places.* **Bromus lepidus**
      Seed length *7-12 mm* (Fig. 24.34); caryopsis mostly concealed within the lemma and palea. Lem-
      ma 7-9-nerved, obovate to elliptic in general outline, rounded dorsally, clearly bifid apically,
      and rather hairy; rarely hairless; awn straight, nearly as long as its lemma, and arising behind
      the apex. *A common annual weed of cultivated ground, waste places, etc.* **Bromus hordeaceus**

36.   Awn *terminal* or *subterminal* 41
      Awn *dorsal* or *basal* 37

37.   Awn *basal* 40
      Awn *dorsal* 38

38.   Length *3 mm* or *more;* seed-stalk *present* 39
      Length *1.2 mm* or less (Fig. 24.35). Lemma and palea and membranous, very blunt, ovate or ob-
      ovate in outline, obscurely 5-nerved, loosely arranged around the caryopsis, hairless except for
      the presence of a very short basal tuft of hairs; the awn bent, arising from the lower half. *Plant
      from heaths, moors, hill pasture, peaty soils; used for lawns.* **Agrostis canina** or **A. vinealis**

39.   Lemma very *blunt* or *short-toothed* (Fig. 24.36); the latter 8.0-16.0 mm in length, firm except for
      thin margins, rounded dorsally, narrowly oblong-lanceolate in side view, 5-nerved, hairless ex-
      cept for short basal hairs, rough upwards. Awn arising from about mid-dorsal area, geniculate,
      firm, twisted. Seed-stalk long-haired. *Frequent from damp alkaline soils.* **Avenula pubescens**
      Lemma apex ending in 2 *bristle-points* or 2 long teeth (Fig. 24.37); outline narrowly lanceolate in
      side-view, 4.5-5.5 mm long, 5-nerved, somewhat loosely arranged, firm except for very thin
      margins, hairless. Awn bent, twisted and dorsal. Seed-stalk hairy. *Common from old pastures,
      roadsides and grassy areas, etc., particularly those on dry alkaline soils.* **Trisetum flavescens**

40. Awn extending *far* beyond the apex of the lemma (Fig. 24.38); lemma glossy, 4-6 mm, membranous, entire or short-toothed apically, short-bearded basally; palea blunt apically; seed-stalk short, hairy. *A frequent plant, mostly from heathlands*, moorlands, etc. **Deschampsia flexuosa**
Awn *not* extending far beyond the apex of the lemma (Fig. 24.39); lemma 3-4.0 mm long, glossy, membranous, very ragged or torn apically, short-bearded basally; seed-stalk short, hairy; palea deeply *bifid* apically. *A frequent plant in rough, damp grassy places.* **Deschampsia cespitosa**

41. Lemma *flat* or *rounded* dorsally 42
Lemma *keeled* dorsally, at least upwardly (Fig.24.40), firm except on the margins, 5-nerved, 4-7.0 mm, s.v. lanceolate, fringed with stiff hairs or at least clearly toothed, asymmetrical; the awn short. Seed-stalk hairless. *A cultivated plant; also occurs in grassy areas.* **Dactylis glomerata**

42. Awn *terminal, i.e.* lemma tapering gradually into an awn 46
Awn *subterminal, i.e.* awn arising just below the lemma apex 43

43. Seed-stalk *dorso-ventrally compressed,* and *adhering* to the palea 45
Seed-stalk more or less *cylindrical,* and *not* adhering to the palea 44

44. Awn *equalling* or exceeding the lemma (Fig. 24.41); the latter rounded dorsally, 6.0-10.0 mm, 5-nerved, firm except for membranous margins, hairless. Awn somewhat subterminal, straight or curved. Seed-stalk cylindrical, rough. *Frequent in damp woodlands, etc.* **Festuca gigantea**
Awn *much* shorter than the lemma, subterminal (Fig.25.01); the latter firm except on the margins 6.0-9.0 mm long, 5-nerved, rounded basally, keeled apically, apex blunt or pointed, hairless. *A very frequent plant from rough grassland, roadsides, old pastures, etc.* **Festuca arundinacea**

45. Width *1.8-2.2 mm;* length 5.0-9.0 mm (Fig. 25.02). Lemma blunt apically, rough on the back, 5-9 nerved, and later becoming swollen and hard. Awn long, up to 20 mm in length, rough. Seed-stalk somewhat tapered basally, smooth, and adhering closely to palea; the latter often separated from the lemma by the protruding caryopsis. *Forage plant or weed.* **Lolium temulentum**
Width *1.6-1.8 mm;* length 5-10 mm (Fig. 25.03). Lemma blunt, somewhat ragged or torn apically, rounded or flat dorsally, 5-nerved, either smooth or minutely rough. Awn is straight, minutely rough, long, arising from behind the apex. Caryopsis not protruding. Seed-stalk smooth. *Common cultivated forage plant; also occurs as a short-lived weed in lawns.* **Lolium multiflorum**

46. Seed-stalk *present* 47
Seed-stalk *absent* (Fig. 25.04). Lemma narrowly lanceolate, or lanceolate-oblong in outline, 5.0-10.0 mm in length, 3-nerved, 2-3-keeled or somewhat 3-angled, rough on the keels, becoming hardened at maturity, terminated with a short straight awn 1.0-3.0 mm in length. Palea blunt and slightly shorter than lemma. *Frequent unpalatable plant from heathlands.* **Nardus stricta**

47. Awn *shorter* than its lemma 49
Awn *equalling* or *exceeding* its lemma 48

48. Width *1.5 mm* or *less* (Fig. 25.05). Lemma 5.0-10.0 mm long, 5-nerved, rounded dorsally, firm, linear-lanceolate, margins often inrolling, smooth or minutely rough, and tapered into a rough awn of at least equal length. *A very frequent plant from dry sandy soils, etc.* **Vulpia bromoides**
Width *2.5 mm* or *more* (Fig. 25.24). Seed widest in the centre, tapered apically and basally, very blunt, rounded or flat dorsally, and nerves not prominent; 6-9 mm long. Ventral surface either wrinkled or smooth; with a narrow longitudinal groove. *Important cereal.* **Hordeum sativum**

49. Lemma clearly *not* more toothed on its upper half 50
Lemma clearly *more* toothed on its upper half (Fig. 25.06); the latter 3-4.5 mm in length, yellowish, finely 5-veined, rounded dorsally, oblong to ovate in outline, ending in a very short terminal awn, and clearly more toothed on its upper half. Seed-stalk short, and somewhat T-shaped. *From grassy places. Used, in the past, in seed-mixtures; used for lawns.* **Cynosurus cristatus**

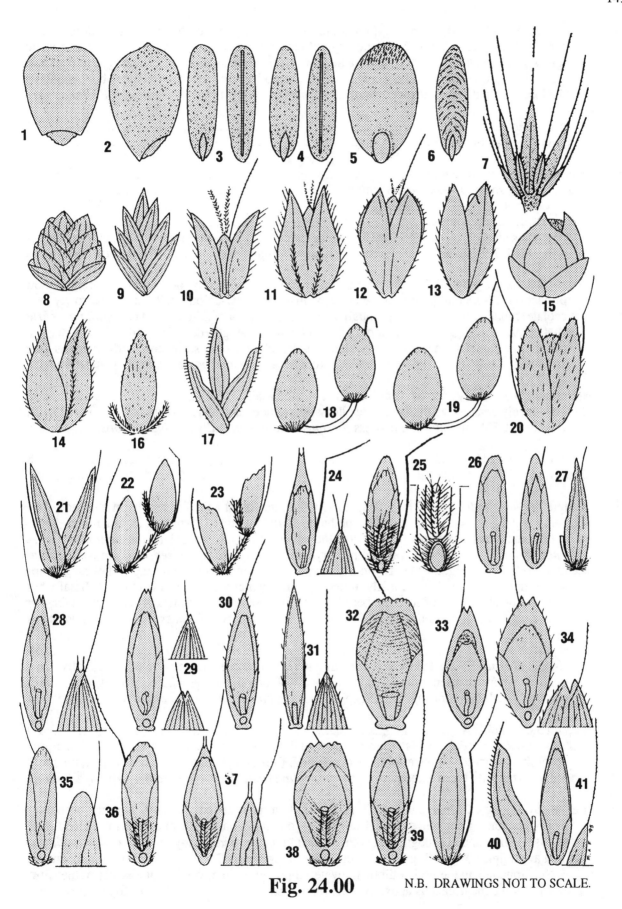

Fig. 24.00   N.B. DRAWINGS NOT TO SCALE.

50. Length *2.75-4.04 mm* (Fig. 25.07). Lemma rounded dorsally, narrowly lanceolate in dorsal view; 5-nerved; awn 0.5-1.5 mm. Seed-stalk cylindrical. *From heaths, moorlands, etc.* **Festuca ovina**
     Length *4.5-7.5 mm* (Fig. 25.08). Lemma narrowly lanceolate in dorsal view, tapered into an awn 1.5-3.5 mm, rounded dorsally, finely 5-nerved, smooth, rough or hairy. Palea about equalling the lemma. Seed-stalk narrow, cylindrical. *Common; many different situations.* **Festuca rubra**

51. Lemma *5-nerved* or less 61
     Lemma *7-nerved* or more 52

52. Hairs *absent* from the lemma 55
     Hairs, at least some, *present* on the lemma 53

53. Lemma apex *not* 3-toothed 54
     Lemma apex clearly *3-toothed* (Fig. 25.09); the latter 5-7.5 mm, smooth, rounded, elliptic, densely bearded, 7-9-nerved; marginal nerves hairy. *Frequent from heaths.* **Danthonia decumbens**

54. Seed-stalk *absent* (Fig. 25.10). Lemma 6.0-15.0 mm, blunt apically but usually bearing a short projection 0.2-0.8 mm long near the apex, whitish or straw-coloured, 7-nerved, narrowly lanceolate, rounded or often keeled, rough, bearded basally. *A dune grass.* **Ammophila arenaria**
     Seed-stalk *present* (Fig. 25.11). Lemma 15.0-25.0 mm in length, densely short-haired, 7-nerved, firm pointed apically, narrowly lanceolate, tough, yellowish or straw-coloured. Palea as long as lemma. Seed-stalk cylindrical, hairy or rough. *A common coastal plant.* **Leymus arenarius**

55. Lemma *neither* deeply concave ventrally *nor* cordate basally 56
     Lemma deeply *concave* ventrally, convex dorsally, rounded apically, cordate basally (Fig. 25.12); 7-9-nerved, 3-4 mm long and hairless. *Frequent plant, especially on alkaline soils.* **Briza media**

56. Width *1.5 mm* or less 58
     Width *1.8-2.2 mm* or *more* 57

57. Length *5-9 mm;* width 1.8-2.2 mm (Fig. 25.13). Lemma blunt apically, rough on the back, 5-9-nerved, and later becoming swollen and hard. Awn long, up to 20 mm in length, rough. Seed-stalk somewhat tapered basally, smooth, and adhering closely to palea; the latter often separated from the lemma by the protruding caryopsis. *Forage plant or weed.* **Lolium temulentum**
     Length *10-22.0 mm* (Fig. 25.14). Lemma greenish or yellowish in colour, tapered or ragged apically, smooth, often glossy, mostly rounded dorsally, nerves effaced dorsally and basally but distinct apically. Base of seed *not* bearded. Outline somewhat oblong. *An Infrequent form of a cereal grain of diminishing economic importance which usually lacks an awn.* **Avena sativa**

58. Seed-stalk *present* 59
     Seed-stalk *absent* (Fig.25.15). The lemma thin in texture, somewhat membranous, laterally compressed, 7-nerved, often minutely hairy, broadly oblong in outline, and very blunt, of a silvery colour, 1.0-2.0 mm long, and keeled apically. *An important cultivated plant.* **Phleum pratense**

59. Length *5.0 mm* or less 60
     Length *5.8-7.6 mm* (Fig. 25.16). Lemma rounded dorsally, blunt or somewhat pointed, margins becoming incurved, elliptical-oblong or oblong, firm except for a thin whitish apex, minutely rough, 7-nerved. Palea 2-toothed. *A common plant from wet muddy ground.* **Glyceria fluitans**

60. Lemma averaging *4.25 mm* in length (Fig. 25.17). Lemma rounded dorsally, margins becoming incurved, blunt or pointed, elliptical-oblong or oblong in outline, firm except for a broad thin, whitish apex, minutely rough, 7-nerved. *Common on wet and muddy ground.* **Glyceria plicata**
     Lemma averaging *3.5 mm* (Fig. 25.18); the latter prominently 7-nerved, minutely rough on the nerves, rounded on the back, elliptic to ovate-elliptic in outline, very blunt apically, rather firm except for the membranous apex. *A common plant in riversides, canals, etc.* **Glyceria maxima**

145

61. Seed-stalk *present* 67
Seed-stalk *absent* 62

62. Length *4.0 mm* or *more* 66
Length *2.25 mm* or *less* 63

63. Seed *dorso-ventrally* compressed 64
Seed *laterally* compressed, and 5-nerved (Fig. 25.19); Lemma thin in texture, somewhat membranous, 5-nerved, usually minutely hairy, broadly oblong in outline, very blunt, of a silvery colour, 1.2 mm, keeled apically. Seed stalk absent. *Important cultivated plant.* **Phleum pratense**

64. Palea *blunt* apically 65
Palea gradually *narrowed* apically (Fig. 25.20). Lemma thin, obscurely 3-toothed, membranous,
finely 3-5-nerved, rounded, short-bearded. *Common from many situations.* **Agrostis capillaris**

65. Nerves, *all,* barely discernable (Fig. 25.21). Lemma very thin, membranous throughout, about 1.5
mm long, very blunt, faintly 5-nerved, outline broadly lanceolate, rounded dorsally, and minutely bearded; caryopsis spindle-shaped, laxly enclosed. *Very common.* **Agrostis stolonifera**
Nerves, *central,* discernable (Fig. 25.22). Lemma thin and membranous throughout, about 1.5 mm
long, very blunt apically, faintly 5-nerved, broadly lanceolate in outline, rounded dorsally, and
minutely bearded; caryopsis spindle-shaped. *Frequent in cultivated ground.* **Agrostis gigantea**

66. Lemma *with a tuft of white hairs* from near the base (Fig. 25.23); the former 6-15.0 mm in length
blunt apically but usually bearing a short projection 0.2-0.8 mm long near the apex; whitish;
5-nerved; round or slightly keeled; minutely rough. *Common dune grass.* **Ammophila arenaria**
Lemma *unbearded* basally (Fig. 25.24); 6-9 x 2.75-4.0 mm. Seed widest in the centre, tapered apically and basally, very blunt, rounded or flat dorsally, and nerves not prominent. Ventral surface wrinkled, and with a narrow longitudinal groove. *An important cereal.* **Hordeum sativum**

67. Lemma *5-nerved* 68
Lemma *3-nerved* (Fig. 25.25); the former tapering gradually towards the apex, hairless, 3.5-6 mm,
rounded dorsally, pointed or blunt apically, smooth. Palea keels clearly visible; the seed base
somewhat rounded. *A common plant from heathlands, moorlands, bogs, etc.* **Molinia caerulea**

68. Seed-stalk *hairless* 69
Seed-stalk *hairy,* long, slender (Fig. 25.26). Lemma *unbearded* basally; the latter 6-9 x 2.75-4.0
mm. Seed widest in the centre, tapered apically and basally, usually very blunt, rounded or flat
dorsally, and nerves not prominent; the latter smooth or toothed. Ventral surface either wrinkled or smooth, and with a narrow longitudinal groove. *Important cereal.* **Hordeum sativum**

69. Dorsal surface *flat, rounded,* or keeled *apically*; seed somewhat dorso-ventrally compressed . 75
Dorsal surface *keeled throughout;* seed somewhat laterally compressed 70

70. Lemma, upper half, *not fringed* with short stiff hair 71
Lemma *fringed* with stiff hairs (Fig. 25.27); the former strongly *keeled* dorsally, firm, pointed or
blunt apically, 5-nerved, 4-7 mm long, fringed with stiff hairs, or at least prominently toothed,
and somewhat asymmetrical apically. Seed-stalk hairless. *Cultivated plant.* **Dactylis glomerata**

71. Nerves, all, *not hairy* 72
Nerves, all, *hairy* (Fig. 25.28). Caryopsis loosely enclosed within the palea and lemma; the latter
semi-elliptic or oblong, membranous with a broad delicate apex, 2.5-4.0 mm in length, keeled
throughout, 5-nerved; a basal web of hairs present or absent. *An ubiquitous plant.* **Poa annua**

72. Marginal nerves *hairless* 74
Marginal nerves *fringed with hairs* 73

73. Palea with *very rough* keels (Fig. 25.29). Lemma blunt or somewhat pointed, finely 5-nerved, firm but margins thin, membranous, keeled, 3.0-4.5 mm; the keel and marginal nerves fringed with silky hairs; basal web of silky hairs often present. *Frequent in dry habitats.* **Poa pratensis**
Palea with *minutely rough* keels (Fig.25.30). Lemma 2.5-3.6 mm in length, keeled, 5-nerved, narrowly oblong to oblong-lanceolate in side view, blunt or slightly pointed apically, firm except for a membranous apex; web present or absent. *Frequent from shaded places.* **Poa nemoralis**

74. Caryopsis *loosely* enclosed within lemma and palea (Fig. 25.31); the latter semi-elliptic or oblong in outline, membranous with a broad delicate apex, 2.5-4.0 mm in length, keeled throughout, 5-nerved; basal web of hairs present or absent. *Ubiquitous; see also couplet no.71.* **Poa annua**
Caryopsis *tightly* enclosed within the firm lemma and palea (Fig. 25.32). The lemma 2.5-3.8 mm long, keeled, narrowly oblong and pointed in side view, 5-nerved; apex and margins membranous; base and lower half of keel fringed with crinkled hairs. *Ubiquitous plant.* **Poa trivialis**

75. Base of lemma and its veins *hairless* 76
Base lemma, or veins, *minutely hairy* (Fig.25.33); the former elliptic to broadly oblong in outline, firm except for a white membranous apex and margins, purplish or whitish in colour, rounded, blunt apically, 3.0-5.0 mm in length. *A frequent plant of salt-marshes.* **Puccinellia maritima**

76. Seed-stalk *not broad-lipped* apically; pale neither yellow nor yellowish in colour 77
Seed-stalk very short and *broad-lipped* apically (Fig. 25.34). Lemma clearly *more* toothed on its upper half, 3.0-4.5 mm in length, of a yellowish or brownish colour, finely 5-veined, rounded dorsally, oblong to ovate in outline, and usually ending in a very short point. *Ubiquitous plant of grassy and waste places; used, in the past, in seed-mixtures for lawns.* **Cynosurus cristatus**

77. Seed-stalk *dorso-ventral* compressed, and *adhering closely* to the palea 82
Seed-stalk more or less *cylindrical*, and *not* adhering closely to the palea 78

78. Seed-stalk *minutely rough* 79
Seed-stalk *smooth* 80

79. Seed widest *towards the base* (Fig. 25.35); the former tapering gradually towards the apex; base somewhat rounded. Lemma 5-nerved, 3.5-6.0 mm long, hairless, rounded dorsally, pointed or blunt apically, smooth. Keels of the palea clearly visible; *see couplet no. 67.* **Molinia caerulea**
Seed widest *about the middle* (Fig. 25.36); the former tapered gradually apically and more abruptly basally. Lemma firm except for membranous apex and margins, 6-9 mm, 5-nerved, rounded or nearly flat basally, often keeled apically; the central nerve ending abruptly behind the apex. Seed-stalk cylindrical. *Frequent from rough grassland, roadsides, etc.* **Festuca arundinacea**

80. Width *1.2 mm* or *less* 81
Width *1.5 mm* or *more* (Fig. 25.37). Lemma flat or rounded dorsally, firm except for membranous margins and apex, 5-nerved, 5-8 mm long, blunt, or bluntly pointed, and hairless or minutely toothed dorsally. Seed-stalk cylindrical. *Frequent from grassy places, etc.* **Festuca pratensis**

81. Length *2.75-4.0 mm* (Fig. 25.38). Lemma rounded dorsally, narrowly lanceolate in dorsal view, and hairy or hairless; 5-nerved. Seed-stalk cylindrical. *Frequent on heaths, etc.* **Festuca ovina**
Length *4.5-7.5 mm* (Fig. 25.39). Lemma narrowly lanceolate in dorsal view, tapered to a point, rounded dorsally, hairy or hairless. Seed-stalk cylindrical; *see couplet no. 50.* **Festuca rubra**

82. Central nerve ending very *suddenly* (Fig. 25.40) - indicating a detached awn. Width 1.6-1.8 mm, 5-10 mm. Lemma blunt, ragged or torn, rounded or flat dorsally, smooth or minutely rough, 5-nerved. Palea keels often toothed; see couplet no.45; *rare weed of cereals.***Lolium multiflorum**
Central nerve *not* ending suddenly (Fig. 25.41). Lemma blunt, somewhat ragged or torn, rounded, dorsally, smooth, width 1.4 mm or less, length 7 mm or less. Keels of the inner palea smooth or minutely rough. *Important pasture and hay plant, abundant in many areas.* **Lolium perenne**

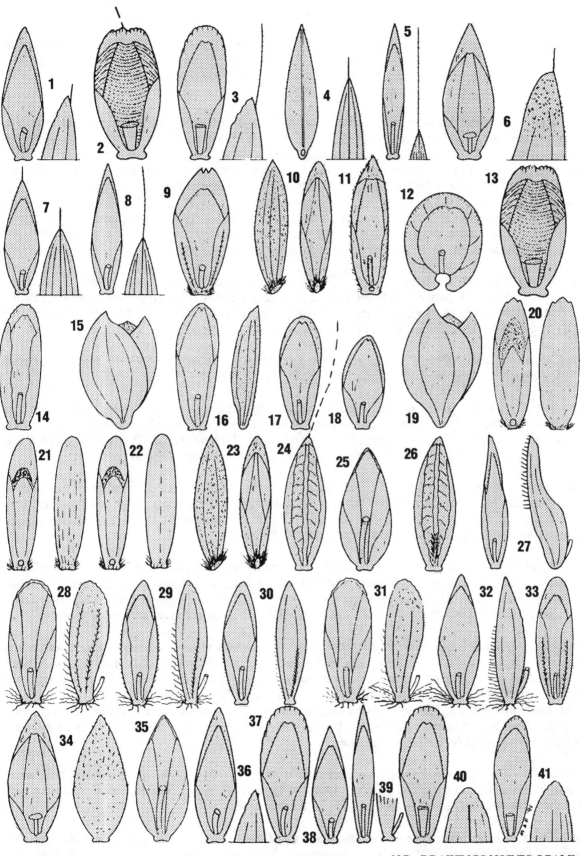

Fig. 25.00

N.B. DRAWINGS NOT TO SCALE

FRUITS

A fruit may be described as a ripened ovary. The ovary wall in the mature end product is known as the fruit wall or *pericarp*. It is generally divisible into 3 zones: (i) the outer layer or *epicarp* or *exocarp*; (ii) a middle area or *mesocarp*; and (iii) an inner zone or *endocarp*. Differences in the texture of the fruit wall, together with a number of other features, determine fruit type.

Classification of Fruits: There are 3 principal fruit types, each of which has many subdivisions. The principal divisions are: 1 *simple*, 2 *aggregate* and 3 *multiple* fruits. Simple fruits, in turn, can be subdivided as 1.1 *dry* fruits and 1.2 *succulent* fruits. Dry fruits, where the pericarp does not increase in size other than to keep pace with developing seed within, can be categorised as 1.1.1 *achenial*, 1.1.2 *capsular* and 1.1.3 *schizocarpic*. Succulent fruits, characterised by having increased fleshy pericarps, may be of the 1.2.1 *drupe*, 1.2.2 *bacca*, 1.2.3 *hesperidium*, 1.2.4 *pepo* or 1.2.5 *pome* type. Simple fruits are the products of a *simple* or *compound* ovary, derived from 1 or more carpels, respectively. Furthermore, the pericarp of some dry fruits may rupture, *i.e. dehisce* in various ways, or remain intact in others, *i.e. indehiscent*. One classification scheme for the many different fruit types could be as follows:

1.1.1.1 Achene: This, of the achenial type, is a dry, indehiscent, 1-seeded fruit where the pericarp remains dry and firm. The pericarp and seed coats do not fuse together. These types are commonly called 'seeds', and, indeed, form the so-called 'seeds' of many crop and weed plants. They are produced, for example, by plants of the families Asteraceae, Ranunculaceae and Urticaceae.

1.1.1.2 Caryopsis: This achenial fruit type, of the Poaceae, is characterised by the pericarp and seed coats fusing together during maturation (see grass 'seeds', page 137).

1.1.1.3: Nut This, a third achenial type, is the term popularly applied to several hard-shelled fruits. It is characterised by having a very firm or hard pericarp, and is produced, mainly, by tree species. Good examples are the fruits of *Corylus* (hazel), *Fagus* (beech) and *Quercus* (oaks). The fruit types of a few herbaceous plants, particularly members of the Polygonaceae, are often described as nuts.

1.1.1.4: Samara A fourth representative is the samara. Again, this is produced mostly by tree species. Most samaras are 1-seeded, for example those of *Ulmus* (elm) and *Fraxinus* (ash), but those of *Acer* (sycamore) are 2-seeded. The latter could be described as a double samara. There is no common example of a samara representing the fruit of a herbaceous plant.

1.1.2.1 Capsule: This is the first example of capsular fruits which are characterised by the presence of many seeds at maturity. They are derived from compound ovaries, and rupture or dehisce in various ways to scatter their seeds. Thus, *Papaver* species shed their seeds through apical pores, those of *Juncus* by means of longitudinal separations, while those of *Plantago* do so by means of transverse ruptures.

1.1.2.2 Legume: A second capsular fruit is the legume. It is characteristic of nearly all members of the Fabaceae. Popularly, it is known as a pod. Legumes or pods are mostly straight, but in *Medicago* they are twisted or curved. The mode of dehiscence is generally along both the ventral and dorsal sutures. In a few exceptional cases the pod may not dehisce, as for example in *Trifolium lappaceum, Onobrychis viciifolia* and *Medicago* species. The pods of *Ornithopus sativus* are described as schizocarpic because they separate into 1-seeded segments. Finally, the fruit or pod of all *Trifolium* species dehisces by splitting transversely, giving, what is termed, a *pyxidium*.

1.1.2.3 Follicle: This is a third example of a capsular fruit. It develops from a single ovary, and, when ripe, ruptures along 1 suture, thereby differing from the legume which opens along both. Examples are found in the Crassulaceae, *Aconitum* species and *Caltha palustris*.

1.1.2.4 Siliqua: This capsular type fruit is characteristic of most members of the Brassicaceae. One distinguishing feature is the fact that it is much longer than broad; it often terminated by a beak. Opinions differ as to whether is arises from 2 or 4 carpels. It opens along both sutures but the central partition remains in place, giving a 3-part structure when dehiscing. One exception is the siliqua of *Raphanus* which is transversely dehiscent into 1-seeded segments.

1.1.2.5 Silicula: This fruit is closely allied to the siliqua but differs in being about as broad as long; it dehisces in the same manner. It occurs in a few genera, particularly *Lepidium* and *Capsella*.

1.1.3.1 Mericarp: This is the first of the schizocarpic fruits. It is characteristic of all members of the Apiaceae. The mericarp arises from a compound ovary which splits into 1-seeded halves when mature. Each half forms a so-called 'seed'. It is invariably terminated with a characteristic swollen glandular structure termed the stylopodium, and generally has 5 longitudinal ribs of varying prominence.

1.1.3.2 Carcerule: The carcerule is another schizocarpic fruit derived from a compound ovary. When mature, it separates, from the base upwards, into individual ovaries which contain 1 seed, but the styles remain attached together for some time. This the fruit type characteristic of the Geraniaceae.

1.1.3.3 Nutlet: The final schizocarpic type is formed from a single, deeply 4-lobed ovary; each lobe or cell, when mature, contains 1 seed. When mature, it separates into 4 1-seeded segments or nutlets which generally have 1 convex and 2 flat faces. This fruit type is found in 3 common families, *viz*. the Boraginaceae, Lamiaceae and the Verbenaceae.

1.2.1 Drupe: The drupe is the first example of a succulent fruit. It is characterised by having a stone-like endocarp, a fleshy or succulent mesocarp and a thin exocarp, *i.e.* the skin. The endocarp encloses a single seed. Drupaceous fruits are found in all members of the genus *Prunus*.

1.2.2 Bacca: Another succulent fruit is the bacca or berry. It is derived from from a compound ovary. Baccate fruits have many seeds embedded in a fleshy endocarp and mesocarp. The exocarp forms the skin. Examples are found in many plant families, but particularly in the Solanaceae.

1.2.3 Hesperidium: This is often listed as a type of berry. It has a thick, leathery rind (peel), with numerous oil glands, and a thick juicy section composed of several wedge-shaped locules. The peel consists of the exocarp and mesocarp, and the juice sacs are outgrowth from the endocarp. Examples of hesperidia are confined to citrus fruits.

1.2.4 Pepo: This is another berry-like fruit. It differs from true baccate fruits in having the rind composed of receptacle tissue which surrounds and encloses the exocarp. The fleshy portion is essentially mesocarp and endocarp. Examples are confined to the Cucurbitaceae.

1.2.5 Pome: This fruit type is characteristic of a subfamily of the Rosaceae which includes, *inter alia*, the genera *Malus, Cotoneaster, Pyrus* and *Sorbus*. Various interpretations have been given as to the morphological nature of the pome, but the distribution of the vascular tissue would suggest that most of the fleshy portion is derived from the floral tube, and that the remainder is derived from the exocarp.

2.1 Etaerio of drupes: Aggregate fruits are derived from several independent carpels of the *same* flower. Thus, in the genus *Rubus*, the many individual carpels give rise to a drupe. However, the drupes remain together in a cluster. These clusters are often incorrectly referred to as berries.

2.2 Etaerio of achenes: This fruit is very similar to the above, but the carpels give rise to achenes instead of drupes. Examples are found in the genera *Potentilla* and *Fragaria*. In the latter, the receptacle forms a fleshy mass of tissue in which the achenes are embedded.

3.0 Multiple fruits: These are formed from the individual ovaries of several flowers. These fruits, taken individually, may be classified as simple fruits but on maturity all remain together as a unit. Examples are the fruits of *Humulus lupulus* (hop), *Morus nigra* (mulberry), *Ficus carica* (fig) and *Ananas comosus* (pineapple). A common example is shown by some forms of *Beta vulgaris* (beet, Fig. 20.09).

SEED TESTING

Historical: The adoption of scientific seed testing, as an aid to practical agriculture, dates from the year 1867, when the first institution for the testing of seeds was established in Germany (Lafferty, 1934). A Herr Nobbe carried out germination tests on agricultural seeds before they were sown in the field. In many areas Nobbe's work was received with enthusiasm, and between 1870 and 1894 seed testing stations were established throughout Europe. However, owing to the absence of standardised testing methods, the data obtained for germination and purity tests at the various stations were of very little value for comparative purposes. Since such data should constitute the basis of International trade in agricultural seeds, the directors of the various stations soon realised the necessity for uniform methods and techniques. Despite attempts in 1906 and 1910 to achieve these objectives, it was not until 1921, in Copenhagen, that an Association, known as the *European Seed Testing Association (ESTA)*, was formed. The principal aim of this Association was to obtain greater uniformity in respect of analysis results at the various stations. At another meeting in Copenhagen, in *1924*, it was decided to enlarge the scope of ESTA, and to extend its activities to all the countries of the world in which the testing of seeds was practised, and to re-constitute it under the name of the *International Seed Testing Association (ISTA)*. A draft form of a set of *International Rules for Seed Testing (IRST)* was discussed at a meeting in Rome, in 1928, and at a Sixth Congress, in Wageningen, in 1931, the rules were accepted for Temperate regions. A meeting, in Dublin, in *1953*, adopted standardised rules for all stations. ISTA now holds triennial meetings, and the International Rules for Seed Testing are revised and updated, if necessary. Tests now carried out in national seed testing stations include those for purity, germination, seed health, verification of species and cultivar, weight and moisture content.

Meanwhile, a Professor Johnson, who had studied seed testing in Germany, established an unofficial station in the Royal College of Science, Dublin, in 1896. In early 1900, a Royal Commission was established to examine the whole question of the quality and sale of agricultural seeds. The commission found that the quality of seeds on sale in Ireland was low in terms of germination and purity. These findings were in agreement with those found in a survey by Land Commission Inspectors in 1895 and 1896 (McCarthy, 1988). It recommended that an official station be established in England. In December 1900, the newly established Department of Agriculture and Technical Instruction for Ireland, made the unofficial station official, and appointed Professor Johnson its director, albeit part-time. This was the first official station in the British Isles. By 1909 sufficient information had been accumulated to show that the quality of certain kinds of seeds imported into Ireland was very inferior. It was decided that, if any improvement was to be achieved, a full-time director, supported by necessary legislation, should be made responsible for seed control. Thus, a Dr Pethybridge was appointed Director in 1909, and he soon achieved a marked improvement in the quality of seeds in circulation.

Legislation

Prior to 1909, the only Act governing seeds was the 'Adulteration of Seeds Act', of 1869, but apart from making it a criminal offence to 'kill' or 'dye' seeds, it did not prevent a merchant from selling seed of very inferior quality. In 1909, the *'Weeds and Agricultural Seeds (Ireland) Act'* became law, and the second part of this Act, which dealt with the sale of seeds, marked the beginning of the end of the inferior seed trade in Ireland. This Act empowered Department of Agriculture Officials, to enter all premises where seeds were offered for sale and take samples for analysis. The vendors were obliged, if requested, to furnish the name and address of the person from whom the seed was originally purchased. Merchants with inferior seed had their names published in national and local newspapers. In addition, public notices were posted in the vicinity of towns in which they did business.

Other Acts followed in 1920, 1933, 1936, 1955 and 1956. These were concerned, *inter alia*, with the production, cleaning, purchase, storage and sale of agricultural seeds. Presently, the sale of nearly all agricultural seeds is subject to Statutory Regulations. The regulations provide that most agricultural seeds may not be offered for sale unless they comply with germination, purity and other standards, as defined in law. Furthermore, the certified seeds must be closed and labelled by an 'Official Certifying Authority'.

Each label carries a code which may be used to identify the 'Seed Assembler', who must be in a position to provide the above information, if requested.

TESTING PROCEDURES

Sampling

The seed on which information is sought is called the *seed lot*. Ordinarily, it is infinitely greater than the quantity on which the actual test is carried out. The first exercise, therefore, is to obtain a sample for transmission to the seed station. Space does not permit a full discussion on the regulations set forth by ISTA in relation to sampling. Persons requiring more detailed information should consult the *Proceedings of the International Seed Testing Association* (1993) - *International Rules for Seed Testing*. Suffice it to say, here, that *all* samples should be taken with great care. Random, unbiased sampling, is of paramount importance, and all material sent to the station should be truly representative of the seed lot. It is important to bear in mind that, irrespective of how accurately the technical work is done in the station, the results can only reflect the quality of the seed on which the tests were carried out.

Generally, each sample size is derived from a nominal 1000-seed weight for each species which, on available evidence, is expected to be adequate for the majority of samples tested. Thus, for small-seeded species, e.g. *Agrostis capillaris*, at least 25 grams is required; for large-seeded types, e.g. *Avena sativa*, at least 1000 grams is necessary. The seed submitted to the station is called the *submitted sample*. In determining the latter, it is generally necessary to draw many individual or *primary samples*. When the primary samples are combined they form the *composite sample*. This is usually much larger than the amount required and, consequently, it must be reduced. The seed on which the actual tests are carried out is called the *working sample*. The working sample, derived from the submitted sample, must be obtained in accordance with the recommendations mentioned earlier. Again, the weight of the working sample is determined by the actual test required and species involved.

The Purity Analysis

The object of the purity analysis tests is to determine: (a) the percentage composition by weight of the sample being tested and, by inference, the composition of the seed lot; (b) the identity of the various species of seeds and inert particles constituting the sample. In accomplishing these objectives, the sample is separated into the following components: (i) *pure seed*, (ii) *other seeds*, and (iii) *inert matter*. Purity tests must be made as a single analysis on at least the weight needed for a working sample, or as duplicate tests on at least half that weight, each independently drawn. In the case of a single analysis, each component should be given to 1 decimal place. Percentages must be based on the sum of the weights of each component. If a duplicate analysis is made of 2 half-samples, the difference between the same components, from each working sample, must not exceed the *permissible tolerance*, as given by IRST. When the difference between the 2 half-samples falls within the permissible range, the average of the 2 figures determines the percentage of each component.

The definition of a pure seed, for purity purposes, may differ depending on the species under examination (see IRST). However, taking *Triticum aestivum* as an example, all mature, undamaged grains of this species fall into this category. In addition, undersized, shrivelled, immature and germinated seeds, provided they can be definitely identified as *Triticum aestivum*, are considered pure seed. Furthermore, pieces of seeds, resulting from breakage, that are more than one-half their original size are also included. Also, diseased seeds, except those altered by fungi to form sclerotia or smut balls, and by galls resulting from nematode infestation, are classified as pure seed.

Seeds are identified on the basis of their morphological characteristics. There are, however, some cases where it is not possible to discriminate with absolute certainty between the seeds of closely plants. For example, the seeds of different forms of *Beta vulgaris* subspecies *vulgaris* are very similar. So, also, are the seeds of different cultivated forms of *Brassica oleracea*. In cases such as above, the percentage of purity, as given in an ordinary report from a seed testing station, is not intended to imply that whole of what purports to be pure seed is necessarily true to name where subspecies or cultivars are concerned. Genuineness or trueness, in such cases, can only be determined by *proving*, *i.e.* by examining, at a suitable stage of development, the plants resulting from such seeds. This is the procedure followed in the production of *certified* seed.

Other seeds, in the present context, would include the seeds of all other plants found in the *Triticum aestivum* sample. With respect to classification as other seeds, the criteria set out for pure seed are applicable. The third component, inert matter, may be a miscellaneous collection of rejects from the other 3, together with remnants of plant material, soil particles, etc.

The Germination Test

Germination tests are carried out under standardised conditions by all member stations of ISTA. Optimum controlled conditions, in terms of appropriate substrata, heat, light, oxygen and moisture, for different species of seed, are set out in the *IRST*. The ultimate object of testing for germination is to gain information with respect to the field sowing value of the seed, and to provide results which can be used to compare the value of different seed lots. Testing under field conditions is normally unsatisfactory as the results cannot be repeated with reliability. Standardised laboratory methods, therefore, have been evolved where some or all of the external conditions are controlled to give the most regular, rapid and complete germination for the majority of samples of a particular kind of seed.

All germination tests are made with seeds from the pure seed component. Four hundred seeds, in replicates of 100, are taken at random and spaced uniformly and adequately apart on a moist substrate. Replicates may be divided into subreplicates of 50 or 25 seeds, depending on the size of the seed and the amount of space needed between them. The duration of the germination test depends on the species under investigation. In the case of agricultural and horticultural (other than flower, spice, herb and medicinal species) seeds, the first count may be made after 4 or 10 days, and a final count after 7 or 35 days. In the case of some tree species, the respective figures may be 35 and 70 days. Generally, 2 counts are made, 1 after a relatively short period and a second some days later. Thus, with respect to *Triticum aestivum*, the first count is made after 4 days and the final count after 8 days. The *percentage of germination* is calculated by counting all normally developed, healthy seedlings, per 100 pure seeds, at the end of the longer period. When 4 100-seed replicates of a test are within the maximum *tolerated range*, the average of the 4, to the nearest full figure, represents the percentage germination. If germination is slow or irregular, or in other ways abnormal and lacking in vigour, it indicates that the seed may be old, badly filled, not fully ripened, possibly damaged by seed dressing or careless storage, injured by attacks of fungi or bacteria. The *germination energy* figure, calculated on the basis of normal healthy seedlings developed after the shorter period, indicates any such defect or signs of deterioration.

Paper substrates: Filter paper, blotting paper or paper towelling may be used as a substrate. The seeds can be germinated on top (TP) of 1 or more layers of paper; this method is used with petri dishes, the 'Jacobsen Apparatus' or with germination cabinets or room-type germinators. In the germinator or room must be maintained at 90-95%. Seed may also be germinated between 2 layers of paper (BT) and kept in closed boxes, wrapped in plastic bags or placed directly on trays in a cabinet germinator. A third method involves placing the seeds in a pleated (PP), accordion-like paper strip. The pleated strips are kept in boxes or directly in a 'wet' cabinet.

Sand substrate: There are 2 methods for using sand as a substrate: (i) in sand (S) when seeds are sown on a uniform layer of moist sand and then covered to a depth of 1-2 cms with sand which is left loose, and (ii) top of sand (TS) when the seeds are just pressed into the surface. The sand must be free from both fine and large particles. Nearly all of the particles should pass through a sieve having perforations of 0.8 mm diameter and be retained on a sieve holes of 0.5 mm. The pH value should be within the range 6.0-7.5. When necessary, the sand must be washed and sterilised before use. The optimum amount of water added should be determined for the main kinds of seeds so that a measured quantity can always be used in routine testing. Cereals, except *Zea mays*, may be germinated in sand moistened to 50% of its water-holding capacity, whereas for large-seeded legumes and *Zea mays* it should be moistened to 60% of its water-holding capacity.

Soil substrate: Generally not recommended, but may be used in certain circumstances. Soil should be of good quality, such as garden loam, and sieved to remove large particles. It must be non-caking and sand should be added if the soil contains clay which would cause it to cake. As with the use of sand, the soil must be sterilised before use.

Moisture and Aeration: The substrate must contain sufficient moisture to meet the requirements for germination, but the moisture must not be excessive. In most cases the substrate should not be so wet that a film of water forms around the seeds. In order to reduce the need for the addition of water after sowing, the relative humidity of the air surrounding the seeds should be kept at 90-95% to prevent loss by evaporation. If distilled water is used aeration may be necessary to increase the oxygen content.

Temperature: Most seeds germinate within the range of 15-30 °C. Some may require a constant temperature of 20 °C. Others germinate best when subjected to alternating temperatures of 20 and 30 °C. In the latter situation, the lower temperature should be maintained for 16 hours and the higher for 8 hours. A gradual changeover, lasting 3 hours, is usually sufficient for non-dormant seeds, but a sharp changeover, lasting 1 hour, may be necessary for seeds which are likely to be dormant. In all cases, the temperature variation should not vary more than + or - 1 °C from that recommended for any given species.

Light: Many species require light for satisfactory germination. Cool white fluorescent lamps are recommended because they have a relatively low emission in the far red and a high spectral emission in the red region which promotes germination. Those seeds for which light is prescribed must be illuminated for at least 8 hours in every 24. Where the seeds are germinated at alternating temperatures, they must be illuminated during the higher period.

Treatments for promoting Germination: Species differ in their readiness to germinate immediately after harvesting. This may be due to physiological dormancy, the hardness of the seed coat or the presence of inhibitory substances. Germination can be promoted by 1, or a combination of, the following: (i) *dry storage* - usually for a short period time; (ii) *pre-chilling* - non-tree seeds may be kept at a temperature of between 5°C and 10°C for an initial period of up to 7 days; tree and shrub seeds are usually treated at a temperature of between 1°C and 5°C for a period, ranging with the species, from 2 weeks to 12 months; (iii) *pre-heating* - replicates for germination may be heated at a temperature not exceeding 30-35 °C with free air circulation for a period of up to 7 days; (iv) *light* - illumination, during at least 8 hours in every 24 hour cycle, and during the higher temperature period when seeds are germinated at alternating temperatures, can enhance germination; (v) *potassium nitrate* - instead of water, 0.2% of KNO_3 solution, is used to saturate the substrate at the beginning of the test; (vi) GA_3 - at strengths varying from 0.2-0.8%, is recommended for some cereal grains; (vii) *sealed envelopes* - some seeds, particularly *Trifolium* species, will show improved germination if retested in a sealed polyethylene envelope.

Seedling Evaluation

Normal seedlings: All normal healthy seedlings are counted at the termination of the germination test. The following is a summary of what constitutes normal and abnormal seedlings. Normal seedlings are those which show the potential for continued development into satisfactory plants when grown in good quality soil under favourable environmental conditions. To be classified as normal, a seedling must conform with 1 of the following: (i) *intact seedlings* - seedlings with all the essential structures well developed, complete, in proportion and health; (ii) *seedlings with slight defects* - seedlings showing certain slight defects of their essential structures, provided they show an otherwise satisfactory and balanced development comparable to that of intact seedlings of the same test; (iii) *seedlings with secondary infection* - seedlings which would have conformed with either of the above, but there being evidence to show that the infection by fungi or bacteria came from sources other than the parent seed.

Abnormal seedlings: Seedlings which do not fall into any of the above categories are considered abnormal. The following seedlings are not included in the final count: (i) *damaged seedlings* - seedlings with any of the essential structures missing or so badly and irreparably damaged that balanced development cannot be expected; (ii) *deformed or unbalanced seedlings* - seedlings with weak development or physiological disturbances or in which the essential structures are deformed or out of proportion; (iii) *decayed seedlings* - seedlings with any of their essential structures so diseased or decayed as a result of primary infection that normal development is prevented.

Hard seeds: Many leguminous seeds do not absorb water or become swollen during the course of the test. The latter remain firm and are referred to as *hard seeds*. Many of these are likely to produce plants under field conditions.

Excised Embryo Test for Viability

This is a comparatively quick method for determining the viability of certain kinds of seeds which germinate slowly or show dormancy under the standard tests. Excised embryos are incubated under prescribed conditions and examined after 5 to 14 days for signs of growth and differentiation. Non-viable embryos show signs of decay.

2, 3, 5-Triphenyltetrazolium Chloride or Bromide

Standard procedures for estimating germination capacity take time. Furthermore, it is not always possible to get an early germination appraisal immediately after harvesting due freshness of seed or dormancy. Biochemical tests enable one to make a rapid estimate of viability, particularly those showing dormancy and those, such as many tree seeds, which take months to germinate using the standard test.

Many workers have been concerned with the staining of seeds with a view to obtaining a rapid estimation of germination capacity (Farragher, 1964). Many different chemicals have been used, but the most successful is the topographical tetrazolium test using *2, 3, 5-triphenyltetrazolium chloride*. Barton (1961) states that the tetrazolium salt is an oxidation-reduction indicator, and that the development of the non-diffusible red colour, in a specific tissue, is an indication of the presence of active respiratory processes in which hydrogen ions are transferred to the tetrazolium chloride. In the presence of viable seed tissue, the colourless solution forms *triphenyl formazan* which is precipitated as a red stain. However, if the results are to be reported on an ISTA International Seed Analysis Certificate, the tests must be carried out under strict conditions. Each test must be carried out on at least 4 replicates of 100 seeds or fruits. All seeds have to be prepared prior to treatment, and the method of preparation varies with different species. An aqueous of 0.1-1.0% of tetrazolium chloride or bromide, within the pH range 6.5-7.5, is used. Mixing with a buffer solution may be necessary to achieve this range. The buffer is prepared by: (i) by dissolving 9.078 gr of KH_2PO_4 in 1 litre of water, and (ii) by dissolving 9.472 gr of Na_2HPO_4 in an other litre of water; the 2 solutions are mixed in a ratio of 1:3, respectively. A 1% solution is obtained by dissolving 1 gr tetrazolium salt in 100 ml of buffer. The seeds are then immersed in the solution, kept at 30 °C, for a period of time varying from 2-24 hours, depending on the species under examination. Subsequently, the seeds are examined for the presence of the red stain. In brief, those having a completely stain embryo are considered viable. However, there are many minor variations in the preparation of different tree seeds, and in the interpretation of what is, and what is not, viable.

Many seed merchants use this test, in an unofficial capacity, to get a quick appraisal of the germination percentage of cereal grains. The test is carried out with the aid of a *vitascope* which controls heat and immersion time. Cereal grains are split longitudinally, and one-half of each grain is discarded. Two hundred halves, representing 200 seeds, are placed on special trays and inserted into the central chamber of the vitascope. In the case of *Hordeum sativum*, the embryos are examined after 20 minutes for the presence or absence of a red colour. Those with completely stained embryos, or with only a small portion of the plumule or radicle remaining unstained, are considered viable.

WEED SEEDLINGS

General: Weed control in cultivated crops is important for many obvious reasons. Maximum control can be effected by the judicious use of herbicides at the pre-emergence or seedling stage. A detailed knowledge of weeds in the seedling stage, and the ability to recognise them accurately is, therefore, a prerequisite to success. It is in this growth-stage that weeds are most easily controlled. After identification, it is then a matter of selecting a herbicide that is effective against the range of species present.

The 'key' presented herewith is intended as a simple guide to the identification of the seedlings (Figs 26.00-27.00) of all the important weeds that are likely to be found in cultivated ground. No attempt is made to classify the seedlings in accordance with botanical taxonomy because species which are closely related can have seedlings differing markedly from each other in form, whereas seedlings which show a close resemblance may not be closely related. The 'key' is therefore compiled without any reference to plant families or genera and is based entirely on the morphological characteristics of the cotyledons, first or second true leaf and the hypocotyl and epicotyl (Fig. 26.01). Identification at this stage is more difficult than that of the mature plant, as the number of criteria at one's disposal is more limited.

Cotyledons are mostly present, though hypogeal germination is shown by some leguminous plants, particularly by *Vicia sativa* ssp. *nigra* (Fig. 26.02) and *Lathyrus pratensis* (Fig. 26.03). In the latter instances, identification is based on leaflet characteristics of the first and second leaf. Cotyledons show a wide range of reasonably consistent variation in length, width, pubescence and shape, though the latter may show a slight overlap between individuals of different species. Thus, the following invariably display long, narrow, sometimes near bristle-like cotyledons: *Coronopus didymus* (Fig. 27.19), *C. squamatus* (Fig. 27.22), *Fumaria muralis*, *F. officinalis* (Fig. 27.23), *Plantago lanceolata* (Fig. 26.17), *Polygonum aviculare* (Fig. 26.18) and *Spergula arvensis* (Fig. 27.18). Many members of the Brassicaceae (Figs 26.33-36), particularly the cultivated and important weed plants, have characteristically large obcordate cotyledons which are mostly wider than long, and deeply emarginate apically.

Other seedlings with shallowly emarginate cotyledons include *Convolvulus arvensis* (Fig. 27.27), *Galium aparine* (Fig. 26.06), *Lapsanna communis* (Fig. 26.32), *Sherardia arvensis* (Fig. 26.05), *Urtica dioica* (Fig. 26.29) and *U. urens* (Fig. 26.30). *Cardamine hirsuta* (Fig. 26.19) is one of the few seedlings with rounded-cordate cotyledons, while those of *Epilobium hirsutum* (Fig. 27.33), *E. montanum* (Fig. 27.39) and *Myosotis arvensis* (Fig. 26.16) are rounded. *Anagallis arvensis* (Fig. 27.39) has bluntly triangular cotyledons. Those of the 2 *Epilobium* species have small peripheral knobs. Seedlings of *Veronica agrestis* (Fig. 27.07), *V. hederifolia* (Fig. 27.04), *V. polita* (Fig. 27.05) and *V. persica* (Fig. 27.06) have bluntly rhomboid cotyledons. Pubescence appears to be reasonably consistent in *Geranium dissectum* (Fig. 26.10), *G. molle* (Fig. 26.11), *Myosotis arvensis*, *Stachys arvensis* (Fig. 27.02), *S. palustris* (Fig. 27.03), *Urtica dioica* and *U. urens*. The 2 species of *Geranium* are further characterised by having asymmetrical cotyledons. *Galeopsis tetrahit* (Fig. 26.27) and *Lamium purpureum* (Fig. 26.28) are the only species with auriculate cotyledons.

The presence or absence of cotyledon stalks, their length relative to the blade portion, and the manner by which the blade portion is tapered towards the stalk, are additional identification criteria. Thus, the 3 principal species of *Cirsium* - *arvense*, *palustre* and *vulgare* can be separated by differences in their cotyledon bases. Cotyledons are clearly stalked in species of *Cardamine flexuosa*, *C. hirsuta*, *Geranium dissectum*, *G. molle* and *Stellaria media* (Fig. 26.20) seedlings.

The usefulness of epicotyls and hypocotyls is limited in seedling identification. They are only important in a few species. They are usually clearly evident in *Euphorbia helioscopia* (Fig. 27.29), *E. peplus* (Fig. 27.42), *Polygonum aviculare* (Fig. 26.18) and *P. persicaria* (Fig. 26.22) where they often have a characteristic colour. Most seedlings have short insignificant hypocotyls and epicotyls.

Variation in the first and second leaf are equally important in identification. *Galium aparine*, *Sherardia arvensis* and *Spergula arvensis* seedlings have whorled leaves. *Galeopsis tetrahit*, *Geranium dissec-*

tum, G. molle, Lamium purpureum, Mercurialis annua (Fig. 26.37), *Stachys arvensis, S. palustris* and *Stellaria media* have the leaves arranged in opposite pairs. Most have the alternate arrangement. Often, due to a very short epicotyl, some of the latter may show what appears to the the opposite arrangement. They can, never the less, be distinguished from the true situation by the unequal size of their leaves; the leaves of seedlings with the opposite arrangement are mostly equal in size.

The outline of the leaves is important. Many, even those with subsequent compound blades, are entire. *Spergula arvensis* has bristle-like leaves; most have relatively broad blades with more or less characteristic shapes. Some, particularly those of *Achillea millefolium* (Fig. 27.11), *Daucus carota* (Fig. 27.09), *Geranium dissectum, G. molle, Heracleum sphondylium* (Fig. 27.10), *Potentilla anserina* (Fig. 26.40), *P. reptans* (Fig. 26.39), *Ranunculus acris* (Fig. 26.38) and *R. repens* (Fig. 26.43) have either pinnate or palmate lobes. Others, such as those of *Galeopsis tetrahit, Lamium purpureum, Mercurialis annua, Stachys arvensis, Urtica dioica* and *U. urens* have toothed margins. Many of the other variations already outlined for the cotyledons are also applicable to the first and second leaf.

One area of difference relates to the type of hairs present. *Capsella bursa-pastoris* (Fig. 26.15) leaves have at least some hairs of the stellate type. *Myosotis arvensis* (Fig. 26.16) seedlings have bulbous hairs, *Atriplex patula* (Fig. 26.13), *A. prostrata* (Fig. 26.14) and *Chenopodium album* (Fig. 26.12) have a 'mealy' appearance due to the presence of papillae, while those of *Urtica dioica* (Fig. 26.29) and *U. urens* (Fig. 26.30) have glandular stinging hairs.

The following 'key' to seedling identification is based on the variation outlined above. All seedlings likely to occur in cultivated ground are illustrated.

KEY TO WEED SEEDLINGS

1.   Cotyledons *absent* 2
      Cotyledons *present* 3

2.   Leaflets 'parallel-sided' in general outline (Fig. 26.02). First true leaflets long, narrow, tapered to fine points; later leaflets much shorter and broader, noticeably mucronate apically, generally glabrous or very nearly so. *Frequent seedling of cultivated ground, etc.* **Vicia sativa** ssp. **nigra**
      Leaflets *without* 'parallel margins' (Fig.26.03). First true leaf with only *one* pair of leaflets; the latter narrowly ovate-lanceolate in general outline, abruptly tapered basally, more gradually tapered apically, glabrous throughout. *Frequent in newly ploughed pastures.* **Lathyrus pratensis**

3.   Leaves *opposite* or *alternate* 6
      Leaves *whorled* 4

4.   Cotyledons and leaves *clearly different* in shape 5
      Cotyledons and all true leaves *very similar* in general appearance (Fig. 26.04); the latter long, narrow, somewhat bristle-like in general outline, initially clustered, glandular, somewhat fleshy. *A very common seedling of cultivated acidic grounds and waste places, etc.* **Spergula arvensis**

5.   Cotyledons *round-oval* (Fig. 26.05); the latter almost sessile, emarginate apically. The true leaves ovate to lanceolate in outline, bristly but not prickly, stiff-pointed apically. Stem guadrangular, pubescent. *Common annual of cultivated ground, lawns and waste places.* **Sherardia arvensis**
      Cotyledons *oblong* (Fig. 26.06); the latter emarginate, abruptly tapered. True leaves somewhat rounded apically, with spiny tips, bristly hairy adaxially, rough with reflexed prickles abaxially and on margins. *Abundant annual; cultivated ground, waste places, etc.* **Galium aparine**

6.   Leaves *without* prickly margins 9
      Leaves *with prickly margins;* cotyledons somewhat *fleshy* 7

7. Cotyledons *clearly tapered* basally 8
   Cotyledons *not* clearly tapered (Fig. 26.07); the latter broadly oblong in general outline, blunt apically, somewhat thick and fleshy, glabrous and practically stalkless. Leaves narrowly elliptical, not clearly tapered apically or basally, margins with *characteristically* long spines. *Abundant biennial of wet places; rarely a troublesome weed of cultivated ground.* **Cirsium palustre**

8. Cotyledons *5.0-6.0 mm* (Fig. 26.08); the latter orbicular-oval in general outline, very gradually tapered basally, very blunt apically, somewhat thick and fleshy, glabrous. Leaves obovate to oblanceolate in general outline, spines very well developed but widely spaced and few in number. *A very common perennial of meadows, pastures, cultivated ground, etc.* **Cirsium arvense**
   Cotyledons *10-15 mm* (Fig. 26.09); the latter noticeably abruptly tapered basally, very blunt apically, somewhat thick and fleshy, glabrous. First leaf obovate to oblanceolate in general outline, with many short, crowded spines; second leaf dull dark green, with dense whitish hairs. A
   common biennial of pastures, meadows, cultivated ground, waste places, etc. **Cirsium vulgare**

9. Cotyledons *symmetrical* 11
   Cotyledons clearly *asymmetrical* 10

10. Second and later leaves *deeply divided*, at least *half-way* to base (Fig. 26.10); the latter palmatifid,
    pubescent. Cotyledons somewhat 2-lobed, on long slender pubescent stalks; lobes of unequal
    size; pubescent. *Common annual; dry pastures; roadsides; waste places.* **Geranium dissectum**
    Second and subsequent leaves *not divided half-way* to the base (Fig. 26.11); the latter palmatifid,
    pubescent. Cotyledons somewhat 2-lobed, on long slender pubescent stalks; lobes of unequal
    size; pubescent. *Common annual; dry pastures, roadsides, waste places, etc.* **Geranium molle**

11. Hairs *completely absent* 61
    Hairs, either *distinct* or *difficult* to see, *present;* or seedling 'mealy' 12

12. Cotyledons and leaves *not* 'mealy', *i.e.* papillae *absent* 15
    Cotyledons and leaves 'mealy', *i.e.* papillae *present* 13

13. Cotyledons *not* reddish-violet abaxially 14
    Cotyledons *reddish-violet abaxially*, fleshy, long-oval, and 'mealy' white (Fig. 26.12). The leaves
    variable, the first oblong-ovate, entire or finely toothed; all often reddish-white, and 'mealy'
    white abaxially. *Abundant annual of cultivated ground and waste places.* **Chenopodium album**

14. Leaves, second pair, *tapered gradually* to the petiole (Fig. 26.13); the latter long-cuneate, mostly
    rhomboid to lanceolate, hastate to almost sagittate, entire or sparsely toothed; first pair elliptic,
    entire. Cotyledons club-shaped. *A common annual of cultivated ground, etc.* **Atriplex patula**
    Leaves, second pair, *truncate* or *shortly* cuneate basally (Fig. 26.14); the latter mostly triangular-
    hastate, irregularly toothed or nearly entire; first pair mostly elliptic, entire; all 'mealy'. Cotyle-
    dons club-shaped. *A frequent annual of cultivated ground and waste places.* **Atriplex prostrata**

15. Laminae of first true leaves *lobed, toothed, distantly sinuate,* or *angular* 27
    Laminae of first true leaves *entire* 16

16. Hairs *not* stellate 17
    Hairs, at least some, *stellate* (Fig. 26.15). The cotyledons small, short-stalked, narrowly orbicular.
    Leaves variable, orbicular to lanceolate, entire to sinuate-toothed, later deeply pinnatifid, most-
    ly dark green. *Common annual of cultivated ground and waste places.* **Capsella bursa-pastoris**

17. Hairs *not* bulbous 18
    Hairs, at least some, *bulbous* (Fig. 26.16). Cotyledons small, short-stalked, and rounded or near-
    ly so. First leaves orbicular-obovate, clearly stalked; later leaves lanceolate, tapered to the pet-
    iole; all entire. *Very common annual of cultivated ground and waste places.* **Myosotis arvensis**

18. Cotyledons *relatively broad* 20
    Cotyledons *extremely narrow* and *bristle-like* 19

19. Cotyledons *grooved* adaxially (Fig. 26.17); the latter long, narrow, and mostly dull blue-green. First leaves broadly lanceolate, 'parallel-veined', lanceolate, not clearly stalked, entire, sparsely pubescent. *Very common perennial; pasture, meadows, waste places, etc.* **Plantago lanceolata**
    Cotyledons *ungrooved* (Fig. 26.18); the latter long, narrow, fleshy. First leaves characteristically broad, elliptical-lanceolate, scarcely hairy, entire. Stipule minute but evident. Hypocotyl long, crimson. *Very common annual; cultivated ground and waste places, etc.* **Polygonum aviculare**

20. First leaves *neither* reniform *nor* cordate 21
    First leaves *reniform to cordate* in outline (Fig. 26.19); the latter entire, short-stalked; later leaves pinnate, frequently pubescent, upper leaflet reniform. Cotyledons small, nearly round, emarginate, long-stalked. *Common annual of cultivated ground and waste places.* **Cardamine hirsuta**

21. Cotyledons and leaves *blunt apically* 23
    Cotyledons and leaves *pointed apically* 22

22. Colour *bright* or *light green* (Fig. 26.20). Cotyledons broadly ovate, tapered to an acute point apically, long-stalked, glabrous. Leaves rather similar, tapered apically, entire, and often hairy in their axils. *A very common annual of cultivated ground, waste places, etc.* **Stellaria media**
    Colour *dull dark green* (Fig. 26.26). Cotyledons mostly ovate, tapered apically to an acute point, short-stalked, pubescent. First leaf broadly ovate, tapered to an acute apex, short-stalked, entire, and pubescent. *Frequent annual of cultivated ground and waste places.* **Solanum nigrum**

23. Sheathing stipule *absent* 24
    Sheathing stipule *evident* (Fig. 26.21). The first leaves rather broad, elliptical-lanceolate, tapered, blunt, entire, scarcely pubescent. Cotyledons ovate-lanceolate, blunt, and tapered. Hypocotyl bright scarlet. *Very common annual; cultivated ground; waste places.* **Polygonum persicaria**

24. Blade of first leaf *abruptly tapered* 25
    Blade of first leaf *gradually* tapered (Fig. 26.22); the latter stalked, ovate, blunt, and entire; later leaves sinuate, scarcely pubescent. Cotyledons narrowly lanceolate. Hypocotyl often purple. *Very common perennial of cultivated ground, old pastures, waste places, etc.* **Plantago major**

25. Leaves *alternate* and clearly *lobed* 26
    Leaves *opposite* (Fig. 26.23); the latter broadly oblong in general outline, short-stalked or nearly sessile, sharply constricted basally, blunt apically, often tinged with purple, usually pubescent, entire. Cotyledons ovate or somewhat rounded, abruptly constricted basally, blunt apically, short-stalked. *Frequent perennial of cultivated ground and waste places, etc.* **Mentha arvensis**

26. Blades of the fourth and fifth leaf *pinnately* lobed (Fig. 26.24); the latter long-stalked, irregularly lobed; earlier leaves entire, toothed or lobed, stalked; all pubescent. Cotyledons narrowly lanceolate, stalked, blunt. *Common annual; cultivated ground and waste places.* **Papaver rhoeas**
    Blades of the fourth and fifth leaf *palmately* lobed (Fig. 26.25); the latter long-stalked, irregularly lobed; earlier leaves entire, toothed or lobed, stalked; all pubescent. Cotyledons mostly oblanceolate, tapered, blunt. *Common annual; cultivated ground and waste places.* **Papaver dubium**

27. Cotyledons *lacking* backwardly projecting basal lobes 29
    Cotyledons *with minute* backwardly projecting basal lobes 28

28. Cotyledons *shallowly emarginate* apically (Fig. 26.27); the latter broadly obovate, short-stalked and with very small backwardly projecting basal, pointed lobes. The first leaves rather short-stalked, broadly ovate, rounded basally, blunt apically, crenately toothed, pubescent adaxially and abaxially. *Very common annual of cultivated ground and waste places.* **Galeopsis tetrahit**

Cotyledons *mucronate* (Fig. 26.28); the latter mostly orbicular, short-stalked, minutely auriculate. First leaves short-stalked, triangular-ovate, cordate basally, blunt, crenate, pubescent, particularly adaxially. *A common annual of cultivated ground and waste places.* **Lamium purpureum**

29. Cotyledons *not* emarginate 38
    Cotyledons *emarginate* 30

30. Cotyledons *other than* oval, oblong or nearly rounded 34
    Cotyledons *oval, oblong* or *nearly rounded* 31

31. Cotyledons *glabrous* 33
    Cotyledons with *stinging* hairs 32

32. Apex of the first pair of true leaves *rounded* (Fig. 26.29); the latter triangular-ovate in outline, margins shallowly serrate, short-stalked, abruptly constricted basally, pubescent; later leaves more distinctly serrate. Cotyledons very small, emarginate apically, short-stalked, mostly obovate, and pubescent. *Very common annual; cultivated ground and waste places.* **Urtica dioica**
    Apex of the first pair of true leaves *acute* (Fig. 26.30); the latter triangular-ovate in outline, margins shallowly serrate, short-stalked, abruptly constricted basally, pubescent; later leaves more distinctly serrate. Cotyledons very small, emarginate apically, short-stalked, mostly obovate, and pubescent. *Very common annual of cultivated ground and waste places, etc.* **Urtica urens**

33. Stipule traces *evident* (Fig. 26.31). First leaves mostly orbicular, scarcely toothed, very blunt apically, cut off square basally, scarcely pubescent; later leaves more toothed. Cotyledons dark green, mostly oblong, emarginate apically, short-stalked. Two species are indistinguishable from each other. *Frequent annuals; dry sandy cultivated ground.* **Viola tricolor** or **V. arvensis**
    Stipule traces *absent* (Fig. 26.32). First true leaves orbicular to broadly ovate, slightly toothed or angular, abruptly stalked, pubescent, and yellowish-green in colour; later leaves more clearly toothed or angular; all pubescent. Cotyledons broadly oval, narrowed into a long stalk, emarginate apically. *Common annual of cultivated ground and waste places.* **Lapsanna communis**

34. Leaves *mostly toothed* 35
    Leaves *deeply pinnately lobed* (Fig.26.33); the former broadly lanceolate in general outline, irregularly lobed, of a dull green colour, clearly stalked and usually covered throughout with short bristly hairs. Cotyledons rather large, blade wider than long, reniform or obovate, widely emarginate apically, stalked. *A frequent annual of cultivated ground and waste places.* **Sinapis alba**

35. Hairs *few, scattered* and *difficult to see* 37
    Hairs *plentiful* and *clearly visible* 36

36. Blades of a *glaucous* colour (Fig. 26.34); the third and later blades variable, often broadly oval in general outline, somewhat lobed, blunt apically, clearly stalked, and covered with short bristly hairs. Cotyledons large, widely emarginate apically, abruptly tapered basally, reniform or ovate, clearly stalked. *A common annual of cultivated ground and waste places.* **Brassica rapa**
    Blades *dark green* or *yellowish-green* (Fig. 26.35); the latter very variable, often broadly obovate, toothed, very blunt; hairs short, bristly. Cotyledons large, reniform or obovate, widely emarginate apically, stalked. *Common annual; cultivated ground; waste places.* **Sinapis arvensis**

37. Third and subsequent blades variable (Fig. 26.34), often broadly oval in general outline, somewhat lobed, of a glaucous colour, blunt apically, clearly stalked, sparingly covered with short bristly hairs. Cotyledons large, clearly stalked, reniform or ovate, widely emarginate apically, abruptly tapered basally. *Common annual; cultivated ground and waste places.* **Brassica rapa**
    Third and subsequent blades *toothed* (Fig. 26.36); the former variable, mostly broadly obovate, toothed, very blunt apically, covered with short bristly hairs. The cotyledons large, reniform or obovate, widely emarginate apically, stalked. *A common cultivated annual.* **Brassica oleracea**

38.   Cotyledons *lacking* yellowish-coloured anastomising veins 39
    Cotyledons with *characteristically anastomising yellowish-coloured* veins (Fig. 26.37); the latter very large, mostly broadly oval, blunt apically, rounded basally, and short-stalked. The leaves mostly broadly ovate, opposite, shallowly and distantly toothed, blunt apically, rounded basally, scarcely pubescent. *Frequent annual; cultivated ground; waste places.* **Mercurialis annua**

39.   First true leaves *not* palmately lobed 45
    First true leaves *palmately* lobed 40

40.   Cotyledons clearly *longer than* broad 42
    Cotyledons nearly *as broad as long* 41

41.   Base of first blade *somewhat cordate* (Fig. 26.38); the latter mostly triangular-ovate in general outline, 3-5 lobed or toothed; second and subsequent leaves more lobed and toothed; all pubescent, petiolate. Cotyledons relatively large, mostly broadly oval, blunt apically, rounded basally, stalked. *Abundant perennial of meadows, pastures, waste places, etc.* **Ranunculus acris**
    Base of first blade cut off *somewhat squarely,* or *abruptly tapered* (Fig. 26.39); the latter mostly ovate shallowly lobed or toothed, blunt apically, petiolate; the subsequent leaves more deeply toothed; pubescent. Cotyledons broadly oval, relatively large, blunt apically, abruptly tapered basally, stalked. *An abundant perennial; cultivated ground, waste places.* **Potentilla reptans**

42.   Seedling *green, greenish, yellowish* or *greyish-green* 43
    Seedling *silvery grey* (Fig. 26.40). First blade mostly ovate in general outline, shallowly lobed or toothed, blunt apically, clearly petiolate; subsequent leaves more deeply toothed; all clearly pubescent. Cotyledons mostly narrowly oval, relatively large, blunt apically, abruptly tapered basally, stalked. *Abundant perennial; cultivated ground and waste places.* **Potentilla anserina**

43.   Seedling *green* or *yellowish-green;* cotyledons relatively *short-stalked* 44
    Seedling *greyish-green* (Fig. 26.41). First blade orbicular to reniform, irregularly toothed or shallowly lobed, bristly; other blades more clearly lobed or pinnate. Cotyledons narrowly lanceolate. *Common biennial of old pastures, roadsides, waste places, etc.* **Heracleum sphondylium**

44.   Cotyledons *somewhat tapered* apically (Fig. 26.42); the latter gradually tapered into a short stalk. Blades variable; the first triangular-ovate, shallowly toothed, abruptly rounded basally; subsequent blades often 3-lobed, lobes toothed; all clearly petiolate, scarcely pubescent. *A very frequent plant of dry pastures, roadsides, sand-hills, gravelly places, etc.* **Ranunculus bulbosus**
    Cotyledons *blunt* apically (Fig. 26.43); the latter abruptly tapered into a short stalk. Blades variable; first mostly triangular-ovate in general outline, shallowly toothed, abruptly rounded basally; subsequent blades usually 3-lobed, lobes toothed; all clearly petiolate, scarcely pubescent. *An abundant perennial of damp cultivated ground, waste places, etc.* **Ranunculus repens**

45.   Blades *alternate,* or opposite *but stems terete* 48
    Blades *opposite* and stems *quadrangular* 46

46.   Cotyledons *orbicular* 47
    Cotyledons *somewhat triangular* in general outline (Fig. 27.01); the latter broadly ovate, blunt short-stalked. Leaves broadly ovate, abruptly constricted basally, blunt apically, opposite, and scarcely pubescent. *Frequent perennial; cultivated ground and waste places.* **Mentha arvensis**

47.   Blades *broadly cordate* to *ovate* (Fig. 27.02); the latter blunt apically, abruptly constricted basally, crenate, pubescent on both surfaces. The cotyledons often orbicular, often broadest basally, rounded at apex, stalked. *Frequent annual; cultivated ground; waste places.* **Stachys arvensis**
    Blades *oval* to *oval-lanceolate* (Fig. 27.03); the latter blunt apically, abruptly constricted basally, crenate, both surfaces pubescent. Cotyledons mostly orbicular, often broadest basally, stalked, apex rounded. *A common perennial of cultivated ground and wet places, etc.* **Stachys palustris**

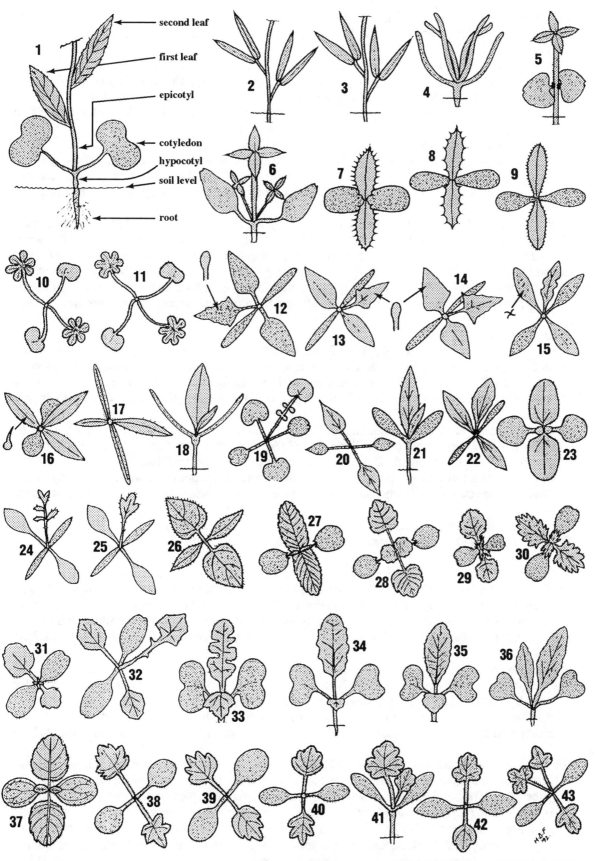

second leaf
first leaf
epicotyl
cotyledon
hypocotyl
soil level
root

Fig. 26.00   N.B. DRAWINGS NOT TO SCALE.

161

162

Cotyledons *abruptly* tapered into a stalk 49

Cotyledon stalk *as long* as or *longer than* the blade (Fig. 27.04); the latter large, broadly ovate, short pointed. Leaves orbicular-triangular, with 1-3 basal teeth on margins, terminal lobe entire, pubescent. *Common annual of cultivated ground and waste places.* **Veronica hederifolia**

First leaves with *about two teeth* (Fig. 27.05); the former short-stalked, frequently broadly ovate, abruptly constricted basally, blunt, scarcely pubescent. Cotyledons small, oval, blunt, abruptly constricted basally. *A common annual of cultivated ground and waste places.* **Veronica polita**

51.   Second blade *shallowly cordate* basally (Fig. 27.06); the latter short-stalked, broadly ovate, blunt, crenate, pubescent. Cotyledons broadly triangular or rhomboid, abruptly constricted basally, blunt apically, stalked. *Common annual; cultivated ground and waste places.* **Veronica persica**
Second blades *cordate basally*, (Fig. 27.07); the latter short-stalked, broadly ovate, blunt, crenate, pubescent. Cotyledons broadly triangular or rhomboid, abruptly constricted basally, blunt apically, and stalked. *A common annual of cultivated ground and waste places.* **Veronica agrestis**

Blades *woolly-pubescent* abaxially (Fig.27.08); the latter orbicular to broadly ovate in general outline, abruptly narrowed into a relatively short stalk, later cordate and long-stalked; the margins with a few fine teeth. Cotyledons narrowly oval to broadly lanceolate, moderately tapered into a short stalk. *Common perennial of cultivated ground and waste places, etc.* **Tussilago farfara**

Cotyledons *long, narrow* and *bristle-like* (Fig. 27.09); the latter very narrowly oblanceolate, several times longer than broad. Blades compound, multi-pinnate, petiolate, of a deep green colour, pubescent throughout; pinnae very narrowly lanceolate, pointed apically. Seedling with a characteristic carrot smell when crushed. *Common wild or cultivated biennial.* **Daucus carota**

Seedling *dark green, yellowish-green,* or *greyish-green* 55

Seedling *greyish-green* (Fig. 27.10). First blade orbicular to reniform in general outline, margins irregularly and broadly toothed or shallowly lobed; subsequent blades more clearly lobed or pinnate; all clearly petiolate, usually bristly pubescent. Cotyledons long-stalked, and lanceolate. *Common biennial of old pastures, roadsides, waste places, etc.* **Heracleum sphondylium**

Cotyledons *unstalked* (Fig. 27.11); the latter mostly oval in general outline, rather small, rounded apically, imperceptibly tapered basally. First leaves compound, pubescent, narrow, with several lateral lobes; lobes sharply pointed, narrow or linear; later leaves more divided. *A common perennial; pastures, roadsides, cultivated ground, waste places, etc.* **Achillea millefolium**

57.   Blade margins *wavy* and irregular (Fig. 27.12); first blades more or less orbicular in general outline, somewhat toothed; the later blades more deeply lobed or narrowly pinnatifid; all scarcely pubescent. The cotyledons narrowly oval, rounded or blunt apically, gradually tapered basally into a stalk. *Common annual of cultivated ground and waste places.* **Sisymbrium officinale**
Blade margins *not* wavy (Fig. 27.13). First true leaves orbicular to broadly ovate, often toothed or angular, abruptly stalked, pubescent, and yellowish-green in colour; later leaves more clearly toothed or angular; all pubescent. Cotyledons broadly oval, narrowed into a long stalk, blunt apically. *A very common annual of cultivated ground and waste places.* **Lapsanna communis**

58. First blades with *basally directed, poorly defined* small teeth 59
First blades with *apically directed* well defined teeth (Fig. 27.14); the former obovate to oblanceolate, tapered into a broad-winged stalk, blunt apically; subsequent leaves often more deeply toothed or lobed; all with a few whitish hairs. Cotyledons small, narrowly oblanceolate and tapered basally. *An abundant annual of cultivated ground and waste places.* **Senecio vulgaris**

59. Blades *tapered* basally 60
Blades *cut-away squarely* at junction with petiole (Fig. 27.15); the former mostly ovate in general outline, thin, shiny, light bluish-green, and abruptly constricted into a long stalk; subsequent blades sinuate-pinnatifid; glabrescent. Cotyledons orbicular, stalked, the apex blunt or sometimes incurved. *Ubiquitous annual of cultivated ground and waste places.* **Sonchus oleraceus**

60. Blades *rather abruptly* tapered to the petiole (Fig. 27.16); the former mostly broadly lanceolate in outline, thin, shiny, light bluish-green, and abruptly constricted into a long stalk; subsequent blades sinuate-pinnatifid; glabrescent. Cotyledons broadly oval, stalked, apex blunt or sometimes incurved. *An ubiquitous annual of cultivated ground and waste places.* **Sonchus asper**
Blades *gradually* tapered to the petiole (Fig. 27.17); the former frequently broadly oval in general outline, thin, shiny, light bluish-green, and abruptly constricted into a long stalk; subsequent blades sinuate-pinnatifid; glabrescent. Cotyledons broadly oval, stalked, apex blunt or sometimes incurved. *An ubiquitous annual of cultivated ground and waste places.* **Sonchus arvensis**

61. Cotyledons *relatively broad* 69
Cotyledons *very narrow, several times longer* than broad 62

62. Blades *relatively broad* 63
Blades, all, and cotyledons *very similar* in general appearance (Fig. 27.18); the former long, narrow, somewhat bristle-like in general outline, initially clustered, glandular, somewhat fleshy. *A very common seedling of cultivated acidic grounds and waste places, etc.* **Spergula arvensis**

63. First leaves *toothed* or *lobed* 66
First leaves *entire* 64

64. Cotyledons *mostly* 'parallel-sided' 65
Cotyledons *club-shaped* (Fig. 27.19); the latter widest and blunt apically, gradually tapered to the base. First leaves mostly ovate, stalked, entire; later leaves clearly lobed; lobes mostly oval; others bipinnate. *Common annual of cultivated ground and waste places.* **Coronopus didymus**

65. Cotyledons shallowly *grooved* adaxially (Fig. 27.20); the former long, narrow, many times longer than broad, bristle-like and mostly dull blue-green. The first leaves narrowly lanceolate in general outline, 'parallel-veined', not clearly stalked, somewhat blunt, entire, sparsely pubescent or glabrous. *A common perennial of pasture, meadows, waste places, etc.* **Plantago lanceolata**
Cotyledons *ungrooved* adaxially (Fig. 27.21); the latter long, narrow, several times longer than broad, bristle-like, fleshy. The first leaves characteristically broad, elliptical-lanceolate, mostly glabrous, rarely with a few short hairs, entire. Stipule minute but usually evident. Hypocotyl long, crimson. *A common annual of cultivated ground and waste places.* **Polygonum aviculare**

66. First leaves *palmately* lobed 67
First leaves *pinnately* lobed (Fig. 27.22); the latter long, narrow, blunt, with a few lobes; subsequent leaves more clearly pinnately lobed. Cotyledons long, narrow, mostly linear-lanceolate in outline, entire. *A frequent annual of tilled ground and waste places.* **Coronopus squamatus**

67. Leaves divided to *about half-way* 68
Leaves *very deeply* divided (Fig. 27.23); the latter long-stalked, pinnately or palmately lobed, and light bluish-green. Cotyledons very long, linear-lanceolate, pointed. Hypocotyl long, pinkish. *Common annuals of cultivated ground and waste places.* **Fumaria muralis** or **F. officinalis**

164

68.	Cotyledons *linear-lanceolate* (Fig.27.24); the latter pointed, tapered. First blades mostly palmately lobed, triangular-ovate, long-stalked; the lobes with 2-3 deep indentations; later blades more deeply divided. *A frequent annual of cultivated ground and waste places.* **Aethusa cynapium**
	Cotyledons *narrowly oblanceolate* (Fig. 27.25); the latter long-stalked, blunt. First blades lobed and toothed, triangular-ovate, somewhat cordate, long-stalked; subsequent blades more deeply divided. *Very frequent annual of cultivated ground and waste places.* **Aegopodium podagraria**

69.	Cotyledons *blunt, rounded* or *pointed* 72
	Cotyledons *emarginate* 70

70.	First leaves *not* reniform 71
	First leaves *generally reniform* in outline (Fig. 27.26); the latter entire, stalked; later leaves pinnate, frequently glabrous; the upper leaflet reniform. Cotyledons small, nearly round, emarginate, long-stalked. *Common annual of cultivated ground and waste places.* **Cardamine hirsuta**

71.	Leaves all *entire* (Fig. 27.27); the latter triangular-ovate; first blades cordate basally, blunt; later blades sagittate, dark green. Cotyledons broadly oval, emarginate, short-stalked, abruptly tapered basally. *Abundant perennial; cultivated ground and waste places.* **Convolvulus arvensis**
	Leaves, all, distinctly *toothed* (Fig. 27.28); the latter variable, often broadly obovate, toothed, and very blunt apically; hairs occasionally completely absent. Cotyledons large, reniform or obovate, stalked; apices widely emarginate. *Common cultivated annual plant.* **Brassica oleracea**

72.	First leaves *entire* 81
	First leaves *toothed, lobed* or with minute *knobs* 73

73.	Leaves *not* emarginate 74
	Leaves *shallowly emarginate* apically (Fig.27.29); the latter obcordate in general outline, gradually tapered apically, usually finely toothed apically, usually glabrous or with a few short hairs, light green. Cotyledons broadly oval, short-stalked, blunt, somewhat rounded. Epicotyl clearly evident. *A very common annual of cultivated ground and waste places.* **Euphorbia helioscopia**

74.	First leaves *relatively broad* 77
	First leaves *long and narrow* 75

75.	Hypocotyl *short*, seedling dark green 76
	Hypocotyl *long* (Fig. 27.30). The first blades narrowly lanceolate, usually few-lobed; subsequent blades with forward directed lobes; glaucous. Cotyledons oblong, rounded apically, tapered basally. *A frequent annual of cultivated ground and waste places.* **Chrysanthemum segetum**

76.	Terminal lobe of first leaf *relatively broad* (Fig. 27.31); the latter pinnately lobed, stalked; lobes entire; later blades more divided with toothed lobes. Cotyledons small, narrowly oval, blunt apically, tapered basally. *A common biennial of waste places.* **Tripleurospermum inodorum**
	Terminal lobe of first leaf *relatively narrow* (Fig. 27.32); the latter narrowly oblanceolate, entire; later blades pinnatifid; lobes toothed. Cotyledons narrowly obovate, rounded apically, tapered basally. *A common annual of cultivated ground and waste places.* **Matricaria matricarioides**

77.	First blades *neither* round *nor* nearly so 78
	First blades *nearly round* (Fig. 27.33); the latter rounded apically and basally, appearing entire but with several small marginal knobs. Cotyledons triangular-ovate, blunt apically, abruptly narrowed basally, short-stalked. *Frequent perennial of damp or wet soils.* **Epilobium hirsutum**

78.	Seedling *bluish-green* or *light-green*, and with *backwardly directed* teeth 79
	Seedling *dark green* (Fig. 27.34). The first blades mostly 5-angled, abruptly tapered; later blades obovate, entire apically, toothed laterally. Cotyledons mostly short-stalked, oval to orbicular, rounded, tapered. *An ubiquitous perennial seedling of several habitats.* **Taraxacum officinale**

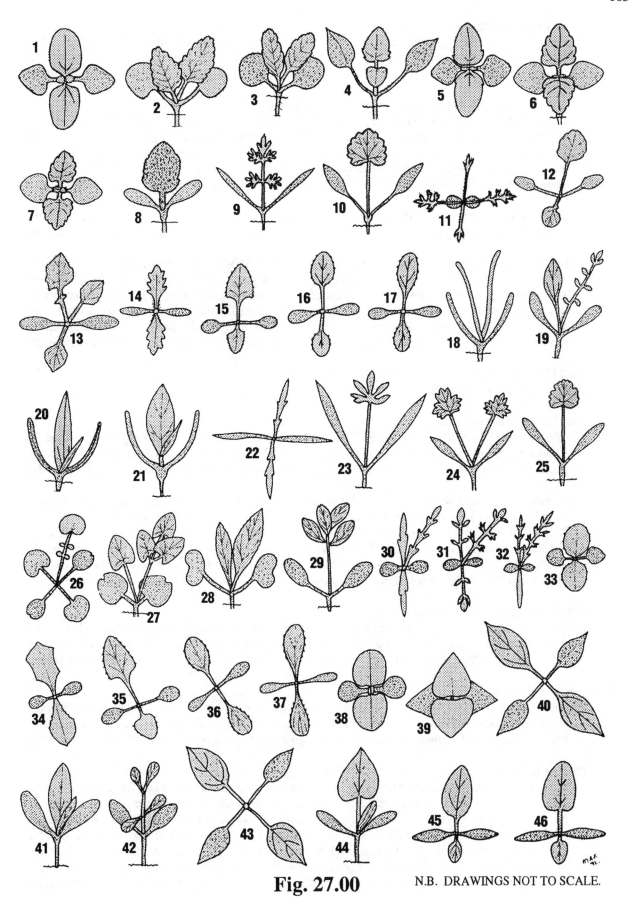

Fig. 27.00

N.B. DRAWINGS NOT TO SCALE.

79. Blades *tapered* basally 80
Blades *cut-away squarely* at junction with petiole (Fig. 27.35), glabrous, mostly ovate, thin, shiny, light bluish-green; petioles long; subsequent blades sinuate-pinnatifid. The cotyledons orbicular, stalked, blunt. *Ubiquitous annual; cultivated ground and waste places.* **Sonchus oleraceus**

80. Blades *abruptly* tapered to the petiole (Fig. 27.36), mostly broadly lanceolate, thin, shiny, glabrous, light bluish-green; petioles long; later blades sinuate-pinnatifid. The cotyledons broadly oval, stalked, blunt. *Ubiquitous annual; cultivated ground and waste places.* **Sonchus asper**
Blades *gradually* tapered to the petiole (Fig. 27.37), light bluish-green, mostly broadly oval, thin, shiny, glabrous; petioles long; subsequent blades sinuate-pinnatifid. Cotyledons broadly oval, stalked, blunt. *An ubiquitous annual of cultivated ground and waste places.* **Sonchus arvensis**

81. Blades clearly *longer than broad* 83
Blades as *broad as long,* or very *nearly so* 82

82. Cotyledons *rounded apically* (Fig. 27.38); the latter short-stalked, very small, and tapered basally. First blades nearly rounded in general outline, abruptly constricted basally, short-stalked, and glabrous. *Very common annual of cultivated ground and waste places.* **Epilobium montanum**
Cotyledons *angular-ovate* (Fig. 27.39); the latter short-stalked, small. First blades somewhat similar to cotyledons in general outline, bluntly pointed, short-stalked, bluntly pointed, and mostly glabrous. *Common annual; cultivated ground, waste places, sand-hills, etc.* **Anagallis arvensis**

83. First blades *either* cordate *or* abruptly tapered basally 86
First blades *tapering gradually* to the petiole 84

84. Cotyledons and leaves *not acuminate* 85
Cotyledons and leaves *acuminate* (Fig. 27.40). Cotyledons broadly ovate, long-stalked, tapered to an acute point apically, glabrous. The leaves rather similar, also tapered apically to an acute point, and opposite. *A very common annual; cultivated ground, waste places.* **Stellaria media**

85. Cotyledons *long, narrow* (Fig. 27.41); the latter linear-lanceolate. First leaves characteristically broad, elliptical-lanceolate, mostly glabrous, entire. The stipule minute but evident. Hypocotyl long, crimson. *Very common annual; cultivated ground; waste places.* **Polygonum aviculare**
Cotyledons *narrowly oval* (Fig. 27.42); the latter rounded or blunt, and tapered into a short stalk. Blades mostly obovate, small, glabrous, and usually of a dull blue-green colour. Hypocotyl and epicotyl distinct. *A common annual of cultivated ground and waste places.* **Euphorbia peplus**

86. Cotyledons and leaves *not acuminate* 87
Cotyledons and leaves *acuminate* (Fig. 27.43). Cotyledons broadly ovate, long-stalked, tapered to an acute point apically, glabrous. The leaves rather similar, also tapered apically to an acute point, and opposite. *A very common annual; cultivated ground, waste places.* **Stellaria media**

87. Hypocotyl *short* 88
Hypocotyl *long* (Fig. 27.44). The blades somewhat triangular, cordate basally, tapered, glabrous, shiny, and short-stalked. Cotyledons narrowly oblong, blunt apically, short-tapered basally. Epicotyl distinct. *Common annual; cultivated ground and waste places.* **Fallopia convolvulus**

88. Base of first leaf *cut-off squarely* (Fig. 27.45); blades broadly oval in general outline, rounded or blunt apically, glabrous, and often of a deep crimson colour; stipule traces frequently evident. Cotyledons somewhat lanceolate, bluntly pointed, short-stalked, rather abruptly tapered basally. *A very common perennial of roadsides, grasslands, waste places, etc.* **Rumex obtusifolius**
Base of first leaf *tapering abruptly* to the petiole (Fig. 27.46); the latter mostly ovate in general outline, bluntly pointed apically, entire, glabrous, and often tinged with a crimson colour; stipule traces frequently evident. Cotyledons narrowly lanceolate, bluntly pointed, tapered into a short stalk. *Very common perennial of roadsides, grasslands, waste places, etc.* **Rumex crispus**

5 POACEAE

General: Grasses are by far the most important group of flowering plants. Members occur in every climatic region. It is impossible to exaggerate their importance. While they rank third in number of genera and fifth in number of species behind other large families they, nevertheless, surpass all other groups as regards number of individuals. It has been estimated that the family - Poaceae or Gramineae - contains over 700 genera and more than 10000 species. Some idea of the importance of grasses may be derived from the fact that grasslands, of various types, extend over approximately 17.75 million square miles and comprise 24% of the world's vegetation. The greater part of the world's food is obtained, directly or indirectly, from grasses and closely related plants.

Grasses have many uses. Apart from the obvious well-known cereals such as wheat, maize, rice, millet, sorghum, oats, barley, rye, and the world's greatest source of sugar - cane sugar - other products derived from grasses include oils, waxes, resins, gums, a wide range of medicines in the form of drugs, starches, cellulose, paints, abrasives, building materials, fibre and newsprint. The list is endless. Sugar cane supplies about 60% of the world's sugar. Wheat is the most important grain crop in the world, and rice feeds more human beings than any other plant produce. Indeed, crops of bamboo grain, though irregular in production, provided a nutritious food which prevented famine in Asian countries on more than one occasion during the past century (Gould, 1968).

Grasses exhibit tremendous variation but the feature common to the vast majority of species is the caryopsis (see page 137). Indeed, so great is the variation within the grass population that completely different types are found in different ecological situations. Thus, some are adapted to warm humid tropical climates. Others are established in polar regions where the growing season is very short and direct sunlight is absent for long periods. Some, hydrophytes, are important elements of marsh and swamp vegetation, while others, xerophytes, inhabit desert regions where the annual rainfall is 12 cm or less. Such plants also play an important role in the prevention of dune and coastal erosion. Grasses are ubiquitous in the temperate zones.

Many researchers now appear to favour the recognition of 6 subfamilies in the Poaceae although unanimity appears to be absent in relation to the placing of certain groups and, to some extent, to subfamily names. The suggestions of Stebbins and Crampton (1961) have received much favourable comment, and indeed have been used as a basis for the formulation of some systems in books on Agrostology. Gould (1968) has modelled his system on these suggestions, and has documented the criteria for the separation of the 6 subfamilies now recognised. The subfamilies (Fig. 28.01) recognised are:

1. **Arundinoideae;**   2. **Bambusoideae;**   3. **Eragrostoideae;**
4. **Oryzoideae;**   5. **Panicoideae;**   6. **Festucoideae.**

Two subfamilies, Bambusoideae and Oryzoideae, contain the bamboo and rice plants, respectively. Several grass species, which are of no great significance to European agriculture, are classified in the Arundinoideae and Eragrostoideae. From the grassland point of view the 2 remaining subfamilies, Festucoideae and Panicoideae, are the most important. Grasses of the temperate regions of the world belong primarily to the former, whereas tropical grasses are populated by tribes of the latter. Temperate grasses have a C_3 biochemical mechanism of photosynthesis in contrast to tropical grasses which have a C_4 pathway. All subfamilies differ anatomically, morphologically and, some, physiologically (Farragher, 1970). The anatomical characteristics of the taxa containing forage grasses are illustrated in Figure 28.02-06, and the taxonomic criteria for the 2 important subfamilies are given later.

Anatomy

In recent times many microscopic structures of the grass plant have been evaluated for grass classification purposes. Row and Reader (1957) were the first agrostologists to describe long and short root-cells in the epidermis of the leaf. They found that long-cells and short-cells can occur together, in which case only the latter produces a root-hair, or that these cells may be of equal size and all capable of producing a root-hair. These 2 forms have been called the 'festucoid-type' and the 'panicoid', respectively, by Booth (1964). It is now generally accepted by many researchers that root-hair type is a useful criterion for grass separation, at least at subfamily level.

The microscopic features of the leaf-blade appear to be more important in taxonomic studies than any other part of the plant. The epidermis is made up of a single layer of cells which are mostly arranged in rows along the long axis of the blade. These cells are of various types and sizes and arranged in different combinations, according to the species involved.

The principal types of cells in the epidermis are 'long-cells' and 'short-cells', the former have long horizontal axes, while the latter are rarely longer than broad. It is the long-cells which form the most conspicuous pattern of the epidermis. The cells in longitudinal lines between the stomata are often somewhat different from the other long-cells, and are usually referred to as interstomatal-cells as a means of separating them from cells not associated with stomata. These are usually shorter than the long-cells. The walls of the long-cells may vary with different species. Variations found include: walls smooth, sinuous, thin, pitted, non-pitted; cells long, intermediate or cubical, and often with end-walls overlapping.

Short-cells are usually fewer in number than long-cells, though in some species, the 2 alternate. They can occur singly, in pairs or rows, they may be confined to the costal or intercostal zones, or they may occur in both zones. They may be absent entirely from some species.

Short-cells are of different types. The most frequent are in pairs, 1 of which deposits suberin and becomes a cork-cell, whereas the other deposits silica and becomes a silica-cell. The silica-bodies so formed often have a characteristic shape. Thus long narrow silica-bodies are described as being rod-shaped and are found, for example, in all *Agrostis* species. Other bodies are described as being dumbbell-shaped; examples of the latter are found in *Molinia caerulea*. Other shapes occur. Finally, the shape of the silica-body in the costal and intercostal zones may differ in a single species. Cork-cells do not exhibit characteristic shapes.

Other short-cells become differentiated as asperities or prickle-hairs which always point towards the distal end of the blade. These asperities, present, can be detected by stroking the blade in a distal-basal direction. It is the experience of the writer, however, that little taxonomic significance can be attached to the presence or absence of prickles, as he has noticed, that, not alone may they be present in some individuals of a species and absent in other individuals of the same species, but they can also vary with different leaves of the same plant.

Yet other short-cells give rise to micro-hairs which are often useful taxonomic criteria at genus level. They are generally present in the Panicoideae and absent from the Festucoideae. These are bicellular and difficult to see, even with the aid of a microscope. The lower cell is thick-walled and clearly different from the distal cell which is tapered and thin-walled. Macroscopic hairs occur commonly in European grasses. They are best exemplified by *Bromus sterilis* and *Holcus lanatus* where they often reach 6 mm in length. They differ from micro-hairs in being unicellular. They are probably the most inconsistent of the epidermal characters and are only of limited use in grass taxonomy. They are difficult at times to distinguish from prickle-hairs or, less often, from micro-hairs (Booth, 1964). Variously shaped projections of the outer wall of the epidermal cells are commonly classed as *papillae*. In some cases the cell projects into a dome-shaped elevation, but also included in this category are projection that are essentially or entirely cell-wall material, or cutin accumulations (Booth, 1964). Papillae occur mostly on long-cells, where they may appear as a single enlargement, or as a number of raised projections. They occur extensively in *Nardus stricta*, *Alopecurus geniculatus* and all *Glyceria* species.

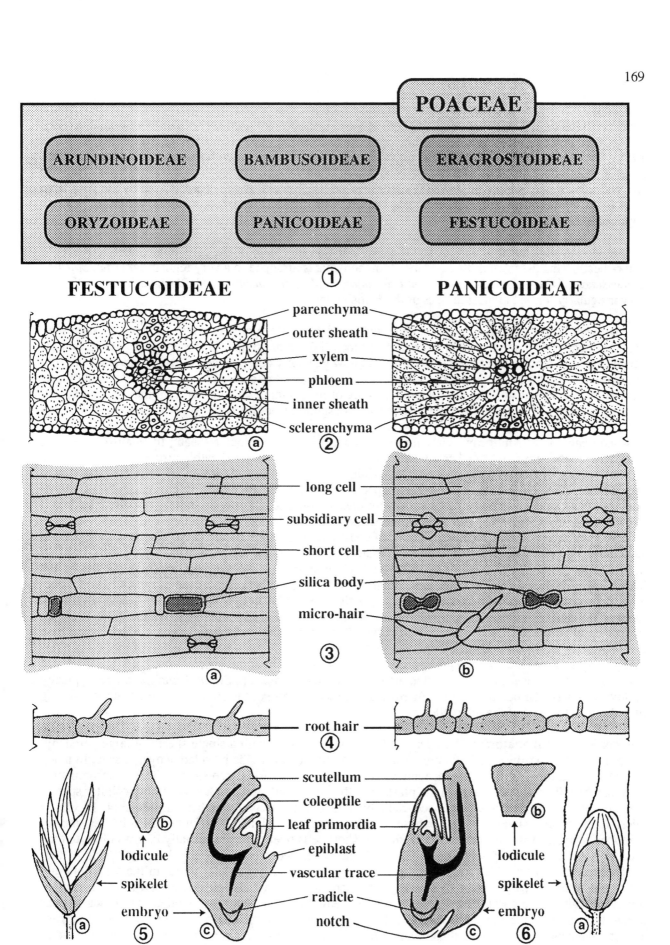

Fig. 28.00

Leaf anatomy, as revealed by blade transverse sections, has been emphasised as a very fundamental character (Gould, 1968). The arrangement of the various tissues are used for subfamily delimitation (Fig. 28.00). The inner or mesotome sheath may be present in some grasses and absent from others (Brown, 1958). Outside the mesotome sheath, when this is present, or in contact with the metaxylem vessels when there is no mesosheath present, is the parenchyma or outer-sheath. The cells of the latter vary greatly in size and thickness of wall. Gould (1868) states that the Eragrostoideae are characterised by the very large cells of this layer. This situation contrasts sharply with those of the Festucoideae which are loosely arranged.

A second anatomical feature of significance is the arrangement of the cells immediately outside the outer-sheath. This tissue, in most cases the chlorenchyma, is arranged in a very regular pattern in *Spartina x townsendii*, tightly packed in *Danthonia decumbens*, *Molinia caerulea* and *Phragmites australis*, and very irregularly formed in most grasses of the British Isles.

Brown (1958) has made a detailed examination of the leaf anatomy of some 100 species, representing 72 genera and, on the basis of his findings, proposed 6 major groups for the Poaceae. These, and the suggestions of Stebbins and Crampton (1968), are now generally accepted as the criteria determining subfamily level. The 6 subfamilies recognised have been mentioned already. Four of these - Arundinoideae, Bambusoideae, Eragrostoideae and Oryzoideae - are of little interest. The criteria for the 2 principal subfamilies represented in the flora of the British Isles are summarised hereunder (Figs 28.02-06):

Festucoideae - Leaf Anatomy: Vascular bundles with an inner and outer sheath; the walls of the inner sheath are characteristically thicker towards the vascular bundles; outer sheath entire or often interrupted abaxially, adaxially, or both, by sclerenchyma Chlorenchyma loosely and irregularly arranged, often with large intercellular air-spaces, *e.g. Glyceria* species. **Epidermis**: Bicellular micro-hairs absent; silica-bodies either round, elliptical, long or narrow, or crescent-shaped; short-cells solitary or paired, seldom if ever in long rows; stomata with low, dome-shaped or 'parallel-sided' subsidiary cells. **Embryo**: The embryo is relatively small, mostly one-sixth to one-fourth the length of the caryopsis; vascular traces to scutellum and coleoptile diverging at approximately the same point; epiblast usually present; lower part of scutellum fused to the coleorhiza; embryonic leaf-margins not overlapping; vascular bundles to the leaves few. **Cytology**: Basic chromosome number, x=7 in most tribes; chromosomes typically large; nuclei not persisting in root-tip cell divisions. **Spikelets and Flowers**: Spikelets with 1-several florets; reduced florets, when present, usually above the fertile; spikelets disarticulating above the glumes, except in species of *Alopecurus* and *Holcus*. **Lodicules**: These are usually elongate, pointed, thick at the base and membranous above, with little or no vasculation. **Roots**: Epidermal cells in region of root-hairs alternately long and short, only the latter giving rise to root-hairs. In some species root-hairs are produced towards the root-tip end of the cell and at an angle towards the root-tip. **Stem**: Culm internodes typically hollow, with vascular bundles in 1 to few rings at the inner margin of the cortex; no meristematic swelling or 'pulvini' at the base of the internode.

Panicoideae - Leaf anatomy: Vascular bundles typically single, with a single sheath of large parenchyma cells; bundle sheaths usually continuous, rarely interrupted on 1 side by sclerenchyma tissue. In many genera, the chloroplasts of the parenchyma sheath are specialised for starch storage. This is accompanied by a loss of ability to store starch in the plastids of the mesophyll, and a reduction in intercellular spaces; the mesophyll cells are often arranged radially around the vascular bundles. **Epidermis**: Bicellular micro-hairs nearly always present, these usually linear or slightly fusiform; silica-bodies mostly dumbbell-shaped, cross-shaped, or saddle-shaped; short-cells over the veins, usually in rows of more than 5 cells; stomata commonly with triangular or tall, dome-shaped subsidiary cells. **Embryo**: This is relatively large, mostly one-third to four-fifths of the caryopsis length; vascular traces to the scutellum and coleoptile separated by a more or less elongated internode; epiblast absent. Lower part of the scutellum separated from the coleorhiza by a notch or groove; embryonic leaf-margins overlapping; vascular bundles to the embryonic leaves numerous. **Cytology**: Basic chromosome number, x=9 or x=10 in most genera; chromosomes are characteristically small to medium sized; nuclei persistent in some, but not in all root-tip cells divisions. **Spikelet and Flowers**: The former have 1 fertile and 1 reduced floret, with the reduced or asexual types attached below the fertile; disarticulation is below the glumes. **Lodicules:** These are usually short, truncate, thick, and heavily vasculated. **Roots:** Epidermal cells in the region of the root-hairs all alike, and all capable of giving rise to root-hairs; root-hairs typically developed near the middle of the cell and perpendicular to the cell axis. **Stems:** Culm internodes predominately solid, with vascular bundles scattered in the ground tissue, or in 2 to a few rings at the margin of the cortex; well-developed 'pulvini' present at base of the internodes.

COMMON GRASSES

There are about 120 grass species in the British Isles, but only about 100 species, mostly of the Festucoideae, occur in Ireland. Of these, about 60 are common or locally abundant. They include the miscellaneous species of grassland and amenity areas, grasses which occur as weeds of tillage, those which are plentiful in hedgerows, woodlands, coastal areas, shallow water, and, of course, the cereal plants. Many people, of diverse interests, need, from time to time, to be able to identify these plants in the (i) *vegetative,* (ii) *inflorescence* and (iii) *'seed'* phase. A detailed consideration of first 2 phases is given hereafter. The third is discussed in *Chapter 4: Seeds-Fruits-Seedlings.* A fourth section deals, in brief, with their morphology and ecology. The principal grain grasses, wheat, barley, oat and rye, are also discussed in some detail.

TAXONOMIC DATA - VEGETATIVE PHASE

The fundamental step in grassland improvement is the identification of the pasture components. Once composition has been determined, improvement is a matter of adopting a treatment that will lead to an increase in the number of the best forage plants and to a diminution of all undesirable species present. Most broad-leaved plants commonly found in pasture are easily identified but some difficulty may be experienced with grasses, particularly in the vegetative phase. This difficulty can be overcome or at least minimised by first of all obtaining a clear understanding of the basic structure of the grass plant, and then by the use of a simple 'key' for identification.

Structure

Germination: Given suitable environmental conditions, the 'seed' absorbs water and germination begins. The first indication of growth is the appearance of the coleorhiza and radicle. The former, a protective covering for the primary root, bursts through the 'seed' allowing the latter to emerge. Two pairs of lateral roots follow in quick succession, although the actual number of rootlets can vary, depending on the species. These rootlets are known as the *seminal* (Fig. 29.13) or *white* root-system, the latter name because of their colour. This root-system is usually of a temporary nature and, in time, its function is taken over by more permanent adventitious roots which develop from the lowermost nodes of the primary shoot and tillers. Meanwhile, the colourless coleoptile pushes upwards through the soil, creating a tunnel, allowing the primary shoot easy travel, within. The tip of the primary shoot remains just above soil level, and forms the *vegetative stem* (Fig. 29.17). Eventually, the first leaf pierces the tip of the coleoptile and unfolds. Other leaves, 1 per node, soon follow and become unfolded in an alternate fashion.

The grass plant differs from all other forage plants in many respects, but principally because of its very short vegetative stem. The presence of this short stem makes grasses ideal grazing plants, as it generally remains undamaged by good cutting and grazing practices. On average, it is about 3 cm in length, and cannot be seen unless the base of the plant is sectioned longitudinally. Its shortness means that all plant material, upwards for about 3 cm from soil level, consists of leaf tissue.

Growth-habit: Determination of the growth-habit is a very necessary step in identification. While it may be 1 of 3 distinct types it is, nevertheless, consistent, with only about 2 exceptions, for any given grass species. The exceptions are *Poa trivialis* and *Agrostis capillaris* which are mostly tufted but, depending on soil conditions, may spread overground.

Tufted: Most grasses in the British Isles have a tufted or *cespitose* growth-habit (Fig. 29.15). This form of growth is characterised by the complete absence of stems other than vegetative stems. When new plants, termed tillers, arise from buds present on the original vegetative stem they remain very close to the parent plant and form a tussock or tuft, hence the term tufted growth-habit. All good forage grasses have this type of growth-habit; no food material is wasted in the formation of secondary stems. *Dactylis glomerata, Lolium multiflorum, L. perenne, Cynosurus cristatus* and *Phleum pratense* are good examples of plants which are always tufted. Many of the 'weed' grasses, such as *Poa annua, Holcus lanatus* and *Arrhenatherum elatius* have a similar growth-habit. All such grasses are characterised by their inability to spread laterally.

A word of caution is opportune here in relation to tufted grasses. Occasionally these grasses, particularly when growing on wet soils or around gateways where there is a lot of animal traffic, become pushed into the soil by livestock or some mechanical means. This results in the vegetative stem elongating and pushing the grass up to soil level once more. The elongated vegetative stem can often give rise to some confusion in that it may show some resemblance to a rhizome, the next growth-habit.

Spreading underground: The next most common type of growth-habit is where 1 or more buds present on the vegetative stem of the parent plant elongate and form long secondary stems which spread horizontally beneath the soil surface. When this happens the grass is said to spread underground (Fig. 29.20), giving a *rhizomatous* growth-habit. The best known example is *Elymus repens*. These underground stems or rhizomes (Fig. 29.18) are easily recognised by being whitish in colour and in having small scale-like leaves (Fig. 29.19). Elongated vegetative stems mentioned earlier differ from underground stems in the near complete absence of scale-leaves and, also, in that they rarely exceed 5 to 16 cm. Furthermore, they differ in having the tillers emanating from the distal end rather than from a lateral position.

Grasses which spread underground have a limited use in agriculture in that a lot of food material is wasted in non-productive secondary stems. Consequently, they tend to occur more as weeds of cultivated ground than as forage plants. They are capable of much greater lateral spread than tufted plants and, therefore, are of use in coastal areas where they stabilise soil and minimise sand movement.

Spreading overground: The third type of growth-habit occurs when 1 or more buds on the parent stem elongate laterally and produce secondary stems which trail on the soil surface. These plants spread overground (Fig. 29.05) and are comparatively few in number. The best known example is *Agrostis stolonifera*. Such grasses have a *stoloniferous* growth-habit. These overground spreading grasses are also considered, by some agriculturalists, to be less desirable than those with a *cespitose* growth-habit.

Leaf

The leaves of most grasses show many consistent variations which are indispensable diagnostic criteria. All grass leaves, with only 1 notable exception, consist of 2 principal parts, the leaf-blade (Fig. 29.04) and the leaf-sheath (Fig. 29.07). Small appendages are present on many leaves.

Leaf-blade: The uppermost part of the leaf is termed the blade (Fig. 29.04). In a comparatively small number of species it is very narrow and bristle-like or *setaceous. Festuca ovina* is a good example. Bristle-like blades are characterised by being tightly inrolled and it is extremely difficult in such cases to distinguish between the upper and lower surface. There is a tendency for grasses of this type to occur in dry areas such as wall-tops and mountain pasture.

The contrast to the bristle-type blade is found where there is a clear distinction between the upper and lower surface, as in *Dactylis glomerata* and *Lolium multiflorum*. Such plants are described simply as having flat blades. Where the blade is flat the question of width comes up for consideration. In most cases it is not a reliable criterion for use in identification. The leaves of many grasses, growing on fertile substrata or in very exposed situations, are invariably wider than those of plants elsewhere. Blade length is also an unreliable criterion. Many grasses growing in shaded conditions produce longer leaves than other plants of the same species growing under more open or exposed conditions.

The shape of the blade can be a helpful feature. Many grasses, for example, *Poa pratensis* and *Lolium perenne*, produce 'parallel-sided' blades. More, such as *Phleum pratense*, are invariably widest towards the base, while others still are widest at a point about one-third of the distance from base to apex, or in the middle. Blade colour is another useful character though, on its own, it is not totally dependable. Thus, the blades of *Dactylis glomerata*, *Phalaris arundinacea*, *Ammophila arenaria*, *Leymus arenarius* and *Elymus farctus* are always bluish-green in colour, those of all *Poa* and *Glyceria* species light green, and *Brachypodium sylvaticum* yellowish and glossy. *Lolium* species are immediately identified by their deep glossy-green blades.

The manner in which the upper surface of the blade is ribbed can provide another useful feature. *Dactylis glomerata*, for example, is ribless (Fig. 29.30) or, at best, only very finely ribbed. It also is the only grass with a single central groove. All *Poa* species and *Avenula pubescens* have a double central groove containing a small *median ridge* (Fig. 29.29). Others are regularly ribbed (Fig. 29.31); dune grasses are prominently ribbed (Fig. 29.32). *Festuca rubra* has alternating large and small ribs.

Youngest blade: The youngest blade always occupies the central position on the plant (Fig. 29.01). When it emerges from within the older leaf-sheaths, it may be arranged in 1 or other of 2 distinct positions. In about 25% of our grasses, one-half of the blade is folded flatly on the other to give the folded arrangement. This condition can be determined with little difficulty by the naked eye but in cases of doubt it is best that a section be taken where the young blade emerges (Fig. 29.02). When viewed with a hand lens the section will show the halves folded on each other (Fig. 29.21). The other situation, found in most grasses, is where one-half inrolls and is overlapped by the other (Fig. 29.22); a few grasses may have both margins inrolled. *Dactylis glomerata* is a good example of a plant where the youngest blade is folded, while *Phleum pratense* always has the youngest blade in the rolled position. *Conduplicate* and *convolute* are the respective technical terms.

Ears: Outgrowths from the base of the leaf-blade are termed *auricles* or ears (Fig. 29.06). They are consistently present in about 20% of grasses in the British Isles and vary widely in prominence. *Lolium perenne* is the only grass species where they may be present or absent. In plants such as *Lolium multiflorum* (Fig. 29.36) and *Hordeum sativum* (Fig. 36.01), they are very prominent. *Secale cereale* (Fig. 38.04), *Lolium perenne*, when they are present (Fig. 29.36) and *Elymus repens* (Fig. 30.01) have small or inconspicuous projections. Two grasses, *Festuca arundinacea* (Fig. 35.10) and *Triticum aestivum* (Fig.34.01), are characterised by having the hairs confined to the ears.

Collar: The collar is the name applied to the area at the junction of the blade and sheath (Fig. 29.16). It is usually lighter in colour than either the sheath or blade. It is of particular interest because in 2 common plants - *Danthonia decumbens* (Fig. 31.10) and *Anthoxanthum odoratum* (Fig. 30.06) - it is more hairy than any other part of the plant.

Leaf-sheath: The second major part of the leaf is the sheath (Fig. 29.07). The portion of the grass plant, from about 2 cm above ground, consists of leaf-sheaths and is sometimes referred to as the pseudostem. The sheath, like the blade, shows a number of consistent variations. In a relatively small number of plants they are said to be entire in that they consist of a complete tubular structure. The opposite to this is where the sheath is split longitudinally and where the margins overlap to a slight degree. One way to decide if it is split or entire is simply to hold the plant firmly in 1 hand and to pull a blade downwards and observe if the sheath ruptures due to the pulling action (Fig. 29.27), or if 1 margin is slightly overlapped by the other (Fig. 29.28). A more definite method is to take a transverse section through the base of the plant (Fig. 29.08) and note the outline of the section with a hand lens. Entire sheaths will appear as in Figures 29.24

and 29.27, while split sheaths will show contiguous or overlapping margins, Figures 29.25 and 29.28. All *Bromus* species and *Dactylis* are good examples of plants with entire sheaths, whereas *Lolium perenne*, *Elymus repens* and *Phleum pratense* show split sheaths. When examining the plant it is desirable to remove the outer 1 or 2 sheaths as these are often damaged or split due to some external factor and may not represent the true nature of the plant.

A relatively small number of plants have strongly compressed sheaths. In the common species - *Dactylis glomerata* and the very rare species - *Poa chaixii*, for example, compression is so strong that 2 sharp ridges, termed *keels*, develop on diametrically opposite sides of the sheath (Fig. 29.23). *Avenula pubescens*, and *Poa* and *Glyceria* species are other examples of plants with a similar sheath type. The opposite situation is where the sheaths are rounded (Fig. 29.24). This is the shape of most grass sheaths, for example, *Elymus repens*, *Phleum pratense*, all *Festuca* species and *Lolium multiflorum*. An intermediate shape is shown by *Lolium perenne* and occasionally *Cynosurus cristatus*. In these the sheaths are said to be slightly compressed in that keels fail to develop (Fig. 29.26). Should any doubt arise as to the shape of the keel, the best way to arrive at a correct decision is to take a cross-section through the base of the plant (Fig. 29.08) and to examine it with a hand lens.

Ligule: The ligule is the term applied to the piece of tissue that is mostly present at the top of the sheath (Fig. 29.03). It is easily seen in *Dactylis glomerata* where it is long, *membranous* and whitish in colour. While it shows considerable variation for different grass species it is, nevertheless, fairly consistent for the individuals of any 1 species. A range of the different types found in grasses of the British Isles shown in Figures 29-33. In a few species the ligule is represented by a fringe of hairs, *i.e.* the *ciliate* type. *Molinia caerulea* (Fig. 32.10) is a good example of a plant with this type of ligule.

Prophyl: When the buds on the original vegetative stem develop into plants similar to the parent the new plants are termed primary tillers. The latter can give rise to secondary tillers; secondary tillers can give rise to tertiary tillers. The first leaf of each tiller differs from all subsequently formed leaves in that the blade fails to develop. Only the basal portion, termed a prophyl or prophyllum (Fig. 29.14), develops. This name is also applied to the outer scale of each bud present on the original vegetative stem or to buds present in the axils of the tiller leaves. Only in 1 case is it necessary to refer to the prophyl to decide the identity of the plant. The only consistent difference between the vegetative phase of *Festuca pratensis* (Fig. 32.11 p) and *Lolium multiflorum* (Fig. 29.36 p) is found in the shape of the prophyl. In the former it is narrow and hairless, whereas in the latter it is broad at the base and has a few short hairs.

Plant-base: A useful sheath feature is the colour of its base. Thus the lower part of the sheath in all *Festuca* species, *Alopecurus pratensis*, *Phleum pratense* and most *Lolium* plants, is mostly dark brown or purplish in colour. *Arrhenatherum elatius* is characterised by having rust-brown or yellowish sheath-bases and roots, while *Cynosurus cristatus* usually has a canary-yellow colour. The pink-coloured veins of *Holcus lanatus* and *H. mollis* are most characteristic.

Phleum pratense is often described as having a bulbous base, but this feature only develops towards the end of the growing season when the food material manufactured in the leaves is translocated and stored in the plant base. *Arrhenatherum elatius* var. *bulbosum* has a number of clearly defined, small, onion-like, rounded bulbs (Fig. 31.11). Most grasses have unswollen bases.

Pubescence: The determination of the presence or absence of hairs, *i.e.* plant *pubescent* or *glabrous*, can provide some problems in grass identification. Caution must be exercised when hairs appear to be absent, in that many grasses have very short hairs that are not easily seen. It is, therefore, very necessary to examine all parts of the plant for the presence or absence of hairs and particular attention should be given to the upper surface of the blades. Two grasses which come to mind here are *Ammophila arenaria* and *Elymus farctus* which have extremely short hairs on the upper surfaces. Some forms of *Arrhenatherum elatius* may also have short hairs. One method to facilitate the determination of the presence or absence of hairs is to loop the blade over the index finger and hold the blade between the eye and some source of light. Using this technique even the smallest of hairs can be seen.

KEY TO THE VEGETATIVE PHASE

1.   Ligule either *absent* or represented by a *membranous* piece of tissue 5
      Ligule represented by a *dense fringe of hairs* 2

2.   Youngest blade *convolute*; collar *glabrous* 3
      Youngest blade *conduplicate*; blades green, up to 60 cm x 10.0 mm, blunt or abruptly pointed api-
cally, finely ribbed, flat or inrolled, firm, mostly *glabrous* though usually with a *tuft of hairs*
on the collar; auricles *absent*; sheath rounded, split often with some hairs; ligule *ciliate* (Fig.
31.10). *Frequent, densely tufted perennial; heathlands, moorlands, etc.* **Danthonia decumbens**

3.   Growth-habit *rhizomatous* 4
      Growth-habit *cespitose* (Fig. 32.10); blades up to 60 cm x 10.0 mm, finely ribbed adaxially with a
rounded abaxial keel, thin, pubescent or glabrous, deep green, contracted basally and tapered
apically into a fine point; the youngest blades *convolute*; sheaths split and rounded; ligule *ciliate*; the roots *cord-like*. *Frequent perennial of bogs, heathlands and moors.* **Molinia caerulea**

4.   Blades *bluish-green* or *glaucous* (Fig. 32.04); the latter up to 72 cm x 45 mm, contracted basally,
long-tapered to a very fine curved or flexuous tip, contracted basally, flat, closely ribbed abaxially, falling early; sheaths rounded, split; ligule a *fringe* of short hairs. *Common rhizomatous
plant of marshy ground, canals, shallow rivers, fens, bog-margins, etc.* **Phragmites australis**
      Blades dull *yellowish-green*; the latter up to 67 cm x 18.0 mm, tapered to sharp hard tips, inrolled
or flat, firm, smooth, closely flat-ribbed adaxially; youngest blade *convolute;* sheaths rounded,
cross-veined, smooth; ligule a *fringe* of short hairs (Fig. 29.35); roots *soft, thick*. *Rhizomatous
plant of coastal mud-flats, estuaries, etc., where it forms large patches.* **Spartina x townsendii**

5.   Blades *flat, not* setaceous 13
      Blades very *narrow*, mostly *setaceous* 6

6.   Growth-habit *cespitose, stoloniferous* or plant *with* trailing shoots 9
      Growth-habit *rhizomatous* 7

7.   Ligule *distinct;* sheath *glabrous* and mostly *split* 8
      Ligule *minute* or *absent* (Fig. 33.12); blades tightly inrolled, 10.0-100 cm x 2.0-5.0 mm, glabrous,
often *bristle-like*, ribbed adaxially, ribs alternating large and small; bluntly keeled abaxially;
pointed; the rhizomes thin, *thread-like*; sheaths entire, mostly *pubescent*, *very variable*. *A very
common rhizomatous plant of wall-tops, roadsides, lawns, dunes, pastures, etc.* **Festuca rubra**

8.   Ligule *minute* and *emarginate* (Fig. 32.01); blades *narrow* and *bristle-like*; the latter up to 30.0
cm, dark green, tightly inrolled, rather stiff; the sheaths rounded, often rough upwards. *A frequent perennial in most mountainous districts, dry heaths, moors, etc.* **Deschampsia flexuosa**
      Ligule *long, tapered, toothed* or *torn* (Fig. 31.04); blades up to 20.0 cm x 1-3 mm, flat or inrolled,
sometimes *bristle-like*, finely ribbed, glabrous; auricles *absent*; the youngest blade *convolute;*
rhizomes slender. *A frequent plant on heaths, hill pastures and peaty soils.* **Agrostis vinealis**

9.   Growth-habit *cespitose* 10
      Growth-habit *stoloniferous* (Fig. 31.03); blades 2-20 cm x 1.0-3.0 mm, flat or inrolled, soft, finely
ribbed, bright green, pointed; auricles absent; the youngest blades convolute; sheaths rounded,
split; ligule 2-5 mm, tapered; stolons slender. *Frequent plant on peaty soils.* **Agrostis canina**

10.   Growth-habit *loosely cespitose*; blades *not rigid* 11
      Growth-habit *densely* and *compactly tufted* (Fig. 29.38); blades *tightly inrolled, bristle-like*, 5-30
cm x 0.5-0.7 mm, very *rigid*, sharp-pointed, green or greyish-green, ribbed adaxially; the ribs
minutely hairy; sheaths erect, *very compact*, the outer short; ligule 1-2 mm. *A densely, tufted
wiry perennial plant, frequent on heaths, moors, hill and mountain pastures.* **Nardus stricta**

11.   Sheath *merging gradually* with the blade 12
      Sheath, top of, *rounded and auricle-like* (Fig. 33.11); the blades 1.0 mm or less wide, *bristle-like*, smooth or rough, green or grayish-green, glabrous; sheath split, rounded; ligule minute or absent. *A very frequent perennial plant of moors, heaths and mountain pasture.*   **Festuca ovina**

12.   Ligule *evident, emarginate* (Fig. 32.01); the blades *narrow* and *bristle-like*; the latter up to 30.0 cm, dark green, tightly inrolled, rather stiff; the sheaths rounded, often rough upwards. *A frequent perennial in most mountainous districts, dry heaths, moors, etc.* **Deschampsia flexuosa**
      Ligule *minute, truncate* or *flat-topped* (Fig. 33.08); blades 5-15 cm x 0.5-2.5 mm, finely pointed, flat or more often inrolled, shortly pubescent adaxially, soft to somewhat firm, finely ribbed, green or yellowish-green; sheaths rounded, smooth; auricles *absent. Frequent annual grass of dry stony or open sandy places, heathlands, hill pasture, waste places, etc.* **Vulpia bromoides**

13.   Youngest blade *convolute* 29
      Youngest blade *conduplicate* 14

14.   Leaf-blades *glabrous* 16
      Leaf-blades *pubescent* 15

15.   Blades with a distinct *central groove* which contains a *small median ridge* (Fig. 32.03); the latter 'parallel-sided', up to 100.0 cm x 2-9 mm, ribbed, keeled abaxially, with *long, scattered hairs*, dark green, rarely yellowish-green; auricles *absent;* sheaths keeled, split, with *reflexed hairs*; ligule *long*, up to 8.0 mm. *A frequent perennial of damp calcareous soils.* **Avenula pubescens**
      Blades *without a distinct central groove* containing a small *median groove* (Fig. 33.07); the latter 'parallel-sided', up to 100 cm x 2-6 mm, often inrolled, ribbed, dull green, keeled, glossy abaxially, finely pointed, with *few, long, scattered hairs*, auricles *absent*; sheaths rounded, entire. *A densely tufted perennial grass, locally abundant on dry calcareous soils.*   **Bromus erectus**

16.   Auricles *absent* 17
      Auricles, *small, present* (Fig. 29.37); the blades up to 25 cm x 2-6 mm, somewhat 'parallel-sided', ribbed adaxially, *deep glossy-green* abaxially and *contrastingly dull green* adaxially, keeled, glabrous, smooth or minutely rough, pointed or blunt; youngest blades *conduplicate*; sheaths split, rarely entire, slightly compressed, but *not keeled*, often purplish basally; ligule short, 1.0-2.5 mm, blunt. *An ubiquitous perennial grass, of great economic importance.* **Lolium perenne**

17.   Leaf-sheaths *with* cross-veins 18
      Leaf-sheaths *without* cross-veins 20

18.   Growth-habit *other than* rhizomatous 19
      Growth-habit *rhizomatous* (Fig. 32.06); the blades up to 80 cm x 20 mm, glabrous, smooth or minutely rough, widest basally, tapered abruptly into a *hooded* apex, clearly keeled, light green; auricles *absent;* sheaths compressed, smooth, entire, soon splitting, keeled; ligule *characteristically abruptly pointed*, 3-7 mm. *A common perennial of shallow waters.* **Glyceria maxima**

19.   Ligule *acuminate*, up to 12 mm (Fig.32.07); blades 5-35 cm x 3-12 mm, often 'parallel-sided', flat, finely ribbed, pointed, distinctly keeled, light or yellowish-green, smooth or rough on the margins, glabrous; youngest blades *conduplicate;* sheaths entire, keeled, smooth; auricles *absent; Frequent perennial of streams, shallow waters and other wet muddy places.* **Glyceria fluitans**
      Ligule *blunt*, 2-8 mm (Fig. 32.08); blades 5-35 cm x 3-14 mm, widest basally, tapered towards the apex, flat, keeled, glabrous, minutely rough or smooth on the margins; youngest blades *conduplicate* light or yellowish-green in colour; sheaths entire, keeled, mostly glabrous; auricles *absent. Frequent perennial of streams, shallow waters and wet muddy places.* **Glyceria plicata**

20.   Plant *rhizomatous, stoloniferous* or *with* trailing shoots 27
      Plant *cespitose* 21

21.   Blades *without* a median ridge 24
       Blades *with* a median ridge 22

22.   Ligules *distinct;* blades *light green* 23
       Ligule very *short, truncate,* entire (Fig.31.02); blades rather short, 4-14 cm, narrow, 1.5-3.5 mm, dark green in colour, glabrous but often minutely rough, flat, weak, finely to *abruptly pointed,* keeled abaxially and minutely rough; the auricles *absent;* sheaths split, keeled, somewhat compressed. *A frequent perennial of dry shaded places, wall-tops, woodlands, etc.* **Poa nemoralis**

23.   Blades *soft,* many *transversely wrinkled* (Fig. 33.01); the latter weak, *abruptly pointed* or *hooded,* always widest at the base, up to 15 cm long and 2-5 mm wide, *yellowish-green,* flat, *glabrous,* minutely rough on the margins; ligule 2-5 mm, blunt, toothed; sheaths strongly compressed, keeled, smooth; auricles *absent. An ubiquitous annual plant of lawns, ground etc.* **Poa annua**
       Blades *firm, none* or *few* transversely wrinkled (Fig. 33.03); the latter flat, 5-30 cm x 2.0-8.5 mm, *abruptly pointed* or *hooded, light* or *yellowish-green,* centrally grooved, keeled, smooth or minutely rough, *glabrous;* ligule 3-10 mm, tapered; sheaths compressed, keeled; auricles *absent. Ubiquitous perennial plant of cultivated ground, old pastures, waste places, etc.* **Poa trivialis**

24.   Sheath-bases *without a purplish colour* 25
       Sheath-bases *purplish* (Fig. 29.37); the sheaths split, very rarely entire, slightly compressed, but *not* keeled; blades up to 25.0 cm x 2-6 mm, rather 'parallel-sided', moderately ribbed adaxially, deep glossy-green abaxially and *contrastingly* dull green adaxially, keeled, glabrous, smooth or minutely rough, pointed or blunt apically; youngest blades *conduplicate;* ligule short, 1-2.5 mm, blunt, entire. *Ubiquitous perennial grass, of great economic importance.***Lolium perenne**

25.   Ligule *very distinct, tapered* or *somewhat tapered* 26
       Ligule rather short, *0.5-1.5 mm, truncate* or *very blunt* (Fig. 30.05). Blades *finely* to *moderately ribbed,* 5-20 cm x 2-6 mm, *dull green* adaxially, *glossy,* poorly or moderately keeled abaxially, tapered apically to a fine roughish tip, widest and *characteristically asymmetrical* basally, glabrous; youngest blades mostly *conduplicate;* auricles *absent;* sheaths rounded or *somewhat compressed* but *not* keeled, smooth, split, the outer *characteristically canary-yellow* in colour. *A common perennial of old pastures, meadows, waste and grassy places.* **Cynosurus cristatus**

26.   Blades *without* a deep central groove, *dull green* in colour, *prominently ribbed,* very rough on the ribs and margins, rather *stiff,* sharply pointed, glabrous, *not* distinctly keeled abaxially, flat or inrolled and up to 80 cm x 2-5 mm; auricles *absent;* ligule very *prominent, tapered,* up to 15.0 mm (Fig. 32.02); sheaths split, smooth or rough upwards, rounded or very *poorly keeled. Frequent, densely tufted, useless perennial, of poorly drained grasslands.* **Deschampsia cespitosa**
       Blades *with* a deep central groove; greyish-green or *glaucous, ribless* or very finely ribbed, 8-110 cm x 2.0-18 mm, firm but *not* stiff, pointed, *distinctly* keeled, sharply pointed, mostly *glabrous* but often somewhat rough on the margins and keel; youngest strongly conduplicate; sheaths *sharply keeled,* the inner whitish, entire, soon splitting, rough, glabrous or with few short hairs; ligule (Fig. 31.01) very *prominent,* up to 12 mm, tapered, bluntly toothed. *A common densely tufted perennial plant of pastures, meadows, roadsides, waste places, etc.* **Dactylis glomerata**

27.   Growth-habit *stoloniferous* or plant *with* trailing shoots 28
       Growth-habit *rhizomatous* (Fig. 33.02); blades very variable, 5-150 cm x 1-8.5 mm, usually much less, 'parallel-sided', with *hooded* apices, glabrous or with a few short hairs on collars, smooth or nearly so, dull grey or glossy-green; the ligule *minute;* sheaths keeled, split, glabrous or pubescent. *A very frequent perennial of dry habitats such as wall-tops, dunes, etc.* **Poa pratensis**

28.   Sheaths *compressed* but *not* keeled (Fig.32.09); blades *whitish* in colour and *rounded dorsally,* 4-8 cm x 2-4 mm, flat or inrolled, *hooded* apically, ribbed, rough or smooth, *whitish* or *glaucous* collar *clearly white;* auricles *absent;* ligule *rounded apically, entire,* 2-4 mm, *often wider* than the blades. *A common perennial of salt-marshes and muddy estuaries.* **Puccinellia maritima**

Sheaths *compressed and keeled*, blades *not whitish* (Fig. 33.03); the latter 5-30 cm x 2.0-8.5 mm, *abruptly pointed* or *hooded, light* or *yellowish-green*, centrally grooved, keeled, smooth or minutely rough, *glabrous*; the ligule 3-10 mm, tapered; sheaths green or purplish; auricles *absent*. *Ubiquitous perennial plant of cultivated ground, old pastures, waste places, etc.* **Poa trivialis**

29. Auricles *absent* 41
    Auricles *present* 30

30. Growth-habit *cespitose* 32
    Growth-habit *rhizomatous* 31

31. Blades *bluish-green* or *glaucous*, up to 70 cm x 22 mm, flat, convolute or involute, sharp-pointed, prominently ribbed, firm, rough on the latter and margins, *glabrous*; the youngest blade *convolute*; the auricles *very prominent* (Fig.29.39); ligule *very short*, about 1 mm, minutely hairy; sheaths rounded, mostly smooth. *A very common rhizomatous dune grass.* **Leymus arenarius**
    Blades *dull grey-green*, up to 35 cm x 20 mm, either *conspicuously* or *inconspicuously* pubescent, flat, soft, *not* distinctly ribbed, finely pointed, smooth or rough; the youngest blades *convolute*; auricles *small* (Fig. 30.01); ligule *absent* or a mere outline; sheaths rounded, split, sparingly or densely hairy. *An ubiquitous, pernicious plant of cultivated and waste ground.* **Elymus repens**

32. Hairs *absent*, or *localised* and *not* plentiful over the entire grass 34
    Hairs *plentiful* over the entire grass 33

33. Sheaths, at least at first, *entire* (Fig. 33.06); the latter mostly rounded, covered with *stiff reflexed hairs*; blades finely ribbed, dull green adaxially, keeled and deep green abaxially, 10.0-80 cm x 5-18 mm, flat, tapered, and usually covered with *long conspicuous hairs* which are often more prominent on the margins than elsewhere; youngest blades *convolute*; auricles *very prominent*; ligule 2-5 mm, toothed or ragged. *Frequent perennial grass; shaded places.* **Bromus ramosus**
    Sheaths *split*; the latter rounded, soft, pubescent; blades up to 25.0 cm x 12.0 mm, *inconspicuously* pubescent adaxially and abaxially, light green, weak, ribbed adaxially, moderately keeled abaxially, tapered, somewhat rough; youngest blade *convolute;* auricles very prominent; ligule (Fig. 29.40) very short, 1.0 mm long. *A frequent annual of waste ground.* **Hordeum murinum**

34. Sheath-bases *not* purplish in colour 38
    Sheath-bases *purplish* in colour 35

35. Auricles *without* very short hairs 36
    Auricles *short-haired* (Fig. 33.10); blades variable, from 10-100 cm x 2-19 mm, firm, dull green adaxially, keeled and glossy-green abaxially, usually strongly serrate on the margins, minutely rough on both surfaces, prominently ribbed, tapered; youngest blades *convolute*; ligule short, 2-4 mm, toothed; sheaths rounded, rough, usually *purplish*; frequently with *short underground tillers*. *A common perennial plant of rough grasslands, roadsides, etc.* **Festuca arundinacea**

36. Auricles very *prominent* 37
    Auricles very *small* (Fig. 29.37); blades up to 25 cm x 2-6 mm, somewhat 'parallel-sided', ribbed adaxially, deep glossy-green abaxially and *contrastingly* dull green adaxially, keeled, smooth or minutely rough, *glabrous*, and pointed or blunt; the youngest blades *convolute*; sheaths split or rarely entire, slightly compressed but *not* keeled, often purplish basally; ligule short, 1.0-2.5 mm, blunt, entire. *Ubiquitous perennial grass, of great economic importance.* **Lolium perenne**

37. Prophyl *narrow, glabrous* and *rounded* apically (Fig. 32.11 p); the blades 10-50 cm x 5-10 mm, light glossy-green in colour and moderately keeled abaxially, bright green, moderately ribbed glabrous throughout, rough or smooth on the margins, and tapered to a fine point; the youngest blades *convolute*; auricles very *prominent*, glabrous; sheaths rounded, split, smooth; ligule about 1 mm long. *Frequent nutritious perennial plant of damp fertile soils.* **Festuca pratensis**

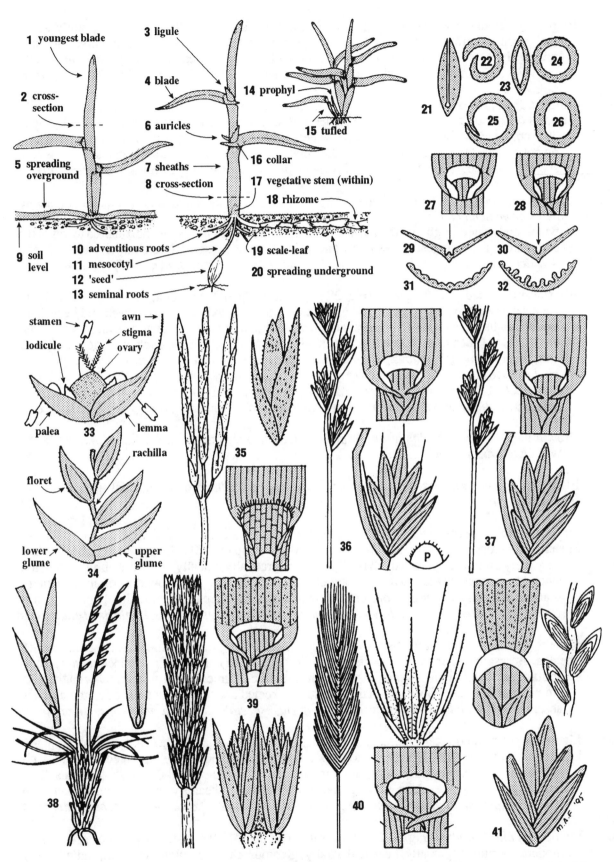

1 youngest blade
2 cross-section
5 spreading overground
9 soil level
10 adventitious roots
11 mesocotyl
12 'seed'
13 seminal roots

3 ligule
4 blade
6 auricles
7 sheaths
8 cross-section

14 prophyl
15 tufled
16 collar
17 vegetative stem (within)
18 rhizome
19 scale-leaf
20 spreading underground

21
22
23
24
25
26
27
28
29
30
31
32

stamen
awn
stigma
lodicule
ovary
palea
33
lemma

rachilla
floret
lower glume
upper glume
34

35
36
P
37
38
39
40
41
M A F '95

Fig. 29.00

N.B. DRAWINGS NOT TO SCALE.

Prophyl *broad* and *with a few hairs* (Fig. 29.36 p); the blades up to 30 cm x 12 mm, rarely wider, dull green and prominently ribbed adaxially, keeled and dull *glossy-green* abaxially; smooth or rough, glabrous, and finely pointed; the youngest blades *convolute*; auricles *prominent*; sheaths rounded, split, smooth or rough. *An annual or biennial cultivated grass.* **Lolium multiflorum**

38.   Auricles *small, not* overlapping 39
Auricles *overlapping* (Fig. 36.01); the blades soft, finely ribbed adaxially, keeled abaxially, up to 30.0 cm long and 12.0 mm wide, rarely wider, twisted *clockwise*, glabrous or shortly pubescent on collar area, tapered to a fine point; the youngest blades *convolute*; the ligule short, 1-2 mm; sheaths rounded, split. *An annual cultivated plant of economic importance.* **Hordeum sativum**

39.   Auricles *glabrous* or with *minute* hairs 40
Auricles *prominently hairy* (Fig. 34.01); blades soft, finely ribbed, up to 30 cm long and 15.0 mm wide, rarely wider, light green, *twisted clockwise*, mostly glabrous, or short-haired towards the collar area; the youngest blades *convolute*; sheaths rounded, split; ligule rather short, 1-2 mm. *A widely cultivated, annual grass, of world-wide economic importance.* **Triticum aestivum**

40.   Abaxial surface *deep glossy-green* (Fig. 29.37); blades up to 25 cm long, rather narrow, 2-6 mm wide, somewhat 'parallel-sided', *dull green* and ribbed adaxially, *glossy-green* and keeled abaxially, glabrous, smooth or slightly rough, and pointed or blunt; youngest blades mostly *conduplicate;* the sheaths split, rarely entire, slightly compressed but *not keeled*; ligule short,1-2.5 mm, blunt, entire. *Ubiquitous perennial plant, of great economic importance.* **Lolium perenne**
Abaxial surface *not* deep glossy-green (Fig.38.04); blades finely ribbed, up to 30.0 cm x 12.0 mm, *twisted clockwise*, minutely hairy or glabrous; youngest blades *convolute*; auricles *characteristically small*; ligule 1.0-2.0 mm; sheaths rounded, split. *Annual cereal grass.* **Secale cereale**

41.   Growth-habit *cespitose, stoloniferous* or plant *with* trailing shoots 50
Growth-habit *rhizomatous* 42

42.   Blades *pubescent*, or at least with a few very *short, appressed hairs* on the adaxial surface 47
Blades *glabrous* 43

43.   Sheaths *without* externally visible cross-veins 44
Sheaths *with* cross-veins (Fig. 30.12); blades *grey-green*, up to 35.0 x 2.5 cm, finely ribbed adaxially, very prominently keeled abaxially, tapered gradually apically and more abruptly basally, smooth; youngest blades *convolute*; auricles *absent*; sheaths split, rounded, often with membranous margins and *conspicuous* cross-veins; ligule *broad*, up to 10 mm, bluntly toothed; rhizomes *thick, extensive, spreading. A common perennial of wet habitats.* **Phalaris arundinacea**

44.   Sheaths, inner, *not* white or whitish in colour 45
Sheaths, at least the inner, *characteristically white* (Fig. 32.05); the latter rounded, *entire*, soon splitting, smooth; ligule *minute*; blades slender, soft, dull green, margins minutely rough, 5-15 cm x 2-5 mm, finely ribbed adaxially, obscurely keeled abaxially, blunt apically; the auricles absent. *A perennial with short rhizomes, frequent, especially on calcareous soils.* **Briza media**

45.   Ligule *not* short and truncate 46
Ligule *truncate* (Fig. 31.05); blades 5-20 cm x 2-6 mm, rarely wider, finely pointed, ribbed adaxially, moderately keeled abaxially, soft or stiff, flat or involute, slightly rough, glabrous, light or dark green; youngest blades *convolute*; auricles absent; sheaths rounded, split, smooth. *An ubiquitous perennial plant; old pastures, meadows, roadsides, heaths, etc.* **Agrostis capillaris**

46.   Ligule *very blunt, toothed* (Fig. 31.06); blades up to 25 cm x 12.0 mm, usually dull or *grey-green*, moderately ribbed adaxially, keeled abaxially, glabrous, finely pointed, flat, firm, and minutely rough; the youngest blades *convolute*; auricles *absent*; sheaths rounded, split, smooth or rough. *A frequent perennial in many open habitats, and especially in tillage fields.* **Agrostis gigantea**

Ligule *relatively long*, up to 5 mm, membranous, the uppermost *tapered, toothed* or *torn* (Fig. 31. 04); the blades relatively short and narrow, up to 20.0 cm x 1.0-3.0 mm, flat or inrolled, sometimes *bristle-like*, finely ribbed, glabrous, smooth or rough; auricles *absent*; the youngest blade *convolute;* rhizomes slender. *Frequent; heaths, hill pastures and peaty soils.* **Agrostis vinealis**

47. Ligule *prominent* 49
    Ligule *minute* or *absent* 48

48. Blades *green* (Fig. 33.12); the latter soft, 10-100 cm x 2.0-5.0 mm, glabrous, ribbed adaxially, ribs alternating large and small; bluntly keeled abaxially; pointed; youngest blades *convolute*; auricles absent; sheaths entire, mostly *pubescent*; ligule *minute* or *absent*; rhizomes thin, *thread-like*; *A very common, variable grass: wall-tops, roadsides, dunes, pastures, etc.* **Festuca rubra**
    Blades *glaucous* (Fig. 29.41); the latter firm, flat or involute, up to 40 cm x 3-8 mm, prominently ribbed adaxially, ribs densely covered with *minute appressed hairs*, smooth and *rounded* abaxially, finely pointed, abruptly *contracted* basally; the sheaths rounded, *split*, smooth; auricles *absent*; ligule minute. *Very frequent perennial of dunes and sandy sea shores.* **Elymus farctus**

49. Sheaths *with* distinct pink-coloured veins (Fig. 31.08); the latter rounded, mostly split, softly pubescent or glabrous; ligule 2.0-5 mm long, hairy, bluntly toothed; the blades mostly *pubescent*, soft, rarely glabrous, 5-25 cm x 5-15 mm, moderately keeled abaxially, ribbed, and *dull grey-green* in colour, flat, pointed, mostly smooth; auricles absent; rhizomes tough, thick, creeping. *A common plant in open woodlands, heathlands, etc.; rarely a weed of tillage.* **Holcus mollis**
    Sheaths *without* pink-coloured veins (Fig. 30.07); the latter split, smooth; blades *glaucous*, firm, up to 75 cm x 6 mm, flat or involute, prominently ribbed adaxially, ribs minutely and densely hairy, rounded abaxially, sharp-pointed; the youngest blades *convolute*; auricles absent; ligule *very prominent*, length up to *25.0 mm*, tapered; the rhizomes very extensive, spreading. *A perennial, and an important sand stabiliser; very common in coastal areas.* **Ammophila arenaria**

50. Growth-habit *cespitose* 54
    Growth-habit *stoloniferous* or plant *with* trailing shoots 51

51. Ligule margins *not* overlapping 52
    Ligule margins *clearly overlapping* (Fig.31.09); the latter usually entire and blunt, or rarely finely toothed; blades short, up to 22.0 cm x 8 mm, ribbed adaxially, keeled, flat, light or yellowish-green, pointed, glabrous but rough on the ribs, smooth abaxially; auricles absent; sheaths split, rounded, the upper often inflated. *A frequent perennial in wet places.* **Alopecurus geniculatus**

52. Ligule *not* tapered 53
    Ligule *relatively long*, up to 5 mm, membranous, the uppermost *tapered, toothed* or *torn* (Fig. 31. 03); the blades relatively short and narrow, up to 20.0 cm x 1.0-3.0 mm, flat or inrolled, sometimes *setaceous*, finely ribbed, glabrous, smooth or rough; auricles *absent*; youngest blade *convolute;* stolons slender. *Frequent perennial; damp or wet lowland peaty soils.* **Agrostis canina**

53. Ligule *truncate* (Fig. 31.05); blades 5-20 cm x 2-6 mm, rarely wider, finely pointed, ribbed adaxially, moderately keeled abaxially, soft or stiff, flat or involute, slightly rough, glabrous, light or dark green; youngest blades *convolute*; the auricles absent; sheaths rounded, split, smooth. *An ubiquitous perennial of old pastures, meadows, roadsides, heaths, etc.* **Agrostis capillaris**
    Ligule *not* truncate (Fig. 31.07); the latter prominent, up to 5 mm, often blunt and finely toothed; the blades up to 18 cm x 5 mm, dull or grey-green, pointed, finely and closely ribbed adaxially, moderately keeled abaxially, glabrous, minutely rough; youngest blades *convolute*; auricles absent; sheaths rounded, split, smooth; the stolons often extensive. *A common perennial in old pastures, roadsides, waste places, hedgerows, salt-marshes, heaths, etc.* **Agrostis stolonifera**

54. Plant-base *not* bulbous 56
    Plant-base *bulbous, i.e.* with a very small *onion-like* swelling 55

55. Sheath-bases *green or chocolate-brown* in colour (Fig. 30.08); the latter rounded, split, smooth; blades widest towards the base, light green, greyish-green or somewhat glaucous, up to 18 cm x 10 mm, pointed, mostly glabrous, flat, firm, inconspicuously ribbed adaxially, clearly keeled abaxially, smooth or minutely rough; the youngest blades *convolute*; auricles absent; ligule up to 6.0 mm, blunt. *A perennial, frequent in pastures, meadows, roadsides, etc.* **Phleum pratense**
   Sheath-bases and roots *yellowish* or *rust-brown* in colour; small onion-like swellings (Fig. 31.11a) invariably present on plant base; ligule (Fig. 31.11b) 2-5 mm long, blunt, often toothed; blades characteristically *dull grey-green* in colour, flat or finely ribbed, moderately keeled abaxially, widest at a point one-third the distance from base to apex, finely pointed, up to 60 cm x 18 mm, mostly with *scattered inconspicuous hairs*, rarely glabrous; auricles *absent. An ubiquitous perennial; rough grasslands, roadsides, waste places, hedgerows, etc.* **Arrhenatherum elatius**

56. Plants with at least *some hairs*, the latter *conspicuous* or *inconspicuous* 67
   Plants *glabrous* 57

57. Sheaths, inner, *not* whitish in colour 58
   Sheaths, at least the inner, characteristically *whitish* (Fig. 32.05); the latter rounded, *entire*, soon splitting, smooth; ligule *minute*; blades slender, soft, dull green, margins minutely rough, 5-15 cm x 2-5.0 mm, finely ribbed adaxially, obscurely keeled abaxially, blunt apically; auricles absent. *Perennial often with short rhizomes, frequent, especially on calcareous soils.* **Briza media**

58. Sheath-bases some colour *other than* purple or chocolate-brown 62
   Sheath-bases *purplish* or *chocolate-brown* in colour 59

59. Abaxial surface *not* deep glossy-green 60
   Abaxial surface *deep glossy-green* in colour; blades up to 25 cm long, rather narrow, 2.0-6.0 mm wide, somewhat 'parallel-sided' for most of their length, moderately ribbed adaxially, *contrastingly dull green* in colour abaxially, moderately keeled abaxially, glabrous, smooth or minutely rough and pointed or rather blunt; youngest blades mostly *conduplicate*; sheaths split, rarely entire, slightly compressed but *not keeled*; the ligule (Fig. 29.37) rather short, 1.0-2.5 mm long, blunt, entire. *An ubiquitous perennial grass of great economic importance.* **Lolium perenne**

60. Ligule *somewhat* rounded apically and 2-6 mm long 61
   Ligule *truncate* (Fig. 30.10); the latter 1.0-2.5 mm long; blades up to 100.0 cm x 12 mm for plant in shaded situations, frequently much less in exposed areas, rough or nearly smooth, glabrous throughout, flat, moderately ribbed adaxially, not prominently keeled abaxially, of a dull green colour; youngest blades *convolute*; auricles *absent*; sheaths rounded, split, colour mostly dark. *A common perennial of old pastures, meadows, grassy and waste places.* **Alopecurus pratensis**

61. Ligule *pubescent* (Fig.30.11); the latter up to 6.0 mm, toothed; blades 5-20 x 0.3-1.0 cm, glabrous, green, smooth or rough, finely ribbed adaxially, keeled abaxially, pointed; the youngest blades *convolute*; auricles absent; sheaths rounded, split, green or purplish. *An annual grass, frequent as a weed of tillage fields, waste places, etc., though rare in Ireland.* **Alopecurus myosuroides**
   Ligule *glabrous* (Fig.30.08); the latter up to 6 mm, blunt; the blades widest towards the base, light green, greyish-green or somewhat glaucous, up to 18.0 x 1.0 cm, pointed, mostly glabrous, flat, firm, inconspicuously ribbed adaxially, clearly keeled abaxially, smooth or minutely rough; youngest blades *convolute*; sheaths *green or chocolate-brown* basally, rounded, split, smooth; auricles absent. *A perennial, frequent in meadows, pastures, roadsides, etc.* **Phleum pratense**

62. Ligule *not* tapered; blades firm but *not* stiff 63
   Ligule *tapered, very prominent*, up to 15.0 mm (Fig. 32.02). Blades dull green in colour, *prominently ribbed*, very rough on the ribs and margins, somewhat *stiff*, sharply pointed, glabrous, *not* distinctly keeled abaxially, flat or inrolled adaxially and up to 80.0 cm x 2-5 mm; auricles *absent;* the sheaths split, smooth or rough upwards, rounded or very *poorly keeled. A very frequent densely tufted, useless perennial of poorly drained grasslands.* **Deschampsia cespitosa**

Fig. 30.00

63.   Sheath-bases *neither yellowish* nor *rust-brown in colour* 64
        Sheath-bases and roots of a *yellowish* or *rust-brown* colour; small outgrowths or protuberances
            invariably present on plant base; ligule (Fig. 31.11) 2-5 mm long, blunt, often toothed; blades
            characteristically *dull grey-green* in colour, flat or finely ribbed, moderately keeled abaxial-
            ly, widest at a point one-third the distance from base to apex, finely pointed, up to 60 cm long
            x 18 mm wide; *appearing* hairless, though very occasionally so; auricles *absent. An ubiquitous
            perennial; rough grasslands, roadsides, waste places, hedgerows, etc. Arrhenatherum elatius*

64.   Ligule margins *not* overlapping 65
        Ligule margins *clearly overlapping* (Fig. 30.09); the latter usually entire and blunt, rarely finely
            toothed; blades short, up to 22 cm x 8.0 mm, ribbed adaxially, keeled, flat, light or yellowish-
            green, pointed, glabrous but rough on the ribs, smooth abaxially; auricles absent; sheaths split,
            rounded, the upper often inflated. *A frequent perennial in wet places. Alopecurus geniculatus*

65.   Ligule *short, flat-topped*; blades *not* twisted anti-clockwise 66
        Ligule *very prominent* (Fig. 38.01); blades variable, *twisted anti-clockwise*, up to 50 cm x 30 mm,
            but often much less, finely ribbed adaxially, moderately keeled abaxially, green or dark green,
            glabrous, pointed; youngest blades *convolute;* sheaths rounded, split, smooth or rough; the au-
            ricles absent. *Annual cultivated cereal plant of decreasing economic importance. Avena sativa*

66.   Sheaths, *outer, characteristically canary-yellow* (Fig.30.05); the latter rounded or *somewhat com-
            pressed*, split; the blades *finely to moderately ribbed*, 5-20 cm x 2-6 mm, *dull green* adaxially,
            glossy, moderately keeled abaxially, tapered apically, widest and *characteristically asymmet-
            rical* basally and glabrous; youngest blades rarely *convolute*; auricles absent; ligule minute. *A
            common perennial of old pastures, meadows, waste and grassy places.   Cynosurus cristatus*
        Sheaths, all, *green*; the latter rounded, split, smooth; blades 5.0-20.0 cm x 2.0-6.0 mm, very rarely
            wider, finely pointed, moderately ribbed adaxially, moderately keeled abaxially, soft or some-
            what stiff, flat or involute, slightly rough, glabrous throughout, of a light or dark green colour,
            and *symmetrical basally;* ligule (Fig. 31.05) short, 1.0-2.0 mm, *truncate;* auricles *absent. An
            ubiquitous perennial plant of old pasture, roadsides, meadows, heaths, etc. Agrostis capillaris*

67.   Blades *neither* yellowish-glossy-green *nor* the keeled abaxial surface *directed* upwards 68
        Blades *yellowish-glossy-green* (Fig.30.04); the latter flat, up to 45 cm x 14 mm, *characteristically
            yellowish-glossy-green, inverted*, pointed, central groove well defined, thin, widest about the
            middle, prominently keeled, and with the *macro-hairs more distinct on the keel and margins
            than elsewhere,* hairs on latter directed apically; youngest blades *convolute*; sheaths rounded
            or slightly keeled, slender, smooth and *entire*; the auricles absent; ligule 1-3 mm, toothed, mi-
            nutely hairy. *A common perennial of woodlands and shaded places.Brachypodium sylvaticum*

68.   Sheaths *split* 71
        Sheaths *entire* 69

69.   Sheaths *lacking* distinct pink-coloured veins 70
        Sheaths with distinct *pink-coloured veins*; the latter rounded, often with *reflexed* hairs, rarely gla-
            brous; blades 25.0 cm x 5-15 mm, usually flat, rather soft, pointed apically, moderately ribbed
            adaxially, moderately keeled abaxially, usually covered with a *tangled mass of matted hairs*,
            rarely with a few hairs, characteristically *grey-green* or somewhat *glaucous*; youngest blades
            convolute; the auricles *absent*; ligule (Fig. 31.09) 1-5 mm, blunt, finely toothed, pubescent. *An
            ubiquitous perennial; old pastures, tillage fields, roadsides, waste places, etc. Holcus lanatus*

70.   Blades *dull greyish-green*; the latter finely pointed, rarely more than 30.0 cm x 8.0 mm, flaccid to
            firm, usually *sparsely* pubescent, flat, minutely rough, ribbed adaxially, keeled abaxially and
            dull green or with purplish tinge; youngest blades *convolute*; auricles *absent*; sheaths rounded
            or very slightly keeled, entire but soon splitting; the ligule (Fig. 33.07) 2.0-5.0 mm long, finely
            toothed. *Annual or biennial, frequent in several habitats with dry sandy soils. Bromus sterilis*

Blades *green* (Fig. 33.04); the latter up to 20.0 x 1.0 cm, soft, finely ribbed adaxially, moderately keeled abaxially, pointed, densely and shortly pubescent, mostl y smooth; the youngest blades *convolute*; the auricles absent; sheaths rounded, rarely somewhat keeled, *entire*, splitting, pubescent, occasionally with light purplish streaks; ligule 2-3 mm, pubescent, toothed. *Common annual or biennial of grasslands, roadsides, waste places, sand-hills, etc.* **Bromus hordeaceus**

71. Hairs *scattered throughout* the entire grass 75
    Hairs *few* and *localised* 72

72. Sheath-bases *not* chocolate-brown in colour 73
    Sheath-bases *mostly chocolate-brown* in colour (Fig. 30.08); the former rounded, split, smooth; blades widest at the base, light green, greyish-green or somewhat glaucous, up to 18.0 x 1 cm, finely pointed, with only a few hairs, firm, very finely ribbed adaxially, prominently keeled abaxially, and smooth or minutely rough; the youngest blades *convolute*; auricles absent; ligule 2-6.0 mm, blunt. *A perennial, frequent in pastures, meadows, roadsides, etc.* **Phleum pratense**

73. Ligule *prominent* 74
    Ligule *minute, truncate* or *flat-topped* (Fig. 33.08); blades 5-15 cm x 0.5-2.5 mm, finely pointed, flat or more often inrolled, shortly pubescent adaxially, soft to somewhat firm, finely ribbed, green or yellowish-green; sheaths rounded, smooth; auricles *absent. A frequent annual plant of dry stony or open sandy places, heathlands, hill grassland, waste areas, etc.* **Vulpia bromoides**

74. Hairs *confined* to the collar area (Fig. 38.01); blades variable, *twisted anti-clockwise*, up to 50 cm x 30 mm, often much less, finely ribbed adaxially, moderately keeled abaxially, green or dark green, mostly glabrous, pointed; youngest blades *convolute;* auricles absent; sheaths rounded, split; ligule up to *7.50 mm. Annual cereal plant of limited economic importance. A vena sativa*
    Hairs present on *collar area* and the *sheath* (Fig. 38.02); the latter rounded, split; blades variable, soft, up to 50 x 3.0 cm, often much less, finely ribbed adaxially, moderately keeled abaxially, green or dark green, twisted *anti-clockwise*; the auricles absent; the ligule up to 7.5 mm. *A frequent, very troublesome annual plant, of many cultivated areas, waste places, etc. Avena fatua*

75. Sheath-bases *neither* yellowish *nor* rust-brown in colour; hairs conspicuous 76
    Sheath-bases and roots of a *yellowish* or *rust-brown* colour; small outgrowths or protuberances invariably present on plant base; ligule (Fig. 31.11) 2-5 mm long, blunt, often toothed; blades characteristically *dull grey-green* in colour, flat or finely ribbed, moderately keeled abaxially, widest at a point one-third the distance from base to apex, finely pointed, up to 60 cm x 18 mm, mostly with *scattered inconspicuous hairs*, rarely glabrous; auricles *absent. An ubiquitous perennial; rough grasslands, roadsides, waste places, hedgerows, etc.* **Arrhenatherum elatius**

76. Sheaths *lacking* pink-coloured veins 77
    Sheaths *with* distinct pink-coloured veins (Fig.31.09); the latter rounded, often with *reflexed* hairs; blades 25 cm x 5-15 mm, finely ribbed, soft, pointed, usually *covered with a tangled mass of long hairs*, rarely with only a few hairs, characteristically *grey-green* or somewhat *glaucous;* youngest blades *convolute*; auricles absent; ligule 1-5 mm, blunt, apex finely toothed, hairy. *A ubiquitous perennial; old pastures, tillage fields, roadsides, waste places, etc.* **Holcus lanatus**

77. Hairs *longer* and *more* plentiful on collar than elsewhere (Fig. 30.06); blades variable, 5-20 cm x 3-15 mm, finely ribbed adaxially, keeled abaxially, pointed, dull green or yellowish-green, *collar with a characteristic tuft of long hairs*, hairs elsewhere shorter, plentiful; youngest blades *convolute;* auricles absent; sheaths rounded, split; ligule prominent, 2-5.0 mm, toothed. *A common perennial of hill and lowland pastures, heaths, peaty soils, etc.* **Anthoxanthum odoratum**
    Hairs *neither* longer *nor* more plentiful on collar area than elsewhere (Fig. 31.12); blades 5-20 x 2-6 mm, finely ribbed adaxially, tapered, green or light green, and clearly pubescent; youngest blades *convolute*; sheaths rounded, split, with an abundance of *reflexed hairs*; ligule minute. *Frequent perennial of grassy places, especially those on calcareous soils.* **Trisetum flavescens**

TAXONOMIC DATA - INFLORESCENCE PHASE

Identification of the second phase is generally much easier as it is not influenced or altered to any great extent by environmental conditions. The only significant variation that may be noticed is a difference between the height of the flowering stem from plants growing on soils of different fertility. Thus, *Lolium perenne* flowering stems can be as little as 3 cm when growing on soils between rock crevices and other similar situations, or up to 60 cm or more when growing on rich moist soils. Little notice should be taken, therefore, of marked differences in flowering stem height. The essential structure of the floral parts remains unchanged by environmental conditions.

Structure

The grass inflorescence differs from most other flowering plants in that there are no colourful floral parts, such as sepals and petals, present. The latter are one of nature's inventions to attract pollinators, but as grasses are mostly wind or self-pollinated, there is no need for colourful floral structures to attract insects to the florets. Instead the essential male and female parts of the floret are enclosed within small leaf-like structures termed flowering pales.

The grass floret: The flower or floret of most grasses contains both male and female parts (Fig. 29.33). The sexual parts show very little variation from one species to another. The male, in the grass family as a whole, has 6 stamens, in 2 whorls of 3. In most forage grasses the number is 3, though in a few species it may be further reduced to 2 or just 1 stamen. The female has a single pistil, consisting of a small ovoid ovary, very short style, and a well-branched stigma. Both essential parts are always enclosed within 2 small leaf-like structures termed *flowering pales*. The term flowering pale is used because they directly enclose the sexual parts.

The female part or *pistil* always occupies the central position. The *feather-like* stigma provides a very large receptive area for wind blown pollen. The ovary, which contains a single ovule, is usually subtended by 2 small, hygroscopic, scale-like structures, known as *lodicules*. These play an important role in the opening of the grass floret to facilitate cross-pollination.

One flowering pale is always comparatively large and is termed the *outer flowering pale* or *lemma*. It is more important than the second for identification purposes because it shows a much wider range variation. The second, known as the *inner flowering pale* or *palea*, is always smaller and has only 2 veins. Because of the consistency shown by the latter pale, from one grass species to the next, it is of little use in grass identification. Grass florets which contain both male and female parts are described as being *bisexual* or *hermaphrodite*. However, the florets of a few species may lack a pistil. When this happens the floret is said to be *staminate*. Some florets of *Phalaris arundinacea* and *Anthoxanthum odoratum* are without sexual organs. These florets are said to be *asexual*.

Flowering pales: Flowering pales of different species differ in many ways. The most obvious variation is the presence in some species of an *awn* or *bristle*, *e.g. Avena fatua* (Fig. 38.02), and the absence of the bristle from others, *e.g. Briza media* (Fig. 32.05). The bristle may be straight (Fig. 24.12) or bent and knee-like (Fig. 24.24). Furthermore, its position on the flowering pale may be significant. It may arise from the base of the pale, from the mid-dorsal area, from near the apex, or directly from the apex. In all cases the bristle is an exsertion of the central vein.

The number of *veins*, also called nerves or vascular bundles, always provide a very important clue towards identification. Thus, in most *Bromus* species and in *Avena fatua* and *A. sativa*, the number is never less than 7, whereas in all *Festuca* species, and most other grasses, the number is consistently 5. The common moor plant, *Molinia caerulea*, is characterised by having only 3 veins (Fig. 32.10). Some difficulty may be experienced in determining vein number. The problem can be eased considerably by pre-soaking in 10% alcohol, or by using reflected light from a binocular microscope. The pale apex, in a small

Fig. 31.00

number of grasses, is very ragged (Fig. 24.38), or has a definite number of teeth. *Danthonia decumbens* is the only grass with a well defined 3-toothed flowering pale (Fig. 25.09).

The presence or absence of hairs, or the location of hairs, can be a further help. *Bromus hordeaceus* has very hairy pales (Fig. 33.04), while those of *Festuca pratensis* (Fig. 25.37), and many others, are hairless. *Poa trivialis* (Fig. 25.32) has the hairs confined to the central vein, whereas in *Poa nemoralis* and *P. pratensis* the hairs are present on both the central and marginal veins (Figs 25.29-30). Species of the latter genus may also have a tuft of silky hairs or 'web' at the base of the lemma. The shape of the pale may be characteristic of a species. *Briza media*, for example, has very distinctive concave pales (Fig. 25.12), while those of *Vulpia bromoides* are very narrow or linear (Fig. 33.08). *Bromus hordeaceus* (Fig. 24.34) has obovate pales. They are *keeled* in *Dactylis glomerata* and all *Poa* species (Figs 25.28-32). Those of the *Lolium* species are *flat* dorsally. Length is not a dependable feature. Texture may be firm, as in the *Lolium* species and many more, or membranous as in the *Agrostis* species and others.

The grass spikelet: The spikelet is the ultimate branch of the inflorescence. It may be located by separating the inflorescence until the end of a branch is found. The essential difference between a spikelet and a floret is that the former is mostly subtended by 2 additional structures termed *empty pales*, collectively known as the *glumes*. The term empty is used because they do not directly enclose the sexual parts of the floret. The inflorescences of the *Lolium* species are characterised by the presence of only 1 empty pale (Fig. 29.36-37) on each lateral spikelet. The spikelet of *Nardus stricta* is distinguished by its very small lower empty pale and the absence of the upper (Fig. 29.38). The number of florets in a spikelet can vary from 1 to many. Spikelets are 1-flowered in *Ammophila arenaria* (Fig. 30.07), *Phleum pratense* (Fig. 30.08), all *Alopecurus* (Figs 30.09-11) and *Agrostis* (Figs 31.03-07) species, and 2 or more-flowered in most other grasses (Figs 31.10-33.12). However, it is important to note that the number of florets, above 2 per spikelet, is not consistent, even for spikelets of a single inflorescence.

Figure 29.34 shows the relative positions of the structures found in a typical many-flowered spikelet. Six florets are shown. These are attached to a central flexuous stalk known as the *rachilla*. Subtending the 6 florets and rachilla are the 2 empty pales. One is attached at a slightly lower level than the other and, accordingly, is known as the *lower* empty pale; the second is known as the *upper* empty pale.

Differences in the empty pales are often used in the identification of a grass inflorescences. They may be about equal in length (*e.g.* Fig. 30.10), or the upper may be nearly twice as long as the lower (*e.g.* Fig. 30.06). Where the number of florets per spikelet is 1 or 2, the empty pales are mostly equal in length to that of the spikelet (*e.g.* Figs 30.08-12). On the other hand, where the spikelet contains many florets, the pales are usually shorter than the spikelet (*e.g.* Figs 33.01-04). Differences in texture, number of veins and pubescence, also occur.

When the spikelets are ripe they shatter readily and the florets or 'seeds' fall away from each other. In most grasses the fracture which enables the 'seeds' to part occurs at the rachilla joints (Fig. 29.04). Where this happens the 'seeds' separate with 1 internode of the rachilla permanently attached. The name rachilla is retained for the internode remaining on the 'seed.' Alternatively, this rachilla internode may be known as the *'seed-stalk'*. In a few species, particularly *Holcus mollis*, *H. lanatus* and *Alopecurus* species, disjoining takes place where the spikelets are attached to the inflorescence branches. This type of disjoining, which is readily recognised by the fact that the entire spikelet comes away with the minimum of force, results in the spikelet falling away to give the 'seed' of some grasses. The mode of disjoining, or *disarticulation*, can be ascertained by gently tapping the inflorescence against some object. Where disarticulation is above the empty pales the florets will disunite leaving the former attached to the branch ends; where disarticulation is below the level of the empty pales the latter will fall off leaving bare branch ends.

In all but 3 grasses the spikelets of the inflorescence are alike. The exceptions are found in *Cynosurus cristatus* and in some *Hordeum* species. The *Hordeum* species can have all spikelets fertile, or more often, 1 of each 3-spikelet cluster fertile and the other 2 asexual or staminate. The spikelets occur in pairs in *Cynosurus cristatus*. One member of each pair shows normal florets, while the second carries all asexual florets (Fig. 30.05). One form of *Festuca ovina* has most unusual spikelets in that many of the florets are replaced by very small plant-like structures.

Spikelet arrangement: The inflorescence consists of the flowers, spikelets and the many branches which unite them. Inflorescence type is consistent for any given species. The main stem of the inflorescence is termed the *rachis* (Fig. 30.03r). The spikelets are attached to the rachis directly, or by primary, secondary, or tertiary branches. Mode of attachment determines the inflorescence type, and the latter is consistent for the individuals of any given species.

Where the spikelets are attached by secondary and tertiary branches the inflorescence is termed a *panicle* (*e.g.* Figs 31.02-31.06). In situations such as this the branches are easily seen. Good examples of a panicle are shown by the *Avena* species, *Arrhenatherum elatius* all *Festuca* and *Agrostis* species, and many more. Many grasses have the spikelets attached directly to the rachis. This type of inflorescence is called a *spike* (Fig. 29.36). *Triticum aestivum, Elymus repens, Lolium* species, and many more, have the spike-type inflorescence. *Hordeum* species, characterised by the presence of 3 spikelets per rachis node, may be said to have a modified spike in that the lateral spikelets are short-stalked. An inflorescence, which is somewhat intermediate between a panicle and a spike, is the *spike-like panicle* (Fig. 30.07). Here the secondary and tertiary branches are very short and covered by the spikelets. This results in a compact panicle, but, since it resembles a spike, it is, for convenience, termed a spike-like panicle. Examples of this are found in *Alopecurus pratensis, Cynosurus cristatus* and several other species.

Two rare types are found in a few common grasses. Thus, in *Spartina x townsendii*, a common plant of coastal mud-flats, what is known as a *compound spike* occurs (*e.g.* Fig. 29.35). Here many spikelets are attached directly to the primary branches. A similar type may be found as an abnormality in *Lolium* species and *Elymus repens*. Finally, in 2 other grasses - *Brachypodium sylvaticum* and *B. pinnatum* - the spikelets are attached to the rachis by a very short primary branch. This type of grass inflorescence is called a *raceme* (*e.g.* Fig. 30.04).

Anthesis: Most grass inflorescences emerge between the months of May and July. A few, particularly *Alopecurus pratensis* and *Hierochloe odorata*, depending on weather conditions, appear late February or early March. One, *Poa annua*, produces an inflorescence throughout the entire year. Other *Poa* species develop inflorescences towards the end of March or early April. *Agrostis* species are the latest of all, their inflorescences generally emerge in the months of June and July. Grasses in areas which are slow to warm, such as in woodlands and on mountains, are usually slower than those in lowland situations. Actual flowering, or *anthesis*, differs with different species.

Most chasmogamic grasses, depending on weather conditions, have regular daily opening times. *Anthesis* or flowering begins with the lodicules swelling and pushing the flowering pales apart. The stigmas separate, spread outwards, and become more feathery in appearance. At the same time the filaments elongate rapidly and push the anthers out of the floret. An abundance of light pollen is released into the air where it is wafted away, much to the annoyance of people who suffer from pollenosis! After a short interval, the filaments fall-off, the lodicules dry-out, shrivel, become reduced in size, and the flowering pales close and interlock once more. An examination of the lodicules at this time provides little useful information. However, they can be of taxonomic importance. Their general features are best studied in a young inflorescence before anthesis takes place.

The majority flower from 4-9 a.m., some about noon, and others in the late afternoon, *i.e.* 3-7 p.m. Two species, *Anthoxanthum odoratum* and *Holcus lanatus*, flower twice daily, from 6-10 a.m. and from 5-7 p.m. A few grasses, such as *Alopecurus pratensis* and *Spartina x townsendii*, lack lodicules. The stigma in these grasses pushes out from the pale tips. Some *Bromus* species and *Danthonia decumbens* exhibit cleistogamic florets. These grasses exhibit self-pollination which takes place within the closed florets. In all grasses, anthesis begins in the middle of the inflorescence, on the main culm and, thereafter, spreads upwards and downwards until all the florets have flowered. Several days are required for anthesis to reach completion. It has been estimated that an acre of *Triticum aestivum* takes 2 weeks from the time the first floret opens until the last floret opens.

KEY TO THE INFLORESCENCE PHASE

1.　Inflorescence a *spike, modified spike* or a *compound* spike 2
　　Inflorescence a *panicle, spike-like panicle*, or a *raceme* 16

2.　Spike *simple* (*e.g.* Fig. 30.01) or *modified* (*e.g.* Figs 24.07 and 36.1a) 5
　　Spike *compound* (*e.g.* Fig. 29.35) 3

3.　Spikelets *more* than 1-flowered; lateral spikelets with 1 glume 4
　　Spikelets *1-flowered* (Fig. 29.35); the latter with narrow, somewhat pointed glumes; inflorescence
　　　　a dull brown colour. *A common plant on coastal mud-flats where it is an aggressive coloniser,
　　　　usually covering very large areas of ground; it is a male-sterile hybrid.* **Spartina x townsendii**

4.　Lemmas *with* awns (Fig. 29.36); the former rounded or flat dorsally, 5-nerved, smooth or minute-
　　　　ly rough, blunt, 5-10 mm. Spikes vary from 10-30 cm. Spikelets up to 3.0 cm, 5-14-flowered,
　　　　narrow edges towards the rachis; all, except the terminal, with 1 glume; the latter either blunt
　　　　or pointed, firm, apices thin, 5-7-nerved. *Abnormal form of couplet no.14.* **Lolium multiflorum**
　　Lemmas *lacking* awns (Fig. 29.37); the former rounded or flat dorsally, 5-nerved, smooth or mi-
　　　　nutely rough, blunt, and 5-10 mm. Spikes 10-30 cm. Spikelets up to 3 cm, 5-14-flowered, nar-
　　　　row edges towards the rachis, all, except the terminal, with 1 glume; the latter blunt or pointed,
　　　　firm but with thin apices, and 5-7-nerved. *An abnormal form of couplet no.15.* **Lolium perenne**

5.　Spikelets carried on *2* sides of the rachis 6
　　Spikelets carried on *1* side of the rachis, in 2 rows (Fig. 29.38), finely pointed, 1-flowered, and all
　　　　very narrow. Lower glumes small, upper mostly absent. Culms erect, wiry, up to 45 cm, rarely
　　　　more. *Widespread plant, common on peaty soils, hill and mountain pasture, etc.* **Nardus stricta**

6.　Spikelets *1* or *3* per rachis-node 7
　　Spikelets *2* per rachis-node (Fig.29.39), 2-5-flowered, wedge-shaped, 20-30 mm long. The glumes
　　　　and lemmas similar, all pointed, firm, and covered with short dense hairs. Spikes firm, com-
　　　　pact. *A vigorous coastal plant, it is common on the seaward-side of dunes.* **Leymus arenarius**

7.　Spikelets *many-flowered*; *1* spikelet per rachis-node 9
　　Spikelets *1-flowered*; *3* spikelets per rachis-node 8

8.　Lemmas of the lateral asexual florets blunt and *lacking* awns (Fig. 36.01). Asexual florets narrow,
　　　　and very small. Lemmas of the central fertile florets longer, wider, and with very long awns.
　　　　All glumes hairy, linear-lanceolate, pointed or awned. *An important cereal.* **Hordeum sativum**
　　Lemmas of the lateral florets *with* long straight awns (Fig. 29.40). The lateral florets staminate or
　　　　asexual, their glumes slightly dissimilar; the central florets hermaphrodite. The glumes linear,
　　　　fringed with hairs, and ending in long awns. *Frequent on waste ground.* **Hordeum murinum**

9.　Spikelets with *narrow* side towards the rachis 13
　　Spikelets with *broad* side towards the rachis 10

10.　Lemmas *not* fringed with long sharp teeth 11
　　Lemmas *fringed* with long sharp teeth (Fig. 38.04); the former 5-7-nerved, keeled throughout and
　　　　ending gradually in long rough awns. Spikelets usually 2-flowered. Glumes narrow, keeled,
　　　　pointed, minutely hairy, 1-nerved, shorter than the spikelets. Paleas sparsely short-haired, and
　　　　very blunt. The caryopsis mostly protruding when mature. *A cultivated cereal.* **Secale cereale**

11.　Glumes and lemmas *symmetrical*; internodes clearly visible 12
　　Glumes and lemmas clearly *asymmetrical* (Fig. 34.01). The latter all strong, firm, 5-nerved, blunt,
　　　　with an asymmetrically placed short blunt beak, and with broad and narrow shoulders. Spike-
　　　　lets 2-5-flowered; the rachis-internodes concealed. *An important cereal.* **Triticum aestivum**

12.   Central spikelets *not* overlapping 2 internodes (Fig.29.41). Spikes stout, 5-15 cm, and readily disarticulating at the rachis-nodes when mature. Spikelets wedge-shaped, large, firm, 15.0-30 mm long, 3-8-flowered, greyish-green or whitish-green, and about 1.5 times as long as the rachis-internodes. Glumes and lemmas firm, blunt, or short-pointed. *A coastal plant.* **Elymus farctus**

       Central spikelets *overlapping* 2 or more internodes (Fig. 30.01). Culms from 30.0-80.0 cm. Spikes compact or loose, slender, straight, 5-25 cm, green or purplish-green;the rachis sinuous, rough, hairy or hairless. The spikelets wedge-shaped, 8.0-25.0 mm, 3-10-flowered, falling as a unit. Glumes and lemmas very similar, all mostly lanceolate, blunt, pointed, or with awns, and firm. *A common troublesome plant weed of tilled and waste ground, roadsides, etc.* **Elymus repens**

13.   Lemmas *awnless* 15

       Lemmas *awned* 14

14.   Upper glume *equalling* or exceeding the florets (Fig.30.02);the former 9-28 mm long, 7-9-nerved, firm, strong, blunt, flat or rounded dorsally, and either rough or smooth; the lower glume suppressed except for the terminal spikelet. Lemmas all firm, rounded dorsally, 5-9-nerved, blunt or toothed apically, and ending in a long subterminal awn. A dark-coloured caryopsis mostly protruding. *Rare grass weed of cereal crops, or a cultivated forage plant.***Lolium temulentum**

       Upper glume mostly *shorter* than the florets (Fig. 29.36). Spikes vary from 10.0-30.0 cm. Spikelets up to 30.0 mm long, 5-14-flowered, with narrow edges towards the rachis; terminal with 2 glumes; the latter blunt or pointed, firm with very thin apices and 5-7-nerved.The lemmas similar to the glumes, rounded dorsally, 5-nerved, smooth or minutely rough, shortly 2-toothed or blunt, 5-10 mm long. *A very important short-lived pasture and hay plant.***Lolium multiflorum**

15.   Lateral spikelets with *1* glume (Fig. 29.37). Spikes varying from 5.0-25.0 cm, erect. Spikelets up to 27.0 mm, 4-14-flowered, with narrow edges towards the rachis, the terminal with 2 glumes; the latter blunt, 5-7-nerved. Lemmas similar to the glumes, all 5-nerved, rounded dorsally,firm with thin apices, and smooth. Very variable, there are numerous cultivated strains. *A most important pasture and hay plant; often abundant in many ecological situations.* **Lolium perenne**

       Lateral spikelets with *2* glumes (Fig.30.03); the latter very unequal, smooth, pointed. The lemmas firm except for thin apices, rounded dorsally, 5-nerved; paleas as long as the lemmas. Spikelets up to 32 mm long, narrow and 6-15-flowered; florets well spaced. *An intergeneric hybrid - Festuca pratensis x Lolium perenne; frequent in many grassy places.* **x Festulolium loliaceum**

16.   Inflorescence a *panicle* or a *spike-like* panicle 18

       Inflorescence a *raceme* 17

17.   Awns *absent* (Fig. 30.03). Lemmas blunt or pointed, firm but for thin apices, rounded dorsally, 5-nerved; paleas as long the lemmas. Glumes 2 per spikelet, firm, very unequal, smooth,pointed. Spikelets up to 32 mm in length, erect or spreading, narrow and with 6-15 well spaced florets; *Intergeneric hybrid, frequent in grassy areas, etc.; see couplet no.15.* **x Festulolium loliaceum**

       Awns *present* (Fig. 30.04); the latter rough and up to 12 mm long. Spikelets *short-stalked,* somewhat cylindrical, 6-16-flowered. Glumes unequal, pointed or with short awns, rounded dorsally, pubescent, lanceolate, firm and 5-9-nerved. The lemmas 6-12 mm, 7-11-nerved, lanceolate, firm, rounded, short-haired, stiff. *A frequent plant in shaded areas.***Brachypodium sylvaticum**

18.   Inflorescence a *panicle;* most branches *clearly* visible 25

       Inflorescence a *spike-like* panicle; most branches very short and usually *hidden* 19

19.   Spikelets *monomorphic, i.e.* all of similar structure 20

       Spikelets *dimorphic,* in pairs, 1 bearing fertile, the other asexual, florets (Fig. 30.05).The inflorescence a spike-like panicle, 1-sided, stiff, 3-10 cm. The fertile spikelets concealed by the asexual; all claw-shaped, 2-6-flowered, 3-8 mm long; the asexual spikelets consist of 10-20 linear, toothed lemmas. All glumes narrow, keeled, pointed, 1-nerved. *Widely distributed over many soil types; common in old grasslands, hedgerows, and waste places, etc.* **Cynosurus cristatus**

20. Glumes *equal*, or nearly equal, in length 21
   Glumes clearly *unequal* in length (Fig. 30.06), the upper about twice as long as the lower. Spike-like panicles of a brown, brownish or greenish colour. Spikelets 3.5-5.5 mm long, 3-flowered, 2 of which are asexual, and with 1 dark brown, hairy, blunt lemma, each; both have long awns; the latter - 1 *straight*, 1 *bent* - arise at different levels. Fertile lemmas smaller, hairless, blunt, and without awns. *A common plant in old pastures, meadows, etc.* **Anthoxanthum odoratum**

21. Awns *present* on either the glumes or lemmas; spikelets usually pubescent 22
   Awns *absent* from the glumes and lemmas (Fig. 30.07). Inflorescence a spike-like panicle, dense, more or less cylindrical. Spikelets 1-flowered, 10-18 mm long. Glumes unequal, pointed, narrow, 1-3-nerved, firm, minutely rough, and exceeding the floret. Lemmas keeled, up to 15 mm, 5-7-nerved, blunt or short-pointed, and with a short tuft of white hairs from near the base. *Perennial, and an important sand staliliser; very common in coastal areas.* **Ammophila arenaria**

22. Glumes *awnless* awns; lemmas *awned* 23
   Glumes *with* short firm awns (Fig. 30.08); the former mostly equal, truncate, keeled, 3-nerved, 3-4 mm long, fringed on the keel with long stiff, white hairs, and exceeding the floret. Lemmas 1.5-2.2 mm long, 5-7-nerved, blunt, membranous, rarely minutely hairy basally, and without awns. Inflorescence a spike-like panicle, dense, firm, up to 20 cm. *A very importantcultivated plant; it is also common on field margins, grassy places and roadsides, etc.* **Phleum pratense**

23. Glumes *not* divided to the base 24
   Glumes *divided* to the base (Fig. 30.09); the former 2-3 mm long, free to the base, 3-nerved, blunt, keeled, membranous, and fringed with soft hairs on the keel; sides also hairy. Lemmas very blunt, 5-nerved, about 2 mm long, and often short-haired on the nerves apically. Inflorescence a spike-like panicle, soft, narrow, and of a glaucous colour. Flowering stems bent at the nodes *A frequent plant in wet muddy places, shallow pools, damp fields, etc.* **Alopecurus geniculatus**

24. Glumes united basally for *one-quarter*, or less, of their length (Fig. 30.10); the latter 4-8 mm long, 3-nerved, gradually tapered apically, firm, and fringed on the keels with soft hairs; sides also hairy. The lemmas membranous, equal to, or less than, the glumes, keeled, bluntly tapered, 4-nerved - the fifth exserted from near the base in the form of a long bristle. Paleas absent. Inflorescence a spike-like panicle, soft, 2.0-14 cm, cylindrical; spikelets compressed, 1-flowered. *A common plant in old pastures, hedgerows, roadsides, woodlands, etc.* **Alopecurus pratensis**
   Glumes united for *one-third*, or more, of their length (Fig.30.11); the latter mostly equal, 4-7 mm, 3-nerved, pointed, and with minutely hairy, slightly winged keels. Lemmas very blunt, keeled, equal to, or less than, the glumes, membranous, smooth, 4-nerved, with the fifth exserted from or near the base in the form of long awns. Inflorescence 3-12 cm long, dense, tapered apically, green, purplish or blackish. *A rare annual weed of tillage crops, etc.* **Alopecurus myosuroides**

25. Spikelets *not* occurring in large clusters 27
   Spikelets *occurring* in large clusters 26

26. Glumes *exceeding* florets (Fig. 30.12); the former equal or nearly so, 3-nerved, keeled, pointed, firm, and minutely rough. Inflorescence a panicle; the spikelets laterally compressed, purplish, greenish or whitish in colour, up to 6 mm long, and with 1 hermaphrodite floret and 2 asexual florets; lateral asexual florets consist of small, short, narrow, hairy, hair-like lemmas; the fertile lemmas 3-4.0 mm, keeled, firm, pointed, 5-nerved, minutely hairy apically. *A common palatable plant in wet places, such as margins of rivers, streams, lakes, etc.* **Phalaris arundinacea**
   Glumes *shorter* than the florets (Fig.31.01); the former strongly keeled, unequal, pointed; the upper 3-nerved, firm, 4-7.0 mm long, and mostly fringed on the keels with short stiff hairs. Lemmas rather similar, 4-7.0 mm, strongly keeled, firm except for membranous margins, 5-nerved, asymmetrical, pointed or short-awned, and fringed with short stiff hairs or short teeth. Inflorescence a panicle; spikelets in dense clusters, claw-shaped, 2-5-flowered, and 5-10 mm long. *A common cultivated plant; it also occurs in grassy and waste places, etc.* **Dactylis glomerata**

Fig. 32.00

27.   Spikelets with *2* or *more* florets (some florets may be reduced to a single awned lemma) 33
    Spikelets *1-flowered*, and very rarely more than 3.0 mm in length 28

28.   Marginal nerves of the lemmas *hairless* throughout 29
    Marginal nerves of the lemmas *hairy* basally (Fig.31.02); the latter 2.5-3.5 mm, 5-nerved, firm but with membranous margins and apices. Glumes very unequal in all spikelets, keeled, pointed, 2-3 mm long, and with rough keels. Panicles rather variable, lax or compact, 5-25 cm, mostly greenish or purplish in colour. Spikelets comparatively small, 1-flowered, length 3.0-5.0 mm. *A common or abundant plant of mostly of shaded places, open woodlands, etc.* **Poa nemoralis**

29.   Lemmas *awnless* 31
    Lemmas *awned* 30

30.   Upper glumes *rough* towards the apex (Fig. 31.03). Flowering stems erect or becoming decumbent; panicles loose and open, becoming somewhat dense, erect or nodding, 5-18 cm in length, and either green, reddish or purplish in colour; branches slender and minutely rough. Spikelets 1.5-3.0 mm long, and all 1-flowered. Glumes exceeding the floret, slightly unequal, narrowed apically, 1-nerved, and membranous; the upper glume rough apically. Lemmas membranous, blunt, about 1.0 mm, short-bearded basally, 4-nerved, the fifth exserted from near the base in the form of a bent awn; paleas small. *A frequent plant of lowland peaty soils.* **Agrostis canina**
    Upper glume *smooth* towards the apex (Fig. 31.04). Flowering stems erect or decumbent; panicles mostly contracted before and after anthesis; green, reddish or purplish in colour; branches clustered, slender. Spikelets narrow, 2-3 mm in length, 1-flowered; glumes unequal, exceeding the floret, pointed, membranous, 1-nerved; the upper smooth apically. Lemmas membranous, about 1.2 mm, blunt, short-bearded basally; 4-nerved, exserted from near the base in the form of a bent awn; paleas very small. *Common plant of heaths, hill pastures, etc.* **Agrostis vinealis**

31.   Lemmas *without* minute teeth 32
    Lemmas *with* 3 minute teeth (Fig. 31.05). Flowering stems erect or spreading; panicles variable, very lax, up to 20.0 cm long, green or purplish in colour;branches many, fine, slender, spreading. Spikelets narrow, 2-3.0 mm long, 1-flowered. Glumes mostly unequal, 1-nerved, pointed, membranous, but with thinner margins, keeled; the lower rough on the base. Lemmas membranous, about 1.5 mm, somewhat 3-toothed, 3 to 5 nerved, short-bearded, and mostly without an awn; paleas small, about 1 mm. *Common plant of grassy places, heaths, etc.***Agrostis capillaris**

32.   Inflorescence branches *spreading* before and after flowering (Fig.31.06); the former a panicle; the latter effuse, up to 30.0 x 18.0 cm, with numerous densely clustered slender branches, green or purplish in colour. Spikelets 2-3.0 mm. Glumes, *both*, with rough keels. Lemmas blunt, about 1.5 mm in length, membranous, short-bearded basally. Paleas about 1 mm. *A frequent plant in waste places, rough pasture, hedgerows, roadsides, cultivated ground, etc.* **Agrostis gigantea**
    Inflorescence branches *erect* before and after flowering (Fig. 31.07); the former a panicle; the latter contracted before and after flowering; colour very variable - green, whitish or purplish; the branches dense and clustered. Spikelets narrow, 2-3.0 mm, 1-flowered. Lemmas membranous, blunt, 5-nerved, about 1.5 mm long, and short-bearded. Paleas about 1.0 mm long. *A common plant; old pastures, roadsides, waste areas, hedgerows, salt-marshes, etc.***Agrostis stolonifera**

33.   Glumes, both, *shorter* than the length of the spikelet 45
    Glumes, at least 1, *equalling* the length of the spikelet 34

34.   Spikelets disarticulating *above* the glumes 36
    Spikelets disarticulating *beneath* the glumes 35

35.   Awns *straight* (Fig. 30.08). Stems erect or spreading; the panicles compact or loose, whitish, pale grey or purplish. Spikelets all 2-flowered - the lower floret hermaphrodite, upper staminate, up to 7 mm long. Glumes unequal, pointed, keeled, thin, the lower 1-nerved, the upper 3-nerved;

both with short hairs on the keels and nerves, rough elsewhere. Lemmas membranous, 2.0-3.0 mm in length, finely 5-nerved, short-bearded basally; the upper have subterminal awns, lower awnless. *Frequent plant of open woodland, heaths, etc.; rarely a tillage weed.* **Holcus mollis**

Awns small and *hook-shaped* (Fig. 31.09). Stems erect or spreading; panicles variable in colour; greenish, pinkish, whitish, or tinged with purple. Spikelets 2-flowered, laterally compressed. 3.0-6.0 mm long. Glumes equal or very nearly so, thin, keeled, somewhat blunt, but with an obscure minute awn, and with short hairs on the keels and nerves, rough elsewhere; the lower 1-nerved, the upper 3-nerved. Lemmas 2.0-2.5 mm long, membranous, faintly 5-nerved, and short bearded basally; the upper with small characteristically hook-shaped awns. *A very common plant in many situations; old pastures, grassy places, waste ground, etc.* **Holcus lanatus**

36.   Lemmas *awned* 38
       Lemmas *awnless* 37

37.   Lemmas *3-toothed* (Fig. 31.10), 7-9-nerved, length up to 7.5 mm, short-bearded basally, rounded dorsally, firm, with hairy marginal nerves. The glumes *5-nerved* or less, often obscurely so. Inflorescence a panicle; the latter clearly branched, though often narrow and compact. Spikelets 6-14 mm, 4-6-flowered, and plump in shape. Glumes equal or nearly, rounded basally, keeled and tapered apically, and 3-5-nerved. *A frequent plant of heaths, etc.* **Danthonia decumbens**
       Lemmas *not* 3-toothed (Fig. 38.01); the latter rather similar, 7-nerved, rounded dorsally, and up to 22.0 mm long; usually lacking an awn. Glumes *7-nerved* or more, unequal, chaffy, 7-9-nerved, equalling or exceeding the florets, yellowish in colour, tapered apically. The inflorescence a panicle; the latter rather loose, spreading, with relatively long rachis-internodes; branches long and arising from 1 side of the rachis. Spikelets large, mostly 2-flowered, up to 25.0 mm long, pendulous, rounded dorsally, and hairless. *A cultivated cereal; often an escape.* **Avena sativa**

38.   Awns, of the lemmas of any 1 spikelet, arising from the *same* level 40
       Awns, of the lemmas of any 1 spikelet, arising from *different* levels 39

39.   Lemmas, *2 hairy*, the *third glabrous* (Fig. 30.06). Glumes clearly *unequal*, the upper about twice as long as the lower. Inflorescence light brown, brownish or greenish in colour. Spikelets 3.5-5.5 mm long, 3-flowered, 2 of which are asexual and consisting of 1 dark brown, hairy, blunt lemma; both lemmas awned - 1 *straight* and 1 *bent*. The fertile lemmas smaller, shiny, *hairless*, blunt, *without* awns. *Common plant in old pastures, meadows, etc.* **Anthoxanthum odoratum**
       Lemmas *pubescent* basally, firm, but with membranous apices, 7-nerved and 7.0-11.0 mm long (Fig. 31.11). Glumes unequal, membranous, rounded dorsally, pointed. Panicles lax; branches in clusters of 1-10. Spikelets 6.0-10.0 mm in length, 2-flowered, the lower hermaphrodite with a bent basal awn, upper staminate and with or without a short, straight, terminal or subterminal awn. *An abundant plant in hedgerows, roadsides, waste places, etc.* **Arrhenatherum elatius**

40.   Lemmas *7-nerved* 41
       Lemmas *5-nerved* 42

41.   Lemmas *not* bearded (Fig. 38.01, with an awn). Panicles well branched, rather loose, spreading, with relatively long rachis-internodes; the branches long and arising from 1 side of the rachis. The spikelets large, up to 25.0 mm, nodding, mostly 2-flowered, rarely more. Glumes unequal, chaffy, 7-9-nerved, equalling or exceeding the florets, yellowish in colour, tapered apically, rounded dorsally, and hairless. *A cultivated cereal plant; see couplet no. 37.* **Avena sativa**
       Lemmas *bearded* at the base (Fig. 38.02). The inflorescence a panicle; the latter very lax, up to 40 x 20 cm; the branches mostly arising in clusters from 1 side of the rachis. Spikelets mostly 2-flowered, rarely more, up to 30.0 mm long, pendulous; the rachillas long-haired. Glumes equal or very nearly so, tapered, exceeding the florets, 7-11-nerved, rounded dorsally. Lemmas 10-23.0 mm long, 7-9-nerved, hairless or covered with dark-coloured hairs, always with a basal, shallow, concave, long-bearded depression; minutely toothed, and with strong, rough, long, geniculate, dorsal awns. *A frequent weed in tillage fields, waste places, etc.* **Avena fatua**

42. Lemmas *not* ending in 2 fine points 43
Lemmas *ending in 2 fine points* (Fig. 31.12). Inflorescence a panicle; stems mostly erect; panicles loose or compact, greenish, yellowish, or of a golden colour. Spikelets 5.0-7.5 mm long, 2-5 flowered, wedge-shaped and gaping when mature; rachillas slender, and hairy. Glumes equalling or less than the florets, pointed, unequal, membranous, keeled. Lemmas membranous, 4.5-5.5 mm long, keeled, 5-nerved, ending in 2 narrow bristle-points, and with dorsal, geniculate awns. *A frequent plant of dry calcareous soils, grasslands, roadsides, etc.* **Trisetum flavescens**

43. Awns *clearly* visible, extending well *beyond* the apices of the lemmas 44
Awns *neither* clearly visible *nor* extending well beyond the apices of the lemmas (Fig. 32.02); the latter with very *ragged* or *torn* apices, membranous, 3.0-4.0 mm long, rounded dorsally, minutely bearded basally, obscurely 4-nerved, the fifth exserted from near the base in the form of short awns. The spikelets 2-flowered, 4-6 mm long; rachillas relatively long, slender, and long-haired. The glumes equalling the florets, or nearly so, membranous, keeled, pointed. The panicles loose or dense, greenish, purplish, or silvery. *A very common, coarse, useless plant from badly-drained, low-lying, rough pastures, moorlands, wet ground, etc.* **Deschampsia cespitosa**

44. Spikelets *6.0 mm in length or less* (Fig. 32.01); the latter all 2-flowered; rachillas small and short-haired. Inflorescence a panicle; the latter lax, purplish, brownish, or mostly silvery, and with spreading flexuous, slender branches. Glumes unequal, membranous, keeled apically, pointed, smooth or minutely rough, and 1-3-nerved. Lemmas membranous, 4-6 mm, entire or minutely toothed apically, short-bearded basally, 4-nerved with the fifth exserted from near the base in the form of bent awns. *Frequent plant of light dry peaty soils in heaths.* **Deschampsia flexuosa**
Spikelets *9.0-22.0 mm* in length (Fig. 32.03), 2-3-flowered; rachillas rather long, with long silky hairs. Inflorescence a panicle; the latter up to 22.0 cm long and 7.0 cm wide, mostly of a greenish colour; branches straight, slender, mostly erect, straight, rarely flexuous. Glumes lanceolate, unequal, membranous, pointed, keeled, the upper equalling the length of the florets (awns excluded). Lemmas 9-17 mm long, obscurely 5-nerved, membranous, broad apically, rounded dorsally, bearded basally, and with very prominent dorsal, bent awns. *Frequent perennial, and occasionally locally abundant, on many low-lying damp calcareous soils.* **Avenula pubescens**

45. Spikelets *not* obscured with a profusion of long hairs 46
Spikelets *obscured* with a profusion of white, brown, or brownish-purple long hairs which protrude freely from between the florets (Fig. 32.04). Panicles large, either lax or dense, up to 45 cm long. Spikelets 2-6-flowered, 9.0-18.0 mm long. Glumes very unequal, thin, pointed. Lemmas awl-shaped. *A common plant of lakemargins, riversides, canals, etc.* **Phragmites australis**

46. Glumes and lemmas *not* showing rounded, concave, hooded apices 47
Glumes and lemmas *showing* characteristically hooded, rounded apices (Fig. 32.05) and cordate bases; the former 7-9-nerved, 3-5.0 mm; the latter 3-5-nerved, 2.5-3.5 mm. Panicles lax, more or less pyramidal in outline, sparingly branched. Spikelets often pendulous, with 5-12 tightly packed florets. *A common plant, especially in old pastures on calcareous soils.* **Briza media**

47. Lemmas *awned* 59
Lemmas *awnless* 48

48. Lemmas *5-nerved* or less 51
Lemmas *7-nerved;* spikelets usually long and narrow 49

49. Lemmas *4.5 mm* or more in length 50
Lemmas *3-4 mm* in length (Fig. 32.06); the latter blunt, 7-nerved, hairless, rounded dorsally, firm but with membranous apices, often of a purplish or brownish colour. Glumes equal or nearly so, 2-4.0 mm long, thin, bluntly pointed. Stems stout, robust. Inflorescence a panicle; the latter profuse, very large, with numerous spikelets, and up to 45.0 cm. Spikelets 5-10-flowered, 6.0-12.0 mm long, hairless. *A common plant of riversides, canals, ditches, etc.* **Glyceria maxima**

50.   Lemmas mostly *6-7.5 mm* (Fig. 32.07); the latter blunt, firm except for broad membranous apices, 7-nerved, rounded dorsally. Glumes unequal, much shorter than the florets, blunt, thin. Inflorescence a panicle; the latter long-branched, open and loose. Spikelets long, narrow, up to 35.0 mm long, 6-16-flowered, hairless. *A frequent plant of wet muddy ground, etc.Glyceria fluitans*
       Lemmas mostly *4.5-5.5 mm* (Fig. 32.08); the latter rather blunt, 7-nerved, rounded dorsally,with broad whitish apices. Glumes very unequal, membranous, very blunt, 1-nerved, rounded dorsally. Inflorescence a panicle; the latter well-branched. Spikelets on relatively long branches; the former long and narrow, up to 22 mm long, 6-16-flowered, hairless. *A frequent plant, often dominating wet muddy ground, bordering streams, ponds, ditches, lakes, etc.* **Glyceria plicata**

51.   Florets *laterally* compressed; lemmas mostly *keeled* dorsally 54
       Florets *dorso-ventrally* compressed; lemmas *rounded* or *flat* dorsally 52

52.   Lemmas *hairless* 53
       Lemmas *with* short basal, silky hairs (Fig. 32.09); the former rounded dorsally, firm with white membranous tips and margins, 5-nerved, 3.5-5.0 mm. Panicles variable, usually grey-green with traces of purple, stiff, erect; branches erect or spreading, occasionally the lower reflexed. Spikelets 5-14 mm, with 2-10 closely packed florets. Glumes unequal, shorter than the florets, rounded dorsally. *Common plant of salt-marshes and coastal mud-flats.* **Puccinellia maritima**

53.   Upper rachilla-internode *one-half, or more,* the length of the lemmas (Fig. 32.10); the latter with a blunt, tapered apex; 4-6 mm long, 3-nerved, rarely 5-nerved, firm. Glumes unequal, membranous, 1-3-nerved, tapered apically. Spikelets nearly terete, laxly 2-4-flowered, 6-9.0 mm. Panicles compact to lax, purplish, brownish, green; *Common plant of heaths, etc.Molinia caerulea*
       Upper rachilla-internode *one-quarter, or less,* the length of the lemmas (Fig. 32.11). The panicles loose, erect or spreading, often nodding; the branches usually in pairs, thin, unequal, the longer carrying several spikelets, the shorter with 1-2. Spikelets parallel-sided or cylindrical, 8.0-22.0 mm in length, and with 5-15 loosely arranged, florets. Glumes much shorter than florets, unequal, apices membranous, 2.0-5.0 mm. Lemmas 5.0-7.5 mm, flat or rounded dorsally, though all 5 nerves often prominent apically, firm with thin membranous margins and apices. *Pasture plant, often abundant in low-lying damp grasslands, and other grassy areas.Festuca pratensis*

54.   Glume keels *hairless* 55
       Glume keels *short-haired* (Fig.31.01); the former very strongly keeled, unequal, pointed; the upper 3-nerved, firm, 4-7.0 mm long, and mostly fringed on the keels with short stiff hairs. Lemmas rather similar, 4-7.0 mm, strongly keeled, firm except for membranous margins, 5-nerved, asymmetrical, pointed or short-awned, and fringed; *see couplet no. 26.* **Dactylis glomerata**

55.   Marginal nerves of lemmas *hairless* 58
       Marginal nerves of lemmas *short-haired* 56

56.   Hairs present on marginal nerves, *other lateral nerves hairless* 57
       Hairs present on *all* lateral nerves (Fig. 33.01). Flowering stems varying in height from 2-30 cm. Inflorescence a panicle; the latter mostly pyramidal-shaped, open, loose, or compact; branches few. Spikelets comparatively large, up to 9 mm long, and with 2-10 florets which readily separate when mature. The glumes unequal; the lower ovate and and 1-nerved, the upper broadly ovate and 3-nerved; both keeled, pointed, and with membranous margins and apices. Lemmas ovate, obtuse, but with characteristically broad membranous margins and apices, 2.5-3.5 mm; basal web mostly absent. *Ubiquitous plant of cultivated ground, waste places, etc.* **Poa annua**

57.   Lower glume *3-nerved* in all spikelets (Fig. 31.02). Glumes unequal, keeled, pointed, 2.0-3.0 mm long, and with rough keels. Lemmas 2.5-3.5 mm, 5-nerved, firm with membranous margins and apices; keels and marginal nerves short-haired. Panicles very variable, lax or compact, 5-25.0 cm, mostly greenish or purplish. Spikelets 2-5-flowered, comparatively small, and 3.0-5.0 mm. *Common or abundant plant mostly of shaded places, open woodlands, etc.Poa nemoralis*

Lower glume *1-nerved* in some spikelets (Fig. 33.02); the latter unequal, pointed, and with rough keels. Lemmas 3.0-4.0 mm long, keeled, somewhat pointed, firm with membranous margins and apices, webbed basally, and with hairy marginal nerves and keels. Panicles very variable, often pyramidal, open or contracted, erect or nodding, and green, purplish or greyish in colour. Spikelets comparatively large, up to 7.5 mm long, and 3-6-flowered. *A widespread plant of old meadows, pastures, wall-tops, roadsides, open woodlands and other dry places.* **Poa pratensis**

58. Keels of lemmas *hairy on their lower halves* (Fig. 33.03); the latter 5-nerved, mostly 3.0-3.5 mm long, pointed, usually with a basal web of silky hairs, firm with membranous margins and apices. Glumes unequal, pointed, 2.0-3.0 mm, with minutely rough keels. Panicles variable, compact or lax, often spreading, purplish, reddish or green, and 5-30.0 cm. Spikelets 2-4-flowered, 3.0-4.5 mm. *Abundant plant of hedgerows, waste ground, cultivated ground, etc.* **Poa trivialis**
Keels of lemmas *hairless* (Fig. 33.01, except for the absence of hairs from side nerves); the latter with blunt membranous apices. Panicle mostly pyramidal-shaped; *see couplet 56.* **Poa annua**

59. Awns, all, attached to the *apex* or just *behind* the apex of the lemmas 62
Awns, at least 1, attached to the *base* or *dorsal* area of the lemmas 60

60. Lemmas *5-nerved* 61
Lemmas *7-nerved* (Fig. 31.11); the latter bearded basally; awns *dimorphic*, arising at *different* levels. The upper glume *about twice* as long as the lower; both somewhat membranous. Spikelets 2-flowered. *A common perennial in grassy places; see couplet no. 41.* **Arrhenatherum elatius**

61. Lemmas ending in *2 fine points* (Fig. 31.12). Panicles of a golden-brown colour. Glumes membranous, unequal, upper shorter than the florets. Spikelets 2-5-flowered, wedge-shaped, gaping when mature; rachillas slender and hairy. *Frequent; see couplet 42.* **Trisetum flavescens**
Lemmas *blunt* (Fig. 33.03); the latter 9-17 mm long, membranous, obscurely 5-nerved, broad apically. Glumes unequal, pointed. Panicles silvery; spikelets 10-22 mm, 2-3-flowered; rachillas long, with long silky hairs. *Common; calcareous soils; see couplet no. 44.* **Avenula pubescens**

62. Lemmas *5-nerved* or *less* 66
Lemmas *7-nerved* or *more* 63

63. Lemmas *keeled,* at least apically 64
Lemmas *rounded* throughout the entire dorsal surface (Fig. 33.04); the latter 7-9-nerved, rounded, characteristically bifid, 8.0-11.0 mm long, hairy, rarely hairless, and with straight subterminal awns. Glumes unequal, pointed, 3-7-nerved, 5-8 mm long, hairy, rarely hairless. Panicles very variable, loose and spreading or compact, and from a few to numerous spikelets; the latter up to 25.0 mm long, mostly ovate, with 2-12 closely packed florets. *Very common annual or biennial of grasslands, cultivated ground, waste places, sand-hills, dunes, etc.* **Bromus hordeaceus**

64. Awns *shorter* than their lemmas 65
Awns *equalling* or exceeding their lemmas (Fig. 33.05); the former up to 30.0 mm long, straight, subterminal. Lemmas linear-lanceolate, 7-nerved, up to 22.0 mm, slightly keeled or rounded dorsally, with membranous margins, usually bifid apically, and hairless. The glumes unequal, very narrow, finely pointed, with membranous margins. Panicles very loose; branches often long, few. The spikelets linear lanceolate when young wedge-shaped when mature, up to 35.0 mm (excluding awns), 4-12-flowered. *Frequent weed of open dry sandy areas.* **Bromus sterilis**

65. Awns *subterminal* (Fig. 33.06); the latter shorter than their lemmas. Panicles very lax, open, nodding, up to 20 x 50 cm; branches long, slender, mostly in pairs, spreading, divided, and rough. Spikelets, at least some, often pendulous, 4-12-flowered, and 15-30 mm long. Glumes unequal, pointed, 1-5 nerved, keeled or becoming rounded. Lemmas 9-14 mm long, 7-nerved, firm with membranous margins, short-haired, keeled, and minutely bifid. *Common plant of shaded situations, it is widespread in open damp woodlands, hedgerows, roadsides, etc.* **Bromus ramosus**

Fig. 33.00

N.B. DRAWINGS NOT TO SCALE.

Awns *terminal* (Fig. 33.07); the latter much shorter than their lemmas. Inflorescence a panicle; the latter 10-30.0 cm long, green, purplish, or reddish, with well rather short, characteristically erect branches. Spikelets subterete to compressed, 'parallel-sided' or slightly broad apically, up to 40.0 mm in length, and 4-16-flowered. Glumes narrow, finely pointed, 1-3-nerved, 6-14 mm long. Lemmas 6-10 mm long, firm except for membranous incurved margins, 7-nerved, somewhat keeled on their upper-halves, rough, mostly hairless, rarely hairy. *Frequent on dry calcareous soils; often locally abundant and dominant on raised roadside banks, etc.* **Bromus erectus**

66.   Lemmas *neither* keeled *nor* fringed with short stiff hairs or teeth 67

Lemmas *sharply keeled* and *fringed* with short stiff hairs, apically (Fig. 31.01). The glumes also strongly keeled, unequal, pointed; the upper 3-nerved, firm, 4-7.0 mm long, and mostly fringed on the keels with short stiff hairs. Lemmas rather similar, 4-7 mm long, firm except for membranous margins, 5-nerved, asymmetrical, pointed or short-awned, and fringed with short stiff hairs or short teeth. The inflorescence a 1-sided, erect, oblong to ovate panicle; spikelets compressed, claw-shaped or wedge-shaped, 2-5 flowered, 5.0-10.0 mm long and arranged in dense *glomerules. A very common perennial plant; see couplets no. 26 and 54.* **Dactylis glomerata**

67.   Awns clearly *shorter* than their lemmas 69
       Awns *equalling* or exceeding their lemmas 68

68.   Panicles *small, dense, 2.0-12.0 cm* long, long-exserted from the uppermost sheath, erect to slightly nodding, lanceolate in outline, loose to compact, 1-sided, and green or purplish. The spikelets (Fig. 33.08) closely overlapping, wedge-shaped, 8-16 mm long (excluding awns), and 4-8 flowered. Glumes unequal, the lower 1-nerved and one-half to three-quarters the length of the 3-nerved upper; both very finely pointed. Lemmas linear-lanceolate in outline, rounded dorsally, 5-9 mm long, firm, 5-nerved. *An annual plant frequent on sandy soils.* **Vulpia bromoides**

Panicles *loose,* erect or more often nodding, lanceolate to ovate in outline, often more or less 1-sided, 10.0-60.0 cm in length and greenish in colour; branches rough, mostly in unequal pairs, the longer with many spikelets, the shorter with 1-3. The spikelets (Fig. 33.09) 10.0-30.0 mm long, 3-10-flowered. Glumes unequal, finely pointed, firm except for broad membranous margins. Lemmas rounded dorsally, 6-10.0 mm long, 5-nerved, firm except for membranous margins, hairless, ending in awns which are at least as long as their lemmas. *A frequent plant in shaded situations, such as damp open woodlands, roadsides, hedgerows, etc.* **Festuca gigantea**

69.   Awns *terminal* 70

Awns *subterminal* (Fig. 33.10); the latter inconspicuous very short; rarely absent. Inflorescence a panicle; the latter up to 50.0 cm long, heavy, loose and pendulous, or contracted, and green or purplish; branches angular and rough. Spikelets up to 18.0 mm, with 4-14 closely packed florets. Glumes unequal, pointed, the lower lanceolate and 1-nerved, the upper broadly lanceolate and 3-nerved; both keeled, margins membranous. Lemmas 6-9 mm, 5-nerved, flat or rounded dorsally, hairless but rough and firm, and with membranous apices. *A very variable robust and vigorous plant, common on rough, low-lying grasslands, roadsides, etc.* **Festuca arundinacea**

70.   Lemmas *mostly 2.5 to 4.5 mm* long, rarely more; the latter rounded dorsally, 5-nerved, hairless or hairy, firm, mostly ending in short awns. Glumes unequal, pointed, hairless or hairy. Inflorescence a panicle; the latter erect, lanceolate to narrowly oblong, 4.0-14.0 cm in length, open in flower, later contracted; green or greenish-purple in colour, and more or less 1-sided; branches erect or slightly spreading. Spikelets (Fig. 33.11) 4-8-flowered and 4-10.0 mm in length; some forms viviparous. *Very variable perennial, frequent on heaths, bogs, moors, etc.* **Festuca ovina**

Lemmas *mostly 4.8-6.5 mm* in length; the latter later firm, rounded dorsally, often with strongly incurved margins, 5-nerved, mostly hairy or hairless, and ending in distinct terminal awns, up to 3.0 mm long. Glumes unequal, firm, pointed, rather narrow, and 1-3-nerved. Inflorescence a panicle; the latter often purplish, erect or slightly nodding, open or contracted, well branched. Spikelets (Fig. 33.12) 5.0-14.0 mm in length, 3-9-flowered. *A very variable plant, there are 4 subspecies and many botanical varieties; it is very common in many situations.* **Festuca rubra**

CEREALS

In terms of food supply, cereals are undoubtedly the most important group of plants. They provide about one-half of man's food calories and a large part of his nutrient requirements (Peterson, 1965). Cereals were probably the first plants to be domesticated by man, and have been grown long before the beginning of recorded history. They are cultivated in all parts of the world except perhaps the very wet tropical regions. Total world cultivation is somewhat in excess of 700 Mha. Wheat accounts for about 31% of this figure. Apart from their usefulness as a rich and relatively cheap supply of carbohydrates, cereals are also the source of raw material for a wide range of products. All are annual, rarely biennial or short-lived perennial grasses grown for their relatively large 'seed' reserves. They include wheat, barley, oats, rye, rice, maize or corn, millet and sorghum.

Maize or corn, *Zea mays* (Fig. 38.03), is cultivated in most warm parts of the world. Annual cultivation, worldwide, is over 130 Mha. It is a high yielding and adaptable cereal but unfortunately, because of it C_4 photosynthesis mechanism, is not productive under Irish climatic conditions, except perhaps in exceptionally warm summers. The 2 principal types are flint maize and dent maize. Their large grains differ in endosperm texture, the former is hard throughout and has a more or less rounded apex, while the latter has a core of soft endosperm which extends to the apex and this, on ripening, shrinks giving an indented apex. There are numerous cultivars of each type. Maize is produced mainly for livestock feed, though in some underdeveloped south American countries it is used for human consumption.

Sorghum or milo, species of *Sorghum* (Fig. 38.06), has a smaller grain-size than maize but is generally the same type of cereal. Approximately 50 Mha are cultivated annually, worldwide. It is mostly cultivated in regions which are too dry and hot for the production of maize. The principal producing countries are the US, Pakistan, India, China, parts of Africa. Its main use is in the supply of animal feed but some is produced for forage and human consumption. Like maize, it has a C_4 photosynthesis mechanism and so is not suitable for cultivation in most European countries.

Millet is another type of small-seeded cereal of lesser importance than sorghum. The term is applied to plants of various genera and species (*e.g. Setaria italica*, Fig. 38.07; *Panicum miliaceum*, Figs 38.08) but the principal genus is *Panicum*. All were originally cultivated by the ancient Greeks, Egyptians and Romans and are still part of the human diet in China, Japan and India. They are similar to maize and sorghum in having a C_4 photosynthesis pathway and are grown in areas that are unfavourable for the economic production of the major cereals. Approximately 40 Mha are cultivated annually, worldwide.

Rice, *Orza sativa* (Fig. 38.05) is second in importance to wheat and is the staple food for millions of people in southeast Asia. Over 90% of the world's supply is grown in China, India, Pakistan, and southeast Asia. Some 25 different plant species are involved. Most rice is grown under water in flooded fields and successful production depends on adequate irrigation. A small amount is produced under dry land conditions. Close on 150 Mha are cultivated annually.

The 4 cereals grown in Europe appear to have arisen from wild ancestral grasses in Asia and southeast Europe some thousands of years ago, though their exact evolutionary development is not fully understood. Most, if not all, of these wild ancestral grasses are still to be found in their original habitats. These wild ancestors are all characterised by their possession of a fragile inflorescence, hulled grains and an articulating mechanism. The fragile inflorescence and the articulating mechanisms are of particular importance to the wild grasses in that they ensure, in the absence of harvesting and sowing practices, the perpetuation of these plants. The fragile inflorescences readily disarticulated into bristle-pointed segments which were easily disseminated or buried in the soil with the aid of long barbed awns. With the advent of attempts to produce food and the introduction of some rudimentary harvesting techniques, these undesirable traits were gradually lost. Harvested plants would have had marginally less fragile inflorescences, while the totally fragile types would shatter on collection. The improvement from the fragile to the firm type was exceedingly slow and took place over several millennia through natural selection. All present day cultivated cereals are relatively large-grained, high yielding, and have firm inflorescences. Some, particularly those used for human consumption, have hull-less or naked grains. A detailed consideration of the 4 principal cereals of importance to European Agriculture is given hereafter.

WHEAT

General: Wheat is the world's most important cultivated crop. A wheat crop is harvested somewhere in the world every month of the year. Total global cultivation, in any given year, is well over 225 Mha. Irish acreage averages 0.075 Mha. It is grown in all temperate and most subtropical regions of the world. About 90% of the crop is produced in the Northern hemisphere. The former USSR is the leading producer, followed by the US, Canada, Argentina, India, Turkey, Pakistan, China, Australia, France, Italy, Spain, and Germany.

There are several wheat species and literally thousands of cultivars. Wheat has many uses. The most important species is *Triticum aestivum*, the common or bread wheat, which is grown principally for the production of flour used in bread making. Of all plant products, only wheat flour contains the particular kinds and combination of proteins necessary for the production of light porous loaves. Wheats grown in dry climates are generally hard types, having a protein content of 11-15% and strong gluten; the wheats of humid regions are softer, with a protein content of 8-10%. The former are better for bread making, while the latter are used mainly in confectioneries.

Historical: Wheat is the world's most important crop and it has been used as a food by man since prehistoric times. It is also possibly the oldest crop in the world. Various dates have been mentioned in the literature for its first appearance as a food plant. Mann (1946), cited by Leonard and Martin (1963), states that it was probably grown in the Middle East as early as 15000-10000 BC. 'Very ancient Egyptian monuments, older than the invasion of the shepherds and the Hebrew Scriptures, show the cultivation already established, and when the Egyptian or Greeks speak of its origin, they attribute it to mythical personages, Isis, Ceres, Triptolemus'. Indeed, there is ample historical evidence to show that it was cultivated in ancient Greece, Iran, Egypt and throughout Europe.

The wild wheat plant or wild einkorn - *Triticum monococcum* var. *boeoticum* - occurs naturally in open herbaceous oak park-forests and steppe-like formations in southwestern Asia, northwestern Iraq and southwestern Turkey. In this area and beyond it, in Transcausia, Syria, central and western Turkey and Greece, wild einkorn is also very common in secondary habitats such as roadsides and edges of fields. This wild wheat and a number of other wild grasses are generally believed to be the progenitors of all cultivated wheats.

Despite many years of intensive investigation it has not been determined accurately when or where wheat was first cultivated. Advances in nuclear physics have helped immensely to fix the dates in history for the beginning of agricultural plants. While plants are alive they absorb from the atmosphere carbon dioxide which contains traces of C_{14}. When plants die the supply of radioactive carbon ceases because carbon dioxide is no longer being ingested. Since C_{14} disintegrates slowly and, since the rate of loss of radioactivity is known, the time of plant death can be established by determining the ratio between the radioactivity of the carbon in living and dead plants. Evidence of this kind places the beginning of cultivated wheat roughly 6000-7000 years ago. Carbonised kernels of wheat were found by American archaeologists, in the middle of this century, at the 6000-7000 year old site at Jarmo, Iraq. Feldman (1976) gives about 6750 BC as the date of the Jarmo site, and about 7000 BC for another at Cayonu, while Feldman and Sears (1981) make reference to the 'genetic variability of cultivated wheat which has accumulated over 10000 years of cultivation'!

TAXONOMIC DATA

General: The present position of wheat is, to say the least, at little confused. The problem is not helped by what appears to be an arbitrary change in the nomenclature of some wild grasses suspected of being involved in the evolutionary history of present-day wheats. This has resulted in the listing of up to 13 diploid species. Originally only 2 such species were mentioned. The change was deemed necessary when it was generally accepted that some wheats originated through hybridization between species of different genera. Under ICBN rules, an intergeneric hybrid cannot be included under the generic name of either parent. A new genus has to be created to accommodate the hybrid, or the generic name of one of the putative parents has to changed to coincide with that of the second putative parent. To add to the confusion some writers have suggested or bestowed varietal or subspecies status on some wheats which are more frequently listed as species.

Irrespective of the uncertainty of the ancestral parents, and of the number of discrete species, all writers place the wheats in the genus *Triticum,* tribe *Triticeae* and subtribe *Triticinae.* Barley and rye belong to the same taxa. There is also unanimity as to the basic chromosome number being x=n=7, as indeed it is for all our important cereals. Pioneering work by Sakamura (1918) and Sax (1922) has established that there are 3 ploidy levels in wheats as a whole. Accordingly, on the basis of somatic chromosome numbers, wheat can be classified into diploids (2n=14), tetraploids (2n=28), and hexaploids (2n=42). It is further accepted that the tetraploids evolved from the diploids, and that the hexaploids evolved from the tetraploids.

The wheat inflorescence is a terminal spike (Fig. 34.1a). The spike shows a wide range of variation in shape and can be dense, medium-dense, or lax. One spikelet is borne at each node of the rachis. Further variation is found in the structure of the spikelet. The floret number varies from 1-9, though most plants have an average of 5. These are enclosed by 2 firm, asymmetrical, coriaceous glumes which show additional variation. The lemma is also firm, strong, asymmetrical and may show the presence or absence of firm awns. The caryopsis, in most cultivated forms, threshes free, but in wild types and some cultivated species, it is described as being hulled due to the persistence of the lemma and palea. Spike and spikelet variation is discussed later in relation to important species and cultivar identification.

Wheat is a self-pollinating plant, though natural cross-pollination can occur to the extent of about 5%. Anthesis starts several days after the spike emerges. The main culm flowers bloom first and the tillers later, in order of their formation. Flowering begins towards the middle of the spike and proceeds upwards and downwards.

Germination begins when non-dormant grains are subjected to suitable environmental conditions. First the root-sheath, the coleorhiza, which contains the primary root, breaks through the pericarp. About 24 hours later the primary root itself breaks through the end of the coleorhiza. A pair of seminal rootlets covered by the coleorhiza appears next above the primary rootlet; these also break through. A second pair soon grows out above the first, giving 5 typical seminal rootlets; a sixth rootlet rarely appears. About the time the first pair of lateral rootlets emerges, the coleoptile begins to elongate, surrounding and protecting the primordial leaves and stem, as growth towards the surface begins.

In the early stages the stem consists of a series of nodes and extremely short internodes. The stem has a leaf attached at each node. At the lower node a small bud develops within the angle formed by the stem and leaf. The first internode usually remains very short, about 1 mm. The second internode elongates within the coleoptile, pushing all the higher nodes upwards until they are close to the surface of the soil. This elongated internode is known as the 'rhizome'. If the grain is buried deeply, several internodes may lengthen to form a long 'rhizome' in order to bring the growing point close to the soil surface. The closely packed nodes, internodes and growing point, which are being raised together, are covered by foliage leaves within the coleoptile. The coleoptile itself lengthens rapidly enough to keep the growing point of the stem enclosed. As the tip of the coleoptile emerges from the soil, or shortly thereafter, the first foliage leaf pushes through the small pore in the coleoptile. Foliage leaves appear in an alternate distichous arrangement later.

The wheat plant, in the vegetative phase, resembles that of rye and barley in having its leaves twisted clockwise. The main difference is the presence of short hairs on its medium-sized auricles.

Classification: The most enlightening dissertation on the evolution of the wheats is given by Feldman (1976). In his discussions, he lists the 3 well established ploidy levels. The most noteworthy feature of his classification scheme is the placing together, into a single species, wheats with homologous genomes. Thus, taking into consideration the change in the nomenclature of the *Aegilops* grasses, he lists 5 diploid species. Two tetraploids - *T. timopheevii* and *T. turgidum* - are recognised because hybrids between these 2 show partial asynapsis and high sterility. One hexaploid - *T. aestivum* - is given. The remaining tetraploid and hexaploid wheats are treated as varieties of the 3 main species. Figure 35.00, based on the system of Feldman (1976), summarises current thinking on the evolution and classification of the wheats and their more important wild relatives. The diagrams, which are self-explanatory, show their probable evolutionary pathways.

It will be noted that the name *aegilopoides* is dropped in favour of the older - *boeoticum* - name. A relatively new wild tetraploid, *T. timopheevii* var. *araraticum,* is mentioned as the likely progenitor of the cultivated form. Also, *turgidum* is given as the species containing most tetraploid wheats, even though the original wheats included under this name, are neither widely grown nor ancestral to the other tetraploids. This is because it was the first name applied to any tetraploid wheat. Some mystery still surrounds the origin of one tetraploid - *T. turgidum* var. *carthlicum.* This produces naked grains, a feature determined by a compound genetic locus, termed the *Q factor*. This factor is not present in the other tetraploids and it is suggested that this wheat may have evolved through hybridization of some *T. aestivum* varieties with *T. turgidum* var. *dicoccum.* Morris and Sears (1967) suggest that the free-threshing characteristic of vars *durum, turgidum* and *polonicum* may be due to the accumulation of mutations that reduced the toughness of their flowering pales to the point that this feature was attained.

Evolutionary history

It is now generally agreed that the wild ancestral grasses involved in the evolution of wheats belong to the genus *Aegilops*. At one time, it was widely accepted that the genus *Agropyron* (now *Elymus*) contributed a species to the evolution of the present day wheats. This connection is no longer deemed credible. Boden (1966) appears to be the first researcher to suggest that the relevant wild grasses of *Aegilops* be transferred to the genus *Triticum,* thereby enabling hybrids of these grasses and wild wheats to be classified under the latter genus. Many subsequent writers have acquiesced with this change.

In hybridization, each parent contributes a set of chromosomes termed a genome. Several genomes are found in the wheat group because several parents are said to have been involved. The genomes are not all homologous, so, for convenience and ease of identification, they have been designated with a letter of the alphabet. Thus, in wheats as a whole, there are genomes labelled A, B, C, D, G, S, S^l and S^b. The source of all but genomes B and G can be traced to particular species. The donor or donors of these 2 genomes have so far evaded conclusive identification. They appear to originate in the tetraploid wheats. Zohary and Feldman (1961) put forward the hypothesis that these wheats, as well as their 'modified genomes,' may be regarded as the products of a characteristic type of hybridization between a restricted number of initial amphidiploids sharing 1 genome in common. Riley (1965) uses cytological, geographical and morphological data to implicate *Triticum speltoides* as the source of B, whereas Kimber (1974) uses cytogenetic evidence to suggest that the same species may have contributed the G genome. Feldman (1976) makes certain assumptions which lead him to state that both B and G would have to be considered modified S genomes. The latter suggestion appears, at this point in time, to be the most plausible in the absence of conclusive evidence to their origin. There are some genome C types derived from other *Aegilops* species which have not played a significant role in the development of the wheats.

IMPORTANT SPECIES

T. monococcum L.: This is the oldest of the wheat species, and it is the one from which, it is claimed, all others have evolved. It has not been subjected to a nomenclature change but is now subdivided into 2 varieties, 1 wild, 1 cultivated. Both are more grass-like in appearance than all other types.

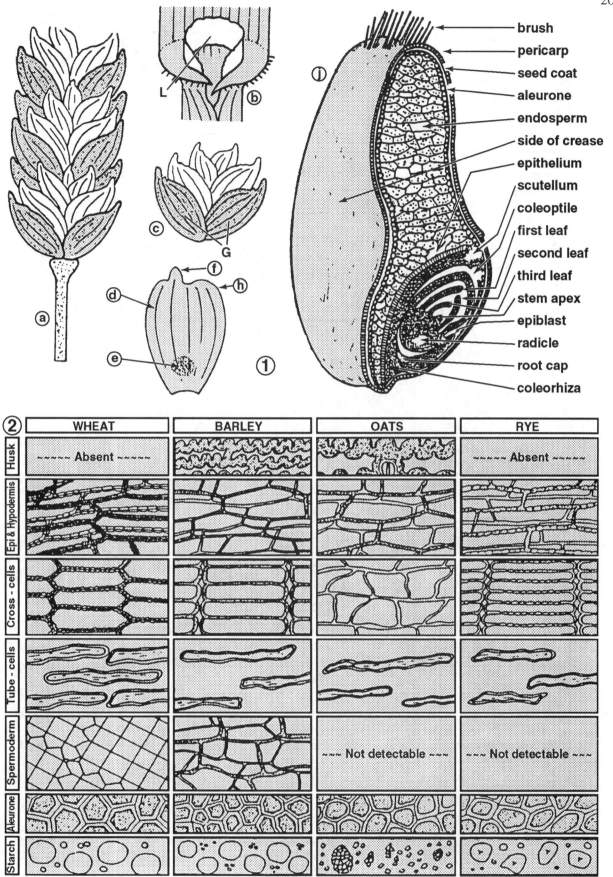

brush
pericarp
seed coat
aleurone
endosperm
side of crease
epithelium
scutellum
coleoptile
first leaf
second leaf
third leaf
stem apex
epiblast
radicle
root cap
coleorhiza

	WHEAT	BARLEY	OATS	RYE
Husk	~~~~ Absent ~~~~~			~~~~~ Absent ~~~~~
Epi & Hypodermis				
Cross - cells				
Tube - cells				
Spermoderm			~~~ Not detectable ~~~	~~~ Not detectable ~~~
Aleurone				
Starch				

Fig. 34.00

206

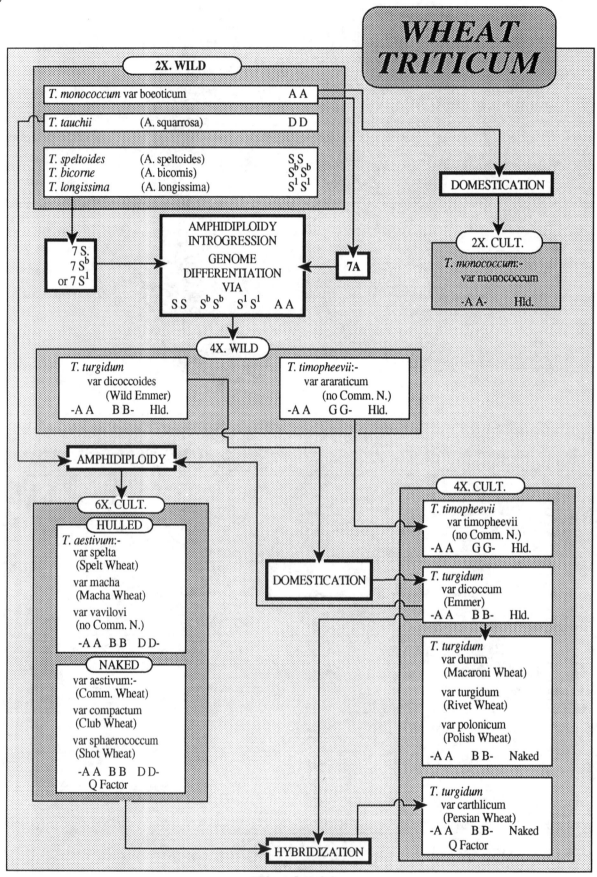

Fig. 35.00

Var. *boeoticum* (Wild Einkorn): This is considered by all researchers to be the wild wheat ancestor of all cultivated wheats. It occurs naturally in open herbaceous oak-park forests and steppe-like formation in southwestern Iran, northwestern Iraq and southwestern Turkey, parts of Syria, Turkey and Greece.

The ears are slender, bearded and narrower across the face than across the 2-row side. Rachis is extremely brittle and the whole rachis breaks up at maturity. Spikelets normally contain 2 florets, usually only 1 is fertile, so that a single 'seed' is formed in each spikelet, hence the German name - 'einkorn'. Grains are small, narrow, flattened and pointed at both ends, grey-blue, olive-green, or light red in colour. Glumes are 2-keeled and each keel ends in a point or tooth. Lemmas awned. Stems slender and either solid or hollow, with thick walls. Synonyms: *T. boeoticum* Boiss.; *T. aegilopoides* Bal.

Var. *monococcum* (Small Spelt or Einkorn): This is an improved form of the above and derived from it in cultivation. On average, it has a shorter, denser spike with shorter awns. The rachis is somewhat less brittle. Spikelets 3-flowered but only 1 grain is produced. Grains are slightly larger. Except for the slight differences mentioned the 2 wheats are very similar. It was cultivated at Jericho and in southwestern Asia from about the seventh millennium BC, and more widespread later, mainly in hilly areas but now almost obsolete. Synonym: *T. monococcum* L.

T. timopheevii Zhukov: This has many features in common with *T. dicoccoides*. However, it displays limited variation and has a very narrow geographical distribution.

Var. *timopheevii* (Timopheevi Wheat): This is possibly the least important of the tetraploids. It is derived from the wild brittle form - var *araraticum* - but only shows minor improvements. Presently cultivated in the mountainous regions of western Georgia, the former USSR, western Iran, northwestern Iraq and eastern Turkey.

The ears are laterally compressed; rachis somewhat brittle at maturity. Glumes are thick, rough and have a single prominent keel. Each spikelet contains 2 grains which do not thresh-out freely. The ear is often shorter than broad. Synonym: *T. timopheevii* Zhuk.

T. turgidum L.: Wheats in this group are the most important and most widely cultivated of the tetraploid group. This name was originally applied to rivet wheat but, under revised classification schemes, now includes 5 additional wheats which were previously treated as distinct species.

Var. *dicoccoides* (Wild emmer): A tetraploid, considered to be the progenitor of the important wheats in this group, it is said to have arisen through hybridization between *T. monococcum var. boeoticum* and one or more species of the old *Aegilops* genus. Rather similar to var. *boeoticum*. It is also said to be morphologically indistinguishable from *T. timopheevii* var. *araraticum*. It grows wild in the countries around the eastern end of the mediterranean sea and in Iran and Iraq.

The ears are laterally compressed; rachis very brittle. Glumes are hard. The terminal spikelet is usually sterile; other spikelets 3-flowered, producing 1-2 grains; the latter are pointed apically and basally. The essential difference between wild emmer and wild einkorn is that the plants of the former tend to be larger in all their parts; usually less compressed due to the increased number of grains. Synonym: *T. dicoccoides* Korn.

Var. *dicoccum* (Emmer): Essentially an improved form of the forenamed and has its general features. However, by contrast, this plant usually has a shorter, denser head, a less hairy and less brittle rachis, a plumper grain and a fertile terminal spikelet. Spikelets 3-4 flowered, though usually ripening only 2 grains. Awned and awnless forms exist.

Emmer is grown to a very limited extent in a few countries bordering the Mediterranean, in Iran, Pakistan, India, and also in the USA; it is adapted to poor, light soils and warm temperatures. It is used mainly for livestock feed. Synonym: *T. dicoccum* Schubl.

Var. *durum* (Macaroni Wheat): The first of the naked tetraploids, it has a wider distribution than emmer. It is grown in all countries bordering the Mediterranean, in the former USSR, Iraq, Pakistan, India, China, Canada, the USA, Mexico, Uruquay, Chile and the Argentine. Said to be descended from emmer but exhibits strikingly different characteristics. Most of its forms have a tough rachis, relatively soft loose glumes, 2-4 free-threshing grains per spikelet and ears, which are often about equal across the face and 2-row sides, less laterally compressed than those of emmer. The glumes have a very distinct single keel from base to apex. Stems are typically tall and either solid or hollow with thick walls. Short-stem forms have been bred. Grains are white, amber, red, or purplish in colour; they are also long, narrow, hard and translucent. On average, it is the hardest of all known wheats. Synonym: *T. durum* Desf.

Var. *turgidum* (Rivet Wheat): Grown extensively in the south of England in the past but is now replaced by common or bread wheats. It is presently grown in most Mediterranean countries, in western Europe and China. In general, it has taller stems, broader leaves, thicker heads, shorter glumes, broader and more mealy grains than the other wheats; it is strongly bearded.

Spikelets usually contain 3-5 grains, which are characteristically soft and mealy. In addition, they have a distinctive dorsal 'hump'; they are usually reddish in colour. The straw, although very long, is strong and wiry. The upper part is semi-solid and bends freely so that the mature ear is pendulous. Synonym: *T. turgidum* L.

Var. *polonicum* (Polish Wheat): Despite its name, it neither originated nor is it cultivated in Poland. It is grown to a very limited extent in Mediterranean countries, in Ethiopia, the former USSR and China. Generally a tall wheat, it has mostly lax, bearded ears. The rachis shows some tendency to shatter. Spikelets have 1-3 grains; the latter are long. The glumes are characteristically long, narrow, equalling or exceeding the lemmas and usually have a papery texture. Synonym: *T. polonicum* L.

Var. *carthlicum* (Persian Wheat): This is the only tetraploid whose origin is not fully understood. It carries a different gene complex, the Q factor, found in hexaploid wheat. Its narrow distribution in Transcausia may indicate that it originated relatively recently, presumably by hybridization of an unknown tetraploid with a hexaploid of the *aestivum* group.

Outwardly, Persian wheat resembles the hexaploids but it crosses readily with most tetraploids to produce fertile hybrids. It is difficult to cross with hexaploids and such hybrids are highly sterile. Like the hexaploids, the ears are broader across the face than the 2-row side; the latter is narrow. The rachis is thin, tough and flexible. Spikelets with 2-5 flowers and 2-4 grains are produced. These are reddish, very flinty and usually with a comparatively high protein content. Ears may be blackish, red or white. Glumes and lemmas are awned. Stems are solid, or hollow with thick walls. Synonym: *T. persicum* Vav.; *T. carthlicum* Nev.

***T. aestivum* L.**: This is the species which contains all the hexaploid wheats. All varieties have evolved from members of the tetraploid group through hybridization with wild grasses. A wide range of variation is shown because of their polyphyletic origin. They are, as a group, the most recently evolved and the most useful. All are cultivated and none is known in the wild. The first 3 mentioned hereunder have thick hard glumes, persistent flowering pales and somewhat brittle rachises, features reminiscent of the more primitive tetraploids. These are also said to be the progenitors of the final 3 varieties - *aestivum, compactum,* and *sphaerococcum.* The 6 important ones are:

Var. *spelta* (Spelt or Dinkel Wheat): Confined to the mountainous regions of Germany and, to a lesser extent, in Spain, France, Italy and Switzerland. Ears are long, narrow and lax. Spikelets 3-4 flowered, with 2-3 grains. These remain enclosed in the lemmas and attached to a rachis internode. Lemmas may be awned or awnless. Synonym: *T. spelta* L.

Var. *macha* (Macha Wheat): Another hulled hexaploid of limited distribution and importance; it is found or cultivated mostly in West Georgia and the former USSR. It has many undesirable features including a brittle rachis, hulled grains and thick, tough glumes. Each spikelet only produces 2 grains. Stems are hollow. Synonym: *T. macha* Dek. & Men.

Var. *vavilovi* (Vavilov Wheat): The last of the hulled wheats is, like the forenamed, of limited importance and distribution. Its cultivation is confined to Armenia and parts of Turkey. The ears are relatively large, with somewhat fragile rachises, and exceptionally long, branched spikelets. It is said to be exceptionally adapted to dry growing conditions. Synonym: *T. vavilovi* Jakubz.

Var. *sphaerococcum* (Shot or Indian Dwarf Wheat): This is a naked wheat grown principally in northern India and Pakistan as a bread wheat. It has very short stout stems in market contrast with most bread wheats, most of which are taller. Other parts of the plant, such as leaves, ears, glumes and grains, are said to be shorter, but the plant tillers profusely. Spikelets have 6-7 flowers which produce nearly spherical or hemispherical flinty grains. The latter are either red or white. Synonym: *T. sphaerococcum* Perc.

Var. *compactum* (Club Wheat): Ears are characteristically short, broad and dense. The rachis is firm. Spikelets are 6-7 flowered and produce 2-5 grains. In addition, they are broad, short, and attached almost at right angles to the rachis because of the extremely short internodes of the latter. Grains are small, plump, usually laterally compressed, red or white and typically soft and mealy, but occasionally flinty. It is grown extensively, particularly in Asia and the Pacific states of the US. Synonym: *T. compactum* Host.

Var. *aestivum* (Bread or Common Wheat): This is undoubtedly the most important variety of all wheats. Cultivation is worldwide. Tremendous variation in is shown in all aspects of the ear, and literally thousands of cultivars have been developed; there are winter, spring and intermediate types.

Spikes are tough, usually long in proportion to their width, although dense or 'squarehead' types are not uncommon. The spike is either square in cross-section or else wider across the face than the 2-row side. Spikelets 5-9 flowered, with up to 5 grains. Glumes and lemmas are very variable; both strong, firm and asymmetrical. Stems generally hollow with thin walls, though in some cultivars, they may be partially or completely filled with pith.

Grains also variable. Generally rounded dorsally and with a deep ventral, longitudinal groove, termed the crease. Apices are often capped with many short hairs called the 'brush'; the extent of the latter varies. Shape is more or less ovoid, and differences in size and weight are common. The position of the embryo is visible on the lower dorsal side and looks somewhat like a 'finger-nail on a finger'. Synonym: *T. aestivum* L. em. Thell.

IDENTIFICATION OF WHEAT CULTIVARS

A cultivar, in this instance, is a wheat which differs from all others in some consistent trait. In other words for a new cultivar of wheat to be recognised as such it must be clearly distinguishable from all other. The differences may be morphological, physiological, or agronomical. Identification at this level is a difficult task because of the vast number involved. Feldman and Sears (1981) have stated that there are up to 20000 cultivars of wheat! Also, the characters used are mostly microscopic and difficult to examine. Gill *et al.* (1980) state that cases have been recorded where wheat cultivars have been distinguishable only by differing iso-enzymes of alpha-amylase. A summary of the criteria used in wheat cultivar identification is given hereunder:

Agronomical: These characters are only observed in the field. They are of limited usefulness because they can be influenced by environmental conditions. The most important ones are:

Germination and establishment: There are real differences between wheat cultivars in regard to the rapidity with which germination and shoot emergence take place but they cannot, in most instances, be availed of as diagnostic characters owing to the influence which soil conditions exert on them. When soil conditions are favourable a good wheat stand may be obtained in less than fourteen days. On wet soils, however, germination may take place very slowly and 2 months may elapse before the young seedlings appear above soil level.

Tiller production: Early after germination wheats undergo what is known as vernalization, a condition induced by low temperatures. It begins shortly after the formation of the third foliage leaf and is necessary for ear and seed development. Winter and spring wheats differ in their cold requirement for vernalization, the former requiring a relatively long, the latter needing a much shorter period.

One of the features of vernalization is the development of tillers. In general, winter cultivars are prostrate and have a high tillering capacity. Early maturing spring cultivars produce fewer, more erect, tillers. While real differences exist between tillers of different wheats in relation to habit of growth, this feature is strongly influenced by environmental conditions, and it cannot express itself to the greatest extent unless seeding is sufficiently thin. Nevertheless, it is a useful diagnostic character.

Morphological: Characteristics of the ear and straw, based mainly on small structural differences, are possibly the most useful and widely used. Caution, however, must be exercised to ensure that 'like-areas' are examined as differences can exist between different parts of the ear and different parts of the straw.

Straw characters: The wheat culm usually has about 6 internodes, of which the upper is the longest. The uppermost may be hollow with a large central cavity and a thin wall, or semi-solid with a small cavity and a thick wall. In general, English cultivars tend to conform to the former, while French cultivars have the latter. Differences in straw or culm length also exist. The trend is towards the production of short-strawed types which do not lodge readily.

Ear: The shape of the ear is mostly used for the identification of the wheat varieties listed in Figure 35.00, though some overlap may occur. Four different shapes are recognised: (i) *oblong* - this is where the ear is the same width from base to apex, and occurs in both dense and lax types. It is mainly associated with *aestivum* wheats; (ii) *elliptical* - or oval shapes are associated with *compactum* cultivars and in crosses between the latter and some common wheats; they are quite common; (iii) *clavate* - or club-shaped occurs with a decrease in the length of the upper internodes of the rachis. The ears are wider apically than basally; (iv) *fusiform* - this is the reverse of the latter, and due to an increase in the upper internodes; the ears are narrower apically than basally. They are found mostly in lax types.

Density is said to be influenced by environmental factors and is, therefore, not regarded as a major character. Neither is it a guide to overall yield. Spike length varies from about 5 to 15 cm but tends, to some extent, to be inversely proportional to density so that all forms have roughly the same number of spikelets, usually about 22-24. Nevertheless, it can be used, under normal conditions, for the separation of cultivars displaying wide differences. It is calculated by dividing the rachis length by the number of internodes present. Three levels of density are recognised: (i) *dense* - where the mean internode length is 3.5 mm or less; (ii) *medium* - mean length varies from 3.6-4.5 mm; (iii) *lax* - is where the length is 4.6 mm or more. Other characters associated with the ear include: cross-section, *i.e.* cylindrical, quadrate or flattened; ear width, *i.e.* narrow or broad.

Glumes: These structures show a wide range of consistent variation and, accordingly, are indispensable in cultivar identification. Some of the variation encountered include:

Colour and pubescence: Glabrous and pubescent forms occur. Glabrous forms are preferred because the latter hold water and favour mould growth. Colour is either dull reddish-brown or white, *i.e.* pale yellow or straw-coloured. Peterson (1965) reports the occurrence of black glumes.

Size and shape: A comparison of the glumes of different ears of the same cultivar shows that the glumes of spikelets at the same level are closely similar. They are longest and widest about the centre of the ear and decrease gradually towards the apex and towards the base of the ear. Because of the above, the ratio of length to width is a good criterion. Length and width, *per se*, are deemed minor features because of the fluctuation within anyone spikelet. As with all comparisons, it is essential that 'like areas' are examined.

Keel: The glumes of all wheats have a keel which stands out as a sharply defined apical ridge. Differences, however, exist between cultivars as regards the expression of the keel, basally. In some the keel is clearly defined throughout. In other cultivars it is lost or effaced in the body of the glume. Careful examination, however, will show that there are marked differences, even in the same ear, for this trait. There is, nevertheless, one aspect of the keel which appears to be consistent. It is consistently straight, inbent or hollow, apically.

Beak: Wheat glumes are asymmetrical with a terminal projection or beak (Fig. 34.1f) which can vary in length. It is long and pointed in awned cultivars but always short, rarely exceeding 2 mm, in awnless types. Other wheats have beaks approximately 1 mm long, while some have humped beaks.

Shoulder: Asymmetrical glumes have 1 narrow and 1 broad shoulder (Fig. 34.1h). The latter may be: (i) *square* - when the upper and lateral margins converge at an angle of approximately 90°; (ii) *oblique* - when they merge without forming an angle; (iii) *inclined* - when the distinction between upper and lateral margins is lost. Instead the lateral margin curves gradually to the base of the beak.

Differences in the contour of the shoulder occur within a single ear so, again, it is very important that comparisons are made from 'like-areas'.

Lateral nerve: The narrow shoulder contains a ridge or lateral nerve (Fig. 34.1d) which is well developed in some cultivars. It may carry a number of short, stiff hairs, basally. The development of the latter and the extent to which the hairs are produced are useful diagnostic criteria.

Inner surface pubescence: Hairs are present on the inner-surface of the glume in many cultivars. These are usually directed towards the glume apex. The hairs are always short, bulbous and thinly but evenly spaced over the lower half. On the upper half, however, they may be replaced by long, white silky hairs. The extent to which these hairs are developed and the area over which they extend, differ in different cultivars. Furthermore, this character is not influenced by environmental conditions and is therefore of major importance. Three levels of pubescence are recognised: (i) hairs few and confined to the keel region; (ii) hairs few but not confined to the keel region; (iii) hairs scattered over the whole of the upper third of the glume.

Watermark or imprint: Some cultivars have glumes with a thin semi-translucent area (Fig. 34.1e) towards the base. This area is known as the *watermark or imprint* and, when present, can only be seen from the inner surface. It may be large or small.

Lemmas: These are less variable than the glumes and consequently are not emphasised in cultivar identification. Variation includes the presence or absence of hairs, and the presence or absence of awns; the latter may be long or short. Normal and inflated lemmas occur. Most cultivars grown in Ireland have normal, awnless and hairless lemmas.

Grain characteristics: Some variation may exist between grains of different cultivars. Thus the grain may be hard or semi-hard, white or red in colour, vary in size and shape, have a narrow or broad, deep or shallow crease. The extent of the brush area, and the prominence of the embryo, can also vary. Caution, however, is necessary as variation for many of these characters may be found within anyone ear of wheat.

Physiological: These are laboratory-determined characteristics and are mostly used to supplement the criteria discussed earlier. One physiological difference, based on differing amylase enzymes, has already been mentioned. Another, commonly known as the *Phenol Test*, involves soaking the grains in 1% aqueous phenol and exposing same to the air. Cultivars are reasonably consistent in their reaction to this chemical. A colour change may be observed after 4 hours, the former varying from unchanged to very dark brown. Alternatively, pieces of grain may be stored on paper impregnated with catecol; this gives a similar reaction. *DNA finger-printing* could be a possibility in the future.

STRUCTURE OF THE GRAIN

The wheat grain (Fig. 34.1j, longitudinal section) is derived from the entire ovary and is, therefore, a fruit rather than a seed. It is classified as a particular type of achenial fruit, a *caryopsis*. The latter is character-ised by being one-seeded, indehiscent and by having the seed coats and ovary wall fused together. It is more or less ovoid with a terminal tuft of hairs known as the *brush*. The fruit wall, which gives rise to the *bran*, is termed the *pericarp*. This is a composite structure consisting of about 6 different layers. These are, from the outside toward the centre, the *epidermis*, the *hypodermis*, remnants of *thin-walled cells*, *intermediate cells, cross-cells* and *tube-cells*. Some or all of these layers occur in the pericarp zones of the other cereals but the wall patterns often differ. These differences, together with differences in starch grain outlines, make it possible to determine the composition of a ground-cereal mixture. Prior to exami-nation, excessive starch should be removed by boiling gently in dilute mineral acid for about 2 minutes and the tissues cleared (see Fig. 34.02).

The epidermis and hypodermis are commonly referred to as the outer pericarp. The former consists of a single layer of cells, while the latter has 1-2 layers. These cells are arranged with their long axes par-allel to the long axes of the grain and are closely arranged; there are no intercellular spaces. Most cells are elongate but some at the brush end are nearly as broad as long. The cell walls of these layers are thick-ened and beaded. The outer epidermal cell walls have a thin cuticle. Many epidermal cells, at the apex of the grain, are modified to form the hairs of the previously mentioned brush. Intermediate cells of the inner pericarp are irregularly shaped and occur mostly at the brush end and in the region of the embryo. Over most of the grain there is a layer of thick-walled, beaded, elongate cross-cells, so termed because their long axes lie perpendicular to the long axes of the other cells. Tube-cells, with a long cylindrical, sinuous shape, form the inner epidermis and underlie the cross-cells. Between the tube-cells are many wide intercellular spaces.

Immediately inside the pericarp is the cuticularised seed-coat. It is firmly joined to the cross-cells on the outside and to the nucellar epidermis on the inside. The seed coat is crushed at maturity and appears as a zone in which cell walls can scarcely be identified. When detectable the cells appear thin-walled and regularly arranged. The term *spermoderm* is often applied to these tissues. Nucellar tissue or *perisperm* may be present or absent at maturity. When present it consists of 1 or 2 layers of irregular cells with thick walls. The crushed cells form a colourless line (Fig. 34.02).

The aleurone layer, usually single in wheat, is the outermost layer of the endosperm. It joins the starchy endosperm except over the embryo; there is continuity between the 2 types of endosperm tissue. Aleurone cells are thick-walled and often appear square or rectangular in cross-section. There are no intercellular spaces. The cells contain protein in the form of aleurone granules. Inside the aleurone layer are the endosperm cells which contain many starch granules embedded in a proteinaceous matrix. They are thin-walled and located with their long axes at right angles to the grain.

The embryo consists of 2 major parts, the *embryonic axis* - which includes the rudimentary root and shoot, and the *scutellum*. The latter, which is considered by some to be the *cotyledon* or first leaf of the embryo, functions as a storage, digestive, and absorbing organ. At the apex of the embryonic axes is the *plumule*, consisting of rudimentary leaves and surrounded by the *coleoptile*. At the base of the axis is the primary root covered with a root cap; both are enclosed by the *coleorhiza*.

The scutellum, attached to the side of the embryonic axis towards the endosperm, forms the largest part of the embryo. It is shaped like a broad shield and fits closely to the endosperm. The epidermis of the scutellum in contact with the endosperm consists of elongated, cylindrical, secretory cells - the *scutellar epithelium.* These secrete enzymes which render soluble the reserve starch stored in the endosperm cells. The *epiblast*, a structure considered to be a vestigial second cotyledon, is a small projection of parenchymatous tissue opposite the scutellum. Between the scutellum and the epiblast lies a young plant complete with rudimentary root, stem and leaves. The rudimentary leaves - the plumule - are enclosed within the coleoptile. The roots are enclosed within the coleorhiza. The stem is extremely short and has a minute cone-shaped growing point. Partially developed conducting tissues form a provascular bundle from the embryonic axis into the scutellum.

Many authorities regard the scutellum, epiblast, coleoptile and the first-leaf as the first 4 leaves of the wheat plant. The alternative disposition of these structures, on opposite sides of the axis, supports this view. Another view held by other botanists is that the scutellum and coleoptile represent the single cotyledon of the wheat plant, while the epiblast is the vestigial second cotyledon.

GROWTH STAGES AFTER GERMINATION

Many workers have claimed to be able to distinguish several stages or phases of growth in the development of the wheat plant after germination. A very useful guide for indicating the stages of development of wheat was devised by Feekes (1941) who established 6 stages. His system was later amplified by English workers. A more recent system devised by Zadoks (1974) and expanded by Tottman, Makepeace and Broad (1979), is known as the *Decimal Code.* This system is likely to gain wide acceptance. The code is based on 10 principal growth stages. Each of these has provision for subdivision into 10 secondary stages. These are denoted by a second series of digits extending the scale from 0.0 to 9.9. Where so many subdivisions are unnecessary the positions are left blank.

The principal advantage of the code is that it should be possible to pinpoint more accurately a particular stage in the development of the plant. Identification of the latter is often necessary to obtain maximum benefit from the application of agrichemicals, e.g. fungicides, pesticides, etc., and to avoid unnecessary damage from same to the plant. The significant feature of a typical plant can be clearly specified and the crop treated when the majority of the plants conform to that description. If the timing of the treatment is critical for crop safety, it may be necessary to state that plants before or beyond certain stages of growth are likely to suffer damage.

When describing the development of a crop or population of plants it will first be necessary to describe each plant, or a random sample of plants, in accordance with the code and then report the mode and range of these observations. For example, a sample of plants from an uneven crop may range from 5 leaves unfolded, with a main shoot and 2 tillers to 7 leaves unfolded, a main shoot with 5 tillers and the first node detectable. The most commonly occurring stage may be represented by a plant with 6 leaves unfolded, a main shoot with 3 tillers, and an erect pseudostem. The Decimal Code, as given by Tottman *et al.* (1979), is reproduced hereunder:

DECIMAL CODE

0 **Germination:**
   0.0 Dry Seed;
   0.1 Start of imbibition;
   0.2 ....
   0.3 Imbibition complete;
   0.4 ....
   0.5 Radicle emerged from caryopsis;
   0.6 ....
   0.7 Coleoptile emerged;
   0.8 ....
   0.9 Leaf at coleoptile tip.

1 **Seedling growth:**
   1.0 1st leaf through coleoptile;
   1.1 1st leaf unfolded;
   1.2 2 leaves unfolded;
   1.3 3 leaves unfolded;
   1.4 4 leaves unfolded;
   1.5 5 leaves unfolded;
   1.6 6 leaves unfolded;
   1.7 7 leaves unfolded;
   1.8 8 leaves unfolded;
   1.9 9 leaves unfolded.

2 **Tillering**:
  2.0 Main shoot only;
  2.1 Main shoot and 1 tiller;
  2.2 Main shoot and 2 tillers;
  2.3 Main shoot and 3 tillers;
  2.4 Main shoot and 4 tillers;
  2.5 Main shoot and 5 tillers;
  2.6 Main shoot and 6 tillers;
  2.7 Main shoot and 7 tillers;
  2.8 Main shoot and 8 tillers;
  2.9 Main shoot and 9 or + tillers.

3 **Stem elongation**:
  3.0 Pseudostem erection;
  3.1 1st node detectable;
  3.2 2nd node detectable;
  3.3 3rd node detectable;
  3.4 4th node detectable;
  3.5 5th node detectable;
  3.6 6th node detectable;
  3.7 Flag-leaf just visible;
  3.8 ....
  3.9 Flag-leaf ligule just visible.

4 **Booting**:
  4.0 ....
  4.1 Flag-leaf-sheath extending;
  4.2 ....
  4.3 Boots just visibly swollen;
  4.4 ....
  4.5 Boots swollen;
  4.6 ....
  4.7 Flag-leaf-sheath opening;
  4.8 ....
  4.9 First awns visible.

5 **Inflorescence emergence**:
  5.0 ....
  5.1 1st. spikelet visible;
  5.2 ....
  5.3 1/4 of inflorescence emerged;
  5.4 ....
  5.5 1/2 of inflorescence emerged;
  5.6 ....
  5.7 3/4 of inflorescence emerged;
  5.8 ....
  5.9 Emergence complete.

6 **Anthesis**:
  6.0 ....
  6.1 Beginning of anthesis;
  6.2 ....
  6.3 ....
  6.4 ....
  6.5 Anthesis half-way;
  6.6 ....
  6.7 ....
  6.8 ....
  6.9 Anthesis complete.

7 **Milk development**:
  7.0 ....
  7.1 Caryopsis water ripe;
  7.2 ....
  7.3 Early milk;
  7.4 ....
  7.5 Medium milk;
  7.6 ....
  7.7 Late milk;
  7.8 ....
  7.9 ....

8 **Dough development**:
  8.0 ....
  8.1 ....
  8.2 ....
  8.3 Early dough;
  8.4 ....
  8.5 Soft dough;
  8.6 ....
  8.7 Hard dough;
  8.8 ....
  8.9 ....

9 **Ripening**:
  9.0 ....
  9.1 Caryopsis hard (difficult to divide);
  9.2 Caryopsis hard (not dented by thumbnail);
  9.3 Caryopsis loosening in daytime;
  9.4 Overripe, straw dead and collapsing;
  9.5 Seed dormant;
  9.6 Viable seed giving 50% germination;
  9.7 Seed not dormant;
  9.8 Secondary dormancy induced;
  9.9 Secondary dormancy lost.

BARLEY

General: Barley, domesticated from wild races found today in southwestern Asia, has been described as the world's fourth most important crop. It is a short season, early maturing plant, grown in nearly all cultivated areas of the temperate zones and in many subtropical regions. Early spring types grow within the

Arctic Circle, farther north than any other cereal. This plant is grown in nearly all cultivated areas of the temperate zones, in many subtropical areas and in the high altitude regions of both hemispheres. It grows in the arid climates of the Sahara, the high plateaus of Tibet, and the tropical plains of India. Barley tolerates alkali, drought and frost, but is best adapted to well-drained, fertile, deep-loam soils with a pH of 7-8; it grows particularly well where the ripening season is long and cool. Growth is enhanced with moderate rather than excessive rainfall. High temperatures are tolerated if humidity is low.

The total world cultivation is approximately 80.0 MHa. In the past, it was used mainly as a human food, especially in areas unsuited to wheat, because it is hardier. It is an important crop in Europe, North America, North Africa, Argentina, Australia, China and Europe. The former USSR is the principal producer. Harlan and Martini (1936) have stated that barley is grown in the New World largely for livestock feed, whereas in the Old World it is produced mainly for human food. Indeed, it would appear that it was cultivated in earlier times for use as the latter when many of the cultivated forms were of the 'naked' type. Such naked barleys are still cultivated in parts of Asia unsuited to wheat-growing.

Presently, in order of importance worldwide, it is used for animal feed, for brewing malts and for human food. Some of the better known derivatives used for human consumption include barley flour and pearled barley. The latter, produced by the removal of the flowering pales or husks and the pericarp, is used for soups and drinks. Barley, however, cannot be used to make a flour that will produce a porous loaf because it contains very little gluten. The flour is used instead to make an unleavened type or flat bread, and to make porridge, especially in North Africa and parts of Asia. Most European barleys are hulled; naked forms exist in Eastern countries. The presence of the husks increases roughage and lowers nutritional value but improves malting quality.

Historical: Barley is considered by many as the most ancient cultivated cereal. Nevertheless, there appears to be some uncertainty as to whether wheat or barley was the first domesticated cereal. Matz (1969) states that it is one of the world's oldest domesticated crops, and that it probably shares, with wheat, the distinction of being the first wild plant form brought under cultivation. Bell (1965) states that barley was a most significant crop in the origin and development of early civilization, and that there is considerable archaeological, cultural and historical evidence for this. Other researchers, however, refer to it as the most ancient grain. Ancient Egyptian hieroglyphics indicate that barley was more important than wheat for human consumption. An indication of its possible earlier existence is shown by the fact that the Sumerians had a god for barley but none for wheat.

Heads, carbonised grains and impressions of grains on vessels have been unearthed in several countries. Archaeological remains of what appear to be wild forms with fragile ears were found at Mureybat, in Syria, dating perhaps to 8000 BC. Similar remains were found at Beidha, in southern Jordan, dated about 6800 BC, and at Jarmo, Iraq, at about the same time. Cultivated barley was found at Ali-Kosh - Iran, Ramad - Syria, Jericho - Jordan, and at Catal Huyuk and Hacilar - both in Turkey, dating between 7000 and 6000 BC. Overall, the above and other publications, indicate that barley is approximately 10000 years old. Doubts remain, nevertheless, as to the identity of the earliest types of barley. Bell (1965) emphasises that most of these in parts of Asia and the Mediterranean region were of the 6-row type. Other writers, however, in the sixties, state that they were all of the 2-row type but that 6-row and naked forms had appeared by 6000 BC. Belief in the identity of the earliest types appears to have had a strong influence on the postulation of different hypotheses on barley evolution in the past 50 years.

TAXONOMIC DATA

General: The genetics and cytology of barley have been extensively studied. Compared with wheat, it has presented fewer taxonomic problems, though the classification of the cultivated forms has presented some difficulties. This is possibly due to its monophyletic, as opposed to wheat's polyphyletic, origin. All barleys, wild and cultivated, are classified in the genus *Hordeum*, and in the same tribe and subtribe as rye and wheat. The inflorescence is also a modified spike (Fig. 36.1a) and the genus is characterised by the presence of a *triad* of spikelets (Fig. 36.1b) at each node of the rachis; the central spikelet is sessile and the laterals are short-stalked. Each spikelet is one-flowered but the laterals may show some reduction.

Barley is a self-pollinating plant, though some cross-pollination can occur to a limited extent. The young seedling differs from that of wheat in some minor details. On germination the coleorhiza is first to emerge from the base of the grain. The main shoot and secondary seminal rootlets follow in close succession. The coleoptile pushes through the seed coat and pericarp, grows up the dorsal side of the grain, and emerges near the apex before pushing through the soil. The first leaf emerges from its tip while, at the same time, the seminal rootlets are growing and extending into the soil. In general there is 1 main root and between 1 and 10 seminal rootlets. The rootlet number is partly characteristic of a variety, but within one variety, there is a wide spread of numbers; small grains tend to have fewer rootlets. Mature leaves are characterised by the presence of prominent glabrous auricles. The former, like those of wheat, are twisted clockwise.

The basic chromosome number is 7 and all cultivated forms are diploids (2n=14). Tetraploids have appeared spontaneously but are a negligible part of the crop. The grass species may be diploid (2n=14) and tetraploid (2n=28), hexaploid (2n=42), but polyploidy is not known in the cultivated forms, or in the 'wild' grain types. The grasses of the genus are genetically isolated from the grain producers in spite of certain similarities of some species with the grain types. No natural hybrids are known between the grain producers and the grass species and artificial hybrids are difficult to produce, resulting in sterile plants when successful.

The number of species is said to be about 25. This number has to be exaggerated because it has been an unfortunate practice, particularly in relation to cultivated barley, to classify each minor variant as a new species. Thus, a review of the pertinent literature will show many different specific names for the different cultivated forms. Many of the more dramatic differences in spike morphology are due to differences in one or a few, often mutagenic, genes. Minor genetic differences are rarely sufficient justification for the bestowal of species status, but are more often the basis for varietal delimitation.

Classification: All cultivated forms and the putative wild ancestors are freely cross-fertile and so, biologically, and in accordance with the ICBN, should be treated as a single species. The appropriate legitimate name for all cultivated barleys would be *Hordeum sativum* Jess., as it appears to be the earliest name used in this sense. Some publications continue, incorrectly, to use the name *H. vulgare* L. *'sensu lato'* to refer to all cultivated forms. The OEEC retains the latter for 6-row barleys and *H. distichum* L. for 2-row types. Other names - previously used for the various cultivated forms and which should now be considered defunct, together with an overall classification system, are given in Figure 37.00. In a revised classification system the defunct specific names could be retained but reduced to varietal status.

Whatever classification system is used, or whether all cross-fertile types forming the cultivated complex should be treated as a single species, 2 wild forms merit special attention. The first of these, mostly termed *H. spontaneum* Koch, or wild 2-row barley, is now widely considered to be the progenitor of all cultivated barleys. The rachis is brittle and shatters easily at the nodes. The central spikelet of each triad is rather narrow, long-awned and bisexual, while the laterals are staminate. The laterals are smaller, awned, short-stalked and, together with the central and 1 rachis-internode, form a very effective disseminating and soil-insertion mechanism called 'trypanocarpy' by Zohary (1960). Wild 2-row barley is concentrated in an arc in the Middle East, with scattered strands over a much wider area from Tunisia to Afghanistan, but is of doubtful occurrence in Morocco and Ethiopia, and not known to extend into Tibet and W China. There are about 5 variants or varieties: *bactrianium, ischnatherium, nigrum, transcaspicum* and *turcomanicum*.

The second wild form of importance is *H. agriocrithon* Åberg. This is the wild 6-row form. It has many of the characteristics of *spontaneum* but differs in having all florets of each triad fertile. It was claimed by many, in the fifties and early sixties, to be the ancestor of the cultivated 6-row barleys. More recent cytological data, however, have refuted this claim. Indeed, its status as a discrete species is also in

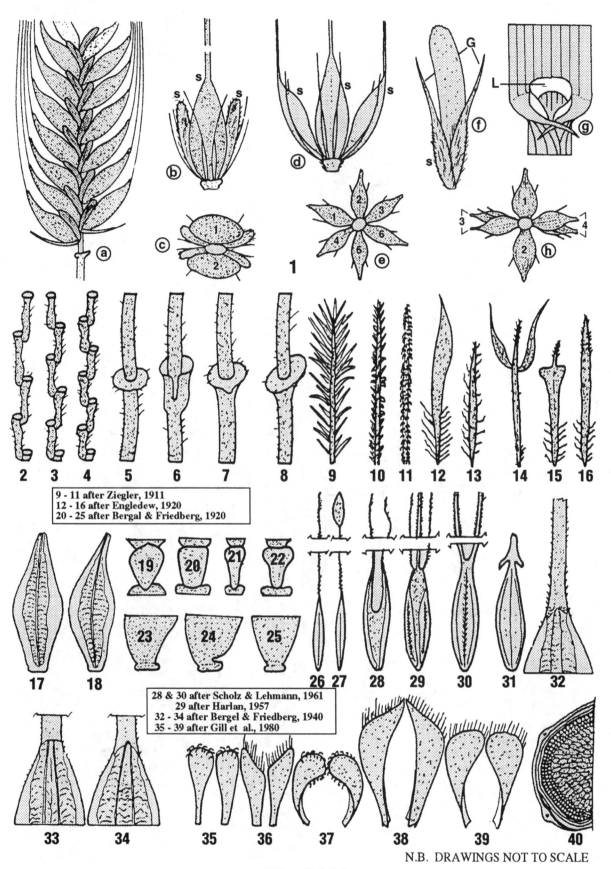

9 - 11 after Ziegler, 1911
12 - 16 after Engledew, 1920
20 - 25 after Bergal & Friedberg, 1920

28 & 30 after Scholz & Lehmann, 1961
29 after Harlan, 1957
32 - 34 after Bergel & Friedberg, 1940
35 - 39 after Gill et al., 1980

N.B. DRAWINGS NOT TO SCALE

Fig. 36.00

218

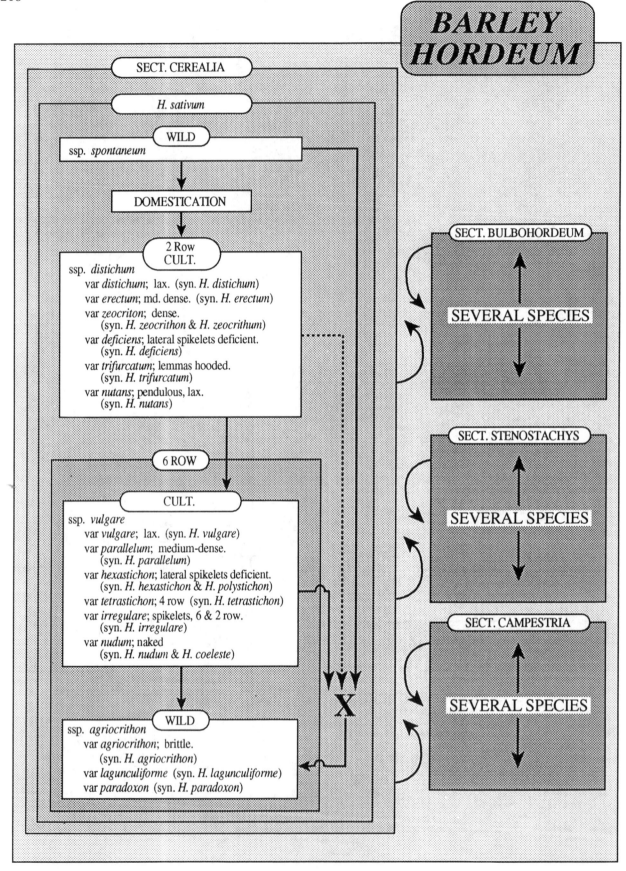

Fig. 37.00

doubt. Plants of this type have a very limited distribution, and are most likely to have arisen through introgressive hybridization involving *spontaneum* and cultivated forms (*see evolutionary history*). Two variants are recorded in the literature: vars *paradoxon* and *dawoense*. Two other wild 6-row forms - *H. lagunculiforme* Bakh. and *H. paradoxon* Schiem. - are also mentioned, but there is no evidence to suggest that they are anything other than forms or variants of *agriocrithon*.

6-row, 4-row and 2-row types: The fertility of the triad spikelets and the length of the rachis-internodes determine the 2 principal divisions of cultivated barley. If all the spikelets of each triad are bisexual and produce a grain, 3 vertical lines of grains will be visible on each face of the ear, giving a total of 6 lines or rows. When the rachis-internodes are short, that is if the ear is dense, the vertical rows will stand out clearly from each other and form 3 lines on each face, or what is known as a *6-row* barley (Fig. 36.1d).

If, however, the rachis-internodes are relatively long, the lateral spikelets do not stand out clearly from the rachis. Instead, the laterals of any triad will take up a position more or less above or beneath those on an adjacent node. The laterals will therefore form 2 vertical rows and these, together with the 2 formed by the central spikelets, constitute the so-called *4-row* (Fig. 36.1h). The 'lateral rows' will contain twice as many grains as the 'central' row. Two types, '4-row' and '6-row', are distinguished only by a difference in rachis-internode length, or density, and indeed grade one into the other. The name *H. tetrastichon* has been applied to the '4-row' forms.

The 2-row types are distinguished by a more dramatic difference in ear morphology. The lateral spikelets of each triad are reduced in size and are staminate or asexual (Fig. 36.1f) and, therefore, cannot produce grains. The grains are produced in the fertile central spikelets, so, considering both faces of the ear, 2 vertical lines of grains or a *2-row* barley is produced (Figs 36.1c). Dense and lax 2-row barleys occur.

Evolutionary history

Leonard and Martin (1963) claim that it is generally recognised that 6-row barley was cultivated earlier than the 2-row form. However, Harlan (1976) suggests that the cultivated 6-row barley was derived from the cultivated 2-row barley under domestication. Contradictory views such as these give some indication of the disagreement that exists relating to the origin or origins of cultivated barley. A review of the copious literature dealing with this topic still leaves the question - which came first, the 2-row or the 6-row? - unanswered. Did the 2-row and 6-row barleys arise separately, or from some other source? What about naked barleys? To date, after careful consideration, one is led to the conclusion that there is no definitive evidence which provides answers to the above questions.

Vavilov (1951), a much acclaimed Russian geneticist, in his *Origin, Variation, Immunity and Breeding of Cultivated Plants*, was the first to suggest what he termed 'centres of origin' for the many cultivated plants. An area which had the greatest diversity of forms for a species was deemed to be the 'centre of origin' for that species. Vavilov lists 2 centres for the barleys. He states that many long-awned, hulled types have come from the Ethiopian and North African centre, and that the hull-less, short-awned, awnless and hooded forms have come from the Chinese, Japanese and Tibetan centre. Wilsie (1951) gives 3 centres for the many cultivated barleys. Claims such as these imply at least a diphyletic, if not a polyphyletic, origin for the different barleys.

The basic wild grain-producing form known for many years is *H. spontaneum* Koch. This, and its variants, are concentrated in an arc in the Middle East, with scattered strands over a much wider area from Tunisia to Afghanistan, but is of doubtful occurrence in Tibet and China. Bell (1965) regards it as the only truly spontaneous form, and also states that it has all the desired characters of the wild type with the disseminating mechanism and means of inserting into the soil. These were the only wild types known until the discovery in 1938, in Tibet, of the brittle-rachis 6-row form, named *H. agriocrithon* Åberg. A similar form was found in 2 centres in Israel, in 1944. This 'species' and its 2 variants were immediately accepted by many authorities as the long sought primitive wild 6-row progenitor of the cultivated 6-row barleys. Briggs (1978), nevertheless, states that the 6-row form has not been seen growing wild in Tibet or Sinkiang, but has been seen growing mixed with 6-row, naked barleys - apparently as a weed. The significance of Briggs' statement will become clear later on. Åberg (1940), however, put forward his alternative views on the origin of cultivated barleys, based on his studies of *agriocrithon,* and postulated a monophyletic theory.

A major study on barley origins was published in 1955 by Takahashi, and received wide attention. His hypothesis of a diphyletic origin was based on the assumption that some primitive cultivated forms of *agriocrithon* were differentiated by mutation from brittle (Bt_2) to tough (bt_2) rachis in Eastern Asia. These differentiated forms or their derivatives spread gradually over various parts of the world. Later, 2-row cultivated forms appeared by mutation from brittle (Bt) to tough (bt) in *H. spontaneum* Koch that grew in southwestern Asia. These 2-row cultivated forms crossed repeatedly with 6-row varieties of genotype Bt Bt bt_2 bt_2 to give new 2-row and 6-row types that became the cultivated forms of Asia and the Occidental countries.

Despite the hypothesis already discussed, doubts persisted as to the credibility of *agriocrithon* as a discrete species. Did it really exist in the wild? If so it would be logical to assume that it was the progenitor of the cultivated 6-row barleys, in the same way as *spontaneum* was deemed to be the wild ancestor of 2-row barley. If it does not exist naturally in the wild, where did it come from, and how did the cultivated 6-row types arise? It would appear that a solution to the *agriocrithon* enigma would provide some answers as to how the cultivated barleys evolved. The *agriocrithon* problem continued to occupy the minds of many researchers over the past 50 years or more.

Bowden (1959) considers that the ancestral stock of the genus was derived from an isospiculate ancestor. This view could possibly be based on the fact that the most ancient record of a 2-row barley, at that time, was only 300 BC. It may also be influenced by the fact that such barleys were said to have been scarce in Europe up until the sixteenth century. It is, of course, a rational assumption when one considers that the trend in plant evolution is towards a reduction in floral parts. More recent archaeological finds of 2-row, non-brittle forms, at Jarmo, Iraq, dated 7000 BC, appear to have laid the basis for more radical thinking on barley evolution in the early sixties. Zohary (1960) maintains that *agriocrithon* is a derived form. Indeed, Bell (1965) states that the barley genus is basically heterospiculate and that the isospiculate condition must be regarded as derived.

Zohary (1960), on the basis of his studies on 'hybrids swarms' in Israel, implies an introgressive origin when he claims that 6-row spontaneous barley represents derivatives of crosses between spontaneous 2-row barley and cultivated 6-row barley. The possibility of genetic introgression from cultivated forms into wild forms and, *vice versa,* is admitted by Briggs (1978) who states that the modern 'wild' forms, for instance these with unusually large grains, are not necessarily similar to the ancestral forms. The principal inference that may be drawn from the foregoing statements is one of doubt as to the credibility of *agriocrithon* as a discrete species. Additional supportive evidence for this conclusion may be obtained from the fact that no truly wild 6-row form, with all the primitive characters expressed, has been identified with absolute certainty. Also, the wild types of Åberg (1940) and Kamm (1954) are fully cross-fertile with all cultivated forms. Therefore, in the absence of definitive evidence to the contrary, it is reasonable to assume that these wild 6-row types have not had a spontaneous origin, but have been derived from another source, possibly a 'wild 2-row cultivated barley complex', through a process of introgressive hybridization.

Indeed, this is the overall view of Harlan (1976) who summarises, most succinctly, the evolution of cultivated barleys. He states that all truly wild forms are 2-row and originated in a limited area, in the Near East. The differences between the various forms are determined by a few recessive mutant genes, and that the 'wild type' is dominant. Thus brittleness of the rachis is controlled by 1 of 2 tightly linked genes - *Bt* and *bt*. Under domestication 2-row forms gave rise to 6-row types. There are 2 genes involved, both with multiple allelic series, but a single recessive mutation, *vv*, is sufficient to cause a 2-row to become a 6-row. Naked barleys are determined by a single recessive gene - *n*.

IDENTIFICATION OF BARLEY CULTIVARS

Barley, on a worldwide basis, is an extremely variable crop despite its diploid nature and possible mono-phyletic origin. Wider variation is shown than in wheat, possibly due to a dramatic difference in the structure of the spike. While there is no information available as to the number of barley cultivars in the world, it can be assumed that it is much less than that for wheat and, possibly, does not exceed a few hundred. Space does not permit an exhaustive account of the wide range of variation encountered in the crop. Briggs (1978), in his book - *Barley* - gives an excellent coverage of the topic. European grown cultivars are a rather homogenous group and many of the characters discussed here may not be applicable to them at this point in time. Nevertheless, a broad discussion is desirable to enable one to get a good insight into the crop. Identification of cultivars is based on the following characters:

Agronomical: These are similar to those of wheat. They have limited usefulness in cultivar identification, as they are often influenced by environmental conditions.

Germination: There are real differences between cultivars in regard to the rapidity with which germination takes place but they cannot, in most instances, be availed of as diagnostic characters owing to the influence which soil conditions exert on them. When soil conditions are favourable a good stand may be obtained in less than 14 days. On wet soils, however, germination may take place very slowly and 2 months may elapse before the young seedlings appear above ground.

Tillering: There are also differences, between cultivars, in tillering capacity. These are the same as those described already for wheat. A further difference, only determined under field conditions, relates to a difference in tillering capacity between spring and winter types.

Morphological: These, as in wheat, are based mostly on morphological differences in the structure of the ear and, to a lesser extent, differences in the straw. They are the most useful and widely used but, in barley also, 'like areas' should be compared.

Plant length: Length, in mm, is measured from ground level to the apex of the upper awn. This feature is very much influenced by environmental conditions and, as such, is not reliable; it may be of some use in the separation of cultivars with very different straw lengths. Plant length varies from as little as 75-100 mm in brachytic type barleys to as much as 1500 mm in taller plants. The stems, as in all cereals, are hollow except at the nodes. In all barleys the uppermost internode, immediately below the ear, is always the longest, but there is considerable genetic variability in length in different types. Thus it may be long or short, straight, curled or curved.

Flag-leaf: Characteristics of the flag-leaves are greatly influenced by environmental conditions. They are, however, smaller in surface area than those immediately beneath them. As a group, 2-rowed barleys have narrower leaves than 6-rowed, and it appears that this criterion is useful in classifying the former. Those carrying brachytic genes have much shorter and more erect flag-leaves. Flaf-leaf features of importance include the following: (i) number of stomata per unit area; (ii) the nature of the wax covering on the adaxial surface. The wax deposits may consist of: (i) flat plates; (ii) knobs - stellate in appearance; or (iii) a mixture of (i) and (ii).

Neck: The neck is that section of the culm between the flag-leaf and the lowermost spikelets. It can show a certain degree of variation and often has arbitrary referred to as: (i) short; (ii) medium; and (iii) long; it may also be: (iv) straight; (v) curled; or (vi) curved.

Collar: The basal node of the rachis is termed the *collar*. This is very variable in shape. Thus, in some cultivars it is flat or of the: (i) *platform* type (Fig. 36.05); in others it may have a raised rim, giving: (ii) a *cup-shaped* effect (Fig. 36.06); others, still, may be described as being: (iii) *saucer-shaped* (Fig. 36.07); or: (iv) *oblique* (Fig. 36.08).

Rachis-internode: The central stem or rachis of each ear is bilaterally symmetrical, with alternating nodes and internodes. The latter can vary in length (Figs 36.02-04), shape (Figs 36.19-22), and in the presence or absence of hairs. Differences in length determine density.

Ear: Both ear emergence and ear type are important features to observe. Ear emergence is a useful period for observation as striking cultivar distinctions are apparent. Many of the characters used for identification are accentuated and more easily recognised. Under spring-sown conditions, ear emergence takes place 2 or 3 days after flag-leaf emergence. Some cultivars, however, may have a longer period, others may have shorter. Reference has already been made to 2-row (Figs 36.1a-c), so-called 4-row (Fig. 36.1h) and 6-row (Figs 36.1d-e) types. There is a dramatic difference between a 2-row and the other types. The former is characterised by having the central spikelet of each triad bisexual and the laterals staminate or asexual (Fig. 36.1b). Extreme forms of this type have the lateral spikelets reduced to a single lemma. All spikelets in the so-called '4-row' (Fig. 36.1h) and 6-row (Figs 36.1d-e) are bisexual. In a mature '4-row' ear the laterals on either side are more or less superimposed giving 2 rows. These, combined with the 2 central rows, give the so-called '4-row'. A true 4-row wood barley - *H. sylvaticum* - where the lateral spikelets are bisexual, but the central staminate, is found in the wild. No true 4-row form exists in cultivated barley. An extreme mutant is known where the outer glumes of the median spikelets contain fertile florets which form grains, giving a rare ten-row type. In addition, the occurrence of supernumerary spikelets and grains is common.

Variations in the length and shape of the rachis and its internodes determine the shape and density of the ear. The rachis, which varies in length from 2.5-12.5 cm, may contain up to 30 nodes. Short internodes (Fig. 36.04) in either a 2-row or a 6-row produce dense ears, while long internodes (Fig. 36.02) give lax types. When the basal internodes are extremely short a 'fan-shaped' ear is formed.

Rachilla: The rachillas (Figs 36.09-16) arise at the basal end of the palea and lie in its ventral groove. They may be present or absent. If present, then the length and shape are important. Also, hairs, if present, may vary. These may be: (i) long and straight (Fig. 36.10), *i.e.* the 'Archer type'; or (ii) short and woolly (Fig. 36.11), *i.e.* the 'Chevallier' type. Furthermore, the wax depositions may consist of: (i) filaments; (ii) filament clumps; (iii) ribbons or (iv) knobs. In extreme cases the rachilla may be replaced by floret-like, lemma-like, or awn-like structures.

Glume: The empty pales or glumes, 2 for each spikelet, are spear-like (Fig. 36.1b). Deviations from these include situations, in some 6-row barley types, where the glumes of the median floret are ovate-lanceolate or boat-shaped. Other unusual types involve the glumes of the lateral spikelets, or of all spikelets, being so-replaced. Some rare 2-row barleys are reported to have lemma-like glumes. Finally, glumes with hooded apices are also known (Fig. 36.27).

Lemma: In most European cultivars the lemma and pales adhere to the caryopsis and form the outer coatings or husks of the grain. They are absent from many Chinese and Japanese cultivars giving 'naked' or hull-less grains. The lemma invests the dorsal rounded side of the grain and its thinned edges overlap the thin edges of the palea. Normal lemmas are ovate or broadly lanceolate and are mostly terminated with an awn which has 3 vascular bundles. The lemmas have 5 longitudinal nerves. Variants are known where the apex is deeply bifid (Fig. 36.28) and ending in 2 awns. Other rare mutants have 3 awns (Fig. 36.29). An astonishing array of appendages is known, ranging from awns carrying slight humps or extensions to highly elaborate structures. Some rare types have simple or lobed hoods (Fig. 36.31). More common features involve the presence or absence of spicules on the median and lateral nerves and pubescence on the inside of the lemma tips. These latter features are all useful in cultivar identification (Figs 36.32-34).

Palea: This is not much used in cultivar characterisation, usually only slight differences in the shape of the tip are encountered (Figs 36.32-34). Differences also exist in relation to the presence or absence of hairs on the sides of its central groove. Mutants, with the palea bearing small awns, or where they are deeply divided nearly to the base, are known. The shape of the apex is a useful feature.

Awn: These (Fig. 36.26-34) are apical extensions of the lemmas, glumes, or both, and carry 3 vascular bundles. The central bundle is the largest. They are roughly triangular in cross-section and contain 2 tracts of parenchymatous photosynthetic tissue. They are mostly straight, rarely curved, or absent. Awns are usually toothed or pubescent, and the size and disposition of the teeth, or their absence, on the median or lateral ridges, or at the apex or base, are characters of importance. Normally the teeth point apically. Smooth-awned cultivars tend to lack sufficient numbers of stigmatic hairs and show a tendency to sterility.

Lodicule: Two hygroscopic scales or lodicules are located between the pales and the caryopsis. The manner of their insertion, their size, their exact shape and hairiness, type of hair, combined with their morphology, is often characteristic and consistent for any one cultivar. Their differing ability to take up stain is a useful diagnostic feature. Gill *et al.* (1980) have identified 5 types based on size, shape and hair type: (i) *small, with woolly hairs* (Fig. 36.35); (ii) *small, with straight hairs* (Fig. 36.36); (iii) *large, with woolly hairs* (Fig. 36.37); (iv) *large, pointed, with straight hairs* (Fig. 36.38); and (v) *large, blunt, with straight hairs* (Fig. 36.39). The small lodicules are sometimes termed the *'parvisquamose'* type, while the large are known as the *'latisquamose'* forms.

To facilitate their examination, the grains should be soaked in a few drops of ammonia for about 10 minutes, followed by washing for several minutes. After washing, they are next placed in a 1% alcoholic gentian solution for about ten minutes, then washed very quickly in water. Following a very rapid wash in water or weak alcohol, they are dehydrated in absolute alcohol, passed through xylol and mounted in Canada Balsam.

Grain: These are mostly composite structures, made up of the flowering pales or lemma and palea, the lodicules, often the rachilla and the caryopsis. The latter may be free and, if so, it alone constitutes the grain. In addition, awns may be present or absent. These structures show a wide range of variation and a discussion of the grain must involve a consideration of all of them.

The grains may differ in shape but are often spindle-shaped, tapering apically and basally, and with a central ventral groove. Differences in symmetry occur. Thus a sample of grain, containing all symmetrical grains (Fig. 36.17), suggests a 2-row origin, while a sample with 2-third symmetrical and one-thirds asymmetrical (Fig. 36.18), suggests a 6-row origin. The base of the grain also varies. Viewed from the dorsal surface, it may show a 'horseshoe', an incomplete 'horseshoe' or a transverse marking or crease. Also, the same, in side-view, may be: (i) of the *falsum* type (Fig. 36.23), *i.e.* tapered sharply or bevelled; (ii) others may have the *verum* type (Fig. 36.24), *i.e.* possess a transverse nick; (iii) finally, others still may show the *spurium* type (Fig. 36.25), *i.e.* a base which is neither nicked nor bevelled. The grain-base, as described above, is not a totally dependable feature.

The base and apex, when the husks are present, are usually damaged, either by the removal of the awn or other apical appendages, or when the grain is freed in the threshing process. The attachment scars left on the grain are of no help in cultivar identification. The surface itself can be wrinkled, semi-wrinkled or slightly wrinkled. Furthermore, the wax formations can be either of: (i) dense filaments; (ii) plate-shaped structures; (iii) a mixture of (i) and (ii); or (iv) consist of dome-shaped protuberances. Length, thickness and width of grain, may also be useful features.

Grain colours are characteristic. Naked grains may appear blue or red-purple, from anthocyanin pigments in the aleurone or pericarp, respectively, or they may appear pale-yellow through various shades of grey to black. Hulled grains may appear greenish if the aleurone is blue and the outer layers yellow. The husk itself may be several shades of yellow or buff, or even partly orange, brown, various shades of grey or black. The nerves may be purple, or the whole husk may be purple.

Caryopsis: The true fruit is known as the caryopsis. It provides very few characters of diagnostic importance. Differences in colour exist. More subtle differences relate to it being widest mid-way or towards the base. The shape of the crease, and the attitude of the caryopses when placed on a horizontal surface, are helpful criteria.

Physiological: Many investigations have been conducted to differentiate, in a chemical or physico-chemical way, morphologically similar cultivars. The phenol and electrophoresis tests are the most useful. The phenol test requires a higher concentration and longer exposure time than that recommended for wheat. An electrophoresis test involves passing a current through a support medium. The current accelerates differential mobility, and consequent separation of compounds such as proteins, according to their molecules. Cultivars with different proteins can thus be separated and identified.

STRUCTURE OF THE GRAIN

The barley grain is somewhat spindle-shaped in outline. One important difference between its structure and that of wheat is the persistence of the husks or flowering pales, *i.e.* lemma and palea, in most cultivated forms of the former (Fig. 36.17-18). Small, but significant, differences exist in the structure of the pericarp. Cells of the lemma are thick-walled and sinuous. The cross-cells of barley have thick unbearded walls, while those of wheat are pitted or bearded (Fig. 34.02). A dramatic difference is found in relation to the aleurone layer. In barley it consists of 3 layers (Fig.36.40), whereas in wheat it is represented by a single layer. The starch grains also tend to be smaller. The structure and distribution of the embryonal tissues are essentially the same as those of wheat and the other cereal grains.

OATS

General: Oats are not significant world crops. Winter and spring oats are not of great importance in Ireland, and rank third behind barley and wheat in terms of acreage cultivated. Indeed, the crop has shown a dramatic decline in popularity over the past 70 years. In world production, oats have consistently ranked fifth behind rice, wheat, barley and maize. Total world production fluctuates somewhat but a figure of just over 30 Mha has been estimated for the early seventies. Leading producing countries include the USA, the former USSR, Canada, France, the UK, Germany, Poland, Austria and New Zealand.

Oat plants require more moisture to produce a given unit of dry matter than any other cereal except rice. As a result, they are especially vulnerable to injury by hot dry weather, particularly from the early heading stage to the production of kernels. It is no surprise, therefore, that the cultivated oats are grown in the cool temperate regions of the world, while the more heat resistant red-oats are cultivated in warm climates. Unlike the 2 other principal cereals grown in this country, oats will grow on soils that are sandy, low in fertility, or even slightly acidic. Oat grains are high in carbohydrates and contain 15 and 7.5%, protein and fat, respectively. They are a valuable feed grain for all classes of livestock, particularly for horses, poultry, young animals and breeding stock; the straw is a valuable roughage for ruminants. Some are processed for human consumption, especially as a breakfast cereal, but oat flour is not used in bread making. Rolled grains and flattened dehusked kernels are used mainly for oatmeals.

Historical: At the dawn of agriculture, a form of wild-oat existed as a weed in widely separated regions of the world. At an unrecorded time, it began to be grown in more or less pure stands, probably because it gave larger and more persistent yields than other grain crops grown under the climatic and soil conditions existing in some parts of northwestern Europe (Matz, 1969). However, opinions differ as to when it first appeared in Europe and, indeed, as to the identity of the first writer to mention it in the literature. Some authors consider Theophrastus (b.c. 372, d.c. 287 BC), the Greek peripatetic philosopher, to be the first to write about oats. Others believe the honour belongs to his compatriot, Dieuches (c. 400 BC).

De Candolle (1882) states that the ancient Egyptians (c. 3000-2000 BC) did not cultivate oats, but that they are now grown in Egypt. However, archaeological finds indicate that they are about 4000 years old. Many writers make reference to their occurrence in the Bronze Age, *i.e. c.* 2000-500 BC. Baum (1977) says that the cultivated oats were known in Europe in that era, but that they have become much more plentiful since the early Iron Age (c. 100 BC), and that archaeological finds are chiefly from Germany, Poland, England, Sweden, Denmark, Czechoslovakia and Switzerland. De Candolle claims that the ancient Greeks (c. 400 BC-AD 200) called it *Bromos*, as the Latins called it *Avenae*. He continues by reporting that these names were commonly applied to species which were not cultivated, and were weeds mixed with cereals. He also adds that there is no proof that they cultivated the common oat.

Authentic historical information on the cultivation of oat plant appears in the early Christian era. It is known to have been cultivated in China as early as 386 BC-AD 200. Pliny (b.c. 23, d.c. AD 79), cited by De Candolle, remarked that the Germans lived on oatmeal, and this remark was interpreted by the latter as implying that the species was not cultivated by the Romans. He adds that its cultivation, therefore, was practised anciently to the north of Italy and Greece, and was diffused later and partially in the south of the Roman Empire. Preceding writers referred to the plant as a weed, forage, or medicinal plant.

TAXONOMIC DATA

General: There are different types of oat plants, scattered in several genera and species. The annuals constitute the important grain plants and all are located in the genus *Avena*, tribe *Aveneae* and subtribe *Aveninae*. The perennials, such as downy oat-grass, yellow oat-grass and false oat-grass, form the indigenous 'oat-grasses' of waste places, and are located in other genera. They are of no economic significance, and all have smaller inflorescences and spikelets.

The basic chromosome number of *Avena* is x=7. A ploidy series, similar to that of wheat, exists. Thus there are diploids (2x=2n=14), tetraploids (4x=2n=28) and hexaploids (6x=2n=42). All are regular bivalent formers. Unlike wheat, however, the manner by which the hexaploids have evolved from the tetraploids, or from the diploids and tetraploids, is far from being satisfactorily explained.

There is no general agreement on the number of species in *Avena*. Hackel (1890) gives the largest number, 50, and has stated that they occur in the temperate zones of the Old World and, to a lesser extent, in the New World. This figure obviously takes into consideration the many perennial oat-grasses that occur throughout the world. Bell (1965) states that 7 species seem basic but other figures of 12, 14, 26 and 27, have been mentioned in the literature. Baum (1977), on the basis of a computer-aided numerical taxonomy study, involving over 5000 samples collected worldwide, lists 27 species.

The annual species of *Avena* are easily distinguished from the 3 other cereals discussed earlier because of clear-cut morphological differences. All have rather loose, spreading panicles, with relatively long rachis internodes, branches and large pendulous spikelets (Fig. 38.1a). The latter are 2-3 flowered, with long, coriaceous, 7-nerved lemmas which many have a stout, twisted, scabrid, geniculate, dorsal awn. The glumes are subequal, coriaceous, 7-nerved or 9-nerved, and exceed the length of the florets (Fig. 38.1c). As in barley, the flowering pales remain attached, except in naked oats, to the ripened caryopsis (Fig. 38.1d). The caryopsis is like that of barley in being somewhat tapered basally and apically, but differs in being hairy all over. The mode of disarticulation differs for different species, and is often used in species identification. Lodicules are lanceolate and acute.

Self-pollination is the general rule, with natural crossing being very limited and seldom exceeding 1%. Anthesis begins in the upper spikelets and 5 to 7 days may be required before the entire panicle has bloomed. On germination only 3 seminal rootlets are produced. The seedling also shows a further difference from that of wheat in the development of a 'mesocotyl'. The mature vegetative phase can be distinguished from that of the other 3 main cereals by the complete absence of auricles, and by the anticlockwise twisting of the expanded blades.

Classification: Various classificatory systems purporting to show, at least to some extent, the phylogenetic relationship of some species of *Avena* have appeared over the years. The species have been divided into various sections, subsections, series and subspecies. Indeed, some writers have labelled the various genomes in a manner similar to that of wheat. No attempt is made here to reproduce any system, because no definitive system exists and, indeed, much work remains to be done before such a system appears. Later comment on important species will be made under the 3 well established ploidy levels.

Evolutionary history

The origin of cultivated oats is lost in antiquity and so remains, and is likely to remain, a matter of conjecture. An exhaustive review of relevant literature leads to the conclusion that the exact time and area may never be known. Nevertheless, there is widespread agreement that it originated in the general Mediterranean area. Thence it appears to have spread northwards and westwards into Europe as a weed of other cereal crops. As environmental conditions became progressively less favourable for the successful cultivation of wheat and barley the latter were superseded by their weed component - oats.

There is much controversy surrounding the phylogenetic relationships of the various species, particularly those of the hexaploid group. Indeed Holden (1976) states, on the basis of the finds of some new and apparently good genetically isolated species, that other distinct entities remain to be identified before the evolution of the cultivated hexaploids is understood, and that differentiation between genetic entities at the same and different ploidy levels is frequently obscured by morphological similarities.

For nearly 100 years it was widely accepted that the wild-oat, *A. fatua* (Figs 38.2a-e), was the progenitor of the cultivated oat, *A. sativa*. Haussknecht, in Germany, in 1885, made the assumption that common wild-oat was the progenitor because of its primitive characteristics. Many writers supported the views of Haussknecht and, indeed, the much-acclaimed classificatory system of Malzew (1930) appears to have been modelled on his assumptions. Malzew's system also indicates that the second important cultivated species - *A. byzantina* - evolved directly from the wild red-oat - *A. sterilis*.

However, in 1946, Coffman pointed out that many researchers, who had studied derived forms of *A. sterilis*, had found plants which were *A. sativa* in type. On the basis of a mass of evidence derived from plant pathology, physiology and genetics, he put forward his theory that the *A. sativa* may be derived from *A. byzantina* - and hence all hexaploid forms from *A. sterilis*. He suggested that *A. fatua* had probably arisen in the transition through the centuries in a manner somewhat similar to the *'fatua-like'* aberrants which were stated to occur by other researchers not too uncommonly in *A. sativa*, and more plentifully in some forms of *A. byzantina*.

This theory is not without its critics. Nevertheless, it is adopted by Holden (1976) who bases his decision on genetic evidence from his own work and that of many contemporary researchers. Holden appears to support his decision by stating that the 2 possible progenitors of cultivated oats - species *sterilis* and *fatua* - have quite different distributions and ecological preferences. The former forms dense stands in relatively stable primary habitats, while the latter is virtually confined to crops and disturbed land and abandoned agriculture and is not known in primary habitats. Acceptance of the Coffman theory means, of course, that the wild-oat, *A. fatua*, must be considered a derived rather than a primitive form.

The original choice of *A. fatua* was a natural one because of the universal acceptance, up to about 40 years ago, that comparable wild plants were the progenitors of cultivated cereals. These wild plants are characterised by their dispersal and articulating mechanisms which enable them to perpetuate themselves in the wild. Plants lacking these features were considered derived as they can only survive in cultivation. Another example of the reversal from the cultivated to the wild has been discussed already in relation to the origin of 6-row barley.

Diploids: The exact number of species with 14 chromosomes cannot be stated with any degree of certainty; there are both wild and cultivated species. All known species in this category are closely related because of their similar genomic constitution. One species, *A. strigosa*, bristle oat, has differentiated sufficiently to develop a distinct form with short broad grains. It is often named *A. brevis*. A naked form is known and this is frequently referred to as *A. nudibrevis*. These 3, together with the closely related *A. hirtula*, are often mentioned as the *strigosa* group. Spikelets of diploid species are smaller than those of tetraploids or hexaploids; yield is also lower.

A. strigosa is frequent as a weed of cereal crops in Ireland, obviously introduced as an impurity in other grain. It can be distinguished from the more pernicious common wild-oat and cultivated oat, by its narrow lemma which terminates in 2 bristle points - hence the common name, bristle oat. The lemma also carries a distinct awn. Both naked and hulled forms were cultivated in western Europe.

Tetraploids: These are of minor importance and are of interest only because of their involvement in the next group. Some of the species frequently mentioned include the wild or weed types: *barbata, magna, murphyi* and *vaviloviana,* and the cultivated *abyssinica.* The *abyssinica* forms *(barbata, vaviloviana, abyssinica)* are often regarded as the tetraploid level of the *strigosa* group. They do not appear, from either cytogenetical or morphological evidence, to have contributed to the putative ancestor of the hexaploids - *A. sterilis.* The diploid ancestry of species *murphyi* and *magna* has not been fully explained. Species *abyssinica* is cultivated in Ethiopia (Abyssinia).

Hexaploids: This is the most important group as it contains the cultivated oats, together with some pernicious weed types. It is often described as containing 2 basic species - *sativa* and *byzantina* - each of which has weed forms, *fatua* and *sterilis,* respectively, and cultivated naked forms. Some current thinking suggests that *sterilis* is the wild ancestor of all 42-chromosome oat plants, and that the other species arose directly or indirectly from it. Indeed, all hexaploid species are fully interfertile. All have the basic set of chromosomes of the diploids and tetraploids. Holden (1976) states that cytogenetic evidence implicates the diploid - *strigosa* - as one donor, and that either *magna* or *murphyi* must be regarded as the tetraploid donor. However, as the diploid ancestry of these latter species is unknown, the full identity of the second and third genomes of the hexaploids remains enigmatic.

IMPORTANT SPECIES

Wild-oat Plants

A. strigosa Schreb. (Bristle Oat): This diploid has been discussed already. The following are hexaploids.

A. sterilis L. (Wild Red-oat): Many researchers now consider this species to be the progenitor of all 42-chromosome oats. It has not been recorded in Ireland, but it occurs as a major component of the herbaceous flora around the shores of the Mediterranean and through Asia Minor to the slopes of the Zagros mountains in the east. It is an aggressive pioneer of disturbed land. The spikelets are large and usually 2-flowered. Lemmas are mostly covered with a dense growth of hairs, and both florets bear long, strong, geniculate, twisted awns. However, the plant as a whole is quite variable. The 2 florets are firmly attached together and fall-off as the diaspore because disarticulation occurs at the base of the lower floret.

A. ludoviciana Durieu (Winter Wild-oat): This species is of no consequence in Ireland but it occurs in England where it is reported to be widespread in southern central areas. It is rather similar to wild red but has smaller spikelets. It is also rather like the common wild-oat but differs in that disarticulation occurs at the base of the lower floret. It is often classified as a subspecies of *sterilis.*

A. fatua L. (Common Wild-oat): This is the most important wild-oat plant. It is a well known weed of tillage fields and broken soils, in many countries. Occasional plants were recorded in the Irish Flora up to about the middle of the present century. Since then it has become widespread. It has been known in England since about 700 BC. However, it is used rather extensively for hay in the interior and coastal valleys of California, where it is estimated that approximately 16000 ha are harvested annually.

Common wild-oat is readily distinguished from other wild-oat plants and from the cultivated species. It is a tall plant and, when present in other cereal crops, it usually attains a greater height. General structure is similar to that of the cultivated plants. However, the spikelet (Figs 38.2a-e) shows a number of well defined characteristic features. Each of the 2 florets bears a twisted geniculate awn and has a basal concavity. This is variously referred to in the literature as a 'sucker-mouth', 'callus' or 'abscission scar'

(Fig. 38.2e). The 'knob' of the rachilla fits into the basal cavity, so that shedding or disarticulation of the grain takes place when the knob shrinks on drying. This characteristic mode of floret attachment is sometimes termed the 'ball and socket' arrangement. Unlike species *ludoviciana* and *sterilis,* all florets are shed separately. Great variation is shown as regards hairiness and colour of the lemmas. As a rule, in Irish plants (Fig. 38.2e), the base of each grain carries a dense tuft of prominent hairs. This species, elsewhere, is often divided into subspecies and botanical varieties.

Two factors, the ease with which the spikelets shatter and its marked dormancy, make common wild-oat a difficult plant to control. Curran (1966) has reported a marked periodicity of germination, a small proportion of seeds germinating in autumn and a higher proportion in spring. Many investigations have been conducted in the United States, where it is a most troublesome weed, to develop cultural methods for its control and possible eradication, but only with partial success.

Cultivated Oat Plants

A. byzantina (Red-oats) and *A. sativa* (Cultivated Oats): Two types of oats are cultivated throughout the world, the cultivated oat of temperate zones and the red-oat of warmer climates. Both are fully interfertile and, consequently, should be placed in a single species. Convention, however, treats each separately, the former in *A. sativa,* the latter in *A. byzantina.* There are numerous cultivars of each. The essential difference between the 2 species is shown by their mode of floret separation. Articulation is lost in both, but the spikelets of cultivated oat are separated by fracture at the rachilla node (Fig. 38.1e), whereas, in red-oat, fracture takes place just above the lower floret, so that the rachilla forms a 'stalk' on the upper floret.

A third species - *A. nuda* - has been cultivated in China and the mountainous parts of southwestern Asia since early times. It is not known elsewhere. This species is characterised by the fact that the flowering pales do not adhere to the caryopsis. In addition, the rachilla internodes are extremely long.

IDENTIFICATION OF OAT (*A. sativa*) CULTIVARS

The general characteristics of the panicle and spikelets have been described already in the introduction to 'Taxonomic data.' Cultivar identification is based primarily on minor differences in general structure of the panicle, spikelet and straw; agronomical features are of little significance.

Agronomical: These are features which can be observed only in the field and are mainly of use in separating winter and spring cultivars. As is the situation with the other cereals, they are highly influenced by environmental conditions. As such, they are not very dependable.

Tiller production: Winter grown cultivars produce many tillers which tend to have a loose growth-habit; spring types produce fewer but are more erect.

Morphological: Characters of the panicle, based mainly on structural differences are, as in other cereals, more useful and more widely used. Some of them may be influenced by environmental conditions.

Straw height: Plant height may be used to distinguish between cultivars that are otherwise similar. This character may be influenced by environmental conditions and time of sowing.

Straw strength: The culms, including the inflorescence branches, differ in diameter and coarseness. Strength is usually described as stiff or weak. Differences in colour may also be evident; most are yellow. All features described here are of minor importance.

Pubescence of culm and leaf: Some cultivars may have consistently developed hairs on the upper node of the culm. Differences in relation to the presence or absence of hairs on the leaf-sheaths and leaf-blades may also be consistent features. The ligule may be absent from the leaves of a few cultivars.

229

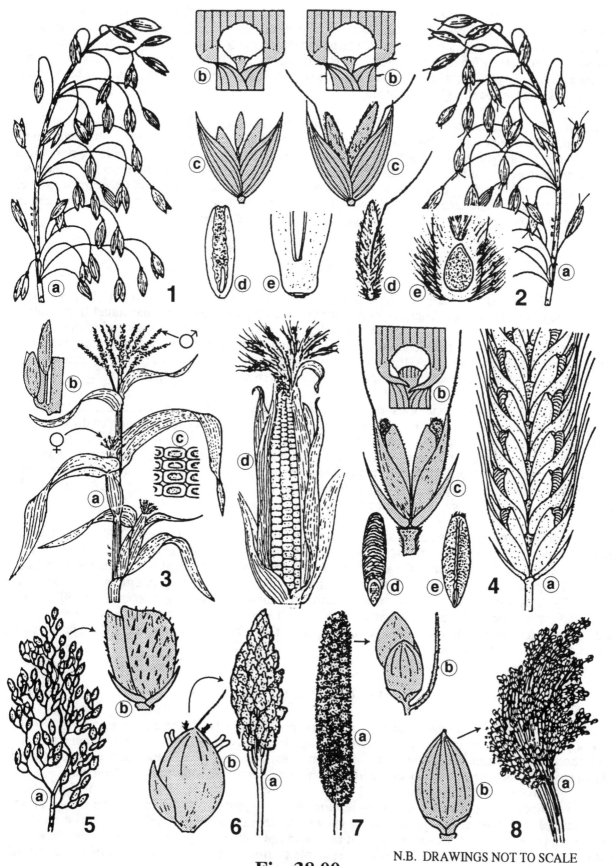

Fig. 38.00

N.B. DRAWINGS NOT TO SCALE

Panicle: This differs in being either spreading or 1-sided and banner-like; the former is frequently described as being 'equilateral', while the latter is known as the 'unilateral' arrangement. Intermediate forms occur. Panicle branches really arise in alternating clusters at the rachis nodes and the overall shape is mainly pyramidal. Differences in panicle size may occur, *i.e.* small, medium or large. Long drooping branches give a 'lax' panicle; when short and ascending in their attitude they are described as being 'compact'. The rachis may be straight or flexuous.

Spikelet: The usual number of florets is 2, or 2 with a small 3rd; potato oat produces only 1 ripe floret. The length, width, number of veins and colour of the glumes have been used in cultivar characterisation. Colour varies from light to dark green and is of value only at time of full heading. Similar observations may be made for the lemmas. Here, however, the colour is whitish in most cultivars but yellowish, brown, grey or even black types are known. Grains may be 'plump' or long and thin. Other grain features include the frequency and length of basal hairs, their presence or absence from the body of the lemma, frequency of awn expression, and the pubescence, length and shape of the rachilla.

RYE

General: Rye is among the least important cereal crops in the world. Indeed, Vavilov (1951) has stated that the peoples of Persia and Afghanistan considered it to be a noxious weed and found it difficult to control. It was considered noxious because it was difficult to separate from the grains of wheat. It is of negligible significance in Irish agriculture, though in the early part of this century, it provided about 70% of the straw used in thatching roofs. Present acreage is so low that it does not warrant a separate entry in statistical records. In world cereal production, in the early seventies, it ranked eighth.

The main advantage of rye appears to be its ability to survive under climatic conditions that are unfavourable for other cereals. For this reason it is an important grain cereal in countries with cold winters and hot dry summers. It will also grow reasonably well on soils of low fertility. It is an important crop in northern and eastern Europe and parts of Asia, and is also grown in Spain, Greece, and Italy. Its main use is in the making of rye bread, though in many countries it is grown as a forage plant. It may be sown alone, or in mixtures with other plants, usually in late autumn, to provide early spring grazing before other less winter hardy forages commence growth.

Historical: Rye, like oats, appears to be of relatively recent origin. De Candolle (1882) states that the ancient Greeks (c. 400 BC-AD 200) did not know it. Nevertheless, it is supposed to have come into cultivation in Asia Minor more than 4000 years ago. De Candolle also states that the grains have not been found in Egyptian monuments, and that the first author to mention it in the Roman Empire was Pliny (AD 23-79). Rye is reported to have become widely distributed during the medieval period (c. AD 500-1450), particularly in Europe. After the Renaissance (yr 1450 +) it was more prominent than wheat. Indeed, it is said to have been the predominant world bread grain until the nineteenth century. Since then, however, it has been replaced more and more by wheat.

TAXONOMIC DATA

General: Rye is classified in the genus *Secale* and species *cereale*. Of all the cultivated cereals, it is the nearest relative of wheat, and the genus *Secale* is considered very close to *Triticum* taxonomically and phylogenetically. Indeed, successful crosses between it and *Triticum*, have been accomplished. Its close relationship with the wheats also places rye in the same tribe - *Triticeae* and subtribe - *Triticinae*.

The number of species within the genus is variously mentioned as 5 to 14. In addition to rye, annual and perennial wild grass and weed species are known. All are diploids with 2x=2n=14. However, there are many reports of accessory chromosomes in both wild and cultivated species (Evans, 1976; Muntzing, 1943, 1944, 1945 and 1946). The genus is comparable to *Hordeum,* and it is an acceptable practice to divide its species into an *Agrestis* (grass) Section and a *Cerealia* (cereals) Section.

The inflorescence is a spike (Fig. 38.4a), with 1 spikelet per rachis node. The spikelets (Fig. 38.4c) are mainly 3-flowered, though the uppermost floret is mostly minute or abortive. The 2 glumes are long, narrow and acute, through shorter than the overall length of the spikelet. The lemmas are relatively broad but taper gradually into long stout awns and bear stiff hairs on their keels. The caryopsis (Fig. 38.4d) is long, narrow, somewhat rugose, greyish in colour, and tapered apically and basally. In the mature inflorescence it tends to protrude from the divergent flowering pales. Unlike most other cereals, the florets are *cross-pollinated*, though some cultivars are rather self-fertile. The vegetative phase can be distinguished from that of the cultivated oat plant by the presence of auricles and the clockwise twisting of its blades, and from wheat and barley by the smallness of its auricles.

Evolutionary history

Rye, like the other temperate cereals, is thought to have originated in southwestern Asia. It shows a strong parallel to the origin of the oat plant in that both appeared first as weeds of other cereal crops. Indeed, both are frequently referred to as secondary crops. As the rye-contaminated wheat and barley spread further and further into Europe, they encountered progressively less favourable conditions and, in time, were replaced in northern and eastern Europe by the more cold resistant rye plants. It is said to have been established, independently, at many locations. How the cultivated form evolved from the wild plants is not fully understood. Nevertheless, there is a wide spectrum of agreement that the immediate progenitor was *S. montanum*, which is a perennial with a brittle rachis. However, *montanum* differs cytologically from *cereale*. They differ by 2 major interchanges, involving 3 pairs of chromosomes. How the change, from the truly wild *montanum* to the agriculturally dependent *cereale* complex occurred, is not clear. There is some disagreement as to whether or not other intermediate species were involved.

MAIZE

General: In terms of worldwide production, *Zea mays* is the third most important grain crop. Most maize is consumed indirectly as feed for livestock rather than directly as human food (Goodman, 1976). Its nutritional value, however, is less than that of most other cereals. Numerous cultivars are recognised, based mainly on the structure of the grain and its usage. The principal areas of cultivation include the US which accounts for nearly one-half of the world's total, the former USSR, Romania, the former Yugoslavia, Hungary, Italy, China, Brazil, Mexico, Argentina, South Africa, Indonesia and India.

Historical: The first reference to maize in recorded history began on November 5, 1492 (Mangelsdorf, 1950). On that day two Spaniards, whom Columbus had delegated to explore the interior of Cuba, returned with a report of a 'sort of grain called maiz which was well tasted, baked, dried and made into flour'. Later explorers to the New World found maize being systematically cultivated by the American Indians, although some other grains were harvested from the wild state. It was introduced into Spain by Columbus in 1494, thence it spread into other European and African Countries. Despite its present-day worldwide importance, its evolutionary history is not fully understood.

Maize differs from the other 4 cereals in that it appears to have originated in the New World, in that no wild forms exist, and in that it is incapable of perpetuating itself without man's intervention. There is no evidence that it was known in the Old World in ancient times. Pollen samples, taken from drill cores collected in Mexico City in the mid-1950s and dated at up to 8000 years old, have been identified as belonging to maize, teosinte or their common ancestor (Goodman, 1976). Teosinte, *Euchlaena mexicana* Schrad., is regarded as an ancestor of maize. Other discoveries, of plants with small very small cobs, in Mexico and New Mexico and dated at about 5000 BC, were thought to be 'wild maize', a postulation that is not unanimously accepted (Mangelsdorf, et al., 1964). From these discoveries it was deduced that wild maize had a 2.5 cm ear with 2 husks borne high on the stalk. At maturity, the husks opened to permit the dispersal of the caryopses. Above the ear was a male spikelet, about 3-5 cm in length. The 'seeds' were rounded and either brown or orange. Repeated selection for larger 'seed' size, more 'seeds' per cob, and other desirable features, led to maize as it is presently known. However, the wild form no longer exists.

Either it was completely displaced by domesticated maize, or cultivated maize hybridised with it, rendering it incapable of independent propagation (Matt, 1969).

Maize, together with teosinte, are classified in the tribe *Maydeae*. A distinguishing feature of this tribe is the presence of male and female flowers in separate spikelets on the same plant. The basic chromosome number is *2n=20*. Maize (Fig. 38.03) is a tall, vigourous annual, with a single stem up to 3 m in height. Leaves are alternate, glabrous or somewhat pubescent, with long lanceolate blades 40-175 x 5-15 cm. Male inflorescences, or tassles, are up to 40 cm long at the apex of the stem. The staminate flowers are carried in pairs in each spikelet, and the spikelets are arranged in groups of 2. One spikelet is sessile, while the other is borne on a short stalk. The glumes are robust but the lemmas and paleas are delicate and membranous. The female inflorescences, or cobs, are carried in the axils of the middle leaves about half-way up the stems. Here, also, the pistillate flowers are 2 per spikelet; 1 flower is sterile. The spikelets, in turn, are also in pairs. The pistillate flowers are sheathed with modified leaves called the husk. The stigmas extend well beyond the bracts, and are often referred to as the *silks*.

There are different types of maize (de Rougemont, 1989): (i) flint maize (Fig. 38.4d) - grains are variable in colour, from white through yellow to red, purple or almost black, have rounded ends and consist mainly of hard endosperm with a little soft central starch; (ii) dent maize (Fig. 38.4c) - grains are larger, also variable in colour, but have a higher proportion of soft starch extending to the apex which shrinks on drying to produce a characteristic dent; (iii) Popcorn - grains are rather small and have a higher proportion of hard endosperm than flint maize. On heating, the steam produced within causes the grain to explode; (iv) sweet corn - grains are shiny and translucent when immature, become wrinkled when dry, but remain broader than flint maize; the endosperm has a relatively high sugar content; (v) waxy maize - this form has the starch consisting entirely of amylopectin which has a waxy appearance.

CEREAL MICROSCOPY

Ground cereal material can be identified on the basis of microscopic differences in pericarp structure. The pericarp consists of about 6 layers of tissue. Some have a distinctive outline for each cereal. In addition, the grains of barley and oats have persistent pales which show epidermal cells with thick sinuous walls. The pericarp tissues form the *bran*, while the flowering pales are known as the *husks*.

Other tissues of diagnostic assistance include the *spermoderm* - derived from the seed coats, the *aleurone cells* and *starch grains*. Spermoderm tissue is difficult to detect in oat and rye but in wheat and barley form distinctive patterns. Starch grains, particularly those of oats and rye, are quite characteristic. All tissues of diagnostic importance are illustrated in Figure 34.02.

Prior to examination excessive starch should be removed and the tissues cleared. The starch may be removed by sieving and then boiling gently in a dilute mineral acid for about 2 minutes. Large quantities of starch may require additional boiling. Clearing is effected by boiling in dilute (0.1%) potassium hydroxide for another 2 to 3 minutes: repeated rinsing in water is then necessary for the elimination of the potassium. Afterwards small pieces of transparent tissue are placed on a drop of glycerine in the middle of a micro slide and a cover slip added. The tissues are then examined under the microscope.

The pericarp tissues consist of a few layers of elongated cells arranged in a 'brick-work' fashion. All such tissues in wheat have beaded walls. Those of barley are straight-walled. In oat the outermost layer is faintly pitted, but the inner layers are mostly straight. This pattern is repeated for rye, though the outer cell walls are more clearly beaded; the inner layer is mostly thin-walled and difficult to detect.

The cross-cells show a more clear-cut difference. All are arranged with their long axis at a right-angle to that of the grain. Their short walls are contiguous. Long and short walls, in wheat, are beaded, while those of barley are thin and unbeaded. They show a rather irregular pattern in oat. The cross-cells of rye are most characteristic; the end walls are unbeaded but the long walls are beaded.

6 BRASSICACEAE

General: The Turnip family constitutes a very important economic group of flowering plants. Storage 'roots,' succulent leaves, modified inflorescences and oil-rich seeds, of many members, are used as animal and human foodstuffs. The seeds of some species are used as condiments. The plants often have a characteristic smell, and many have stellate hairs. Members show a remarkably uniform floral structure, yet classification of some cultivated forms poses major difficulties for taxonomists. Most are found in the northern hemisphere, though some are grown in the tropics at high altitudes or where some months of the year are cool. The family contains approximately 350 genera and 3000 species (de Rougemont, 1989), of which about 46 and 140, respectively, are represented in the flora of the British isles (Stace, 1991). Representatives are annuals, biennials or perennial herbs, though some woody forms are known. This taxon is 1 of 6 exceptions which may be legitimately referred to under 2 different names, *viz.* Cruciferae and Brassicaceae. The former is the older and is derived from the characteristic crucifix arrangement of the petals (Fig. 39.1b). The latter name is relatively new and is based on the type genus - *Brassica*.

TAXONOMIC DATA

Stems: These are mostly erect, herbaceous and either smooth, bristly or ridged; a few are woody throughout; fewer still have woody bases and herbaceous apices; others have fleshy stems.

Roots: All members have a well developed tap-root system; adventitious roots may arise on prostrate stems. Modified 'roots' occur in some cultivated plants.

Leaves: The leaves are alternate, exstipulate, and variously divided. Some species have entire leaves; others have coarsely toothed or even lobed blades; most are pinnately divided, being either pinnatifid or pinnate; leaflets are variously shaped. The leaves are petiolate in most species. In some exceptional cases a plant may have expetiolate or amplexicaul blades.

Flowers: These (Fig. 39.01) are often conspicuous, mostly actinomorphic, and show the hypogynous arrangement. The inflorescence is racemose, mostly a simple raceme; the latter rarely a corymbose; bracts few or absent.

Calyx: The calyx, which is polysepalous, consists of 4 lanceolate, erect or spreading sepals. These occur in 2 whorls of 2; the outer whorl is often pouched.

Corolla: The corolla is polypetalous, of 4 characteristically shaped petals; flowers are rarely apetalous. Each petal has an erect, narrow, basal claw and a broad, flat, mostly spreading, limb.

Androecium: There are usually 6 stamens, rarely 4 or 2. The inner 4, which have filaments of equal length, arise from 2 primordia, and each primordium divides after formation, to give 2 stamens. When the number is 6, the outer 2 are mostly shorter than the inner 4. The relative length of the outer to the inner, and their shape, are important criteria in the identification of the 3 principal species of the genus *Brassica*. Nectaries are present in the filaments bases.

Gynaecium: The pistil or gynaecium is syncarpous, with a superior ovary derived from 4 fused carpels; 2 of these form a partition, giving a bilocular ovary; this partition is sometimes termed the replum or septum. The style is short and terminated with a small bilobed stigma. Placentation is parietal (Fig. 39.1a). Pollination is effected by insects, such as bees and beetles; self-pollination can occur.

Fruit: The fruits are many-seeded capsules. The principal type is known as a siliqua, which is characterised by being long and narrow and mostly terminated with a short beak (Fig. 39.1e). Some members have fruits which are about as broad as long, *e.g.* the silicula of *Capsella bursa-pastoris* (Fig. 39.1d). Indehiscent and schizocarpic types are found in a few genera.

Seeds: Cruciferous seeds (Figs 23.35-40) are non-endospermic but rich in oils. Glucosinolates are present in many species and these, on hydrolysis with thioglucosidase, yield thiocyanates, isothiocyanates and nitriles. These thio compounds are responsible for the pungent flavour of many plants of this family. They may be found in all parts of the plant but are often at their highest concentration in the seeds (Cooper and Johnson, 1984). Seeds of some species, *e.g.* mustard plants, are used as condiments, but can be unpalatable to livestock and may prove poisonous (Gill *et al.*, 1980).

Cotyledon and radicle arrangement within the seed is of systematic importance. Thus, the embryo of *Capsella bursa-pastoris* seed has the radicle folded through an angle of 180° and lying on the flat cotyledons to give the incumbent arrangement. The seed of *Brassica napus*, *B. oleracea* and *B. rapa*, and the seed of some closely related plants, show a somewhat similar arrangement but the cotyledons are folded around the radicle, *i.e.* the conduplicate *C. flexuosa* have the radicle lying close to the narrow edges of the cotyledons, giving the accumbent form. Cotyledons and radicle arrangements have a direct bearing on seed shape.

Germination: Germination is epigeal. There is a wide range of variation in cotyledon shape. The seedlings of all *Brassica* species and of some closely related plants are characteristically obcordate (Fig. 25.01); *Capsella bursa-pastoris* - broadly lanceolate (Fig. 25.15); *Cardamine hirsuta* and *C. flexuosa* - cordate (Fig. 25.19); and those of *Sisymbrium officinale* - oblong (Fig. 26.28). The first leaves, even those of species with mature pinnate blades, are practically entire.

IMPORTANT SPECIES

The Genus *Brassica*

Practically all plants of agricultural importance in the family Brassicaceae are contained within the genus *Brassica*. Other genera of minor importance include *Armoracia, Lepidium, Raphanus,* and *Nasturtium* which contain plants grown for salads or condiments. The genus *Sinapis* is of interest in that it contains 2 species - *arvensis* (charlock) and *alba* (white mustard) - which occur as weeds of tillage and waste places; the former is particularly prevalent. Both species were included, at one time, in the genus *Brassica*, but such a classification is at variance with taxonomic evidence now available.

The genus is remarkable for the diversity of forms which it displays, and there is no other group of cultivated plants of the temperate regions which offers such a variety of specialised habit of growth and edible parts (Curran, 1966). The forms cultivated in the British Isles represent but a small portion of the great range of cultivated types available. Many more are grown in India, China and Asia Minor. Most cultivated types appear to have a Mediterranean origin (Wilsie, 1951).

All *Brassica* species have large, conspicuous, yellow-coloured flowers. One type, *B. oleracea ssp. balearica,* has white petals (Curran and Brickell, 1969). Most have pinnately-divided or lyrate leaves; the leaves of *Brassica oleracea* var *capitata* are one exception. Fruits are cylindrical and of the siliqua type; they are terminated with a distinct beak (Fig. 39.1e). The more or less spherical, reticulate-coated seeds occur in a single row in each loculus, and vary in size and shades of purple. *Brassica* species are mostly cross-pollinated, and infraspecific forms are normally inter-fertile; genetic incompatibility factors prevent the satisfactory growth of pollen tubes following self-pollination (Gill *et al.*, 1980). Pollination is usually effected by small insects.

Cytology: The genus contains both monogenomic and polygenomic species. The former form a series with haploid chromosome numbers of 8, 9, 10, 11 and 12 (Curran, 1970). Each haploid number can be identified from the nature and behaviour of its chromosomes, and it has become customary to designate each with a letter of the alphabet. Thus the monogenomic species have genomes designated A, B, C, D, E and F. In species such as these the haploid number (n) equals the basic number (x), and the term genome may be equated, in its crudest interpretation, with either (Stace, 1980). Accordingly, polygenomic species of *Brassica* will have the genomic constitution of 2 monogenomic species. An example of the former would be *B. napus* which has genomes A and C.

The genus is clearly different from the genera which contain the wheats, barleys, beets, potatoes, etc., in that it does not contain a readily identifiable basic chromosome number. This has lead to some disagreement, not alone as to what the number should be (Catchside, 1934 and Thompson, 1976), but also as to the genomic constitution of some monogenomic species (q.v. Haga, 1938 and Sikka, 1940). Catchside (1934) was one of the first researchers to suggest that 6, *i.e.* x=6, may be the ancestral chromosome number of the genus. Sikka (1940), on the basis of his work with 3 *Brassica* species, interpreted secondary associations as indicating a basic number of 5. The occurrence of 2 bivalents in haploid kale (Thompson, 1956), and the recognition of 6 different chromosome types in *Brassica* genomes (Robbelen, 1960), lend some support to the 5 and 6 hypotheses, but do not resolve the problem. Accepting one or other of the above numbers would suggest that at least some monogenomic species lost 1 or more pairs of chromosomes during their evolutionary development, making them modified amphidiploids derived from a cross between 2 primitive ancestors (Thompson, 1976).

The genus contains about 40 species (Clapham *et al.*, 1962), but less than a half-dozen are of agricultural importance. The overall number, however, varies depending on the taxonomic treatment given to the cultivated members (q.v. McNaughton, 1976a; Ollson, 1954 and Oost, 1986). Three species - *oleracea, rapa* and *napus* - contain the cabbages, brussels sprouts, cauliflowers, kales, rapes, turnips and swedes. Considerable confusion, which may be attributed in part to difficulties in identification and overlapping common nomenclature, has existed in relation to the classification of the various forms. Taxonomic studies in cytology, serology, morphology and geographical distribution, on the species and their hybrids, have lessened the confusion and led to a better understanding of the evolutionary history of the genus. At present the principal remaining problem, at species level, is the acceptance of *B. rapa* or *B. campestris* for the turnip group of plants. Problems also remain in relation to the classification of the many infraspecific forms.

Classification: Several classificatory systems have been devised by many researchers (*cf.* Crane, 1943; Curran, 1962; Oost, 1986; Thomas and Crane, 1942; Wellington, and Quartly, 1972 and Yarnell, 1956). An interesting proposal for the infraspecific classification of *B. rapa* has been advanced by Oost (1984). In it cultivars are grouped together on the basis of having shared characters, but those of origin, phylogenetic relationships or common ancestry are omitted. This system appears to agree, in general terms, with that suggested by Wellington and Quartly (1972). The system given here follows closely that of Curran (1962) but some changes in nomenclature are adopted. It is based on the biological species concept (*q.v.* Heywood, 1986 and Solbrig, 1970) in that all interfertile 20-chromosome forms are treated as a single species. *B. tournefortii* is excluded because it is not interfertile with the other 20-chromosome types. Such a treatment was first advanced by Ollson (1954), and appears to have received a wide level of approval (*q.v.* Li, 1981; McNaughton, 1976a; McNaughton, 1976b; Nishi, 1980; Oost, 1986 and Prat, 1960). A similar treatment has been adopted for the wheats (Feldman, 1976), and is recommended for the barleys by the present writer.

A perusal of relevant literature with show at least 4 different scientific names for the cultivated turnip (*q.v.* Berrie, 1979; Curran, 1962; Gill *et al.*, 1980 and McNaughton, 1976a). The 2 most frequently used names are *B. campestris* and *B. rapa*, but these are generally believed to be conspecific. *B. rapa* is widely accepted as the oldest name (Gill *et al.*, 1980; Oost, 1986 and Wellington and Quartly, 1972) and, in accordance with ICBN procedures, is retained here. This species contains at least 10 subspecies and 30 botanical varieties (Oost, 1986), but only the more relevant types are included in the present system (Fig. 39.02).

There are 3 species of agricultural importance grown in the British Isles. Each species has many different plant forms. A third species, *Sinapis alba*, is cultivated for its seeds which, on being crushed, give a pungent condiment - white mustard. Other mustard plants are classified within the 3 principal groups. The cultivated forms are frequently termed the Turnip Group (*B. rapa* forms), the Cabbage Group (*B. oleracea* forms) and the Swede Group. (*B. napus* forms).

The Turnip Group

Brassica rapa ssp. *campestris* (Wild Turnip): This plant, the putative ancestor of this group, is an annual with a non-bulbous root, and is common as a weed of arable and waste land. It is easily identified by its glaucous, bristly, amplexicaul leaves. Two distinct plant types have evolved from this source. Wild turnip is often mistaken for charlock, another common annual, but the latter has a deep-green colour, and lacks amplexicaul leaves. The 2 plants are often found growing together.

1. *B. rapa* ssp. *rapa* vars *annua* and *oleifera* (Turnip Rapes). These have been selected for its much-branched stems and constitute the turnip rape plants. These, in European countries, are grown for their leaves, *i.e.* the turnip forage rapes, or their seeds, *i.e.* the turnip oil rapes. There are 2 types of oil rapes - winter and summer - and, while they are hardier than the swede rapes, are not as productive. Turnip rapes are annuals and do not produce a swollen 'root'. The summer type is distinguished at an early stage of development by the rapid elongation of the second and higher internodes, and early flower formation. The winter type is very similar to the true turnip in the first year in that a rosette of leaves is formed above the hypocotyl, and in that there is little extension of the first few internodes.

2. *B. rapa* ssp. *rapa* var. *rapa* (Turnips): The second distinct type is characterised by its enlarged 'root' or storage organ and gives the true turnips. These are important as forage for sheep and cattle, and are also eaten as a vegetable in many parts of the world (McNaughton, 1976a). They show a great range of shapes. Most, however, are globe or semi-tankard shaped. Two types are recognised, *i.e.* the white-fleshed and the yellow-fleshed. The first are quick maturing with a low dry matter content of 7-8%. They are not frost hardy and do not store well, and must be used as soon as the storage organ reaches its maximum size (Gill *et al.*, 1980). The yellow-fleshed have a higher d.m.c., up to 9%, while they are slower to mature, they store better.

The corymbose inflorescence of the turnips and turnip rapes can be distinguished from those of the cabbage or swede group in that the opened flowers clearly overtop the unopened flower buds (Fig. 39.3). In addition, the 2 short stamens (Fig. 39.3b) of each flower curve outwards at the base and are approximately one-half the length of the inner. The upper stem leaves are somewhat glaucous, usually glabrous, and clearly clasping the stem; lower leaves are grass-green and bristly; both are variously divided.

The mature 'root' (Fig. 39.1f, and Fig. 39.1g without neck portion) is a composite structure derived from the entire hypocotyl and partly from a small portion of the true root. An anatomical examination of the true root shows a diarch stele and cortex identical with that described later (page 259) for sugar beet (Fig. 40.10). The development of the primary cambium in the root zone is also identical. In the hypocotyl and stem regions fascicular cambium arises from undifferentiated cells. The periderm arises in the inner layers of the cortex.

The mode of secondary thickening in the turnip is totally different to that of the beet, as the massive increase in size is due to the activity of a single cambium. A microscopic examination of the lower part of the 'root' shows the squashed primary xylem occupying the central position. A broad band of radiating secondary xylem, consisting mainly of xylem parenchyma and interspersed with ray parenchyma, surrounds the primary xylem. There is very little lignified tissue present. A much smaller zone of secondary phloem, mainly parenchymatous, is located outside the cambium. The secondary xylem and phloem is enclosed within a thin periderm (as in Fig. 40.19).

The Cabbage Group

Brassica oleracea L. var. *sylvestris* (Sea or Wild Cabbage): This plant, the putative ancestor of the cabbage group of plants, is the sea or wild cabbage which occurs sparingly as a maritime plant in parts of the British Isles and more plentifully in northwestern Europe and the Mediterranean region. It has more or less fleshy leaves and a thick decumbent, irregularly curved stem which becomes woody and covered with leaf-scars. Doubts have been expressed as to whether all the cultivated forms of this group evolved from this wild plant. There are suggestions that other wild diploids, particularly species *cretica, insularis* and *rupestris,* may have been involved (Gill *et al.,* 1980 and Yarnell, 1956). The inflorescence (Fig. 39.4a) of all the cultivated forms can be distinguished by its extended racemes where the unopened flower buds stand well above the opened flowers. In addition, the 2 side stamens (Fig.39.4b), of each flower, are straight and practically equal the inner 4; the flower colour is pale yellow, rarely white. All leaves are glaucous and glabrous. There are many cultivated forms:

1. *B. oleracea* vars *fruiticosa* and *fimbriata* (Kales): Plants of this section are used for animal and human feed, and are the closest to the wild form in habit. Young stems and loosely arranged leaves form the edible portions. There are different types. One, known as thousand-head kale, is characterised by its slender woody stems which produce leafy succulent shoots with flat or moderately wrinkled leaves. Another, marrow-stem kale, resembles the above plants but have rather thick edible stems.

2. *B. oleracea* var. *capitata* (Cabbage): These forms are characterised by the development of a compact head of thick closely overlapping leaves which surround the terminal bud on a short unbranched stem. Stem elongation stops at an early stage in its vegetative growth and a rosette is formed by the fully expanded sessile leaves. The inner more fleshy leaves remain completely unfolded. Large-typed cabbages are grown for cattle feed, while those for human consumption are smaller and more compact. One type, savoy, has large rugose, less compact leaves (var. *sabauda*).

3. *B. oleracea* var. *gemnifera* (Brussels Sprout): Plants of this group may be looked upon as minute cabbages which arise not as terminal but as axillary buds, where all the leaves remain compactly folded. The sprouts are globular, very compact, and carried on short stems. There are many different types.

4. *B. oleracea* var. *gongylodes* (Kohlrabi): This plant is unique among the cabbage group in that it is the only one which produces a 'bulb'. The 'bulb', however, consists entirely of stem tissue, and so is quite different from the swede or turnip. Anatomically, it shows a large pith which forms the greater part of the 'bulb', and a narrow ring of vascular tissue.

5. *B. oleracea* var. *botrytis* (Cauliflower): The edible portion is a terminal curd which develops on a short thick stem at the end of the growing season. The curd represents a condensed terminal inflorescence consisting of a compact mass of short, thickened, white peduncles, pedicels and bracts with the flower initials undifferentiated (Dark, 1938). They are protected by loosely arranged leaves and vary in colour, but white or pale yellow types are the most popular. One winter type is known as broccoli.

6. *B. oleracea* var. *italica* (Sprouting Broccoli): The chief difference between these and the cauliflower or broccoli is that the curds, which are much smaller, are borne on the ends of short axillary stalks.

The Swede Group

Brassica rapa x B. oleracea (Swede Types): Members of this group differ from the preceding 2 in that the parent species - *napus* - is not known in the wild. *B. napus* has long been demonstrated experimentally to be an amphidiploid of species *rapa* and *oleracea* (U, 1935). Indeed, the morphological characteristics of the swede are somewhat intermediate between those of its parents. The inflorescence (Fig. 39.5) is more corymbose than those of the cabbage group, and the flower buds and opened flowers attain more or less the same level. The 2 side stamens are longer than those of the turnip group but shorter than those of the cabbage flowers. Mature swede leaves are like those of the cabbage group in being glaucous and glabrous, but somewhat bristly, like those of turnips. One feature worth noting is that the swedes, unlike the cabbages and turnips, are mostly self-pollinating. There are different types:

1. *B. napus* vars *annua* and *biennis* (Swede Rapes): As in the turnip group, there are 2 types of rape plants with small leaves and unthickened roots, the oil-seed rapes and forage rapes. Oil-seed rapes are an important source of vegetable oils for various uses. There are summer and winter types. Forage rape, on the other hand, is grown for animal feed.

2. *B. napus* var. *pabularia* (Swede Kale): Plants of this section differ from the preceding in having larger leaves and thicker stems. They are not as popular as the kales of the cabbage group.

3. *B. napus* var. *napobrassica* (Swede): Swedes differ from the rape and kale plants mentioned earlier in that the plants produce an enlarged storage organ or 'root'. They are, therefore, very similar to the 'roots' of turnips. The essential difference between the 2 'root' types is the presence of a short neck - derived from the lower portion of the true stem - in the swedes. Anatomical structure and mode of secondary thickening are also similar. Swedes differ in shape, and flesh and skin colour (Figs 39.1f-g).

Other Crop Plants

Raphanus sativus L. (Garden Radish): This is grown mainly as salad vegetable or, more rarely, as a forage plant. It often occurs as an escape in waste places and dumping grounds. This is an annual or biennial, bristly-hairy plant with a swollen reddish, white-fleshed tap-root. Leaves are pinnately lobed, often lyrate. Flowers are either white or mauve.

Garden radish is a very variable plant, and there are many distinct cultivars. Larger types of the plant are important food crops in China, Korea, India and Japan, including the large white radishes or Daikon, and others with 'roots' up to 20 kg in weight (de Rougemont, 1989). The 3 main radish groups include: (i) salad radishes - these are cultivars with white, white and red, or red 'roots' which may be globose, intermediate, or cylindrical, which are picked when small or immature for use in salads; (ii) winter radishes - these are mostly oriental cultivars with firm-fleshed 'roots' which do not become pithy and hollow when mature, and which can be stored for winter use; (iii) fodder radish - in this group, some types have large whitish 'roots', while others have no fleshy hypocotyl but are capable of rapid leaf production.

Nasturtium officinale R. Br. (Water-cress or Green Water-cress): This species occurs naturally, often abundantly, as a perennial in ditches, marshes, streams, etc. It has procumbent to ascending, hollow, angular stems, up to 1 m in height, which root freely at the nodes. Leaves are shiny, dark green, pinnate, with 3-9 pairs of mostly entire leaflets. Flowers are rather small and of a whitish colour. The fruits are slender, small, up to 20 mm in length, and somewhat curved. Different forms are intensely cultivated as a salad vegetable. Water-cress has a distinctive, pleasant, fresh and slightly pungent flavour. The tips of the leafy stems make an excellent salad or garnish for different meats (de Rougemont, 1991).

Lepidium sativum L. (Garden Cress): This cress is an erect, delicate annual up to 30 cm, with a single or branched stem. It differs from other species of the same genus in that the stem leaves are not amplexicaul. Basal leaves are entire or lobed, whereas the upper leaves are 1-2-pinnate; some cultivars have crisped and curled leaves. Overall, the vegetative stage resembles *Petroselinum crispum*. Garden cress is commonly used in the cotyledon stage for salads, garnishes, sandwiches, etc. In some continental countries it is often grown to near-maturity and used as a salad or vegetable soup.

Armoracia rusticana P. Gaert., B. Meyer & Schreb. (Horse-radish): This is a coarse, robust perennial with stems up to 1.2 m in height; the 'roots' are long, fleshy and pungently aromatic, and irregularly and transversely striate. Basal leaves are long-stalked, ovate or oblong in outline, somewhat dock-like, shiny dark green, lobed or entire and with serrated margins. Flowers are small and white. It occurs in waste grassy places. The 'roots' of horse-radish are grated and used as a condiment for many dishes.

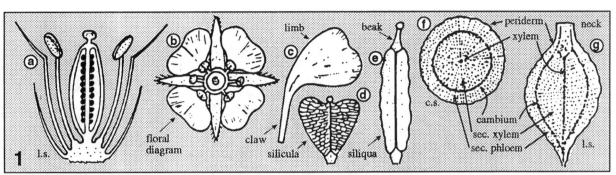

2 CLASSIFICATION: *Brassica* species

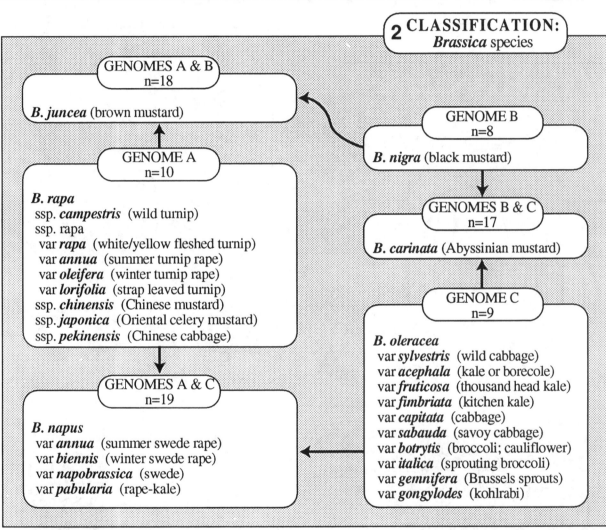

GENOMES A & B
n=18

B. juncea (brown mustard)

GENOME A
n=10

B. rapa
 ssp. *campestris* (wild turnip)
 ssp. rapa
 var *rapa* (white/yellow fleshed turnip)
 var *annua* (summer turnip rape)
 var *oleifera* (winter turnip rape)
 var *lorifolia* (strap leaved turnip)
 ssp. *chinensis* (Chinese mustard)
 ssp. *japonica* (Oriental celery mustard)
 ssp. *pekinensis* (Chinese cabbage)

GENOMES A & C
n=19

B. napus
 var *annua* (summer swede rape)
 var *biennis* (winter swede rape)
 var *napobrassica* (swede)
 var *pabularia* (rape-kale)

GENOME B
n=8

B. nigra (black mustard)

GENOMES B & C
n=17

B. carinata (Abyssinian mustard)

GENOME C
n=9

B. oleracea
 var *sylvestris* (wild cabbage)
 var *acephala* (kale or borecole)
 var *fruticosa* (thousand head kale)
 var *fimbriata* (kitchen kale)
 var *capitata* (cabbage)
 var *sabauda* (savoy cabbage)
 var *botrytis* (broccoli; cauliflower)
 var *italica* (sprouting broccoli)
 var *gemnifera* (Brussels sprouts)
 var *gongylodes* (kohlrabi)

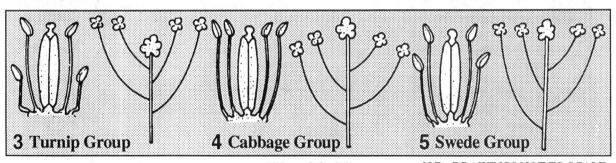

3 Turnip Group **4 Cabbage Group** **5 Swede Group**

Fig. 39.00   N.B. DRAWINGS NOT TO SCALE.

240

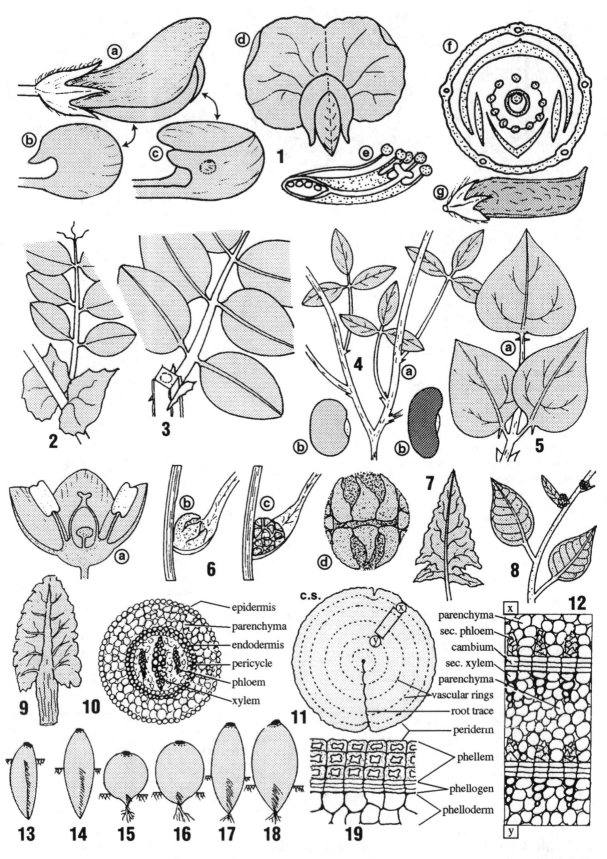

Fig. 40.00

N.B. DRAWINGS NOT TO SCALE.

7   FABACEAE

General: In terms of important cultivated plants the family Fabaceae, commonly known as legumes, is second only to the Poaceae. It is a source of both grain and forage plants and, less importantly, many ornamental species. The importance of the family is due to the nutritional value and the ability of many members to assimilate free atmospheric nitrogen, vital to nearly all major crop plants. Without them the amount of nitrogen circulating within the biosphere would be much less than it is to-day. This ability is made possible through a symbiotic relationship with certain soil bacteria of the genus *Rhizobium* which manifest their presence in the form of nodules on the roots. Legumes, therefore, because of this peculiar relationship, are a most valuable natural substitute for expensive synthetic nitrate fertilisers.

This family represents a very large and natural taxon. The number of genera has been estimated at 750, while the number of species is said to be between 16000 and 19000. In importance, as already stated, it is second only to the Poaceae, and in size only smaller than the Orchidaceae and the Asteraceae. The taxon includes trees, shrubs, woody climbers, annual and perennial herbs with a great variety of habit, including aquatics and xerophytes. Distribution is cosmopolitan in tropical, subtropical and temperate zones. Despite their widespread distribution, however, legumes very rarely form plant communities in which they dominate. The family contains some enormously important forage plants such as lucerne, clovers, sainfoin and serradella, and important grain plants such as the various beans and peas, lentils, peanuts, etc.; the seeds of some species are used as spices. Some shrubs are an important source of grain.

TAXONOMIC DATA

Leaves: The leaves are generally alternate, pinnately compound and stipulate. However, there are many exceptions. In *Ulex europaeus* they are simple and very small, while in *Cercis siliquastrum* they are simple, very large and reniform. The number of leaflets in pinnate leaves can be of taxonomic significance. Many species have leaves which can alter their position at night, a phenomenon which usually involves the folding of the leaflets. Many members have climbing aids in the form of tendrils or hooks. Great variation is shown in stipule size, shape and texture. Most are foliaceous, a few are spinous.

Inflorescence: The inflorescence is usually an erect or pendulous raceme, though a few members have spikes or contracted racemes. The flowers are mostly zygomorphic and hermaphrodite. There are 5 sepals which are often fused basally. Again, exceptions occur where 2 or 3 sepals are totally united to form lobes. The petal number is also 5 but great variation, in relative petal size and arrangement, is frequent. Stamens vary in number from 10 to many. They may be free or united in various ways.

Members of the family are characterised by their fruit which is generally a many-seeded *pod* or *legume*. Some members have pods with transverse constrictions between the seeds. Others have fruits which open explosively. Fruits of *Trifolium* species split transversely. Very few members possess indehiscent fruits. Legumes are mostly dry, rarely fleshy, either inflated or compressed, winged or unwinged, greenish or brightly coloured, and range in size from a few mm to 30 or more cm. Seeds vary in number from 1 to many and often possess a hard or tough coat. They contain a large embryo with little or no endosperm. The food reserves are contained in the embryonic cotyledons.

Classification: Most taxonomists treat the legumes as a single family, either the Fabaceae or the Leguminosae. The former is a relatively new name, coined in conformity with the *International Code of Botani-*

cal Nomenclature, so that all family names would have the suffix - aceae. The old family name is still deemed legitimate because it has been in use for a very long time.

This taxon is very variable and, accordingly, is usually subdivided into 3 subfamilies: (i) the Mimosoideae; (ii) the Caesalpinioideae; and (iii) the Papilionoideae. The classification problem is further confused by some taxonomists who bestow family status on the forenamed subfamilies, *viz.* the Mimosaceae, Caesalpiniaceae and the Papilionaceae. Also, the Papilionoideae are frequently termed the Faboideae or the Lotoideae, particularly in American texts. The Mimosoideae and the Caesalpinioideae are composed mainly of tropical trees and shrubs, including a few economic plants such as Cassia, Tamarind, and many ornamentals such as Flamboyant, *Bauhinia* species, *Amherstia* and mimosas. The Faboideae, on the other hand, contain members which grow mostly in the temperate zones. Most members of economic importance occur in the latter subfamily.

Based on the taxonomic information available at this time, an acceptable classification system for the legumes would be Fabaceae for the family name, and Caesalpinioideae, Mimosoideae and Faboideae for the 3 subfamilies. Such names as Lotoideae, Papilionoideae, Papilionaceae and Papilionatae are best treated as synonyms for the Faboideae.

The main coordinating characteristic is fruit derived from a single carpel whose ventral suture is posterior, *i.e.* directed towards the dorsal surface of the flower. It is unilocular with a single style and stigma, and 2 to numerous ovules arranged along the ventral suture. The fruit is characteristically a *pod* or *legume*, containing 1 to numerous seeds. Variations to the typical fruit type occur.

KEY TO SUBFAMILIES

1. Corolla *actinomorphic*; stamens mostly numerous; leaves *bipinnate*, rarely pinnate .. **Mimosoideae**
   Corolla *zygomorphic* 2

2. Wing petals *external* to the standard .... **Caesalpinioideae**
   Wing petals *internal* to the standard .... **Faboideae**

MIMOSOIDEAE

This subfamily contains only a small number of well known ornamental plants; there are no cultivated species of any importance. Members are either shrubs or small trees with an erect or climbing growth-habit, and with characteristically *bipinnate* leaves; the stipules of many species are often converted into *thorns*. Stem tissue is rich in tannin sacs and gum-passages.

Flowers are generally relatively small, aggregated into stalked globose heads or cylindrical spikes. They differ from the other subfamilies in being *actinomorphic*. The calyx and corolla are valvate. The stamens may equal or double the petal number, or, in some instances, be many times the petal number. They often form the most colourful and conspicuous part of the flower. The 2 most important genera are *Acacia* and *Mimosa*. These genera are regarded by many taxonomists as being congeneric.

CAESALPINIOIDEAE

Again, there is no economic species. This taxon also contains some ornamental shrubs, trees and woody climbers. One of the most widely used species is *Cercis siliquastrum* L. The leaves may be simple, pinnate or bipinnate. Flowers are either large and conspicuous or small and inconspicuous; they are *zygomorphic* and occur in racemes, or the inflorescence may be spicate. The calyx is generally polysepalous; occasionally the upper 2 sepals may be united. The corolla is very varied. Petal size varies in some genera. Other genera may have partly reduced or totally reduced petals. Stamen number is 10.

FABOIDEAE

This is the most important subfamily. It contains many members of economic and ornamental importance. Representatives are shrubs, trees and herbaceous plants, with an erect, climbing or prostrate growth-habit. Practically all leguminous plants of the temperate zones are contained within this taxon. In addition, it contains other crop plants not grown in European countries. These include *Pachyrrhizus erosus* (yam bean), grown for its edible tubers, *Crotalaria juncea* (sunn hemp), a fibre plant used in rope making and *Derris* the source of rotenone, used as an insecticide and for poisoning fish.

Characteristics

Leaves: These are mostly pinnately compound and alternate, though some may have the opposite or whorled arrangement. *Ulex* is one exception because of its small, simple leaves. The leaves resemble those of the of the family Rosaceae, but the latter have mostly serrated leaflets. Leaflet number varies widely and, consequently, the number can be of diagnostic value. Thus, *Ononis repens* (Fig. 06.18) has a single leaflet. *Laburnum anagyroides* (Fig. 17.02), *Medicago lupulina* (Fig. 06.02), and *Trifolium* species (Figs 06.21-23) have 3, *Lotus corniculatus* (Fig. 06.24) has 5, while *Vicia sepium* (Fig. 06.17) and *Anthyllis vulneraria* (Fig. 06.20) have several pairs. *Anthyllis* is further characterised by its very large terminal leaflet. In *Lathyrus* (Fig. 06.13) the number is 2; these are also of the distinctive sagittate type. Many genera (Figs 06.13-17) have leaflets ending in tendrils. The leaflets of most members of this subfamily have entire leaflets. In 1 genus - *Medicago* - they are mucronate, *i.e.* the central vein extends beyond the leaflet margin to form a short projection. This is a very useful diagnostic feature in distinguishing between species of *Medicago* and *Trifolium* that are otherwise very similar, morphologically.

Stipules: All members are stipulate, but the stipules vary in size, texture and shape. They are spinous in *Robinia* but foliaceous in most species. Many are small and insignificant but some species and genera are readily identified by their stipule type. Thus, *Pisum sativum* (Fig. 40.02) is recognised by its stipules which are larger than its leaflets, while those of the *Lotus* species (Figs 06.23-24) are minute. In *Lathyrus* (Fig. 06.13) they are distinctively sagittate. Species of *Medicago* (Fig. 06.22) are separated from those of *Trifolium* (Figs 06.25-28) by their serrated stipules.

Root-system: The root-system of leguminous plants is a tap-root. It differs from the majority of families of flowering plants in bearing root nodules. These are lateral outgrowths from the roots, mainly parenchymatous, but with a vascular supply connecting with the vascular system of the roots. They are caused by the presence of particular bacteria, *Rhizobium* species, which enter the root hairs. They multiply and form an infection-thread, a slender filament of bacterial cells which become surrounded by a tube-like sheath secreted by the root cells. This thread penetrates the cortical cells; in the inner cortex the sheath breaks down, setting the bacteria free in the host cells. Here they multiply and stimulate the host cells to divide, thus causing the proliferation of the inner cortex and the development of a nodule. *Rhizobia* are capable of utilising the free nitrogen of the air and, if conditions are favourable, a state of symbiosis is set up in which the bacteria supply nitrogenous compounds to the plant while the plant provides the bacteria with carbohydrates. If satisfactory symbiosis occurs, the host plant becomes independent of combined nitrogen in the soil. This depends on on soil conditions. For example, nitrogen fixation does not take place in waterlogged soils or in the absence of bacteria.

Flowers: The flowers are clearly *zygomorphic* (Fig. 40.1d), occur mostly in racemes (Figs 06.14-15), though spikes and contracted racemes are represented (Figs 27.1-4).

Calyx: The calyx (Fig. 40.1a) consists of 5 sepals which are united basally into a tube and free apically; the apical portions, or teeth, may be equal or unequal in length. In a few members, particularly in *Ulex* species, the 2 upper and the 3 lower may be united in the form of a 2-lipped arrangement.

Corolla: The corolla is typically *xygomorphic*, with 5 unequal-sized petals. The free uppermost posterior petal is the largest and is termed the *vexillum*; the 2 lateral petals are usually free, claw-shaped and, collectively, form the wings or *alae*; the 2 anterior petals are usually united to form the keel or *carina* which encloses the pistil; the keel petals are also claw-shaped (Figs 40.1a-c).

Androecium: The androecium consists of 10 stamens which may be free or, more often, united. In *Lathyrus pratensis, Lotus corniculatus, Medicago lupulina, Melilotus altissima, Pisum sativum, Trifolium repens* and *Vicia sepium* the *diadelphous* condition, where 9 stamens are united by their filaments and the 10th is free (Figs 40.1a-c), is represented. In other species - *Genista pilosa, Laburnum anagyroides, Lupinus arboreus, Cytisus scoparius* and *Ulex europaeus* - all 10 stamens are united, giving the *monadelphous* arrangement.

Gynaecium: This feature is superior, sessile, rarely if ever stalked, and consists of a single carpel (Fig. 40.1e). The ovary is 1-celled but in a few members may be 2-celled or more due to the development of cross-partitions.

Fruit: The fruit is typically a *pod* or *legume* (Fig. 40.1g), with the seeds forming a single row. Splitting, which is sometimes explosive, is along the full length of both sutures. In *Trifolium* species the fruit is a capsular *pyxidium*. Other modifications include the 1-seeded indehiscent fruits of *Onobrychis viciifolia*, the fruits of *Melilotus altissima* which open along 1 suture and the *lomentum* of *Ornithopus sativus* which has constrictions between the seeds, and which breaks into 1-seeded *mericarps*. *Medicago lupulina* and *M. sativa* plants have coiled fruits.

Seeds: The seeds are mostly numerous but in 1 economic genus - *Trifolium* - there is only 1 seed per fruit. Seeds are non-endospermic. The embryo consists of 2 oval-shaped cotyledons with starch and protein food reserves, a large radicle and a small plumule. They are covered with a thick testa with a conspicuous hilum; the latter varies in colour, shape and length, and is often of diagnostic value. In some species the testa may become impermeable to water, giving what are known as *hard* seeds. These are unable to absorb water and swell, so that when placed on a suitable substratum, germination is delayed. The embryo, however, if alive, and if the testa is cracked or broken by abrasion or frost and water enters, germination will take place. A large percentage of hard seeds is usually a disadvantage in crop plants as it results in poor initial germination. Germination is mainly *epigeal* but in a few genera, particularly *Vicia* and *Lathyrus*, it may be hypogeal.

Pollination: The flowers are usually cross-pollinated, although there are several examples of self-pollinating plants. Many species are almost completely self-sterile. Most of the cross-pollinating plants exhibit special mechanisms which facilitates the transference of pollen from 1 flower to those on another plant. Pollination is usually effected by bees. Indeed, in the production of commercial seed, particularly that of lucerne or alfalfa in the United States, it is frequently the practice to locate hives of bees throughout the crop. Nectar is secreted in diadelphous flowers within the base of the staminal column and can be reached through the slit between the free stamen and those adjacent to it by insects with a long proboscis. In some genera, *e.g. Medicago, Trifolium* and *Vicia*, the pollination process involves the sudden and dramatic release of the staminal column.

In some species of the above genera the staminal column is contained and restrained within the 2 keel petals (Fig. 40.1c). The restraining capacity of the keel petals is supplemented by the wing petals (Fig. 40.1b). A finger-like projection from the latter extends into a pocket-like depression (Fig. 40.1c) on the former. This interlocking of petals keeps the staminal column, and the style and stigma within, completely shielded from an outside pollen source. When an insect visits the flower it invariably alights on the wing petals, causing the finger-like processes to be withdrawn from the keel petals. The net result is a lessening of the restraining capacity of the latter and the staminal column snaps out with some considerable force. This sudden release is called *tripping*.

On the release of the staminal column the stigma and anthers strike the insect on the thorax. Some pollen grains, because of their heavy sticky nature, adhere to the insect. When the insect visits another flower, on another plant, some of the adhering pollen grains will come in contact with its stigma and cross-pollination will be effected. The sudden release of the staminal column also causes a rupturing of the stigmatic membrane and this, in turn, facilitates the germination of the pollen grains and the penetration of the stylar tissue by the pollen tubes.

In other genera, what is termed the piston effect, is encountered. In *Lotus* species the ovary and staminal tube tend to curve upwards but are held in place by a rigid keel. Dehiscence of the anthers takes place in the bud stage and the stigma is thus covered with pollen. When an insect alights on the flower and depresses the wing petals, the staminal column pushes forward, striking the insect.

Seed production in cross-pollinated legumes is largely dependent on an adequate number of bees to effect pollination. These plants are highly heterozygous and agricultural stocks are viable strains rather than distinct uniform varieties. In self-pollinating species well defined uniform varieties exist.

KEY TO IMPORTANT GENERA

1. Leaves *palmately* divided .... *Lupinus*
   Leaves *not* palmately divided 2

2. Tendrils *present* 3
   Tendrils *absent* 4

3. Stipules *larger* than the leaflets .... *Pisum*
   Stipules *smaller* than the leaflets .... *Vicia*

4. Terminal leaflet *absent* .... *Faba*
   Terminal leaflet *present* 5

5. Leaflets *mucronate* 6
   Leaflets *not* mucronate 7

6. Stipules lanceolate or broad, *serrated* .... *Medicago*
   Stipules linear, *entire* .... *Melilotus*

7. Terminal leaflet *larger* than the laterals .... *Anthyllis*
   Terminal leaflet more or less *equalling* the laterals 8

8. Leaves *pentafoliate* .... *Lotus*
   Leaves *trifoliate* 9

9. Stipels *present* .... *Phaseolus*
   Stipels *absent* 10

10. Leaflets *emarginate* .... *Trifolium*
    Leaflets *not* emarginate .... *Glycine*

IMPORTANT SPECIES

There are about 140 species, native and introduced, in the British Flora. Most are herbaceous plants and of minor significance. Some arborescent or suffruticose types, such as *Laburnum anagyroides*, are used as ornamentals. The economically important members include the genera *Trifolium* (clovers), *Medicago* (medicks), *Glycine* (soya beans or soybeans), *Lotus* (trefoils), *Onobrychis* (sainfoins), *Pisum* (peas), *Ornithopus* (serradella), *Vicia* (vetches), *Faba* and *Phaseolus* (various beans).

Several genera, particularly *Acacia, Cercis, Cytisus, Genista, Laburnum, Lupinus, Mimosa, Robinia* and *Wistaria*, are ornamental plants; some of these are also noted for their poisonous properties. Apart from the above economic genera, which are of importance in the temperate zones, there are many more cultivated in warm-climate countries. Included in the latter category are the genera *Cicer* (chick peas), *Vigna* (cowpeas), *Arachis* (groundnuts, earthnuts or peanuts), *Lens* (lentils), *Ceratonia* (carob beans), *Trigonella* (fenugreek) and *Cajanus* (pigeon peas).

Clovers

There are about 250 species worldwide, mostly native to the temperate regions of the northern hemisphere, but a few are indigenous to South America and Africa. The main centre of origin is believed to be Asia Minor and northeastern Europe. True clovers are among the most important cultivated plants used for forage and soil improvement in the temperate zones. They are annual or biennial herbs, mostly adapted to cool moist climates. In regions with hot and dry summers their growth, in the absence or irrigation, is confined to autumn, winter and spring. They grow best on soils rich in phosphorous, potassium and calcium. The basic chromosome number is $x=7$ or 8. There are about 30 species in the British Flora. Some of these are suspected of having poisonous properties. They contain a number of compounds that can have an adverse effect on animal health. The substances present include *cyanogenetic glycosides, goitrogens* and chemicals which can cause *photosensitivity*.

Trifolium pratense L. (Red Clover, Fig. 06.28): This species is one of the most widely grown of all the true clovers. Its soil improving properties have been recognised for several centuries, and it was this species which, from about 1730, was included in 'Turnip' Townsends' Norfolk four-course rotation as a fertility restoring crop (Langer and Hill, 1991). Its centre of origin is southwestern Asia and regions south of the Caspian Sea. It is said to have been cultivated in Europe in the third century. The basic chromosome number is $x=7$; tetraploid forms have been produced.

Characteristics

Stems: The primary stem remains short and bears a rosette of radical leaves. The buds in the axils of the leaves develop into secondary branches which terminate in flower heads. Tertiary branches are also produced and these also terminate in flower heads.

Leaves: These are trifoliate or ternate; leaflets are variable in outline but are mostly ovate with near entire margins, strongly pubescent, and have crescent or horseshoe-shaped markings. Stipules are broad, long-pointed, somewhat membranous, entire or finely serrated and often red or purple-veined.

Inflorescence: The racemes are contracted to form spherical flower heads; the latter are subtended by 2 almost sessile leaves. Peduncles are very short. There are approximately 30 flowers per cluster. These are self-sterile and tripping is obligatory. The corolla is usually reddish-purple in colour, occasionally white. Calyx is gamosepalous; 1 of the 5 teeth is much longer than the remaining 4. The stamens are in the diadelphous, 9+1, arrangement. Red clover is a 1-seeded *pyxidium*.

Types: Three principal types of red clover are recognised. These differ, not so much in morphological characteristics, but in such features as persistency, resistance to disease, time and duration of growth, and yield of fresh matter. The principal types are:

1. Wild Red: A long-lived perennial, it flowers before any other form. The stems are semi-prostrate, short, slender, wiry and solid. The internodes are long so that the leaves are few. Though more persistent than any other form, it is a low yielder and gives little or no aftermath.

2. Early-flowering or Broad Red: An introduced type, it is early flowering, short-lived, high yielding and erect in general habit; it is sometimes referred to as double-cut cowgrass. The leaves are larger and the leaflets are always longer and broader. It starts growth early in spring and flowers early. The vegetative buds do not all develop at the same time; this enables the plant to recover rapidly after cutting. The young shoots grow rapidly to produce an abundant aftermath or a second 'hay' crop. The development of a succession of buds during the first season means that few dormant buds remain to continue growth in the following year. Early-flowering red is, therefore, a relatively short-lived plant, rarely persisting in quantity beyond the second year.

3. Late-flowering Red: The true late-flowering reds are longer-lived plants than the early forms, persist up to 4 to 5 years, start growth later and flower up to 4 weeks later. The leaves are smaller; the leaflets of the leaves produced in early spring are almost circular, although those produced later in the season show little or no difference from those of early-flowering red. It produces more, though weaker, stems than the other types, but these do not become weaker until after flowering. Late-flowering red produces excellent 'hay' but poor aftermath. All stems tend to elongate at the same time, so there are few or no short shoots in active growth at the flowering period. The plants are thus slow to recover from cutting. Numerous dormant buds remain to continue growth the following year.

Trifolium repens L. (White Clover, Fig. 06.25): A stoloniferous perennial, it forms small tufts in the early growth stages. It is essentially a long-lived plant that is ideally suited to grazing management. Some of the large white forms have limited use as 'hay' plants. The creeping habit of white clover allows it wide vegetative spread and it can, therefore, increase rapidly if conditions are favourable. The conditions needed are lime, phosphate and the absence of shade. Thus, if soil conditions are right, it will spread under hard grazing management which prevents it being over-shadowed by taller plants. Some forms of white clover are suspected of having poisonous substances of minor importance.

White clover shows considerable variation in leaf size, persistence, in the amount of adventitious root production, in its response to different degrees of shading, manurial treatments and in the extent to which it is cyanophoric. Its importance lies, not in its yield, as in its effect on companion grasses, although the high protein content of its foliage makes it a valuable grazing plant. It is superior to any other leguminous plant in its ability to withstand hard grazing and in its persistence. White clover is a tetraploid with $2n=32$.

Cyanophoric Types: Some forms of white clover contain the cyanogenetic glycosides - *lotaustralin* and *linamarin* and the enzyme system - *linamarase* - that can hydrolyse them. If a seedling or leaf is crushed the glycosides and enzyme come together and react to produce *hydrogen cyanide gas*. The reaction, in the laboratory, can be hastened by the addition of a small amount of chloroform, and it can be demonstrated by the fact that yellow sodium picrate paper is turned red by the cyanide.

The test is usually made as follows: 1 gramme of picric acid and 10 grammes of sodium bicarbonate are dissolved in 200 cc of water. To make the test a seedling or leaf is placed in a narrow test tube with a strip of picrate paper, a few drops of chloroform are added and the tube is then sealed. Gentle heat, as obtained by holding the test tube tightly in the hand, hastens the reaction. Plants giving a positive reaction will colour the paper *red*; such plants are said to be *cyanophoric*. When no colour change takes place the plants are described as being *acyanophoric*.

Characteristics

Stems: The stems are mostly solid; those of the Ladino types are sometimes hollow. They trail along the ground giving off branches which also trail, resulting in a somewhat mesh-like branched system. This branching and twisting present some difficulties in a botanical analysis. White clover's stoloniferous growth-habit makes it an ideal grazing plant as the young buds, or potential plants, are situated close to ground level and are unlikely to be damaged by grazing animals.

Leaves: The leaflets show variation in outline but are mostly ovate or heart-shaped, deeply emarginate apically, glabrous, finely serrated and have horseshoe or crescent-shaped markings. The petioles are *much longer* than those of red clover. Stipules are small, somewhat membranous and tapered apically.

Inflorescence: The flowers occur in contracted racemes or clusters. They are mostly white, glabrous, and borne on *very long* peduncles; the latter are *longer* than the petioles. Occasionally the colour is pinkish or rose-coloured. Calyx is glabrous and 10-ribbed; 5 ribs are continued into lanceolate teeth which are half as long as the corolla. The 2 upper teeth are slightly longer than the lower 3. The structure of the corolla, staminal column and fruit is similar to that of red clover.

Types: Different morphological and physiological types of white clover are recognised:

1. Wild White: This form is native to Europe and is widely naturalised elsewhere. It is the characteristic form of closely grazed pasture, has rather small, dark green leaves and numerous slender branched stolons with characteristically short internodes. Copious adventitious roots are produced on most nodes so that the stolons are firmly attached to the ground. It commences growth very early. Flowering takes place towards the end of June. Flower heads are relatively small and carried on rather short peduncles. Wild white is usually cyanophoric.

2. Large White: This type consists of a number of forms which may be distinguished from wild white by leaf and stolon size, and by their flowering habit. Generally, they have fewer, stouter stolons with longer internodes, and are less well rooted. They are mainly cyanophoric. Their greater size means that their yield is considerably higher. They are also sufficiently tall to be of some value in a 'hay' crop where they are not over shadowed by grasses species.

3. S.100: There are 2 forms: (i) ordinary S.100 and (ii) S.100 no-mark. The latter is similar to the ordinary but readily distinguished from it by the absence of the genetically dominant leaf-mark. S.100 plants are vigorous and robust, establish readily, and recover quickly after cutting or grazing. They have a strong upright growth-habit and so are not readily suppressed. Because of their greater vigour and speed of recovery, they are capable of almost continuous growth through the life span of the plants.

4. Dutch White: Introduced from Holland, it is a short-lived form, lasting only about 2 years, more erect than S.100, and produces few stolons. It gives an abundance of large flower heads about 1 week earlier. Dutch white is mainly acyanophoric; it is not now used to any great extent.

5. Ladino: This is a distinct form of Italian clover. It is the largest of the white clovers, but is not suited to the climatic conditions prevailing in the British Isles; it is neither tolerant of hard grazing nor of intense competition. These forms are mainly acyanophoric.

Other Clovers

Trifolium hybridum L. (Alsike, Fig. 06.26): This is similar in growth-habit to that of red clover. It produces erect tubular, mostly solid stems. The leaflets *lack* 'horseshoe' or 'crescent-shaped' markings, are glabrous, more or less broadly ovate, and borne on *long* petioles. Stipules are very characteristic in that they are long and finely tapered and have greenish veins. Flower heads occur in axillary positions but, unlike red clover, there are no terminal heads and, thus, the stems continue to grow and become semiprostrate. Peduncles are relatively long. There are no subtending leaves beneath the heads. Colour is white or pale pink, and the flowers shrivel and become reflexed when mature.

Alsike is named after a village near Uppsala in Sweden where it was first recorded. It is not a hybrid as the epithet would suggest but a species in its own right. It is diminishing in importance, although it is better adapted than the previous species to wet soils and cold climates. It grows well on fairly acid soils and tolerates more alkalinity than most clovers (Langer and Hill, 1991).

Trifolium incarnatum L. (Crimson Clover, Fig. 07.01): This annual species is also known as Italian clover. The leaves are softly hairy; leaflets are very broad. The stipules are fairly distinctive, being broad, somewhat truncate or rounded, and frequently red or purple-edged. Growth-habit is erect. The inflorescence is distinctive; the flower heads are *cylindrical* in outline, about 5 cm long, and of a deep crimson colour, rarely white. Crimson clover is usually sown in mixed herbage as a winter annual for forage, particularly for sheep grazing. It is not, however, very hardy, and it has been replaced, to a large extent, by more productive clovers. In the United States it is still an important clover.

Medicks

Members of this group have many features in common with clovers. The vegetative phases can be distinguished by their *serrated* stipules and *mucronate* leaflets. All leaves are trifoliate with the central vein exserted as a short projection or *mucro*, and the laterals extended as teeth. The flowers occur mostly in racemes and are either yellow, purple or a combination of both. Petals are often caducous; sepals are fused basally but free apically in the form of 5 nearly equal-sized teeth. The fruits are characteristically *coiled* or curved, often spiny and usually indehiscent.

Medicago sativa L. (Lucerne or Alfalfa): This legume is considered to be the world's most important forage plant. The total annual world cultivation is well in excess of 35 hectares. As a producer of high quality, palatable, nutritious fodder, this plant has no equal. Not only is it rich in proteins, minerals and vitamins and low in fibre content, it is also an excellent yielder under favourable management conditions (Farragher, 1969). It is not uncommon for lucerne or a lucerne-grass combination to yield as much as 15 tons per hectare.

Lucerne or alfalfa is an autotetraploid (*2n=32*). Its origin, however, is not clear, though is is said to have been derived from a diploid native to the Black and Caspian Seas and which grows wild in these areas. Indeed, wild lucerne is scattered over central Asia, and the cultivated species is believed to have been domesticated in southwestern Asia, possibly Iran. Some of its relatives, such as *M. lupulina*, occur as weeds of tillage and grasslands in the British Isles. Another species, *M. falcata*, is suspected of having contributed to some of the known hundreds of cultivars. The cultivated species is well adapted to a wide range of climates and soil conditions. It performs best on deep loam soils that are well drained in climates where rainfall is moderate, winters cool and summers warm.

As with most legumes, the first leaf is simple but all subsequent leaves are trifoliate with the central leaflet slightly raised on a short petiolule. Leaflets are mostly ovoid, serrated on the upper halves, and the central vein forms a mucro. The stipule is characteristically serrated. Flowers, which are mauve to blue in colour, occur in racemes. Pollination is by bumble or honey, and tripping is obligatory. When sowing, particularly on soil to which the plant is new, the seed should be dressed with an inoculum of an appropriate strain of *Rhizobium*. Application of the organism is best achieved by mixing the inoculum with skim milk before pouring the liquid on the seed, taking care to obtain a uniform spread.

Lucerne has never been a very successful crop in the British Isles, especially in Ireland (Farragher, 1968 and 1969). Experiments by the writer, involving several hundred cultivars, have shown that very few cultivars are suited to the moist climatic conditions of Ireland. Most cultivars showed a high plant mortality rate after the first or second year and only those which originated in countries, or parts of countries, such as northern France, with somewhat similar climatic conditions, persisted for a few years.

Medicago lupulina L. (Black Medick, Fig. 06.22): This is not a high yielding legume and, at best, can only be regarded as a poor substitute for red clover. Its principal interest here is that it frequently occurs as a weed of tillage and grasslands. It is an annual or biennial with weak trailing much-branched stems. The small leaves are trifoliate; leaflets are ovate, mucronate, usually somewhat hairy and serrated. It is very similar, morphologically, to *Trifolium dubium* but may be distinguished on the basis of its leaflets and stipule type. The flowers, like those of the latter species, are small, yellow and in compact clusters.

Trefoils

Trefoils are mostly herbaceous plants, rarely undershrubs, though often woody at the base, erect or prostrate and annual or perennial in habit. The principal members are located in the genus *Lotus* which contains 100-200 species (Allen and Allen, 1981). Species of the genus *Anthyllis* are often included with the trefoils, for general discussion, because of their close relationship.

Lotus corniculatus L. (Common Bird's-foot-trefoil, Fig. 06.23): A native of temperate Europe and Asia, it is grown in the Americas and Europe for both grazing and 'hay' production. It is the most important of the trefoil species. In the past, when it was the custom to have many species in a seed-mixture, it was invariably included. This legume is a deep rooted, herbaceous perennial, occurring naturally in dry pastures, grassy places and other dry situations. Stems are prostrate or semi-erect. Leaves are small, *pentafoliate*, glabrous, with entire leaflets. The stipules are *minute* or *absent*. Flowers occur in axillary, long-stalked, *umbel-like* clusters, and are yellow or yellow and orange in colour.

Lotus uliginosus Schkuhr (Greater Bird's-foot-trefoil, Fig. 06.24): This second trefoil is very similar to the last species but has bigger leaves, wider spreading stems and shorter rhizomes. A most important distinction is its *hollow* stems. The flowers have a similar structure, but the calyx has somewhat *spreading* teeth. It is adapted to wet or marshy ground and areas of high rainfall and, because of its ability to tolerate acid conditions and low phosphate levels, it has potential as a pioneer legume in areas of land reclamation.

Anthyllis vulneraria L. (Kidney Vetch, Fig. 06.20): This vetch is a native of Europe, and is rather plentiful on calcareous grasslands, especially in coastal areas. At one time it was a frequent constituent of seed-mixtures but is now rarely sown except to improve sheep pastures or in 'hay' meadows where conditions are too dry for clover (de Rougemont, 1989). It tends to be short-lived and only gives 1 'cut' per season. This legume is easily identified by its *pinnate* leaves which have relatively *large terminal* leaflets. The flowers occur in yellow-coloured, globose or head-like racemes.

Vetches

This group contains several indigenous, but only 1 cultivated, species. Several species are indigenous to the flora of the British Isles. All are climbing or scrambling annual or perennial herbs. Leaves are pinnate, mostly with several pairs of lateral leaflets, but have branched or simple leaflets.

Vicia sativa L. ssp. *sativa* (Common Vetch or Tare): This is the only cultivated species of this genus; it is derived from the next species. Sown alone or in a mixture with cereals, it grows rapidly, producing a high yield of green fodder. It is an annual with long straggling, angular stems and large leaves ending in branched tendrils. Leaflets are lateral, numerous, *mucronate,* obovate or oblong. Stipules are lanceolate, pointed, often serrate with *dark* central markings. Flowers are short-stalked, solitary or in pairs, red-purple or very rarely white in colour. Seeds are nearly circular, slightly laterally compressed and vary from pale brown with darker mottling to dark purple; hilum *elongate*; germination *hypogeal*.

Beans

Beans are a valuable source of protein for human and animal consumption. There are different types of beans, classified in different species and in different genera. In addition, there are distinct types of beans within species. Plant growth-habit is very varied.

Vicia faba L. (Broad Bean or Field Bean, Fig. 40.03): Contained within this species are leguminous plants which are important in much of the northern temperate zones and sub-tropical regions at high altitudes in the cool season. While they have a wide distribution, China produces over one-half of the world's supply (Bond, 1976). *V. faba* is a diploid with $2n=12$; no polyploids are known. Neither is there a known wild ancestor. Beans are partially self-pollinating. There are 2 distinct types - *broad beans* and *field beans* - within the parent species. Many taxonomists, particularly in recent times, have classified the parent plant in the genus *Faba* and have applied the epithet *vulgaris*. Irrespective of the taxonomic treatment, the 2 types of beans are erect annuals with stout, square, slightly winged stems. Leaves are large with 2 or more pairs of broadly-ovate leaflets and serrated stipules. No tendrils are present but the mid-rib ends in a *short point.* Flowers are white, occasionally lilac or purple and have black blotches on the wing petals. Pods are large, straight, fleshy when young, and with the inner surface downy. There are many phenotypes, mostly based on seed size. The different types are :

1. Broad Beans: These are often referred to as var. *major*. They are grown as a vegetable for human consumption. Seeds are large, up to 25 mm, flat, mostly with pale white or pale green testae. They are usually grown for picking in the unripe seed stage, shelled and cooked fresh, or dried for future use. A number of well defined cultivars exist and these generally fall into 2 groups: (i) *long pods* - these are sufficiently hardy to allow for autumn or very early spring growing under favourable conditions; they are commonly known as winter beans; the pods occur in clusters and contain 4-6 seeds; (ii) *windsors* - these are less hardy and are only suited to spring sowing; pods are usually shorter but broader and are mostly borne singly; each pod contains 2-5 large seeds.

2. Field Beans: These are usually classified under the varietal name - *minor*. They may also be used for human consumption but are more often grown as a source of food for livestock. Seeds are smaller, rarely more than 15 mm and often nearly spherical or cylindrical in general outline, or somewhat laterally compressed. The colour is dark brown. They are grown for harvesting ripe, dried, ground, and mixed with cereal grain. Cross-pollination occurs to the extent of about 50%. There are 3 main groups: (i) *winter beans* - these are hardy and slow growing; stems usually branched; seeds mostly short and cylindrical; (ii) *tick beans* - they are less hardy, spring sown only, less branched basally, but quicker growing; seeds are similar to those of winter beans; (iii) *scotch* or *horse beans* - these are similar to the last group but the seeds are larger and usually flattened.

Glycine max L. (Soya or Soybean, Fig. 40.04): This plant is considered to be the world's most important grain legume as regards total production and international trade (Hymowitz, 1976). World production is in excess of 60 million tons. It is the source of more than 100 commercially important products, ranging from adhesives and plastics to fire-fighting foam and water-proofing preparations (*Encyclopaedia Britannica*, 1984). Most soya beans are made into meal and oil. On a world basis, this crop produces 15% of all edible oils and ranks ahead of butter (Langer and Hill, 1991). It is thought to to have originated in north China. As stated, the oil is used principally for food, although sizeable quantities are used in paint, chemical and other industries. The seeds contain 13 to 25% oil, while the meal has a protein content of up to 50% which can be made into artificial meat and many other products.

The soya bean succeeds on most types of soil but does best on fertile or sandy loams. It is an erect, hairy, branched annual with an extensive root-system. Cultivars range in height from 30 cm to more than 2 m. Leaves are pinnate-trifoliate and vary greatly in size and shape. As a rule they are shed before the seed matures. Stipules are small and pointed. The white or purple inconspicuous flowers are borne in the axils of the leaves. Seeds vary widely in colour; they may be creamy-white, yellow, green, brown, black, or bi-coloured. The most common bi-colour patterns are green or yellow with a saddle-like patch of black or brown extending down on each side of the hilum.

Soya bean is classified in the genus *Glycine* which contains over 200 species, subspecies and taxonomic varieties (Hymowitz, 1976). It is divided into 3 subgenera, and the species of importance - *G. max (2n=40)* - occurs in the subgenus *Soya*. Only 1 other species, *G. wightii*, appears to have any agricultural importance. The latter is a minor forage legume in the tropics. The evolutionary history of the soya bean remains to be fully elucidated. It does not occur in the wild. Based on chromosome number and size, morphology, geographical distribution and electrophoretic banding patterns of seed proteins, *G. max* is most probably the ancestor (Hymowitz, 1976). *G. max* differs from its putative ancestor in having an increased seed size, increased oil but decreased protein content in the seed, an erect growth-habit, in being a larger plant and in having reduced shattering of mature seed from the pods.

Phaseolus vulgaris L. (French Bean or Kidney Bean, Fig. 40.05): This bean species is one of the most widely cultivated in the world. Stems are slender, twisted, twining, angled and ribbed, more or less square in cross-section, and often streaked with purple. The alternate pinnate-trifoliate leaves are large with the terminal leaflet subtended by a pair of small *stipels*; the lateral asymmetrical leaflets are subtended by 1 each. The flowers, which occur in racemes, are white, pink or purple; the standard petal is *reflexed*. The keel petals are *coiled*. French beans are self-pollinating plants. Pods are slender, 10-20 mm long, straight or curved and terminated with a *beak*; they contain 4-6 seeds, rarely more. The seeds are non-endospermic, vary greatly in size and colour from small black 'wild types' to large, white, brown, red, black or mottled black. Germination is epigeal. Other names by which this plant is known include: haricot bean, black bean, navy bean, white bean, red kidney bean, pinto bean and wax bean.

Phaseolus coccineus L. (Runner Bean or Scarlet Runner Bean, Fig. 40.05): This species is less popular than the above named species. It is normally a perennial though usually grown as an annual; it has an erect growth-habit. The leaves are similar to those of *P. vulgaris*. Flowers are white or scarlet and produced in racemes which are longer than the leaves; the number of flowers per raceme varies from 20-30. Pods are broad, rough and green. Seeds are large, mostly pink with black mottling, or white in white-flowered cultivars. Plants are self-pollinating; germination is hypogeal.

Beans of the genus *Phaseolus* are another very important source of grain legumes. All are a very valuable source of proteins, but some, particularly *P. vulgaris*, can have small insignificant amounts of poisonous properties (Cooper and Johnson, 1984). There are up to 100 species. The 3 important species include the above 2 - *vulgaris* and *coccineus* - and *P. lunatus*. The latter species, like the others, is of South or Central American origin. It is rather demanding in its climatic requirements and is really not suited to European agriculture. All are valued grain legumes or pulse crops in many tropical countries and are usually consumed as dry beans. In temperate countries, although some are grown as dry beans, the 2 major species are used for their dry seed, their immature and fleshy pods, or processed as frozen vegetables. All are diploids with $2n=2x=22$.

Peas

The pea crop constitutes one of the most important grain legumes in the world with an estimated total in excess of 6 million tons. They are classified in the genus *Pisum* but there is some uncertainty as to the number of species. Some writers (Allen and Allen, 1981) suggest there are 6, but others (Davis, 1976) claim that the number should be 2. All cultivated forms are generally classified in the genus *Pisum* and species *sativum* although, occasionally, some are described under different specific names. However, some current thinking suggests that only 2 species - *sativum* and *fulvum* - should be recognised. Both are diploids with $2x=14$ and are self-pollinating.

Pisum sativum L. (Pea, Fig. 40.02): All forms of the cultivated pea are climbing, glabrous, glaucous annuals with slender cylindrical stems. The leaves have 1-3 pairs of ovate leaflets and end in branched tendrils. Stipules are characteristically *larger* than the leaflets. Flowers are large, white, pink, red or purple and are borne in short axillary racemes. Cultivated peas fall into 2 categories:

1. Garden Peas: These are sometimes described under the name *P. saccharatum*, a name which should now be considered obsolete. A more acceptable classificatory system is *P. sativum* ssp. *sativum*. Garden peas are characterised by the absence of a lining membrane in the pod, though there may be some exceptions. They are white-flowered with blue-green or near white seeds; the testae are colourless; plants are weak and have to be supported; stipules are unmarked. Germination is hypogeal. Garden peas are further divided into 2 groups:

(a) Green Peas: These are grown as a field or garden crop for picking green, and may be used fresh, frozen or canned. Height varies from 30-150 cm; they mature in 10-15 weeks. Some cultivars have spherical seeds with simple starch grains; these have a low water content. Sugar peas are a distinct form in which the inner fibrous layer of the pod fails to develop; the immature pods are cooked whole.

(b) Dry Peas: Dry peas are grown on a field scale without supports for harvesting ripe and threshing for human consumption as packeted, processed, or split peas. The principal types are: (i) *marrowfats* -

these have wrinkled seeds and green cotyledons; (ii) *blue peas* - this group is characterised by having round seeds and green cotyledons; (iii) *white peas* - the third type has round seeds and yellowish cotyledons; they are grown for use as split peas.

2. Field Peas: The obsolete specific names *pachylobum* and *arvense* are often applied to these. A more acceptable classificatory system is *P. sativum* ssp. *arvense*. They have a well developed fibrous pod lining. Flowers are usually bi-colour, purple and red, with brownish or mottled seeds owing to colouration of the testae. The plants are hardy and have stipules with purple markings. Field peas are grown for livestock feed, for ripe seed, or cut green for forage or silage. The principal types are: (i) *dun peas* - these have large seeds and dull brown testae; (ii) *maple peas* - the seeds of this group are smaller and have speckled testae; they are used for animal feed; (iii) *grey peas* - the final category is characterised by having small spherical, pale green, speckled-violet seeds.

Other Legumes

Onobrychis viciifolia Scop. (Sainfoin): There are about 100 species in the genus *Onobrychis* but only the above has any worthwhile agricultural value. Sainfoin originated in southern Europe and temperate Asia, but is now widely grown on dry calcareous soils in many countries. It is a valuable forage plant in some European countries, particularly France, but in the latter it is now being replaced by lucerne. In England, sainfoin cultivation is confined to the southern parts. There are 2 distinct types: (i) common - the more persistent, and (ii) giant sainfoin. It is used particularly for sheep grazing, and as a 'hay' plant, it mostly used for racehorse feed. It is an erect leafy perennial, reaching up to 50 cm in height. Leaves are pinnate; stipules are characteristically broad, pointed, chaff-like, red and clasping the stem. Flowers, which are mostly pink with red veins, occur in many-flowered racemes which are conical in shape. The fruit is an indehiscent 1-seeded pod.

Arachis hypogaea L. (Peanut): Peanuts are valuable annual herbs whose principal characteristic is the production of underground fruits. There are many distinct cultivars. All grow to a height of about 50 cm. The lower parts of the calyx, corolla and filaments are fused to form a tubular structure. The mature pods are oblong, about 10-ribbed, indehiscent and 3-4 cm long. In addition, there are constrictions between the seeds. Peanuts are highly nutritious, being rich in vitamins B and E, contain about 30% protein and up to 50% oil. They may be used for human consumption, though the crop is grown mainly for its vegetable oil. The residue remaining after the oil has been extracted is a very valuable livestock food. Other names by which peanuts are known include: groundnuts, earthnuts and monkey nuts.

Lens culinaris Medikus (Lentil): This is one of the oldest grain legumes grown. The seed is high in protein content, and is considered one of the most tasty and nutritious pulses. The principal growers are India, Ethiopia, Pakistan and countries bordering the Mediterranean where large quantities are produced. The mature seeds may be split to make 'dhal' which is used in soup, or or the young pods may be eaten as a vegetable. Lentil is a small, light green, annual herb, rarely more than 40 cm, with well branched square stems. The leaves are alternate, pinnate, with about 7 pairs of leaflets and a terminal tendril. The standard petal is white with blue markings, whereas the wings and keel are white throughout. The pods are oblong, broad, somewhat flattened and have 1 or 2 seeds.

Vigna radiata (L.) Wilczek (Mung Bean): These beans are common vegetables in many oriental dishes. The seeds contain up to 25% protein, and both the protein and carbohydrate fractions of the seed are highly digestible and of good nutritional quality (Langer and Hill, 1991). It is a very diverse crop with many distinct phenotypes. Extensive cultivation is carried out in southeast Asia and in the Indian sub-continent. Early taxonomic treatments classified these legumes in the genus *Phaseolus*.

The mung bean is an erect to sub-erect, pubescent annual plant which grows to a height of 1.5 m. Leaves are trifoliate with ovate leaflets which can be as large as 12 x 10 cm and carried on long petioles. Inflorescences occur in axillary racemes on a long peduncle which carries many flowers. Flower colour is predominantly yellow. Pods are rather small, covered with many fine hairs, rarely more than 0.5 cm, and contain many small seeds. The latter are variable in shape, may be globular to square in cross-section, and may be yellow, black or green in colour. There are many distinct races.

Vigna unguiculata L. (Cowpea): Another ancient pulse crop, it is used in a variety of ways ranging from the use of the young green seedlings as a vegetable crop, through cover crops and forages for livestock, to its consumption as dry beans. They are sometimes referred to as *long beans* because some forms produce pods up to 100 cm long. One particular form is known as *black-eye bean*.

This legume, like the previous crop, is a highly diverse species, and indeed many taxonomic treatments give specific rank to some forms. The species mentioned, for the different forms, include *V. sinensis* and *V. sesquipedalis*. The plants are annuals and range from being erect, sub-erect, prostrate or twining. The leaves are trifoliate and carried on large petioles up to 15 cm. Leaflets are usually entire, ovate or rhomboidal with sharp apices; the lateral leaflets are set obliquely onto the petiole. Inflorescences are in the form of small axillary racemes borne on long peduncles. Colour varies from white, yellowish or violet. Pod size, seed number per pod and seed size, are also very variable.

Cicer arientinum L. (Chick Pea): Chick peas are important pulse crops in many Middle Eastern regions, African states and some Mediterranean countries. They are small annuals, rarely more than 60 cm, with erect well branched stems. All parts of the plants are covered with glandular hairs. Leaves are trifoliate, pinnate, with up to 15 pairs of ovate, obovate or elliptical leaflets. Flowers, which are rather small, are mostly solitary and usually carried on peduncles or pedicels up to 5 cm. Colour varies from white, green, pink or blue. Seed colour and shape also vary from white through yellow, reddish-brown and black and from smooth to very rough and wrinkled.

Trigonella foenum-graecum L. (Fenugreek): This legume is grown in warm climates for two uses, firstly, for use as a fodder or forage plant and, secondly, for its seeds which are used as a spice. The seeds are very mucilaginous, bitter to the taste, contain up to 10% aromatic oil and are frequently used as a component of curry powder. Fenugreek is a small medium-sized annual, usually erect, up to 50 cm and pubescent. The leaves are alternate, trifoliate and light grey-green in colour. Flowers are small, white, axillary, solitary or in pairs and carried on short peduncles or pedicels.

Cajanus cajan L. (Pigeon Pea): This hardy pulse crop is grown almost entirely on poor land, and is an extremely important crop in some tropical countries. Almost 25% of the crop is cultivated in India where the hulled seeds are milled to produce red gram. Lesser amounts are grown in the West Indies where the immature pods are picked and processed as a substitute for peas. Pigeon pea is a very diverse crop. It is adapted to a wide range of soils and climates. The dry matter consists of about 20% protein but the latter is deficient in some essential amino acids. The plant is a shrub varying in height from 0.5-5.0 m. Stems are angular and covered with fine hairs. Leaves are pinnate-trifoliate with the central leaflet on a longer petiolule than the laterals. The flowers, which are mostly white, occur in either terminal or axillary racemes.

Ornithopus sativus Brot. (Serradella): This species is a native of north Africa and southwest Europe, particularly Spain and Portugal. It occurs as a casual in the southern parts of England. It is grown to a very limited extent, alone or in mixed herbage, as a forage plant or for green manure in parts of southern and central Europe. Serradella is a weak annual with semi-erect stems up to 50 cm in height. The leaves are pinnate with 6-10 pairs of pubescent leaflets; the terminal is narrowly lanceolate. The flowers, which are white in colour, occur in small heads on long axillary peduncles. The fruits are long schizocarpic pods; each short oval segment contains a single seed.

Lablab purpureus L. (Lablab Bean): Also referred to under the name *Dolichos lablab*, this legume is a tropical annual or short-lived perennial (Langer and Hill, 1991). It has erect or climbing stems and large trifoliate leaves. Flower colour is varied; they may be white, blue or purple, depending on cultivar. The oblong, curved pods have a distinct beak. Lablab requires a high temperature and, consequently, its distribution is limited to the warm tropical or subtropical regions of the world, such as East Africa, Sudan, Brazil, the Philippines and parts of Australia. It is used as both a forage and grain plant.

8   CHENOPODIACEAE

General: The beet or goosefoot family is a comparatively small and uniform taxon with well defined characteristics. The latter name is derived from the genus *Chenopodium* which means the foot of a goose, alluding to the shape of the leaves of some species. There are about 100 genera and 1,400 species, of worldwide distribution, but with centres in xerophytic and halophytic areas, especially on the prairies and plains of North America, the pampas of South America, the shores of the Red, Caspian and Mediterranean seas, the central Asiatic basin, the South African karroo, and the salt plains of Australia (Lawrence, 1951). There are about 9 genera and about 45 species in the flora of the British Isles (Stace, 1991). However, despite the large number of genera and species in the family, the important cultivated plants are confined to 1 or 2 species. Most members of the family are associated with a particular ecological situation, *viz.* soils with a high salt content. Plants adapted to such soils - *halophytes* - generally possess characteristic morphological features. In this family, these features are manifested in the succulence and the possession, in many species, of leaves covered with *mealy* glands, or *papillae*. One physiological feature of halophytes is the plant's ability to withstand high levels of inorganic ions such as those of nitrate and nitrite compounds (Berrie, 1977). These ions can give rise to digestive upsets, particularly in the young. Members consist of herbaceous annuals, biennials and perennials; arborescent forms are rare.

TAXONOMIC DATA

Leaves and **stems**: The leaves are simple, mostly alternate, rarely opposite, sometimes succulent, exstipulate, petiolate or sessile, entire, toothed or lobed. Also, they are usually covered with bladder-like hairs or *papillae* which gives them a 'mealy' appearance. The stems are erect, prostrate or decumbent, herbaceous or woody, and either cylindrical or angular.

Inflorescence: The family is quite distinct from many other dicotyledonous families in having no colourful flowers. The flowers (Fig. 40.6a) are usually small and inconspicuous. They are actinomorphic, mostly hermaphrodite, though unisexual in *Atriplex* species where the plants are also monoecious. True dioecious forms occur in the species *Spinacia oleracea* (spinach). The flowers may occur singly, or, as in *Beta vulgaris* (beets), in axillary clusters. The perianth consists of 1 whorl of tepals. These segments are generally considered to be sepals. They can vary in number from 3 to 5. The male flowers of *Atriplex patula* and *A. prostrata* have 3 sepals, but the female flowers are subtended by 2 somewhat triangular warty bracts; the latter may be present or absent in other genera and species. The petals are said to be absent in all species.

Androecium and **gynaecium**: The androecium consists of 5 stamens in the flowers of most members. A few may show a reduction to 2 or 3. Filaments are slender but the anthers are small and 2-lobed. The gynaecium may show of 2-3 united carpels with the hypogynous arrangement. The ovary is superior in many species. The genus *Beta*, however, is characterised by having a semi-inferior ovary (Fig. 40.6a). The unilocular ovary contains a single ovule in the distinctive *campylotropous* arrangement. Styles are 2-5-lobed; the stigmas are linear or feathery.

Fruit and **seeds**: The flowers occur in clusters in most cultivated members and each flower gives a single seed which remains enclosed in a dry woody-textured pericarp. In addition, all the 2-5 fruits in each glomerule, become fused together so that the end result is a collection of fruits (Figs 40.6c-d). The only exception is found in genetical *monogerm* 'seed' of *Beta vulgaris* subspecies *vulgaris* (Fig. 40.6b) where the fruits occur singly. The true seed, because it is derived from a campylotropous ovule, has a characteristic lenticular shape in most species. Sometimes the testae are characteristically sculptured. The main food reserve consists essentially of perisperm.

The Beet family contains few members of agricultural importance. The important plants are found in genera *Beta, Spinacia, Atriplex* and *Chenopodium*. The 2 last genera are of interest because they contain plants of secondary importance to western agriculture, though, in the tropical American highlands, some *Chenopodium* species are cultivated. The most important of these is the *C. quinoa* plant (Simmonds, 1976a). Another is *C. album* (fat-hen), an ubiquitous weed of cultivated ground, waste places, etc. At one time it was grown as a garden plant and used in the same way as spinach. One species of *Atriplex* is grown, to a limited extent, in some European countries as an ornamental or vegetable plant. Others species, *A. patula* and *A. prostrata*, are common weed plants. Apart from these genera there are several others whose species are indicators of saline habitats. The principal plants in this category are the *Salicornia europaea* (glasswort), *Salsola kali* (saltwort), *Suaeda maritima* (annual sea-blite) and *Halimione portulacoides* (sea-purslane). Most are frequent or common in coastal areas of the British Isles, and have the fleshy leaves which are characteristic of the family.

IMPORTANT SPECIES

Spinacia oleracea L. (Spinach, Fig. 40.07): An annual, short season crop, it is suited to cool weather conditions, where the plant takes 6 to 10 weeks to mature. Its nutritious leaves are eaten as a green vegetable, or the crop is grown for canning and freezing. The flowers are borne in clusters in the axils of the upper leaves. They differ from those of beet in being unisexual. In addition, the plant is dioecious, though monoecious forms exist. The male flowers are subtended by 4-8 perianth segments and contain 4-5 stamens. The female have a 2-4-lobed perianth surrounding a 1-seeded ovary which has 4-5 short styles. Flowering occurs readily when the days are long. Low temperatures also have an important influence on flower initiation. Resistance to bolting is an important factor. New Zealand spinach, *Tetragonia expansa*, is a different species, and it belongs to the family - Aizoaceae.

There are 2 main types - the round-seeded or summer spinach, and the prickly-seeded or winter spinach. Round-seeded cultivars are sown in spring and summer for picking as soon as the leaves are ready, before the plants run to seed. They are also sown in late summer for picking from October to May. Prickly-seeded spinach, as the name suggest, has spiny seeds. It forms a more spreading, branching plant, bearing broadly triangular leaves. It was formerly thought to be hardier than the round-seeded types but this has been disproved and the latter are now preferred even for winter use. Spinach plants quickly run to seed, in summer, producing a leafy stem about 60 cm high, with small green flowers.

Atriplex hortensis L. (Cultivated Orache): Occasionally grown as a substitute for spinach or sorrel, the young leaves are eaten in the same way as spinach. Cultivated orache is a native of western Asia and southeast Europe, where it has been grown since ancient times. It was introduced into western Europe in the Middle Ages, and was widely cultivated in kitchen gardens until the eighteenth century (de Rougemont, 1989). It is still grown to a small extent in France and central Europe, but not on a commercial scale. It is a tall annual plant which grows to 2 m in height. Leaves are up to 12 cm in length and carried on long petioles. The plant is dioecious. There are red-leaved, yellow-leaved and pale green cultivars. The coloured forms are also grown as ornamental plants.

Chenopodium bonus-henricus L. (Good King Henry or Mercury): This plant was formerly cultivated for use as a garden vegetable, and used in the same way as spinach. Although rarely grown today, the plant has become naturalised in some localities, usually near old gardens or buildings (Harrison, Masefield and Wallis, 1969). Good King Henry is a perennial plant, 30-50 cm in height. The leaves, which are rather distinctive, are broadly triangular, have acute basal lobes and wavy margins, and are up to 10 cm long and 8 cm broad. The flowers are similar to those of the common species - *C. album. Chenopodium bonus-henricus* may have been introduced from the eastern Mediterranean in the Middle Ages. Like cultivated orache, it was formerly cultivated in kitchen gardens, but is not widely grown today, and is never grown commercially.

Chenopodium quinoa L. (Quinoa): This is the most important of the cultivated species of *Chenopodium*. It makes a significant contribution to the local food supply in some tropical American countries (Simmonds, 1976a). Unpigmented forms bear a general resemblance to *C. album*. However, it is extremely variable in stature, rate of maturity, plant and seed pigmentation and seed size. It contains variable amounts of saponins which are washed out before consumption. Quinoa is an annual plant, cultivated for its fruits which are roughly comparable with wheat in nutritional quality. They are generally about twice as large in linear dimensions as those of *C. album*. Quinoa flour is made into bread, biscuits, porridge, etc. Comparison of quinoa with *C. album* suggests that human selection for large fruits, non-shattering inflorescences and low seed dormancy was practised (Simmonds, 1976a). Here, there is a striking analogy to the evolution of cultivated cereals.

Beets

The various cultivated beets belong to the genus *Beta*. They include sugar beet, red or garden beet, fodder beet, mangolds or wurzel-mangel, and a variety of leaf beets. Most are grown for their 'roots' which have varied uses. 'Roots' are composite structures, consisting partly of hypocotyl and partly of true root. The proportion of one to the other varies with the different types and, indeed, within some types. Thus, in the garden beet most, if not all, of the end-product is derived from the hypocotyl. Different mangolds have varied amounts, a feature which directly contributes to the existence of a variety of forms. It is interesting to note that the more hypocotyl involved in the formation of the 'root', the greater the ease with which it can be removed from the soil.

Beta contains about 12 species, in 4 sections. Cultivated members are in species *vulgaris*. The 4 types grown for their 'roots' are in ssp. *vulgaris*, while the leaf forms are in ssp. *cicla*. Other species are of little interest though some have extremely high resistance or immunity to various plant ailments. The 'root' forms are diploids with 2n=2x=18, and are normally cross-pollinated. Wind appears to be the effective agent in the transportation of pollen, with insects playing a minor role. The overall classification of the 12 species, as given by Poehlman (1959), is given below:

Species	*Chromosomes*
Section 1 Vulgaris	
B. vulgaris	(2n=18).
B. macrocarpa	(2n=18).
B. patula	(2n=18).
B. atriplicifolia	(2n=18).
Section 11 Carollinae	
B. macrorhiza	(2n=18).
B. trigyna	(2n=36-52).
B. foliosa	
B. lomatogona	(2n=18, 36).
Section 111 Nanae	
B. nana	
Section 1V Patellares	
B. patellaris	(2n=18).
B. procumbens	(2n=18).
B. webbiana	(2n=18).

Species *vulgaris* is divided into 3 subspecies as follows (Campbell, 1976):

1. *B. vulgaris* ssp. *maritima* .... sea beet;
2. *B. vulgaris* ssp. *cicla* .... leaf-beets - spinach and seakale beets;
3. *B. vulgaris* ssp. *vulgaris* .... beetroot or garden beet or red beet;
                                                 mangold or mangel;
                                                 fodder and sugar beets.

Beta vulgaris ssp. ***maritima*** (L.) Arcang. (Sea Beet, Fig. 40.08): This wild beet is deemed to be the progenitor of all cultivated forms. It occurs as a seashore plant throughout Europe and western Asia. The plants are perennials with branched prostrate stems arising from a woody base which have rather small, fleshy, angular, bright-green leaves. The inflorescences are like those of the cultivated forms but the stems on which they are carried are more or less prostrate; the flower clusters are smaller.

Beta vulgaris ssp. ***cicla*** (L.) Arcang. (Leaf-beets, Fig. 40.09): These plants are grown for their fleshy leaves and used as vegetables. The 'roots' are woody and only slightly swollen. There are different types:

1. Spinach Beet. Also known as perpetual spinach, spinach beet resembles beetroot but does not have a swollen root. The leaf-stalks are usually green, though in 1 cultivar they are red; they are used as a green vegetable in a manner similar to that of spinach, which it resembles in flavour. At best it is a poor substitute for the latter. It may be sown in spring, for picking in summer and autumn, or in late summer, for picking in late winter and early spring. Like the beet, it is a biennial and forms seed in its second year.

2. Seakale beet. This is also known as Swiss chard. It is characterised by having thick, white fleshy petioles and mid-ribs, up to 8 cm in width, and usually much reduced leaf-blades. The roots are not swollen. It is used in the same way as spinach. To those who find spinach too acid, the milder flavour of spinach beet and seakale beet may be more acceptable (Harrison, Masefield and Willis, 1985).

Beta vulgaris L. ssp. ***vulgaris***: Four important 'root' plants are classified in this subspecies:

1. *Beta vulgaris* L. ssp. ***vulgaris*** (Fodder Beets, Fig. 40.14): These are similar in anatomy to that of sugar beet but have a greater amount of hypocotyl in the 'root'. They can be said to be intermediate in character between sugar beet and mangolds. They show a continuous range of variation from forms which are selections of sugar beet to those which differ little in appearance from mangolds. They are more useful than the latter in that they can be fed fresh, and in that they have a higher food value.

2. *Beta vulgaris* L. ssp. ***vulgaris*** (Mangold, Figs 40.16-18): Also known as mangel or wurzel-mangel, it may be described as a beet which is grown especially for cattle feed. They are larger than the sugar beet though they have a similar anatomical structure. Weight varies from 2-3 kg. Only one-third of the 'root' is below ground level and, consequently, they are relatively easy to harvest. They are not frost resistant and must be harvested and stored before winter. The shape of the 'root' varies depending on the cultivar. Colour varies from red, orange or yellow. Mangolds have the disadvantage in that the fresh 'roots', due to the presence of nitrites and nitrates, are mildly toxic to livestock, and should be stored in field clamps for 2 to 3 months before use.

3. *Beta vulgaris* L. ssp. ***vulgaris*** (Red Beet or Garden Beet, Fig. 40.15): Beet of this type has very little lignified tissue. Like the sugar beet, described later, the 'root' is a composite structure, consisting partly of hypocotyl tissue and partly of true root. More hypocotyl than root is involved. It has a distinctive colour because of the presence of anthocyanin pigments in the cell sap. Selection is aimed at producing 'roots' of even dark-red colour with as little difference as possible between the rings of the vascular tissue and the intervening parenchyma (Fig. 66.14). Garden or red beets are a low calorie vegetable.

4. *Beta vulgaris* L. ssp. ***vulgaris*** (Sugar Beet, Fig. 40.13): The sugar beet is the most important cultivated form, and is a product of plant breeding. It is the only major crop that was not known in prehistoric times (Poehlman, 1959). Sugar beet is a herbaceous, dicotyledonous biennial, characterised by small, greenish, bracteolate flowers. The latter are perfect, regular, without petals, and carried singly or in axillary clusters. The leaves are glossy-green, alternate, somewhat fleshy, puckered, undulate and roughly toothed. In the first year of growth, it develops a large succulent 'root', in which much sugar is stored. During the second year, it produces flowers and 'seeds'. Prolonged cool periods may cause bolting in the first year or if sown too early (Harrison, Masefield and Willis, 1985). Some cultivars bolt more readily than others at higher temperatures (Martin, 1967). Maximum yields are obtained between latitudes 30 and 40° and decline rapidly between 40 and 50°. Under favourable management conditions yields can average 55 tonnes per hectare, and the sugar content can average 16% of gross weight. The sugar is extracted by first washing and then cutting the 'roots' into strips. These are put into hot water which soaks the sugar. The solution is purified, filtered and boiled to form crystals.

The 'root' is a composite structure derived partly from the true root and partly from the hypocotyl. The amount of hypocotyl involved is small. The 'root' has been developed by selection and breeding from fodder beets. Present forms have a distinctive conical shape, are of a whitish colour, and have an average weight of 1 kg (Fig. 40.13).

True Root and **'Root'**: Most of the 'root' is derived from the upper part of the true root. The distribution of the tissues in the true root is identical with that shown in Figure 40.10. Primary xylem occupies the central position. On either side of the xylem, but separated by a small amount of thin-walled parenchyma, are 2 strands of primary phloem. A small amount of parenchyma separates the pericycle from the metaphloem. These 3 tissues are encircled by a single layer of regular-sized cells called the *pericycle*. The innermost layer of the cortex - the *endodermis* - surrounds and is in immediate contact with the pericycle. Endodermis cells are regular-sided and are characterised by having casparian strips on the radial walls. The greater part of the *cortex* consists of 3-7 rows of elongated barrel-shaped cells with large intercellular spaces. The epidermis, is made up of regular or uniform thin-walled cells.

Primary Cambium: The primary cambium arises partly in the parenchyma tissue between the xylem and phloem and partly in the pericycle. The parenchyma tissue divides repeatedly and form 2 strands of cambia, 1 on either side of the xylem. A few pericycle cells, immediately opposite the protoxylem arms, divide tangentially into daughter cells. The outer daughter cells increase in size and remain part of the original pericycle. The inner cells undergo further division and eventually form 2 arcs of cambia which unite with the ends of the already formed cambial strands. Thus, a complete zone of cambial tissue is formed around the primary xylem but inside the primary phloem. The primary cambium differentiates secondary xylem centripetally and secondary phloem centrifugally.

Supernumerary cambia: The increase in size of the beet 'root' is unique in that it is due to the activity of supernumerary cambia. Up to 30 cambia may be involved. They are derived from the pericycle, or partly from the pericycle and partly from undifferentiated tissue between the pericycle and metaphloem. The cells of the undifferentiated tissue enlarge and divide and so interpolate a widening zone of parenchyma between the pericycle and metaphloem. It is not, however, of uniform width as some of the phloem groups remain closer to the pericycle than others. It is within this zone that the secondary cambium originates, but at the protoxylem arms, it can only arise as did the primary cambium, within the pericycle itself. In the upper region of the hypocotyl the cambium arises entirely in the pericycle, but in the lower region and the root proper, it is confined to the parenchyma and pericycle.

The cambial tissue, formed as above, divides tangentially. The inner daughter cells form the first true secondary cambium and differentiates secondary xylem centripetally and secondary phloem centrifugally; medullary ray tissue, *i.e.* parenchyma, is formed opposite the protoxylem arms. The outer daughter cells increase in size and divide. The inner cells resulting from this division form the second true secondary cambium which behaves exactly as that described for the first true secondary cambium. The outer cells also increase and divide. The inner cells of this last division behave as the third secondary cambium, while the outer once again increase and divide and behave as did the outer cells of previous divisions. The net result of these repeated divisions is the formation of several annular zones of tissue (Fig. 40.11) which give the beet 'root' its characteristic cross-section appearance.

The majority of the cambia are formed in rapid succession, so that a beet no thicker than a pencil contains all the annular zones of growth developing simultaneously. The degree of development of each annular zone may vary markedly in different beets, but, in all cases, the outer zones are less differentiated than the inner. In fact the outermost zones, next to the skin of the 'root', may consist of no more than multiseriate cambial tissue with a few scattered companion cells (Hector, 1936). As soon as the beet seedling begins to show an increase in size due to cambial activity the cortex ruptures longitudinally. Continued increase in size eventually results in the sloughing-off of the cortex. The last of the secondary cambia behaves in a different manner to that of its predecessors. Instead it acts as a *phellogen*, forming a *periderm* or skin (Fig. 40.19), which acts as an insulating layer.

'Seed': Reference has been made already to the structure of the fruit and 'seed' types occurring in cultivated members of the family. The normal 'seed' type is in fact a cluster of corky fruits (Figs. 40.6d). These clusters constitute what is known as the *multigerm seed*, in that they contain several true seeds, each of which can give rise to a beet seedling. This type of 'seed' has a number of disadvantages. Germination within the cluster may be prolonged and, of course, several seedlings may be produced at each station which necessitates singling.

Serious interest in *monogerm seed*, *i.e.* single fruits containing 1 true seed (Fig. 40.6b), appears to have been initiated in America, in 1950, with the accidental discovery of a few plants which had solitary flowers in the leaf axils (Savitsky, 1950). Earlier attempts to breed monogerm seed were unsuccessful, and the first such seed released did not yield as well as contemporary multigerm cultivars. High field emergence has to be an important consideration for monogerm seed, and current interest in breeding is on the production of improved cultivars with the monogerm character.

Historical: Prior to the beginning of the nineteenth century, sugar cane was the only world source of crystalline sugar. Today it only accounts for about 60%, the remainder is extracted from sugar beet. Sugar cane is a large, perennial, tropical grass, grown for its tall thick stems from which 'cane sugar' is extracted. Sucrose was first extracted from this plant more than 1000 years ago, in India. Sugar cane gives its best yields in the tropics, but it is also grown in the subtropics as far north as Louisiana in the US and southern Spain, and as far south as New South Wales. The crop is planted using stem cuttings, and about a year later, in tropical lowlands, the first harvest of stems is ready to be taken. After cutting, the plants throw up successive crops of stems called *ratoons*, which take about the same period to mature as the original crop. The yield, however, declines slowly with each ratoon, and after 2 or 3 ratoon crops have been harvested replanting is usually undertaken (Harrison, Masefield and Wallis, 1985).

Sugar Beet, on the other hand, is an annual or biennial European 'root' crop, developed as an alternative cheap source of sugar less than 200 years ago. Total world cultivation at present is just over 8 Mha. Annual Irish acreage is about 0.04 Mha.

The use of 'beets' probably dates back to prehistoric times. Aristotle mentioned red chards and Theophrastus mentioned light-green and dark-green chards used in the fourth century BC (Campbell, 1976). The Romans used 'beet' as feed for animal and man and, by the sixteenth century, it was widely used for this purpose. The history of the sugar beet, however, may be said to date back to 1747 when a German chemist, Andreas Sigismund Margraff, discovered that the kind of sugar in fodder beets was identical to that of sugar cane. The sugar content of the first beets was only 6.2%. Napoleon encouraged its development as a means of boycotting cane sugar from the British Colonies. Fifty years after Sigismund's discovery, Franz Carl Achard perfected a method for extracting sugar from beets on a commercial scale. In 1801 Achard built the first beet factory, and is now recognised as the father of the beet industry (Coons, 1955). In addition to developing methods for commercial extraction, purification and crystallization, he also established practices for the culture of Sugar Beet and the selection of improved types (Poehlman, 1959). Sugar beet rapidly became an important plant in world trade, and in less than 100 years after the first beet factory was constructed, was responsible for a large proportion of the world's sugar needs. One by-product is molasses which is used as animal feed and for making industrial alcohol. The filter cake, left behind when the juice is purified by filtration, is also used to feed animals either as a wet pulp or dried and mixed with molasses to make a valuable animal feed.

9 ROSACEAE

General: This is a very large family of great economic importance. Members are worldwide in distribution, but are more frequent in the temperate regions. Some difficulty is experienced in defining the general features of the family as extreme variation is shown in plant form and most characters. Species range from prostrate herbaceous plants, such as *Potentilla reptans* (creeping cinquefoil), to tall woody forms, such as *Malus pumila* (apple). Many are of considerable economic importance. This is especially true of the fruit-producing members from which are obtained the apple (*Malus*), pear (*Pyrus*), cherry, prune, nectarine, plum, almond, apricot, sloeberry and peach (*Prunus* spp.), loquat (*Eriobotrya*), blackberry, raspberry and loganberry (*Rubus* spp.), medlar (*Mespilus*) and strawberry (*Fragaria*). Others are widely used ornamental plants. The principal species in this regard are cotoneaster (*Cotoneaster* spp.), spiraea (*Spiraea*), pyracantha (*Pyracantha*), hawthorn (*Crataegus*), rose (*Rosa*), kerria (*Kerria*), flowering quince (*Chaenomeles*), mountain ash, common whitebeam, Swedish whitebeam and rock whitebeam (*Sorbus* spp.), cherry laurel (*Prunus*) and the shrubby cinquefoils (*Potentilla* spp.).

The Rosaceae represents a fairly large family with approximately 150 genera and 3000 species, distributed over most of the world. Throughout the wide range of forms there is a great similarity in floral structure. However, there are several fruit types, including berries, pomes, achenes and drupes. Some taxonomic treatments of the family recognise 6 subfamilies. Many authors, however, afford family status to some of the subfamilies, a view rejected by most phylogenists.

Apart from the well documented morphological criteria, there are certain anatomical features which are more or less characteristic of the family, or at least of the woody members. Cork cambium in 1 subfamily is either epidermal or hypodermal. The primary cortex normally has a collenchymatous hypodermis, and crystal sacs often occur in the parenchyma tissue. Stone-cells are usually absent from the primary and secondary cortex, though present in some fruits. Medullary rays are broad in 2 subfamilies, Rosoideae and Prunoideae, and narrow in the other 4. Prickles and other hard epidermal outgrowths are common in certain genera, *e.g. Rosa* and *Rubus*, while branch thorns are produced commonly in *Crataegus* and *Pyracantha*. The wood is very fine-grained and durable. Pear wood, for example, is one of the smoothest known, and is used extensively for carving and the manufacture of drawing instruments.

TAXONOMIC DATA

Leaves: The leaves are *petiolate*, mostly *alternate,* very rarely opposite or subopposite, and either *simple, pinnate* or *palmate.* Good examples of plants with simple leaves would be *Cotoneaster, Malus, Pyracantha* and *Pyrus* species; *Potentilla reptans* (creeping cinquefoil) has palmate leaves, while *P. anserina* (silverweed), *Filipendula ulmaria* (meadowsweet) and *Geum urbanum* (wood avens) are examples of plants with pinnate leaves. Plants with pinnate leaves, resemble in the vegetative phase, leguminous species with a similar leaf-type. However, the leaflets of Rosaceae plants are usually serrated, while those of the Fabaceae are mostly entire. The number of leaflets, their size, particularly in relation the to terminal leaflet, and the presence or absence of secondary leaflets, are diagnostic criteria of great importance.

Stipules: Leaves are invariably *stipulate*, very rarely exstipulate. The stipules are generally small. In some species they are caducous, but in these situations small scars are left at the base of the petioles, indicating the position of the stipules. In the case of woody plants, it is always best to examine young twigs for their presence or absence. Other species may show stipules adnate to the petiole.

Flowers: These (Fig. 41.01) are mostly *actinomorphic*, rarely zygomorphic; they occur in various dense types of *racemose* and *cymose* inflorescences; rarely solitary. Furthermore, they are usually *hermaphrodite*, very rarely unisexual, and with either the *perigynous* or *epigynous* arrangement.

Calyx: The sepals are typically 5 in number and either distinct or very slightly united basally. An *epicalyx*, which is an additional whorl of smaller sepals, is present in some genera, particularly in *Aremonia*, *Fragaria*, *Geum*, *Potentilla* and *Sibbaldia*.

Corolla: The corolla is always *polypetalous*, and the petals usually equal the sepals in number. The petals alternate with sepals, usually imbricate, or very rarely convolute. In addition, they are often of a large size and brightly coloured. White, red or yellow are the predominant colours. However, they are absent in a few genera as, for example, in *Acaena, Alchemilla, Aphanes, Poterium* and *Sanguisorbia*.

Androecium: Stamens are usually numerous and in several whorls of 5; rarely 5-10; commonly *free*, rarely monadelphous. Anthers are small, 2-celled, and dehisce longitudinally, rarely transversely.

Gynaecium: The number of carpels constituting the gynaecium varies from 1 to many. The carpels may be united and form a *syncarpous* pistil, as for example in the genera *Malus* and *Pyrus*, or free and form an *apocarpous* pistil, as in the genera *Fragaria* and *Potentilla*. Different fruit types are derived from apocarpous and syncarpous pistils. Either pistil type may be free from or united with the hypanthium. The ovary may be *inferior* (*e.g. Malus*) or *superior* (*e.g. Fragaria*). Ovules are usually 2, sometimes 1 or more. The styles, in a syncarpous pistil, usually equal the carpel number. The former may be united or free, and the degree to which they are united can be an important taxonomic factor. For example, the styles are free throughout in *Pyrus* (pear) and connate basally in *Malus* (apple).

Fruits: There is a wide range of fruits types. Achenes, follicles, pomes, drupes, or an aggregate of achenes or drupes, are found throughout the family. Capsules are very rarely represented. Both the fruit of *Malus pumila* and *Fragaria x ananassa* are often referred to as false fruits because the juicy succulent edible portions are not derived from the ovaries.

Classification: This extensive family is generally divided into several subfamilies. The taxonomic data, used to differentiate the 6 recognised subfamilies, are evident in the 'key to subfamilies and genera' which is given later. Some confusion as to the overall classification, however, has arisen because many authors have afforded family status to the various subfamilies, though the family, as a unit, is a more natural assemblage than some other large families (Lawrence, 1951). Thus, the Spiraeoideae are referred to as the Spiraeaceae, the Maloideae as the Malaceae or the Pomaceae, the Rosoideae as the Poteriaceae, the Prunoideae as the Amgydalaceae, and the Chrysobalanoideae as the Chrysobalanaceae. Two subfamilies, *viz.* Chrysobalanoideae and Neuradoideae, are of little significance in temperate areas. Members of the Rosaceae are characterised by the usual presence of stipules, the general pentamerous arrangement of the floral parts (gynaecium excluded), the presence of a hypanthium in most genera, and the near absence of endosperm. Fruit type plays a major role in the delimitation of subfamilies.

KEY TO SUBFAMILIES AND GENERA

1.   Ovary *superior* **2**
     Ovary *inferior* or semi-inferior .... **Maloideae**

> This taxon is sometimes referred to as the Pomoideae in some texts. It is a very distinct subfamily. Besides its characteristic morphological characters, the the basic chromosome number is *x=17*, and is not found in any other subfamily. Members are usually trees or shrubs, with long and short shoots, the latter often modified as sharp thorns. Leaves are simple or pinnate. Small stipules are present, at least in young wood; they are often caducous, leaving a small stipular scar. The hypanthium is concave; it encloses the carpels, and is fused to them (Fig. 41. 3a). Consequently, the ovary is *inferior* or at least semi-inferior. Flowers are 5-merous; sta-

mens are more than 10; carpels 1-5, with 1-2, rarely numerous, ovules. The fruit is mostly a *pome*. There are several genera of importance; most are grown as ornamentals.

a. Ovules or seeds *8* or more .... **b**
    Ovules or seeds *1-5* .... **c**

b. Leaves *serrate;* styles *connate* basally (ornamental) .... ***Chaenomeles***
    Leaves *entire*; styles *free* basally (ornamental or cultivated) .... ***Cydonia***

c. Flowers arranged *other than* in racemes .... **d**
    Flowers in *racemes* (ornamental) .... ***Amelanchier***

d. Leaves *entire*, at least basally .... **e**
    Leaves *other than* entire .... **f**

e. Carpels *free on the inner side* (ornamentals) .... ***Cotoneaster***
    Carpels *almost completely united* (ornamental) .... ***Mespilus***

f. Leaves *serrate* .... **g**
    Leaves *lobed* (ornamentals) .... ***Crataegus***

g. Flowers in *compound corymbs* .... **h**
    Flowers in *simple corymbs* .... **i**

h. Plant *thorny* (ornamental) .... ***Pyracantha***
    Plant *not* thorny (ornamentals) .... ***Sorbus***

i. Styles *connate basally* .... ***Malus***
    Styles *free* .... ***Pyrus***

2. Fruit *indehiscent* **3**
    Fruit *dehiscent*, mostly a *follicle*, rarely a capsule .... **Spiraeoideae**

Members are mostly unarmed shrubs, very rarely herbs. Leaves are mostly small, and either stipulate or exstipulate. Flowers invariably small, numerous, and occur in panicles, or simple or compound corymbs. Hypanthium cup-shaped or saucer-shaped, *not* enclosing the carpels, though fused with them basally (Fig. 41.07). Stamens more than 10; the carpels 3-5. Fruit dry, consisting of a group of *follicles*, very rarely capsules or achenes. Basic chromosome number, $x=8$ or 9. There are about 80 species; many are grown as garden ornamentals.

Flowers small, white, crowded in large terminal panicles (ornamentals) .... ***Spiraea***

3. Gynaecium *syncarpous* or of a *single* carpel **4**
    Gynaecium *apocarpous*, at least apically **5**

4. Corolla *zygomorphic* or somewhat so; style *basal* .... **Chrysobalanoideae**
    Corolla *actinomorphic*; style *terminal* .... **Prunoideae**

Members of this subfamily are trees or shrubs. The leaves are simple, stipulate, and serrate to crenate. Flowers occur solitary or in racemes; sepals 5; petals 5; carpels *1*, very rarely 2; ovules 2; stamens more than 10; the hypanthium concave and *not* enclosing the gynaecium and attached to it only at the base (Fig.41.5a). The fruit (Fig. 41.5b) is a *drupe*. Basic chromosome number, $x=8$. There is but 1 genus of importance.

Carpels *1;* style terminal .... ***Prunus***

5.   Carpels *connate* basally and adnate to the hypanthium .... **Neuradoideae**

         Carpels *distinct* and *free* from the hypanthium .... **Rosoideae**

This subfamily is represented by trees, shrubs and herbaceous plants. Leaves are simple, pinnate, palmate or trifoliate, stipulate, entire, serrate or lobed. Flowers may be solitary, in few-many flowered corymbs or in flat-topped panicles. Flowers 4-6-merous; epicalyx often present; carpels mostly usually numerous and spirally arranged, free from each other and from the receptacle; ovules 1 or 2, but seeds always 1; stamens mostly numerous. The receptacle either *concave* (Fig. 41.6a) or *convex* (Fig. 41.2a). Fruits often an *aggregate* of drupes or achenes, sometimes borne on a succulent receptacle, or surrounded by a dry or succulent hypanthium. Basic chromosome number, $x=7, 8$ or 9. There are several species, some are grown as ornamentals, others are cultivated for their fruits.

a.   Receptacle *concave* .... **b**

         Receptacle *other than* concave .... **c**

b.   Plant *thorny* (ornamentals) .... ***Rosa***

         Plant *not* thorny (ornamental) .... ***Kerria***

c.   Epicalyx *present* .... **d**

         Epicalyx *absent* .... ***Rubus***

d.   Stamens *10* or more .... **e**

         Stamens *5*; petals small, inconspicuous (ornamental) .... ***Sibbaldia***

e.   Fruits *fleshy* .... ***Fragaria***

         Fruits *dry* (ornamental and wild) .... ***Potentilla***

IMPORTANT SPECIES

There are about 35 genera and 235 species in the British Isles (Stace, 1991). Most are herbaceous and of minor significance. Some arborescent or suffruticose types are used as ornamentals. The principal species in this regard are those of *Amelancher, Chaenomeles, Cotoneaster, Crataegus, Cydonia, Kerria, Potentilla, Pyracantha, Sorbus, Spiraea* and *Rosa*. The economically important members, grown for their fruits, include species of *Pyrus* (pears), *Rubus* (blackberry, raspberry and loganberry), *Malus* (apples), *Prunus* (cherries) and *Fragaria* (strawberry). Several other members are important fruit plants in warm climates. The latter are *Sorbus domestica* (service tree; fruits used in jam making or alcoholic drinks), *Mespilus germanica* (medlar; edible fruits), *Crataegus azarolus* (arazole; eaten fresh or used for jams and spirits), *Cydonia oblongata* (quince; eaten when cooked or as a vegetable) and *Eriobotyrya japonica* (loquat; eaten fresh or made into jellies). A few species such as *Rubus caesius* (dewberry; used in the same way as the blackberry), *R. phoenicolasius* (wineberry; jams and desserts) and *R. chamaemorus* (cloudberry: jams or stewed) are grown in cold northern climates.

Cydonia oblongata Mill. (Quince): This may be used as an ornamental or grown for its fruits. The ornamental form is a small deciduous shrub with simple entire leaves. The latter are ovate to elliptical in outline, alternate, 5-10 cm, pale but turning dark green adaxially and grey-tomentose abaxially, and with *glandular hairy* stipules. Flowers mostly *solitary*, 2-5 cm in diameter, with numerous stamens, and white or pink in colour. The fruits are 3-5 cm in wild forms but up to 15 cm in some cultivated types. Apart from its ornamental value, its cultivated fruits, when cooked, have a fine flavour. Its principal use is in jams and jellies. The raw fruit, however, is hard and quite unpalatable. Quince is probably a native of northern Iran and Turkestan, but is naturalised in parts of the near East and southern Europe. It is of ancient cultivation in Europe, and was much used by the Romans. It is grown on a fairly large scale in Portugal and Spain. In more southern countries, it is of minor importance (de Rougemont, 1989).

Malus domestica Borkh. (Cultivated Apple): This plant is self-incompatible and can be hybridised readily with most other species of *Malus*. It is thought to have originated by chance cross-pollination between wild species in western Asia (de Rougemont, 1989), though some European species are also said to have contributed to the evolution of the present-day apple (Walkins, 1984). While there are about 25 species in the genus (Rehder, 1954), there appears to be some confusion as to the correct specific name. On this basis of the 'biological species concept', it would appear that too many species have been assigned to the genus. Future taxonomic research may lead to a substantial reduction in the number of species and the apple, in time, may be reclassified. At present, various specific names are given in the literature. These include *M. pumila* Mill., *M. communis* L., *M. sylvestris* ssp. *mitis* (Wallr.) Mansf. and *Pyrus Malus* L. All of the forenamed may, for the time being, be considered synonyms.

Apple trees are *non-thorny* and extremely variable in height. The general habit is upright with spreading branches. Leaves are simple, ovate to oval in outline with more or less broadly rounded or narrowed bases and coarsely serrated margins. They are pubescent on the *abaxial* surfaces; stipules are small. Flowers occur in simple *corymbs*. Petals, narrowed *abruptly* into a small basal *stalk*, are mostly white, though invariably showing various shades of pink. The sepals, which are often persistent in the fruit, are *pubescent* on the *outside*. Carpels vary from 3-5 and have *inferior* ovaries; ovules 2 per ovary; styles are *connate* basally (Fig. 41.3c). The fruit is a *pome* (Fig. 41.3b). The pericarp is *cartilaginous*, and the 'fruit' itself is *without* stone-cells. There are different opinions as to the origin of the 'fruit'. It is often described as a *false fruit* because it is not derived directly from the ovaries. One school of thought suggests that it originates from the *hypanthium*, while another postulates that it is derived from the *base* of the *floral tube*. Certainly, the distribution of the vascular tissue within the fruit supports the latter theory.

Apples are classified into 4 main economic groups (de Rougemont, 1989): (i) ornamental, (ii) dessert, (iii) cooking and (iv) cider apples. Dessert apples, by far the most important category economically, are medium-sized. Cooking apples are usually larger, mainly green, with a higher acid content, requiring the addition of sugar when cooked. Cider apples, usually small fruits, are classed as sweet, sharp, bittersweet or bitter-sharp according to their flavour; these different types are blended to determine the the type and quality of the cider required. Other products include unfermented apple juice, liqueurs and pectin.

Malus sylvestris (L.) Mill. (Crab Apple): This species scarcely qualifies as a crop, although its fruits are sometimes used to make jellies. Its main interest is that it is considered to be the native European parent, together with Asiatic species of *M. domestica*, and that it is commonly used as a rootstock for cultivars of the latter. It is native to northern and central Europe. Very similar to *M. domestica*, it differs in being more intricately branched, and usually *thorny*. Leaves are ovate to elliptical and *glabrous* when mature. Pedicels and outside of calyx are pubescent; flowers pinkish-white; fruits small, about 2-3 cm.

Pyrus communis L. (Pear): The general features of *Pyrus* are similar to those of *Malus*. The principal differences relate to the styles, the shape and characteristically gritty flesh of the fruit which is due to the presence of *stone-cells*. The styles are *free* throughout (Fig. 41.3d), and the pome is *not* indented basally. The flowers of the cultivated pear are whitish, occur in *simple corymbs*, have 2-5 carpels with each having 2 ovules; ovaries *inferior;* stamens 20-30. Leaves are simple, serrate or crenate and glabrous or nearly so; stipules small. The bark is dark brown or greyish, finely and deeply fissured into very small squares.

It is generally accepted that the primary centre of origin of both *Malus* and *Pyrus* is within the region which includes Asia minor, the Caucasus, Soviet Central Asia and Himalayan India and Pakistan (Wilcox, 1962). *Pyrus communis* is not a true natural species, but the result of ancient chance and selected hybridisations between wild species which probably included *P. pyraster. P. salvifolia, P. nivilis, P. syriaca* and *P. cordata*. Pears are an old crop; Homer records their cultivation in Greece about 1000 BC.

The principal cultivars can be categorised on the basis of their uses (de Rougemont, 1989). The most important are the best dessert pears, which can be eaten fresh when fully ripened. Other pears may be canned. Stewing pears are types that are not acid, but hard and lacking in flavour; many of those are also canned after cooking. Perry, a fermented juice comparable to cider, but less common because it does not keep well and is less digestible, is made from certain pear cultivars, mostly in the same areas as cider.

Rubus fruticosus L. *sensu lato*: The genus *Rubus* is an extremely complex taxon. It is usually divided into sections and series. *R. fruticosus sensu lato*, under which the **bramble** or **blackberry** is generally listed, comprises a widespread and polymorphic group of distinct, largely polyploid microspecies, most of which are pseudogamus apomicts (self-pollinating) but occasionally reproduce by cross-pollination giving rise to sterile and fertile hybrids and new apomicts, all capable of spreading vegetatively by root-tips (Scannell and Synnott, 1987). The latter list 78 microspecies in the Irish Flora. De Rougemont (1989) refers to this taxon as an aggregate of some 400 species and subspecies.

Many taxonomists refer to this species under the name - *R. saxatilis*. The numerous microspecies constituting *R. fruticosus* differ in the disposition and type of spines on the stems, leaf shape, flower colour, fruit shape, colour and flavour, and hair and gland distribution. Identification is made even more confusing by the numerous hybrids which occur. In general, the plant is a scrambling shrub, with arched *angled* stems bearing prickles, spines and spines. The leaves mostly palmate, with 3-5 serrated leaflets. Flowers are panicles with pinkish-white petals; usually 5-merous; epicalyx absent. The hypanthium is *flat* with the receptacle usually extended upwards in *convex* structure (Fig. 41.4a). Stamens and carpels generally *numerous*; ovaries *superior*. The fruit is invariably an *aggregate* of *drupes*; *black*. The latter may be eaten raw, or used in tarts or for making jam. Bramble or blackberry is a tetraploid, with *2n=28*. It also of interest because it was once thought to be one of the parents of the loganberry.

Rubus idaeus L. (Raspberry): This species is essentially like *Rubus fruticosus*. The stems are erect, *rounded*, with few to many, slender, weak, straight prickles. The leaves often have more leaflets, 3-7, they are green adaxially, white-tomentose abaxially. Flowers are few, in *racemes*; white. The fruits are *red* in colour, rarely white or yellow.

The raspberry is one of the most important, economically, of all cane fruits. They are usually consumed in the same way as blackberries, but their popularity and distinctive taste have led to a wider range of uses, particularly in flavouring ice-creams, sweets, confectioneries and liqueurs. The plant is a diploid with *2n=14*, and is said to be one of the parents of the loganberry.

Rubus x loganobaccus Bailey (Loganberry): The loganberry is described as a very vigorous plant, producing stems several metres in length which have numerous prickles and which arch and root at the tips. The leaves have 5 ovate leaflets with white-tomentose abaxial surfaces. Flowers are in panicles. Fruits, which are *dark purplish red,* are used for jams, stewing, freezing, canning and juice and wine making. The loganberry was once thought to have arisen as a hybrid between *R. fruticosus* and *R. idaeus* (Eldin, 1967) but is now considered a hybrid between the latter and *R. vitifolius* (Stace, 1991). It was first discovered by a Judge Logan, after whom it was named, in California, at the end of the last century.

Rubus chamaemorus L. (Cloudberry): A rhizomatous plant, it is comparatively small, reaching only about 20 cm in height. It is clothed with *hairs*. The leaves differ from those of the previous species in being *simple*, shallowly palmately lobed and white-tomentose. Flowers are *solitary*, white, terminal; the plant is *dioecious*. The aggregate fruits have an *orange* colour. Cloudberry scarcely qualifies as a cultivated plant, but it is has been a valuable source of rare fruit, for jam making, in northern European countries, where it originated.

Rubus caesius L. (Dewberry): This plant is very similar to blackberry, but has stems which are slender and prostrate, or at most low-arched, has moderate prickles but no hairs, and is bluish green in colour. Leaves are *trifoliate*, with lanceolate stipules; leaflets more or less glabrous to pubescent. Flowers few, white, and in *corymbs*. A native of northern Europe, its *black* drupaceous fruits are used as blackberries.

Rubus phoenicolasius Maxim. (Wineberry): A distinctive species, the long calyces envelop the fruits until they are nearly ripe; the latter are *red*. In addition, the stems, calyces and petioles are covered with red hairs, while those on the flowers being distinctly club-shaped (de Rougemont, 1989). Stems erect and spreading, sparsely spiny; the plant forms long arching canes which must be supported. Leaves with 3-5 ovate, white-tomentose leaflets. The pink flowers occur in racemes. Wineberry, a native of China and Japan, has fruits which are are sweet and juicy, not too acid, and make fine desserts and good jams.

Prunus persica (L.) Batsch (Peach or Nectarine): All *Prunus* species are classified in the subfamily Prunoideae, and the taxonomic criteria given already for the latter are essentially those of this genus. There are about 77 species (Rehder, 1954), scattered in various sections and subspecies. Some of the cultivated species are diploids, tetraploids, hexaploids and octaploids. The first species of importance is the diploid peach or nectarine, described as a small tree reaching 6-7 m in height. Leaves are narrow, lanceolate, thin, usually broadest at or beyond the middle, finely serrate and almost glabrous. Flowers are *solitary*, small, rose-pink, rarely pale pink or white. Fruits of the peach are *globose*, 5-12 cm in diameter, with a *velvety* skin, pale to golden yellow in colour and flushed flushed with deep red on exposed side; sometimes entirely red. Flesh colour is either greenish-white or golden yellow. Nectarines, which are really a variety of peach, have a richer colour, with smooth skin like that of a plum.

The peach did not originate in Persia (now Iran), as the specific name would suggest, but in China. At the present time, it is cultivated in warm-temperate areas. Peaches are said to be the most delicious of all *Prunus* species. However, they do not keep well, so a large proportion of all crops is processed.

Prunus armeniaca L. (Apricot): Another diploid, the apricot is probably native to China, whence it spread to southern Europe. Commercial production requires a warm temperate climate with a hot summer. Spain, Hungary, Italy and France are the main European producers. The apricot tree reaches about 10 m, and generally has dense upwardly spreading branches. Leaves are broadly ovate to almost round, thin, smooth and borne on long stalk. The fruits are smaller than the peach, *velvety*, orange-yellow when ripe, and with a dry flesh which detaches easily from the endocarp.

Prunus amygdalus L. (Almond): The almond, a diploid, and closely related to the peach, became established in an area extending from central to western Asia. It is described as a small tree, 4-6 m, or more in cultivation, with a flattened globose crown. The bark is nearly black, and deeply fissured into small squares. Leaves are finely serrate, 7-12 cm, and ovate-lanceolate in outline. The large flowers appear before the leaves from February to March. The fruits resemble a small *ovoid, flattened* peach, yellowish green or dark brown with age, and covered with *dense felted* hairs. Almonds have many uses; some are cultivated for their bitter and sweet oils which are used for flavouring many dishes, while the seeds of others are ground or eaten whole after their removal from the endocarp.

Prunus spinosa L. (Blackthorn or Sloe): A native diploid, it is a dense bushy shrub with intricate *black* branches ending in *thorns*. It attains a height of up to 4 m. Young twigs downy with numerous side shoots that become *long straight thorns*. Leaves obovate to oblanceolate, *small*, 1-3 cm, dull, serrated. The flowers, which appear *before* the leaves, are nearly sessile, solitary or in pairs. The drupes are 1-2 cm in diameter, *blue-black* with a *dense bloom*. Sloes are too sour to be of much value. However, they are used in some European countries to make sloe wine or sloe gin. This plant is of interest because it is deemed to be one of the putative parents of *P. domestica*.

Prunus cerasifera Ehrh. (Cherry-plum): An introduced tetraploid (*2n=32*) plant, it is used for hedging and as an ornamental. Apart from the ornamental value of some forms, it is also of importance in that it is considered to be the second parent of *P. domestica*. It is a shrub or tree up to 10 m, *often* thorny. Leaves 3-7 cm, ovate to obovate, serrate-crenate, green or purplish, and borne on *pubescent* petioles. Flowers *solitary*, white, bigger than those of the previous species, and appearing *with* or just before the leaves. Fruits *globose, red*, rarely yellow or bronze.

Prunus domestica L. (Plum): A hexaploid (*2n=48*), it is the result of hybridisation between 2 species, *P. spinosa* and *P. cerasifera*. It forms a very complex aggregate, and many different types of plums are known. Stace (1991) divides this species into 3 subspecies: (i) ssp. *domestica* - plums, small trees with sparsely pubescent spineless twigs and usually large fruits with *very flattened* endocarps; (ii) ssp. *insititia* - wild plums, bullace and damsons, generally small shrubs with *densely pubescent*, often *spiny* twigs and fruits with *less flattened* endocarps; (iii) ssp. *italica* - greengages, which are more or less intermediate between (i) and (ii). Some plums are grown as dessert fruits or used in certain sweet dishes, others, particularly large, oval, black-skinned cultivars, are dried and called prunes. The more acid or astringent types are used for cooking or jams. Other products include liqueurs and wines.

The general characters of *P. domestica* are those of *P. cerasifera*, but size, habit, leaf shape, etc., are very variable (de Rougemont, 1989). The fruits have an external longitudinal groove along 1 side, and usually a bloom which is particularly noticeable on the blue-skinned cultivars.

Prunus avium (L.) L. (Wild Cherry): A large native tree up to about 30 m. Leaves are 6-15 cm, obovate to elliptical in outline, pointed, serrated, soft and drooping when young, and pubescent abaxially. Flowers white, occur in small umbels, on relatively long, slender pedicels, have a *shallowly* cup-shaped hypanthium, and appear just before the leaves; sepals *reflexed*. Fruits are *red, black* or *yellow*. The sweet or cultivated cherry, which belongs to the same species, is a smaller tree. Wild cherries are used for cooking and jam making, while sweet cherries are mostly used as dessert fruits.

Prunus cerasus L. (Dwarf Cherry): Of little importance but, in some countries, the sour *bright red* fruits are used for flavouring jams, jellies, liqueurs and spirits. It is a shrub or small tree, up to about 8 m. The plant differs from *P. avium* in its *saucer-shaped* flowers which are *fewer* per umbel; the pedicels are also *shorter*. It is an introduced plant in the British Isles, being found in hedgerows and copses.

Fragaria x *ananassa* Duchesne (Garden Strawberries): These represent one of the world's important soft fruit crops. They are grown extensively in most temperate and subtropical countries. The name 'strawberry' is derived from the practice of laying straw under the fruiting shoots to protect the fruit from damp and dirt. There are over 40 species, though many may not be truly distinct (Jones, 1984). Some of the earliest cultivated species were *F. viridis, F. moschata* and *F. vesca*. All species are very tasty, and have been grown for centuries. One of the above species, *F. vesca*, or wild or woodland strawberry, is widespread throughout Europe, but its fruit are small and production is negligible. Fruits, probably of this species, were tended in gardens by the Romans in 200 BC, and possibly earlier (Jones, 1984). The fruits of garden strawberries are much larger, and some of the modern cultivars are claimed to yield 100 tonnes per acre (Berrie, 1977). Mainly fresh dessert fruits, they are also used for jams, flavouring ice-cream, confectioneries, sweets, liqueurs and many processed dessert foods.

The species of *Fragaria* form a polyploid series, from diploid to octoploid, with a basic chromosome number, $x=7$. The many cultivars of the garden strawberry are ultimately derived from 2 American species - *F. chiloensis* Duchesne and *F. virginiana* Duchesne. Each appears to be an autopolyploid, though possibly with some regulation of chromosome pairing. The former occurs discontinuously along the Pacific coast, commonly on dunes, from the Aleutian Islands to California, and then to Chile where it also grows inland. The former grows in open woodland and hill pastures in North America, from the east coast to the Rocky mountains, and from Mexico to Alaska. However, hybridisation is thought to have occurred in Europe after these 2 New World species were brought together during the period of plant importations of the eighteenth century (Berrie, 1977). The hybrid, *F. x ananassa*, has a flavour like pineapple, hence the specific epithet. It is an octoploid, with $2n=56$.

The garden strawberry is basically similar to *F. vesca*, but is bigger in all its parts. It grows as a perennial, propagating by runners. In reality it produces 2 types of axes, a short shoot system which branches to develop a 'crown' and a long shoot system which is in fact the runner. The inflorescences are borne on the short shoots. The leaves are *trifoliate* or sometimes unequally imparipinnate, *i.e.* with a pair of much smaller leaflets below the normal types. Leaflets coarsely serrate, but entire at the more or less wedge-shaped bases; relatively small. Abaxially, they have fairly long hairs, but are glabrous and blue-green adaxially. Petioles are mostly long and channelled above; stipules are adnate to the petioles. Adventitious roots are formed and the terminal portion of the runners become independent of the mother plant. The whole process of establishment of the runners takes place after flowering during late summer. Flowers white or flushed with pink, and occur in cymes on more or less leafless stems arising from the axils of the leaf-rosettes; 5-merous. Epicalyx *present*. Stamens 20 or less; filaments mostly shorter than the receptacle. The pistil is *apocarpous*, consisting of an indefinite number of carpels. They are carried on a strongly *convex* receptacle; the other floral parts are borne on the lip of a more or less *flat* hypanthium. The strawberry is often referred to as a *false* or accessory fruit (Fig. 41.2c), since most of the juicy, edible portion is obtained from the *receptacle*. The true fruits, or *achenes*, are derived from the individual carpels, and are distributed in a peripheral zone. They are frequently but, incorrectly, termed seeds.

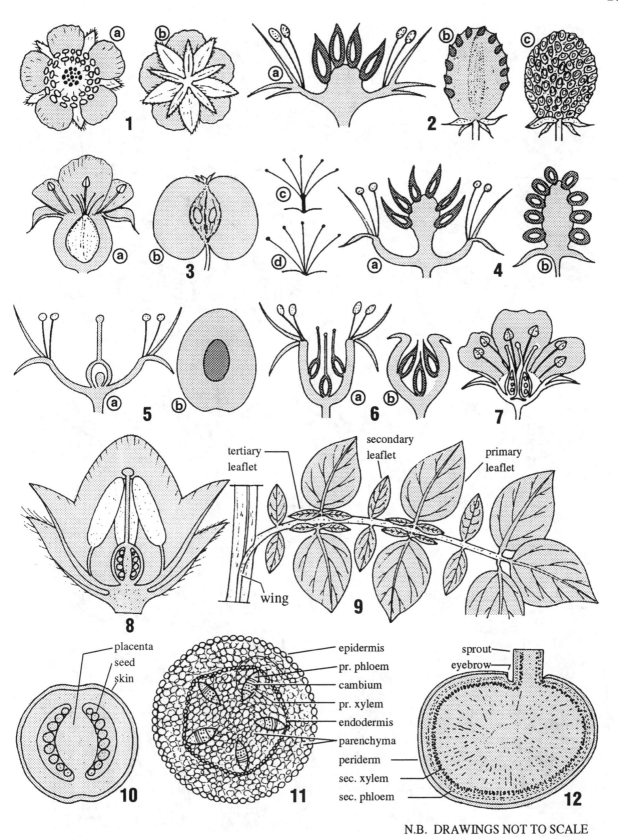

N.B. DRAWINGS NOT TO SCALE

Fig. 41.00

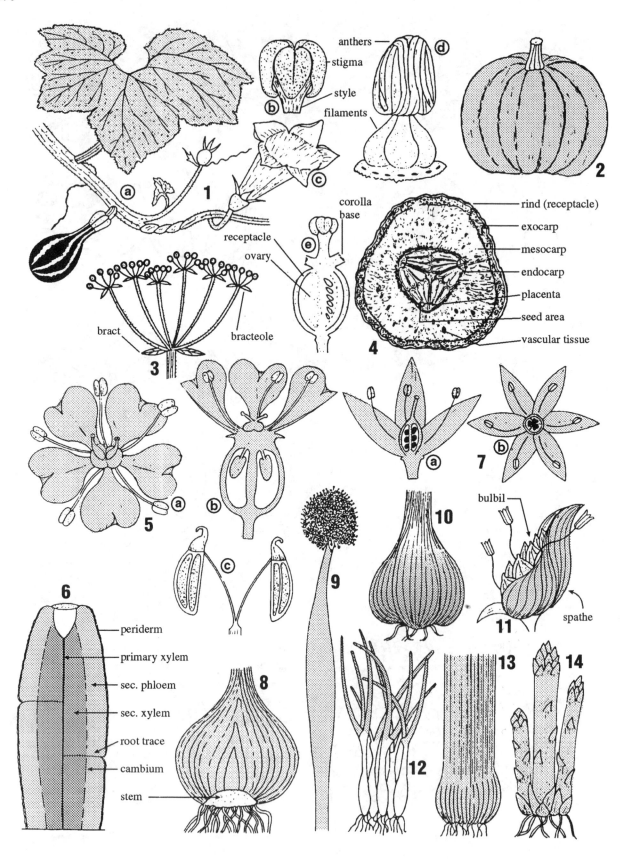

N.B. DRAWINGS NOT TO SCALE

Fig. 42.00

10 SOLANACEAE

General: The potato family - Solanaceae - is one of the most important group of economic flowering plants. Members are herbs, shrubs, trees, often climbing or creeping. There are up to 85 genera and 2200 species worldwide. Only 4 genera and 7 species are represented in the Irish Flora. Distribution is throughout the temperate and tropical regions of the world, though the main centres are in the Americas. It is a rather uniform taxon with well defined morphological and anatomical features. Nearly all members have hispid hairs and are strong smelling; the smell becomes more pronounced when the plant is crushed. One very distinctive feature is the presence of *bicollateral vascular bundles* in the stems. This feature is shared with only 1 other economic family - Cucurbitaceae.

Many members are a source of poisonous alkaloids and saponins. Such drugs as *atropine, belladonna* and *hyoscine* are derived from Solanaceous plants. Indeed, one of the most poisonous indigenous plants, *Atropa belladonna* (deadly nightshade - Fig. 19.16), belongs to this family. In addition there are 3 other poisonous members (Figs 19.07-08 and Fig. 19.23), which are frequent throughout the British Isles. Other poisonous plants, which are used medicinally, include *Datura stramontium* (thorn apple) and *Mandragora officinarum* (mandrake). The alkaloidal glycoside - solanine - is found in *Solanum* species.

There are very few ornamental or weed plants in the British Isles. Despite the large number of species in the family, no more than a half-dozen have any worthwhile economic importance. These include: *Solanum tuberosum* (potato); *S. melongena* (aubergine or egg-plant); *Lycopersicon esculentum* (tomato); *Nicotiana tabacum* (tobacco) and peppers which are used for spices. Most of the latter are in the genus *Capsicum*. All of the latter are native to the 'New World', though some are now cultivated in temperate areas. Thus, the red, cayenne, green, paprika, and tabasco peppers are obtained from the berries of various species and chillies. These peppers owe their pungency to the presence of a volatile phenolic compound, capsaicin, which is distributed throughout the plant, but tends to be concentrated in the placentae.

TAXONOMIC DATA

Stems: These are mostly erect and herbaceous; rarely woody, climbing or prostrate. Most have longitudinal ridges which may be so well developed that the stem is winged (Fig. 41.09). All are characterised by their possession of bicollateral vascular bundles.

Leaves: The leaves are simple or compound, spirally arranged or alternate, though sometimes appearing opposite towards the ends of the stems, petiolate and exstipulate. They are compound in most cultivated members. The petioles are adnate with the stems so that they diverge some distance above their point of insertion. The adnate petioles form wings or ridges on the stems. The former vary in prominence, and are of use in potato cultivar identification. The compound leaf of the potato may have secondary and tertiary leaflets (Fig. 41.09).

Inflorescence: The flowers occur in cymes or dichasia, though solitary flowers are found in some members. The former, in *S. tuberosum*, are easily recognised by their leaf-opposite arrangement.

Flowers: The individual flowers (Fig. 41.08) are conspicuous, often showy, and some members of the family, for example *Petunias*, have been selected as ornamentals. They have a characteristic structure. All are hermaphrodite, mostly actinomorphic and have the hypogynous arrangement. The calyx, which is gamosepalous, consists of 5 sepals, rarely more or less, and often persists in the fruit. The corolla is gamopetalous with 5 petals which are stellate, rotate or tubular in outline. The androecium has 5, rarely 6, epipetalous stamens, with characteristic connivent anthers. Staminodes are present in some members. The gynaecium is superior and consists of 2 fused carpels. The ovary is bilocular, rarely more through the development of cross-partitions; it contains numerous ovules in the axial arrangement. There is 1 style with a bulbous stigma. The fruit (Fig. 41.10) is typically a berry, rarely a septicidal capsule. Seeds are small and are non-endospermic. The best example of a berry is the fruit of the common tomato. Small tomato-like fruits or berries are often found on potato plants in late Autumn. The potato fruit, or 'potato ball' is smooth, globose and green in colour.

Many tomato cultivars show some degree of proliferation of floral parts and also fasciation within the developing inflorescence. This leads to flowers having more than 5 members to each of the whorls. Furthermore, this leads to some having up to 10 fused carpels which give rise to multilocular fruits.

IMPORTANT SPECIES

Potatoes

The potato and tomato plants are the 2 most important member of the family. The former constitutes a major foodstuff of great importance, even though by reason of its relatively low dry matter content it is not as important as wheat. It contains about 80% water and 20% dry matter. Starch accounts for about 85% of the dry matter and the remainder is mostly protein. Nevertheless, small amounts of minerals, such as calcium, iron, magnesium, potassium, sodium and sulphur, are present. They also contain many vitamins, including niacin, riboflavin, thiamine, and Vitamin C. Other constituents include fat at 0.1%, soluble sugars at 1% and crude fibre at 0.6%. They are frequently served whole, in chip form, or mashed as a cooked vegetable; they may also be ground into a flour which is used in baking or for thickening sauces. Some products, whose ingredients can include potatoes, are alcoholic beverages and certain starches used in industry.

When a tuber is exposed to sunlight for several days, either before or after digging, the skin of most cultivars develop a green colour as a result of chlorophyll development, though some can turn purple. Along with this change, increased quantities of solanine are formed in the cortex. Solanine is an alkaloidal glycoside, bitter in taste, and poisonous when ingested in sufficient quantities. The poisonous alkaloid, in solution, is called solanidine. The quantity of solanine in sun-burned tubers may be more than 20 times that in normal tubers. Most of the poisonous chemical disappears if the affected tubers are kept in the dark for about 2 weeks, or is removed when the tuber is peeled.

Potatoes are cultivated in all cold-climate countries, with the former USSR and eastern Europe the leading producers. They are classified in the tribe Solaneae and genus *Solanum* which has a worldwide distribution. It is a very large complex genus with many species. Only a relatively small number, however, are tuber-bearing, and these are found only in the 'New World'. The latter fall into 2 clear groups - wild species and cultivars.

Classification: As already stated *Solanum* is a large complex genus with up to 1500 species. All species have a basic chromosome number of 12, and diploids, triploids, tetraploids, pentaploids and hexaploids occur. *Solanum* is located in the tribe Solaneae and is divided into 6 subgenera. The tuber-bearing members are placed in the subgenus *Pachystemonum*. This taxon is further divided into 25 sections, of which Tuberarium is 1. Tuberarium, in turn, is separated into subsections Basarthrum, in which the articulation of the pedicel is at its base, and Hyperbasarthrum where articulation is above the base. Plants belonging to the latter subsection have multicellular hairs, are mostly tuber-bearing, upright, and have annual stems. Plants of the first subsection have bicellular hairs, are probably without tubers, and are perennials with straggling stems. Finally, Hyperbasarthrum is further divided into 26 series. The tuber-bearing species relevant to our cultivated potatoes are grouped in the series Tuberosa. Series Tuberosa is a euploid with

2n=24, 36, 48 and 60. European cultivars are listed in the species *andigenum (andigena* in some texts) and subspecies *tuberosum*. The complete classification system for the European potato may be summarised as follows:

Family ... Solanaceae
Tribe ... Solaneae
Genus ... *Solanum*
Subgenus ...Pachystemonum
Section ... Tuberarium
Subsection ... Hyperbasarthrum
Series ... Tuberosa
Species ... *andigenum*
Subspecies ... *tuberosum*
or ... *Solanum tuberosum*

Historical: The genus *Solanum* appears to have originated over a very extensive area. Berrie (1977) suggests that the region extending from New Mexico and Arizona down through central America to Argentina, keeping mainly to the higher altitudes of the Cordilleras, is probably the centre of origin of the genus. A narrower area is indicated by Wilsie (1962) for the important species who gives Peru, Ecuador and Bolivia as the centre of origin for *Solanum andigenum,* one of the putative ancestors of European potatoes. A further narrowing of the area of origin is suggested in the writings of Dodds (1965) and Simmonds (1976b). These writers claim that domestication is most likely to have occurred where wild and cultivated diploids show their maximum degree of variability and, to this end, indicate the altiplano region around Lake Titicaca, of Bolivia - Peru. The diploids are still there but have been largely superseded by cultivated tetraploids, but diploids, triploids and pentaploids also occur.

All the original wild diploids had a bitter taste and the first stage in the development of the potato was probably the selection of non-bitter types. It is not known when this happened but, in the general historical-archaeological context, it was possibly in the period 5000-2000 BC. Neither is the identity of the original diploids known but, on systematic grounds, several species of series *Tuberosa* have been mentioned. These include *S. brevicaule, S. leptophyes, S. canasense, S. spegazzinii* and *S. sparsiplum*. Cultivated diploids evolved, under domestication, from these and possibly other species, and they developed a wide range of variability in tuber-shapes, colours and foliage characters. The most important of these appears to be *S. stenotomum*. Autotetraploids evolved from the cultivated diploids, possibly from *S. stenotomum* and the most important tetraploid appears to be *Solanum andigenum*. This species is divided into 2 subspecies - ssp. *andigenum* and ssp. *tuberosum*. The latter represents most of the South American potatoes, while the former covers all European potatoes and, of the 7 or so cultivated species, is the only 1 grown on a worldwide scale. The scientific name is invariably shortened, for convenience, to *Solanum tuberosum*.

There are well defined morphological and physiological differences between *S. andigenum* and *S. tuberosum*. *S. andigenum* is a short-day plant with well divided leaves, while *S. tuberosum* is a long-day plant with less well divided leaves. The original European introductions were possibly all of the *andigenum* type, and probably did not perform well. The *tuberosum* forms most likely evolved from these over a period of 100-200 years. Present day forms developed a wide range of variability in tuber shapes and colours, and in foliage characteristics. The differences in photoperiodic response and leaf shape is controlled by a very unstable set of genes and these respond readily to natural and artificial selection. This means that the European potato could easily have evolved in its present form from *S. andigenum* (Simmonds, 1976a).

The potato was introduced to Europe after the Spanish conquests of the Americas in 1570. It spread to Italy and thence to Austria where the first botanical description of the potato was made by Clusius, a botanist, in 1588. It was introduced independently into England in 1586, probably by an expedition of Sir Francis Drake, though the latter's involvement lacks unanimous agreement. It was most likely introduced to Ireland a few years afterwards by Sir Walter Raleigh who owned an estate near Youghal, County Cork. By the end of the seventeenth century it was a major crop in Ireland, and by the end of the eighteenth was a major crop in continental Europe. The Irish economy became dependent upon the potato. The crop, from the time of its introduction into the British Isles in Elizabethan times until about the year 1845, was not affected by serious disease. In that year, however, it was attacked by a fungus, *Phytophthora infestans*, which, to the present day, is responsible for the well-known disease - *Blight*. Blight first appeared in the Isle of Wight but soon spread to the English mainland and Ireland. The attack was so severe that nine-tenths of the crop rotted in Ireland, and was directly responsible for the disastrous famine of 1845 and 1846. Thankfully, at the present time, the disease can be kept under control with the judicious use of a wide range of effective fungicides.

Characteristics

Stems: The stems of potatoes are of 2 types - aerial and underground. The aerial are annual in habit while the underground are perennial. The latter may be subdivided into: *(i) rhizomes* and (ii) *tubers*. The aerial stems are at first erect and rounded but later become triangular with winged edges (Fig. 41.09). These are solid at first but later become hollow at maturity. Growth-habit may be erect, spreading, or compact; branches are few to many. Growth-habit is best studied on well-spaced plants. The subterranean stems arise in the underground leaf axils. They grow horizontally outwards and later become swollen at the apices to develop tubers. All roots are adventitious; the root system may have a lateral spread of 60 cm and a penetration of 60-150 cm.

Leaves: The leaves (Fig. 41.09) are pinnately compound with a phyllotaxis of 5/13. The petioles are winged and the wings are decurrent on the stems. The prominence of the wings can vary. The terminal leaflet is large and ovate; the laterals are short-stalked and more or less in opposite pairs. Secondary and tertiary leaflets may also occur.

Inflorescence: The inflorescence is a cyme with few flowers. They have the typical family structure, already described. Colour varies from white to mauve or blue. The fruit is a 2-celled berry (Fig. 41.10) resembling a small unripened tomato. The seeds are non-endospermic, pointed-oval, about 2 mm long, flattened, slightly winged, pale brown and slightly rough.

Floral Development: Flower and seed production are very reduced in the potato plant. This is possibly due to natural selection where the principal mode of reproduction is vegetative. Pollen sterility and ovule degeneration are likely factors. Abscission occurs easily at high temperatures or with unbalanced nutrition. The flowers are shed early in many cultivars, but if persistent, pollination takes place and the ovary develops into a globular bilocular berry with numerous seeds borne on fleshy placentae (Fig. 41.10).

Germination: True seeds are of interest to plant breeders who wish to produce new cultivars. They are of no interest as regards the production of a crop of tubers. Germination is epigeal and the cotyledons are small and narrowly-oblong. The first few leaves are simple. Cotyledonary buds grow out horizontally as rhizomes, turn downwards and enter the soil. Other buds present on the cotyledonary ends of these rhizomes behave in a similar fashion, giving 2 to 6 rhizomes in all which originate above soil level. These rhizomes, on entering the soil, produce branches. The tips of the primary rhizomes and their branches swell to form small tubers. The first crop of tubers are quite small and usually reach a diameter of only 1-3 cm. These are harvested in the year of sowing, stored, and replanted the following year to produce rather larger tubers in the normal manner. The harvesting and replanting operations are repeated and full sized tubers are produced in the third and fourth year.

Tuberisation: A normal crop of potatoes is obtained by planting 'seed' tubers whole or partial. Under favourable conditions the buds or *eyes* elongate and form aerial stems at the expense of the vast food reserves held in the 'seed'. Normal compound leaves are produced. Shortly after the emergence of the leaves the young plants are 'earthed-up', a practice designed to encourage the plentiful development of rhizomes from the underground portions of the aerial stems. Tuber formation commences under the influence of a hormone and favourable environmental conditions, usually at about the time of flower bud development. Practically the entire crop of tubers is formed within a few days. Growth is very rapid during the early stages. Differences in size are attributed to unequal growth rather than a difference in age. Tuberisation begins much earlier under short-day conditions but early and second early cultivars commence tuber formation sooner under long-day conditions than main crop cultivars.

Anatomical Development: On warm, sunny days, an abundance of photosynthate is produced by the leaves. The tuber is formed as a result of the photosynthate being translocated to, and accumulating in, the distal ends of the rhizomes. The number of tubers per plant varies from a few to more than 20, depending on cultivar, weather and soil conditions. A detailed consideration of the anatomy of the rhizome, prior to swelling, is necessary for a complete understanding of the structure of the tuber.

The anatomy of the young rhizome is best revealed by preparing a thin cross-section (Fig. 41.11). An examination of the distribution of the vascular tissue reveals typical stem structure. Characteristic bicollateral vascular bundles are scattered in a peripheral zone and surround a parenchymatous pith; the bundles are separated from each other by thin-walled parenchyma. Surrounding the vascular tissue is the *endodermis* or innermost cortex layer. External to the endodermis are several layers of regular parenchyma cells. Intracellular spaces are plentiful. The *epidermis*, which is the outermost layer of the cortex, consists of thin-walled, regular-shaped cells.

Photosynthate is translocated from the leaves and stored mainly in the phloem parenchyma. The rhizome increases in size to accommodate the ever increasing photosynthate. The increase in size is due essentially to a *proliferation* of primary tissues, principally phloem and xylem parenchyma. These tissues become meristematic and undergo repeated divisions. There is neither cambial activity nor secondary thickening and, consequently, there is little or no fibrous tissue in the mature tuber. The relatively neat arrangement of the vascular tissue in the young rhizome is soon lost with its elements being scattered laterally and radially (Fig. 41.12). There is more phloem than xylem tissue in the young stem and, accordingly, more phloem in the mature tuber. Indeed, the bulk of vascular tissue present is mainly medullary phloem, consisting mostly of parenchymatous tissue, but some sieve-tubes and companion cells are also present. The amount of xylem is small and very few lignified cells are present.

As soon as the rhizome apices begins to swell, the epidermal and hypodermal cells become meristematic and form a *phellogen* which produces a *periderm* or skin (Fig. 40.19). The periderm increases in pace with the developing tuber and forms a continuous covering or protective layer. At intervals throughout the periderm groups of *cork* or *phellem* cells remain loosely arranged to form *lenticels*. The latter are essential to allow for the exchange of gases. They are rather inconspicuous but may increase in size to form raised white dots on tubers grown in partially waterlogged soils.

Secondary Periderm: In addition to the ordinary phellogen which produces the tuber skin, more phellogens may arise in response to wounding or cutting. The most obvious example is where large tubers are sectioned prior to planting. A secondary periderm, or wound periderm, forms on the cut surface under favourable environmental conditions to protect the tuber from the invasion of pathogens and consequent decay. Healing of the cut or wounded surface commences by the formation of a fatty deposit of suberin. This deposit is formed by the oxidation and condensation of fatty substances present in the sap of the surface cells. The process is favoured by a low pH and the presence of air and moisture, and is normally accomplished within 48 hours. Cut-potatoes for 'seed' are usually covered with sacking to minimise moisture loss and create a suitable environment for periderm development. Suberisation is followed by the division of the underlying parenchyma cells which ends in the formation of a phellogen. Under favourable conditions this phellogen differentiates a secondary periderm on the wound or cut surface. It has an identical structure to that of the original periderm.

Over exposure of the freshly cut surface to strong sunlight or to a dry atmosphere may result in the desiccation and consequent killing of the cells. Should this happen no worthwhile phellogen formation is possible. Furthermore, the planting of unhealed tubers in dry soils can have a similar effect. It is therefore advisable to store cut tubers in a moist atmosphere, *i.e.* by covering with sacking, prior to planting.

IDENTIFICATION OF POTATO CULTIVARS

Growth-habit: Early cultivars tend to have a low spreading growth-habit; main crop types are mostly tall and erect; second early cultivars are intermediate. This feature is best studied on well-spaced plants.

Stems: These show a wide range of variation. They vary in height, frequency, thickness, degree of branching, colour, and in the presence or absence of hairs. In addition, the wing may be prominent or poorly developed, straight or wavy.

Leaves: Leaf features show inter-cultivar differences, but they may also differ with age. Thus reference should always be made to fully expanded mature leaves. The size and frequency of primary, secondary and tertiary leaflets determine the openness or closeness of the leaf (Fig. 41.09. Leaves may be long, short, broad or narrow, rigid, drooped or arched. The leaflets may be light, dark, or grey-green, dull or glossy, flat, folded or arched, and rugose or smooth. Colour distribution in the mid-rib and petioles is another very useful feature. One potato mutant is known where the leaflets have coarsely dentate margins, like those of a tomato plant.

Leaf-area Index: This is measured by obtaining the following ratio:

$$\frac{Max.\ Width}{Max.\ Length} \times 100$$

for the top left-hand lateral leaflet. The ratio is fairly constant for most cultivars.

Inflorescence: The flowers are borne in compound, terminal cymes or dichasia, with long peduncles. In the potato this type of inflorescence is readily recognised by its leaf-opposite arrangement. Colour is mostly white. Other colours include blue, lilac, purple, rose or reddish, though the petal tips and veins of such flowers may lack colour.

Tubers: These vary as regards colour from pink, reddish-brown or reddish-white to a deeper colour. Flesh-colour normally ranges from white to yellow, but purple-fleshed types are known. Shape and weight also vary. The latter may be up to 300 gr., though higher weights are possible. The well-known cultivar, Golden Wonder, is immediately recognised by its elliptical shape. The depth and distribution of the *eyes* are other useful features. The *eyebrows* (Fig. 41.12), *i.e.* minute scale-leaves, which may be prominent or indistinct, are spirally arranged.

Sprouts: These features (Fig. 41.12) are best examined on normally developed tubers. Sprouts may be bulbous, cylindrical or tapered, depending on the relative development of the base, middle and tip. Colour distribution and the frequency and distribution of hairs are also useful.

Tomatoes

The tomato plant, which is the second most important food plant in the potato family, has a widespread area of cultivation. It is an annual vegetable, grown for its fruits or berries. Despite the popularity of its fruits, the vegetative parts are poisonous! These contain solanine but rarely cause poisoning, as they are not normally eaten by livestock. The concentration of this chemical in the fruit is too low to cause any harm to humans. The feeding of pigs with green side-shoots has caused acute illness and death (Cooper and Johnson, 1984). The reasons for its popularity are many; it is a well known source of vitamins and adds flavour and colour to the human diet. Many cultivars have been bred or selected to suit different environments with fruits suitable for different uses. Botanically, the edible portion is a baccate fruit though, colloquially, it a frequently referred to as a vegetable. Tomatoes are rather intolerant of low temperatures and in cool regions have to be grown under glass. They are closely related to the potatoes, and it has been suggested that the potato genus - *Solanum* - is ancestral to the tomato genus - *Lycopersicon* (Rick, 1976). Indeed, the tomato at one time was known as *S. lycopersicon* The taxonomic distinction between the 2 genera is based on pollen-shedding characteristics. In the potato flower pollen is shed through apical pores in the anthers, while in tomatoes pollen is shed through longitudinal slits. Evidence of the close relationship of the 2 genera may be gleaned from the existence of a potato mutant with tomato-like leaves, and *vice versa*.

Most writers indicate that the tomato originated in the same Andean region as the potato (Cobley, 1976 and Wilsie, 1951). Rick (1976), however, suggests that the bulk of the historical, linguistic, archaeological and ethnobotanical evidence favours Mexico, particularly the Vera Cruz-Pueblo area, as the source of the cultivated tomatoes that were first transported to the Old World. Its introduction into Europe seems to follow closely that of the potato. The putative wild ancestor, var *cerasiforme*, is found in Mexico, central America and parts of South America, but it is a weedy aggressive coloniser, a tendency which obscures its true native origin (Rick, 1976).

The tomato plant is technically known as *Lycopersicon esculentum* Mill., and is classified in the tribe Cestreae. There are less than a half-dozen species in the genus, all native to the Andean region of South America, and have morphologically similar chromosomes (2n = 2x = 24). It seems likely that the modern tomato had a complex ancestry and, after it was domesticated, diversity accumulated in it partly as a result of hybridization with several other species which now grow wild in the New World tropics. The wide range of diversity is manifested in the recognition of several distinct botanical varieties and cultivars, based on differences in fruit size and shape, leaf size and, to a limited extent, growth-habit:

L. esculentum var.	*commune* - common tomato
"	*grandifolium* - large-leaved tomato
"	*validum* - upright tomato
"	*cerasiforme* - cherry tomato
"	*pyriforme* - pear tomato

Characteristics

Stems: These are annual or short-lived perennials. The young stems are round, soft, brittle, much branched and covered with glistening, reddish-yellow, glandular, as well as pointed non-glandular, hairs. They become angular, hard and woody with age. Cultivars may have different growth-habits. Erect determinate types have fairly thick, solid stems with numerous long lateral branches. Trailing or semi-climbing, indeterminate forms, have weaker, thinner stems which usually require some form of support.

Roots: Young seedlings form tap-roots. These, however, are often damaged in transplanting operations. The primary root-system is then replaced by an extensive, copiously branched, adventitious system which arises from the base of the stem. Stem cuttings also produce adventitious roots.

Leaves: Numerous spirally arranged compound leaves are produced on the stems. They are unevenly pinnate with variously indented and lobed margins. The main leaflets are more or less opposite each other along the petiole. Smaller leaflets, which vary in size and number, are found between them. One botanical variety - *glandifolium* - has entire leaflets. All leaflets are stalked and pubescent.

Inflorescence: Inflorescence branches towards the base of of the stems are often monopodial, while those produced higher up are sympodial and vegetative growth continued by the development of a bud in the axil of a leaf below the apex. Consequently, the small cymose inflorescences appear to be opposite the leaves, or opposite but slightly above or below them.

Flowers: The flowers (Fig. 41.08) are small and carried on short pedicels, have a central construction, and often fall early. They are usually aberrant in having 6 petals and 6 sepals; some cultivars are pentamerous. The petals are stellate. The persistent calyx enlarges with the fruit and is covered with glandular and other hairs. Six epipetalous stamens with short filaments alternate with the corolla lobes, and have

278

long coloured, more or less convergent, anthers with longitudinal introse dehiscence. The tips of the anthers are prolonged into sterile beaks. Self-pollination occurs in some cultivars, cross-pollination may occur in others.

The gynaecium is syncarpous, consisting of 2 carpels, and the superior ovary is surmounted by a single style. False septae or partitions may produce 3-20 loculi in the ovary which contain numerous ovules on a fleshy axial placenta (Fig. 41.10). These fleshy placentae form the bulk of the edible portion. The endocarp area of the pericarp form an insignificant proportion of the overall size. The epicarp forms the thin skin. The fruit or berry is reddish or yellowish with a smooth skin. There is a lot of variation in the size and shape, in the thickness of the fleshy mesocarp and in the development of the placentae. Fruit colour is due to the presence of the pigments lycopersicin and caroten which occur in different concentrations in fruit of different colours. A predominance of lycopersicin produces reddish fruits, while a predominance of the latter produces yellowish fruits.

Tomato seeds (Fig. 21.05) are flat, about 2.5 mm in diameter, mostly obovate, light brown and densely covered with small adnate hairs. Germination is epigeal.

A second species - *L. pimpinelliforme,* the cherry or currant tomato - is an important crop in some areas, though not as extensively grown as the common tomato. At one time it was considered a likely candidate as the progenitor of *L. esculentum.* Indeed, breeding experiments have succeeded in restoring the large fruit size of *esculentum* cultivars. It is of some importance because it has been used for crossing with *esculentum* to produce disease resistant forms.

Peppers

The word 'pepper' is used in English to describe a group of spices which are derived from 2 quite different kinds of plants. 'Red' or 'cayenne' pepper is obtained from bushy plants of the genus *Capsicum,* and 'white' or 'black' pepper from the climbing vine *Piper nigrum* (Harrison, Masefield and Wallis, 1969). *Capsicum* has about 5 cultivated species. There are approximately 20 wild species, mostly confined to South America (Heiser, 1984). *C. annuum* and *C. frutescens* are the 2 most common species throughout Asia and Central America for the production of chillies. The other 2 species, *C. baccatum* and *C. pubescens* are grown in the Andes region of South America, and produce extremely hot fruits. Peppers owe their pungency to a phenolic substance, *capsaicin,* which is concentrated in the placentae.

Capsicum annuum L. (Peppers and Chillies): These are erect annual, shrub-like herbs, sometimes woody at the base, and grow to a height of 150 cm; they are strongly branched. The leaves which may be small or large, are extremely variable in shape but remain simple. Flowers are invariably solitary. The pendulous fruits are borne singly, and are extremely variable in length, width, colour and pungency. Some of the many cultivars of this species include the sweet peppers, paprikas, chillies and wrinkled peppers.

Capsicum frutescens L (Bird Chillies): This plant is somewhat similar to species *annuum* but is a perennial, and has smaller leaves. The flowers and fruits are borne in clusters of 2-5, the latter erect, usually red when ripe, elongate or conical, and extremely pungent. This species is used in some of the hottest exotic dishes, and forms the basis for fiery chilli sauces such as tabasco (de Rougemont, 1989).

Piper nigrum L. (Black Pepper): A climbing vine, it is a native of India. It needs the support of living trees, stakes or a trellis. The fruits, which occur in hanging spikes, are called 'peppercorns', and turn red when ripe. Both white and black pepper are made from them.

Solanum melongena L. (Aubergine): A native of Tropical Asia, it is also known as the egg-plant. It is a favourite vegetable for inclusion in lasagnes and other dishes. The berries are variable in size, up to 6-20 cm, globose, ovoid to oblong, sometimes curved, usually with a shiny epicarp which is usually dark purple, though creamy white, greenish and yellowish cultivars are known. It is a much branched perennial, though usually grown as an annual. The leaves are large, alternate, simple, up to 20 cm, with wavy margins and a dense covering of hairs. The purplish flowers are solitary or in small leaf-opposite groups.

11 CUCURBITACEAE

General: The family Cucurbitaceae is of wide geographic origin, and is represented by about 100 genera and 850 species, worldwide. The species are almost equally divided between the New and Old Worlds. Cultivated cucurbits are thought to have arisen in the Western Hemisphere, Africa and Asia. Their fruits, whose flesh consists of about 90% water and only about 5% sugar, are usually eaten raw, pickled and preserved, or used in soups. They are an important source of edible products and useful fibres in the tropics, subtropics, and milder areas of the temperate zones of both hemispheres. In these areas, the cultivated cucurbits are a useful source of carbohydrates when cooked (squash, pumpkin, marrow and chayote), when used as dessert or breakfast fruits (water melon and musk melon), as ingredients of salads (cucumber and gherkins), or as pickles (cucumber and gherkins). There are other minor uses. The fruits of white-flowered gourd (*Lagenaria siceraria*), before the advent of pottery, were indispensable to primitive peoples - the rinds were used as different types of containers for the storage and transportation of liquids in many primitive cultures, while the fibrous material was used for scouring, filters, etc. Some, because of their spreading, tangled vines, are used as a cover crop in plantations.

Cultivated members appear to be diploids with a basic chromosome number, *x=7, 11, 12, 13, 14* or *20*. All are fast growing, trailing or climbing annuals, rarely perennials, with mostly soft, pubescent, prickly or smooth stems. The latter, like those of the Solanaceae and the Gentianaceae, have bicollateral vascular bundles. Leaves are typically large, more or less lobed and spirally arranged. No species is frost resistant, so they can only be grown during the summer months or under glass. However, some species, commonly called pumpkins, develop a very hard exocarp which allows them to be stored for considerable periods of time, thus providing an important source of winter vegetables in some countries. Furthermore, thanks to the endeavours of plant breeders, many cultivars have now spread well into temperate environments in the summer time (Langer and Hill, 1991). Unfortunately, there is no specific English term for the individual species. This, together with the indiscriminate use of such vernaculars as 'pumpkin' and 'squash' has led to some confusion. The problem is further compounded by the fact that some species give more than one product. A very detailed description of the general characteristics of the family, summarised hereafter, is provided by Whitaker and Davis (1962), unless otherwise stated.

TAXONOMIC DATA

Roots: The root-system of all cucurbits is extensive but shallow. The central tap-root may penetrate to a depth of 180 cm, and at a rate of 2.5 cm per day. Numerous horizontal secondary roots develop, but branching of the tap-root is not extensive below 60 cm. The secondary branches are *complexly* and *minutely rebranched* until there is a remarkably efficient network that completely occupies the upper 30-60 cm of soil. Growth rate may be up to 7 cm per day. The root-system of a single plant of some curcurbits may occupy as much 28 m³.

Stems: All cucurbits are alike in their general stem morphology. The main axis is short; from 3-8 branches arise from near the base of the main axis. These stems are prostrate, trailing, hirsute to scabrous, thick and angled in cross-section. They have to be supported on wires, twines or trellises. The main axis is a sympodium, *i.e.* at each node a lateral branch continues the main axis and, by its growth, displaces the terminal branch so that the latter occupies a position on the *opposite* side of the axis from the leaf from which it arises. In most species, the stem grows to a length of several cm, and in a few species of *Cucurbita* the stems may reach a length of 1.2-1.5 m. Species of the above genus also have a tendency to produce adventitious roots at the nodes. Anatomically, all stems have *bicollateral* vascular bundles.

Leaves: These (Fig. 42.1a) are *exstipulate, simple*, petiolate, mostly *palmately* 3-lobed or 5-lobed, rarely entire, *palmately* veined, *spirally* arranged with a phyllotaxy of 2/5. Species of *Citrullus* have pinnatifid leaves. Indeed, there is much variation among species and cultivars, in terms of shape and depth of lobes. The leaves of *Cucumis sativus* are alternate, simple, palmately 5-lobed and somewhat angled when young, but become subcordate when mature. Tendrils are invariably present, and these arise in the *axils* of the leaves, though there is much controversy as to their exact origin. They are mostly *branched*, but unbranched in the above species. The plants, except those grown in greenhouses, are not usually provided with supports and are allowed to grow over the surface of the soil.

Flowers: The flowers (Figs 42.1b-e) of different genera and species vary considerably in size and colour, but are similar in general morphology. They are mostly *unisexual*, very rarely hermaphrodite. Various patterns of distribution are found; they may be *solitary*, but they mostly occur in *cymose* or modifications of cymose or racemose inflorescences. All flowers show the *epigynous* arrangement. Plants are *monoecious, dioecious* or rarely andromonoecious.

Calyx: The sepal number is 5. Sepals of the staminate flowers are linear and alternate with the corolla lobes; those of the pistillate flowers are slender and awl-shaped, and often shorter. Both the calyx and corolla are fused, giving a well-defined tube.

Corolla: The corolla is *gamopetalous*, very rarely polypetalous; petal lobes are large, showy, distinct and 5 in number. Yellow and yellow-orange are the most commonly encountered colours in most species. In addition, each corolla is actinomorphic and campanulate, rotate or salverform.

Androecium: Male flowers may contain a rudimentary style. The basic number of stamens in the family is 5. These can be considered as having arisen from 5 primordia but there is much irregular fusion during development, the anthers not only fusing but also becoming contorted and proliferated (Berrie, 1977). However, the number in most genera appears as 3, 1 of which is unilocular, the other 2 are bilocular. Each bilocular stamen is thought to have arisen from the fusion of 2 such features. Each stamen is inserted below the middle of the perianth tube. They bend towards the centre of the flower and end in the thickened connectives, which fit together closely by their angled surfaces. The outer surface is almost occupied by the anthers which are fringed with hairs. After the stamen primordia appear, 3 small lobes become visible near the bottom of the receptacle, and below and within the stamen lobes. They alternate with the stamens and constitute what is known as the *pistillodium* (Fig. 42.1d).

Gynaecium: Each pistillate flower has a *syncarpous* gynaecium or pistil with an *inferior* ovary, the carpels primitively 5, rarely 4, though usually 3 in most cultivated species; the locule usually 1 with parietal placentation, or sometimes 3, with axile placentation; ovules numerous. Pistillate flowers may also contain 3 rudimentary stamens (Fig. 42.1e).

Fruit: The fruit is essentially a type of *berry*, technically referred to as a *pepo* (Fig. 42.04; c.s.). This fruit type is characterised by the presence of a hard *rind*, derived from the *receptacle*, and a fleshy mesocarp which is the edible portion. Pepos, which show great variation in size, shape and colour, are usually indehiscent, rarely explosive. They are very succulent and, in desert regions, can substitute for water, for example the edible portion of the water melon contains 95% water (Berrie, 1977). In addition, those of the cultivated forms are among the largest fruits found in plant kingdom. Cucurbit seeds are usually flattened and contain no endosperm. Some cultivars are seedless.

IMPORTANT SPECIES

This is a relatively small family with only 3 genera of major importance. The important species, cultivated in temperate regions, are classified in the genera *Cucumis (x=7 or 12), Cucurbita (x=20)* and *Citrullus (x=11)*, in which the fruits are the major items of trade. The genera *Luffa, Benincasa, Trichosanthes, Sechium* and *Momordica* are of minor importance except in Asian countries. The nutritional value of their fruits is much lower than that of the principal species. They are used mainly to enliven rice-based dishes. For reasons stated earlier, it is not possible to give a common English name for the latter.

Citrullus lanatus (Thunb.) Matsum. & Nakai (Water Melon): A variable plant, this cucurbit differs from other members of the family by having *pinnatifid* leaves; the central lobe forms most of the leaf. It is a much-branched trailing annual, up to 1 m, with slender, angled and grooved stems, covered with white woolly hairs when young. Flowers unisexual, *solitary*, axillary, pale yellow in colour; corolla deeply divided into 5 lobes. The fruits are large, up to 70 cm long, *globose* or *oblong*, with hard, smooth rind variously patterned in green and cream, or different shades of green; the flesh very watery, typically *deep pink*, sometimes white, yellow or greenish.

Water Melon is a native of tropical Africa, where a wild form with bitter fruits still occurs. It was cultivated in Ancient Egypt whence it spread throughout the Mediterranean and eastwards to India in prehistoric times. To-day, it is cultivated in all the tropical and subtropical countries, and in temperate countries with a continental climate. In Europe, it is grown in all the Mediterranean countries and in central and eastern Europe where hot summers prevail (de Rougemont, 1989).

Cucumis melo L. (Melon): A diploid, $x=12$, it is a very variable species. Its products include the musk melon, winter melon, ogen melon and cantaloupe. The plants are annual vines covered with soft hairs. Stems are up to 1 m, rather hispid, and ridged. Leaves are large, orbicular, ovate to reniform, angular or shallowly lobed; tendrils simple or branched. Sexual expression is complex, and the plants may carry both female and male flowers, or may have hermaphrodite and male flowers. Male flowers are borne in clusters while hermaphrodite and female are solitary.

Cucumis melo is a native of tropical Africa. Its fruits are very variable in size, shape, colour, type of rind and flavouring. De Rougemont (1989) groups them into 4 types: (i) *cantaloupe* - these are medium-sized or ovoid fruits with a thick rough rind, often with deep longitudinal grooves suggesting segments, and usually yellow or orange in colour; (ii) *ogen melon* - fruits are very small, 15 cm or less, globose, bright orange-yellow, with vertical green stripes, and green flesh with a fine flavour; they are derived from the cantaloupe; (iii) *musk melon* - this type is also known as netted melon because of the network of whitish lines which covers the yellow, pinkish, orange or green rind; the flesh is green to salmon pink and very aromatic; (iv) *winter melons* - these are of inferior quality to the above 3 types, but are harder-skinned and, consequently, keep longer; the fruits are globose or, more often, ovoid to strongly elongate, medium to large, with slightly bumpy or ridged rinds of a uniform pale or dark green colour, and pale green flesh.

Cucumis sativus L. (Cucumber and Gherkins): This species is anomalous in being the only species in the genus with $x=7$ (Whitaker and Bemis, 1984). It is described as a monoecious, annual, trailing or climbing vine with 4-angled, hirsute or scabrous stems. Leaves are entire or shallowly 3-5 angled, finely serrate, and somewhat cordate basally. Flowers are yellow, deeply divided into 5 hairy, crinkled petals; corolla, as in all species, somewhat *rotate*. More male than female are produced. Male flowers are carried in axillary clusters, while the female are solitary and borne on a stout pedicel.

Cucumis sativus is a native of India where it has been in cultivation for many centuries. Its fruits are very variable, mostly green, ovoid to elongate, cylindrical, much longer than broad, often slightly curved and prickly, especially when young. Gherkins are described as small fruited cultivars, picked when very young for pickling.

Cucurbita moschata Poir. (Pumpkins and Winter Squashes): An example of the confusion that exists, in relation to the ambiguity of English names, is shown by the fact that pumpkins and winter squashes are applied to the products of 3 species of *Cucurbita*. The distinction between pumpkins and squashes is not clear, but it appears to be one of texture, in that fine-grained fruits with a mild flavour are described as squash, while the more coarse-grained and more strongly flavoured are called pumpkins. All species have the basic chromosome number, $x=20$. *C. moschata* is a monoecious, annual, trailing, non prickly but softly hairy vine with round or smoothly 5-angled stems. Leaves are also softly hairy, large, shallowly lobed, often with white spots along the veins and somewhat cordate basally. Flowers yellow, and corolla, as in all species, somewhat *campanulate*; calyx tube of staminate flowers short or absent, the lobes often foliaceous; corolla with 5 widely spreading, mostly reflexed lobes. The pedicels, like the stems, mostly smoothly 5-angled, and *expanded* or *flared* at the point of attachment to the fruit. The fruits are very variable, usually large, globular, cylindrical or flattened. They are often divided into 3 categories, mainly on the basis of colour, texture of rind and flesh, size and shape.

Cucurbita maxima Duchesne ex Lam. (Pumpkins and Winter Squashes): This species is said to have been cultivated first in Peru where seeds 700 years old have been found (de Rougemont, 1989). It is very similar to the *C. moschata*. It is also monoecious, mostly trailing or climbing vine, rarely bushy in habit. Leaves are more reniform in shape, not lobed or only obscurely so, finely serrated, and more deeply cordate basally. Flowers yellow; corolla lobes usually reflexed; calyx lobes short and narrow. The pedicels are *spongy*, cylindrical, soft and *corky*.

The fruits are very variable, and de Rougemont (1989) groups them into 5 categories, based on type: (i) *turban squashes* - these are readily identified because the ovaries protrude prominently from the apex of the fruit, shaping it like a turban sitting on a head or an acorn in a cup; size is typically small to medium, with smooth skins, often red or mottled; (ii) *mammoth* or *show squashes* - these are characterised by their large size, up to 50 kg (Fig. 42.02), globular in outline, with soft thin orange-coloured shells or, in some cultivars, with pale brown stripes; all have thick yellow flesh; (iii) *banana squashes* - fruits of this group are much smaller, only 5-6 kg, shaped like a large courgette, with yellow flesh and grey-green skins becoming pink in storage; (iv) *hubbard* - these are pear-shaped fruits, 30-40 cm long and weighing 4-6 kg, with very hard, green shells and dull yellow flesh; (iv) *boston marrow* - these are very similar to the last type, but are more globular in outline, less constricted on the stalk end, and have a softer, orange shell.

Cucurbita mixta Pangalo (Pumpkins and Winter Squashes): This plant was treated, in the past, as a variety of *C. moschata* to which it is very similar. It originated in Mexico or central America, and is intolerant of cool temperatures. Stems are softly hairy, hard, not rough, and clearly 5-angled. Leaves rather large, shallowly to moderately lobed, with or without white blotches. Flowers similar to those of *Cucurbita moschata*. However, the pedicel is *not* expanded or flared at the fruit end but the diameter is greatly *increased* by the addition of firm *warty* cork. Fruits, as in all species of the genus, are very variable. They may be either hard or soft shelled, usually dull in colour, with moderately dry, white to pale tan-coloured flesh.

Cucurbita pepo L. (Summer Squashes, Pumpkins and Marrows): This is one of the oldest cucurbits. It has been cultivated in its native Mexico and the southern United States for at least 8000 years. It is described as an extremely variable plant in terms of size, reproductive characters and in being either slightly bushy or, more usually, with long trailing stems; the latter are hard, often 5-angled, harsh and prickly to the touch. Leaves are broad, somewhat triangular, finely serrated on the margins, mostly deeply lobed and cordate basally. The flowers are similar to those of the other species. The pedicels are 5-angled, but *not* flared or expanded at the fruit end. Fruits are very variable.

Several types are recognised. (i) *vegetable marrows* - plants here are either bushy or trailing with large cylindrical or rounded fruits with green or cream rinds, and greyish-white, rarely yellowish, flesh; *courgettes* or *zucchini* are names applied to small 'baby' vegetable marrows; (ii) *pie pumpkins* - these are produced on trailing plants; they are either large, 8-12 kg, globular or slightly flattened, pale orange with pale cream flesh; another type is more oblong in outline, with drier, dark yellow flesh; others, smaller, 2-5 kg, with dull orange rinds and sweet firm orange flesh, are known; (iii) *fordhook* - plants of this category are either trailing or bushy, with fruits 20-25 cm long, cylindrical in outline, and with whitish rinds mottled with yellow; (iv) *summer crook-necks* and *straight-necks* - the former are club-shaped and curved between the stem and middle, 20-35 cm long, with ribbed, warty, orange yellow or white rinds. The latter are similar but curved in the basal half. Other fruit types, of lesser importance, are also known.

12 APIACEAE

General: Members are mostly biennial or perennial herbs, occasionally suffrutescent, rarely shrubs. Two family names - Apiaceae and Umbelliferae - are used. The former is a relatively new name based on the type genus - *Apium*. The latter is the old name for the taxon and it is still deemed legitimate because it has been in use for a very long time. The family contains about 250 genera and between 2000 and 3000 species, growing in the northern hemisphere though, in the tropics, members are confined to mountains tops. It is often divided into 3 subfamilies, the Hydrocotyloideae, Saniculoideae, and the Apioideae; the distinction is based on differences in leaf and fruit structure, and in the distribution of oil glands on the latter. With the exception of a few genera, the family is readily identifiable by its rather distinctive floral structure. The leaf bases of practically all members have a characteristic shape.

Native members occur mostly as weeds of pasture and waste places (Figs 08.01-13). Some, particularly *Conium maculatum*, are of importance because of their well known poisonous properties (Figs 18.05-08). Others, both native and introduced, are grown either as vegetables or for their 'seeds' which are used for flavouring bread. The latter, which are not of sufficient importance to warrant detailed discussion here, include: caraway - *Carum carvi* L., anise - *Pimpinella anisum* L., fennel - *Foeniculum vulgare* Mill., coriander - *Coriandrum sativum* L., chervil - *Anthriscus cerefolium* (L.) Hoffm., turnip-rooted chervil - *Chaerophyllum bulbosum* L., lovage - *Levisticum officinale* Koch, sweet cicely - *Myrrhis odorata* (L.) Scop., cumin - *Cuminum cyminum* L. and dill - *Anethum graveolens L.*

TAXONOMIC DATA

Leaves: These are mostly alternate, but may be opposite at the base of some stems. They are mostly compound and exstipulate, but have a characteristically flattened petiole base which resembles a stipule. In stipulate leaves, however, there is a clear distinction between petiole and stipule. This is not the situation in plants of this family. The leaflets vary in size, shape, in being entire, toothed or lobed, and in the presence or absence of hairs; all these features are important in identification.

Stems: Most members have herbaceous, erect stems which usually have a large pith that shrinks or dries at maturity with the internodes becoming hollow. Only 1 member - marsh pennywort - has a weak creeping stem. All are distinctively grooved externally.

Inflorescence: The inflorescence (Fig. 42.03) is generally an umbel; it may be simple or compound. It is reduced to a single flower in marsh pennywort. One or more bracts, constituting the involucre, may be present at the base of each umbel. Bracts present at the base of a single umbel, in a compound umbel, are known as the *involucel*. The terms *bract* and *bracteole*, respectively, are also used. Both structures show a wide range of variation and, consequently, are of great taxonomic importance.

Flowers: The flowers are small and mostly actinomorphic (Figs 42.5a-b). In a compound umbel, however, the outer flowers may be zygomorphic, the petals directed towards the exterior being longer than those directed towards the interior. Petal arrangement is unique in that they arise directly from the ovary. In most cases the flowers are hermaphrodite but exhibit protandry. In this respect one could easily be misled into believing that some flowers are unisexual, *i.e.* pistillate, as the stamens drop off when the pollen is shed. Protandry is so marked that selfing is prevented. Nevertheless, different flower types occur. Thus in the marsh pennywort the flowers are almost homogamous, while the andromonoecious condition is seen in the umbel of *Anthriscus*. In the South African genus - *Arctopus* - the dioecious arrangement is shown. Cross-pollination is effected by the activity of very small insects which visit the stylopodium for nectar.

Calyx: The sepals are generally absent; if present they are reduced to small teeth (Fig. 42.5a).

Corolla: The corolla is composed of 5 petals, each either obcordate or obovate; they are often 2-lobed or mucronate with an incurved mucro. The colour is mostly white, occasionally yellow, green or pinkish (Fig. 42.5a).

Androecium: There are 5 stamens with thread-like filaments which curve inwards when the flower is young (Fig. 42.5a).

Gynaecium: The pistil or gynaecium consists of 2 united carpels. The inferior ovary is bilocular and each loculus contains a single pendulous, anatropous ovule, suspended from the medium septum with its micropyle directed upwards and outwards. The pistil is characterised by the presence of a swollen glandular structure, the *stylopodium*, which surmounts the ovary. It may be absent or inconspicuous in a few species. The style is small, bifid, with either straight or outwardly curved arms (Fig. 42.5b).

Fruit: The fruit is dry, indehiscent, laterally or dorsally compressed, and usually schizocarpic. When ripe it splits along the line of union, the *commissure*, of the 2 carpels into 2 1-seeded, indehiscent, partial fruits or fruitlets. Each partial fruit is called a *mericarp* and forms the 'seed' which is characteristic of this family. After splitting each mericarp remains attached for some time to a central stalk, the *carpophore*, which also splits into 2 (Fig. 42.5c).

Each carpel or 'seed' usually bears, on its outer or dorsal surface, up to 9 more or less well defined lines or ridges. Five of these (Fig. 42.5c) are described as primary ridges or *costae*. Two are close to the commissure, the other 3 are dorsal. The 2 marginal ridges may be similar to, or different from, the dorsal costae. Four furrows or *valleculae* are located between the 5 ridges. Occasionally 4 secondary ridges or ribs occupy the spaces. In a few species, *e.g. Daucus carota*, these are as prominent as, or larger than, the primary ridges. All ridges may be continuous raised lines or consist of hairy, prickly, or knob-like projections. Sometimes the prickles occur throughout the entire surface of the mericarp.

Longitudinal canals or *vittae* are frequent in the pericarp (Fig. 20.4c). They are generally situated in the grooves between the primary ribs or below the secondary ridges, if these are present. A similar canal may occur on either side of the carpophore. Each mericarp, as a consequence, may have 6 of these oil ducts but additional ducts may also occur. The canals contain secretions of various balsams, resins, and volatile oils which impart to some fruits or 'seeds' their characteristic odour and taste.

Each mericarp, or half-fruit, contains 1 true seed which completely fills the cavity and is usually adnate to the pericarp wall. The outer surface of the seed has the same outline as the dorsal surface of the partial fruit: the inner surface is either flat or concave. A small embryo is embedded in a copious oily endosperm. The 'seeds' show a wide range of variation in shape, size, prominence of ridges, depth of sini, width of commissure, and in the presence or absence of teeth or prickles.

IMPORTANT SPECIES

Daucus carota L. ssp. *sativus* (Hoffm.) Arcang. (Carrot): This is a cool weather plant (Fig. 42.06) grown throughout the world in temperate climates in spring, summer and autumn but, in subtropical climates, cultivation is confined to the winter period (Banga, 1976). It is derived from the wild carrot, *Daucus carota* subspecies *carota*, which is common throughout Europe and southwestern Asia, Australia and America. Other subspecies - *commutatis, gummifer, fontanesii, hispanicus, major* and *maritimus* - of lesser importance are known from the Mediterranean region and countries eastwards of same. The important cultivated forms are located in subspecies *sativus*. All are diploids with $2n = 2x = 18$.

Leaves: The leaves are tripinnate with small lanceolate, pointed, pubescent segments or lobes and form a dense rosette in the first year.

Stems: The stems remain short in the first year's growth but lengthen in the second up to a height of 90 cm. They are ridged, solid and bear numerous compound umbels. These are flat or slightly convex in flower, but deeply concave in fruit, giving a 'bird's-nest' appearance. The bract number varies from 3-6; each is divided into narrow pointed lobes. The individual flowers are small and white; the central umbels often have a reddish colour.

'Seed': The 'seed' (Fig. 21.37) is characterised by having long hooks or prickles on the primary and secondary ribs, those of the latter being characteristically larger than the former. The presence of prickles makes drilling of the 'seed' difficult. Prior to sowing, the mericarps are rubbed to remove the projections but, nevertheless, some remain. They are roughly hemispherical in shape and about 3.5 mm long.

Germination, which is only about 70%, is epigeal. The cotyledons are strap-shaped and elongated. The first foliage leaves are deeply divided and about as broad as long. The hypocotyl and upper part of the primary root become swollen; the epicotyl remains short.

'Root': The young carrot has a slender rap-root which passes imperceptibly into the hypocotyl. Contraction of the root takes place in most cultivars and, as a consequence, the hypocotyl, which is at first above ground level, is drawn close to or even below soil level. Thickening of the true root and epicotyl then takes place. Adventitious roots develop on the root and hypocotyl so that all external distinction between the 2 disappear.

The relative amount of hypocotyl and true root contributing to the thickened areas varies in different cultivars. In every case, however, the lower portion of the true root remains unthickened and cord-like. The shape of the 'root' thus produced varies. Three distinct types are recognised: *Taper-pointed:* In this type the 'root' tapers uniformly from crown to the tap-root. *Premorse:* This type is characterised by the 'root' ending abruptly and the roots emanating from the nearly flat or dome-shaped base. *Cylindrical:* The third type has cylindrical 'roots' for about two-thirds of their length, thereafter they taper gradually.

Secondary Thickening: The distribution of the primary tissues in the young root is similar to that shown in Figure 40.10. The primary xylem occupies the central position of the true root. On either side, and separated by thin-walled parenchyma, are 2 strands of primary phloem. The primary xylem and phloem, and the thin-walled or procambium parenchyma, is encircled by a single layer of thin-walled, regular cells, termed the *pericycle*. This in turn is enclosed within a single layer of cells, called the *endodermis*. External to the latter are several layers of loosely arranged parenchyma cells and these, in turn, are circumscribed by a final layer of thin-walled, regular cells, known as the *epidermis*.

Almost as soon as the first leaf is expanding, secondary thickening begins in the hypocotyl and root. The thin-walled parenchyma between the xylem and phloem becomes meristematic to form 2 strands of cambium, 1 on either side of the primary xylem. A few cells of the pericycle, immediately outside the protoxylem arms, divide tangentially into daughter cells. The outer daughter cells increase in size and remain part of the original pericycle. The inner cells increase in size, undergo further repeated divisions and form connecting bands of cambium for the 2 cambial strands already formed. Thus a continuous, more or less circular, zone of cambium is formed around the primary xylem (Fig. 42.06).

The cambium differentiates secondary xylem internally and secondary phloem externally. The xylem consists of thin-walled vessels together with delicate, wide-celled xylem parenchyma; no fibrous tissue is formed. The phloem is made up of parenchymatous cells with scattered sieve tubes. As thickening continues, the cortex is ruptured and eventually is sloughed-off in the soil. The pericycle becomes active at an early stage and, after repeated divisions, forms a cork-type cambium termed the *phellogen*. The latter differentiates cork or *phellem* tissue externally and live thick-walled cells - *phelloderm* - internally. The 3 tissues - phellem, phellogen, and phelloderm - constitute the *periderm* or skin (Fig. 40.19). The periderm consists mainly of dead suberised cells which act as an insulating layer.

Some plants may bolt, *i.e.* produce an inflorescence in the first year and then die back, but the majority behave as biennials. If the plants are replanted or allowed to grow in the second year further secondary thickening takes place. The xylem produced in the second year is heavily lignified and the feeding value of the 'root' decreases markedly.

Most cultivars produce an equal amount of secondary phloem and xylem (Fig. 42.06). In mature 'roots' the 2 zones are easily recognised because of the lighter colour of the former or *core*; secondary phloem is also known as *bast*. A high proportion of phloem to xylem is desirable as it is in the former that the greatest amounts of sugar and other nutrients are stored. Carrots can be white, yellow or red, but cultivars for human consumption are all of the latter colour.

Pastinaca sativa L. var. *hortensis* Gaudin (Parsnip): The cultivated form (as in Fig. 42.06) is derived from the wild plant var. *sylvestris* (Mill.) DC., which occurs on waste places throughout Europe and western Asia. Annual and biennial forms exist but the latter are higher yielders. It differs from the carrot in its larger, simply-pinnate leaves with large ovate segments, and taller hollow flowering stems which lack bracts and bracteoles. Like the carrot the succulent 'root' is a composite structure, derived partly from the epicotyl and partly from the hypocotyl. The colour, however, is white and the main food reserve is starch. The 'seed' or mericarp is totally different. It is very strongly dorso-ventrally compressed, flat and disc-like in outline, and has the marginal ribs expanded in the form of a peripheral wing. There are neither secondary ribs nor prickles present (Fig. 21.35). Vittae extend over the entire surface of the seed.

Apium graveolens L. var. *dulce* (Mill.) DC. (Celery): This plant is mostly grown for its succulent petioles which are used as vegetables. It is derived from the wild celery - *Apium graveolens*. Some plants are cultivated for blanching. These are technically known as var. *dulce* (Mill.) DC. Others have swollen edible roots - var. *rapaceum* (Mill.) DC. The latter is commonly known as celeriac. The short stem and upper part of the root are greatly swollen to form a spherical structure 7-14 cm in diameter, with a transversely wrinkled surface. Leafy types are referred to as var. *secalinum* (Mill.) DC. The wild form occurs as a waterside plant both in Ireland and throughout the rest of Europe.

A biennial with coarsely bipinnate leaves, it has greenish-white flowers, short-stalked umbels and entire petals. The fruit is c. 1 mm, oval, and laterally compressed (Fig. 21.47). Each mericarp has 5 low, corky, primary ribs, and 1-3 vittae on the commissural side. The broad petioles mostly consist of parenchyma with a series of vascular bundles, of varying sizes, extending throughout. Collenchyma occurs immediately inside the lower epidermis in association with each vascular bundle. 'Stringy' celery is the result of excessive development of the collenchyma under unfavourable growing conditions.

Petroselinum crispum (Mill.) A.W. Hill (Parsley): This plant (Fig. 21.51), unlike the 3 forenamed, is a perennial with tripinnate leaves and wedge-shaped segments. Some forms have curled leaves. Others have a small swollen tap-root resembling a small parsnip. All have greenish-yellow flowers, flat umbels, with both bracts and bracteoles present. There are 3 principal cultivated forms. Curled parsley, the most common, is a compact plant with short leaf-stalks and very small leaflets. The second, uncurled parsley, is a larger in all its parts. The third form, Hamburg parsley, is rather similar but has a large swollen tap-root resembling a small turnip.

Foeniculum vulgare Mill. (Fennel): This is a tall variable, aromatic perennial, with solid stems up to 120 cm in height. The leaves are finely divided into long awl-shaped segments. Flowers, which are yellow, occur in glaucous compound umbels, of 13-20 rays; no bracts are present. There are 2 forms. One, Florence fennel, with swollen leaf bases, is used as a pot-herb; the other, sweet fennel, is grown for its seeds. They contain the essential oil, anethol, and are used for flavouring or as a a spice.

Carum carvi L. (Caraway): An annual or biennial much-branched herb, caraway has slender, finely furrowed, erect stems which attain a height of 30-80 cm. The leaves are also finely divided into feather-like segments. Umbels are compound, 2-4 cm in diameter, and have 8-15 unequal rays; bracts and bracteoles 1 or absent. The small brown mericarps are aromatic, and used for flavouring food, particularly bread. The young leaves are often used in salads, while the thick tap-root may be eaten as a vegetable.

13   LILIACEAE

General: This is a family of worldwide distribution, and is represented over most of the vegetated land areas of the Earth and are especially abundant in the warm temperate and tropical regions (Lawrence, 1951). There are approximately 240 genera and 4000 species. There has been some considerable confusion in relation to the classification of these species and, indeed, as many as 12 separate families are mentioned in the literature. Members are mostly herbaceous perennials or occasionally small evergreen trees, shrubs or climbers. The herbaceous forms frequently have corms, rhizomes or bulbs.

Most members are of no significance, but several are of ornamental value. Fewer still are grown as economic crop plants. Practically all such plants are in the genera *Asparagus* and *Allium*. Species of the latter are cultivated mainly for their characteristic alliaceous, pungent flavour, a feature which is said to be due to the presence of organic sulphur compounds (Jones and Mann, 1969).

TAXONOMIC DATA

Stems: These are generally glabrous, smooth, erect or climbing, though sometimes modified into fleshy subterranean storage organs such as *corms, bulbs* and *tubers*. Corms are described as shortened vertical, subterranean stems. Bulbs, on the other hand, differ from corms in that food reserves are stored in scaly leaf bases. The stem portion is small, *disc-like,* and has at least 1 central terminal bud which will produce a single upright stem. In addition, there is at least 1 axillary bud that will produce a bulb for the subsequent year. Some species produce highly modified branches, termed *cladodes*. These are so strongly dorso-ventrally compressed as to be leaf-like and, indeed, function as photosynthetic organs. Other species form horizontal, underground stems or rhizomes.

Leaves: The leaves are very variable. They may be either radical or cauline and alternate, or both on the same plant; very rarely whorled or, as in *Scolyopus*, opposite. Mostly laminate, rarely reduced to small scales; outline varies from long, narrow and cylindrical to broadly elliptical. All are exstipulate, simple and entire, though occasionally fleshy, fibrous or with prickly margins. Most have parallel veins, but a few species exhibit parallel-reticulate venation. Sheathing bases may be present or absent.

Flowers: These are variously arranged. The inflorescence may vary from inconspicuous 1 or few flowers, to simple or compound umbels, simple or compound cymes, spikes or terminal, lax or dense racemes. Some inflorescence types may be subtended by a large leaf-like bract termed a *spathe*. Most flowers show the *hypogynous* arrangement, but the epigynous pattern is found in 2 subfamilies. Plants of most species carry *hermaphrodite* flowers but a few are functionally dioecious.

Perianth: The corolla and calyx are not clearly differentiated, so both are considered together and are generally referred to as *tepals*. There are usually 6 segments, in 2 whorls of 3, rarely 4, or in 2 whorls of 4-6 each, and often brightly coloured; these may be free throughout or variously united, especially towards the base. Symmetry is mostly *actinomorphic*, rarely zygomorphic. Furthermore, a long or short funnel-shaped or collar-shaped *corona* is sometimes present within the rows of tepals.

Androecium: The stamen number is usually 6, rarely 3, 4 or as many as 12, hypogynous or adnate to the perianth; the filaments are usually free, rarely connate; anthers 2-celled, extrorse or introrse, basifixed or versatile, usually dehiscing by longitudinal slits or rarely by a terminal pore.

Gynaecium: The gynaecium or pistil is *syncarpous*; the ovary is mostly *superior* or inferior in subfamilies Amaryllidoideae and Alstroemerioideae, generally *trilocular* with axile placentation, rarely unilocular; ovules mostly numerous; styles usually 1, rarely 3 or 0; stigmas usually 3, or 1 and trilobed.

Fruit: This is either a septicidal or loculicidal *capsule;* rarely fleshy. Some members of the subfamily Lilioideae produce berries. Seeds are small, numerous and contain an abundance of endosperm.

Classification: The taxonomic position of this group of plants is a matter of some controversy. Indeed, a perusal of relevant literature reveals several classificatory systems. The cultivated members, in particular, are variously placed in the Liliaceae or the Amaryllidaceae. Taking the group as a whole, up to twelve extra families are sometimes recognised *viz.* Hyacinthaceae, Amaryllidaceae, Asparagaceae, Colchicaceae, Asphodelaceae, Alliaceae, Alastroemeriaceae, Hemerocallidaceae, Ruscaceae, Convallariaceae, Trilliaceae and Melanthiaceae. However, in recent times many taxonomists have adopted a much wider concept of the Liliaceae, and many or all of the above families have been treated as subfamilies. The system, given hereafter, is based on that of Stace (1992).

IMPORTANT SPECIES

Despite the large number of genera, approximately 240, only 2, *Asparagus* and *Allium*, are of economic importance. Most of the food or flavouring plants, and a few poisonous species, belong to the latter. The cultivated types include Chinese chives, chives, rakkyo, common onion, shallots, Welsh onion or Japanese bunching onion, potato or multiplier onion, garlic, leek, kurrat and great-headed garlic. More poisonous plants, bog asphodel and meadow saffron, are in the genera *Narthecium* and *Colchicum*, respectively. Several other genera contain plants of ornamental value, some of the best known being red-hot-poker (*Kniphofia uvaria* (L.) Oken) and lily-of-the-valley (*Convallaria majalis* L.).

KEY TO SUBFAMILIES

1. Ovary *superior*; or *very rarely* semi-inferior **3**
    Ovary clearly *inferior* **2**

2. Spathe *present*. Plants *bulbous*. Leaves radical, linear to narrowly elliptical. Inflorescence 1 flower or a terminal *umbel*; spathe *present*; styles 1; stigma simple or slightly 3-lobed; fruit a loculicidal *capsule*, or dehiscing irregularly; the tepals white to yellow or orange, rarely pink in colour; the corona *sometimes* present. Members are mostly of ornamental value.**Amaryllidoideae**
    Spathe *absent*. Plant with *tuberous* roots. Leaves all cauline, lanceolate. Inflorescence a *terminal* simple or compound *umbel without* a spathe; flowers *somewhat zygomorphic*; style 1, with a 3-lobed stigma; ovary *inferior*; fruit a loculicidal *capsule*; tepals orange with dark markings, free; corona *absent*. There is 1 genus, but it has no economic importance. **Alstroemerioideae**

3. Leaves *other than* Iris-like, *i.e. lacking* 2 identical faces **4**
    Leaves strongly compressed and Iris-like, *i.e.* with 2 *identical* faces, mostly radical. *Rhizomatous*. Inflorescence of a *terminal raceme*; styles 1 or 3; ovary *superior*; fruit a loculicidal *capsule;* tepals yellow to greenish-white, free throughout, and *without* a corona. There is only 1 species of importance, *Narthecium ossifragum* (L.) Huds., a frequent poisonous plant.**Melanthioideae**

4. Cladodes *absent* **5**
    Cladodes *present*. Plant *rhizomatous, dioecious*. Leaves all *cauline* and reduced to *small scales*. Inflorescence an *inconspicuous cluster* of 1-few flowers, in axils of scale-leaves on the main stem or cladodes; the styles 1 or nearly absent, stigma capitate or 3-lobed; the ovary *superior*; fruit a *red berry;* tepals greenish to yellowish-white, free or united basally, and *without* a corona. There are but 2 genera, the ornamental *Ruscus* and economic *Asparagus*. **Asparagoideae**

5. Flowers and leaves present at the *same* time **6**

Flowers, 1-few, arising from *tubers* and appearing *after* the leaves have *died off*; styles 3; ovary *superior*; fruit a septicidal *capsule;* tepals pinkish to pale purple, united basally into a long tube, and *without* a corona. Plant with a *corm*. Leaves *mostly* cauline. There is but 1 species of importance, *Colchicum autumnale* L. (meadow saffron), a rare poisonous plant. **Wurmbaeoideae**

6. Spathe *absent* **7**

Spathe *present*. Leaves radical, cauline, somewhat cylindrical, linear-oblong or elliptical. Inflorescence a *terminal umbel*, scarious spathe *present*; rarely reduced to a single flower or replaced by *bulbils*; styles 1; stigma capitate or 3-lobed; ovary mostly *superior*; fruit a loculicidal *capsule*; tepals variously coloured; corona *absent*. The only important genus is *Allium*. **Allioideae**

7. Plant *bulbous* or with *tuberous* roots; or rhizomatous *but* flowers in long terminal racemes **8**

Plant *not* bulbous basally. Plant *rhizomatous* or with *swollen* roots. Leaves all, or nearly all, radical, linear. Inflorescence a *raceme* or *terminal compound cyme*; styles 1; ovary *superior*; fruit a loculicidal *capsule*; the tepals variously coloured but not blue, and *without* a corona. There are about 4 genera in this subfamily, but none is of great economic importance .. **Asphodeloideae**

8. Flowers mostly *single*, rarely spikes or *lax* racemes. Plant *rhizomatous* or *bulbous*. Leaves all radical, all cauline, or both, linear to ovate or elliptical. Inflorescence a *single* flower, or a *spike* or *raceme;* styles 1 or 0; ovary *superior*; fruit a *berry* or a loculicidal *capsule*; the tepals variously coloured but *not* blue; *without* a corona. This subfamily has about 10 genera. **Lilioideae**

Flowers mostly in *dense*, often corymbose, racemes; ovary *superior*; styles 1, often with a 3-lobed stigma; fruit a loculicidal *capsule*. Plant *bulbous*. Leaves all radical, linear or nearly so; tepals variously coloured, free or united basally, and *without* a corona. There are c. 20 species, some, particularly the 'Star-of-Bethlehem' plants, are of some ornamental importance .... **Scilloideae**

The Genus *Allium*

The genus *Allium* contains more than 600 species (Jones, 1990). They are widely distributed over the warm-temperate and temperate zones of the northern hemisphere and, more sparsely, in tropical and subtropical regions (Hanlet, 1990). A region of especially high species diversity occurs in Turkey and the Irano-Turanian floristic region. A second region occurs in western North America. The principal domesticated forms, which are grown for food, number only 7 main species, though there are a number of others, of minor importance, grown in some parts of the world. In the genus, as a whole, there are 3 basic chromosome numbers, $x=7, 8$ or 9.

Allium cepa var. *cepa* L. (Bulb or Common Onion, $n=8$): The onion is probably the major culinary herb of the world, and all cultures seem to utilise these pungent bulbs to enhance the flavour of their food (Langer and Hill, 1991). *Allium cepa* is a very variable species, with many different varieties and cultivars. The bulb onion (Figs 42.08-09) has a long history of domestication as onions were depicted in ancient Egyptian murals and have been found in coffins of mummies (Langer and Hill, 1991). It is not found in the wild. Practically all of the commercially important bulbing onions come from *A. cepa*. Cultivars differ widely in pungency, flavour, bulb size, shape and colour, storage quality, in their response to temperature and photoperiod, and many other characteristics.

All parts of the plant have a characteristic odour and are lachrymatory. The plants are biennials, though commercial crops are usually grown as annuals. Leaves (Fig. 42.08), which are produced in succession from the bulb, have cylindrical blades which are at first solid, later becoming hollow, glaucous, slightly to markedly flattened adaxially. When conditions become favourable, photosynthate is deposited in the leaf-base to produce the typical onion bulb. The inflorescence, which is a *terminal umbel*, is borne on a scape and may be subtended by a *scarious spathe*. Hermaphrodite flowers are carried on slender pedicels with inconspicuous white to greenish tepals. Six stamens occur in 2 whorls of 3. The tri-locular ovary is superior. The *capsular* fruit contains many small, angular, black, striated seeds.

Allium cepa var. *ascalonicum* Strand (Shallots, *n=8*): These are smaller plants than onions, and produce a cluster of slightly asymmetrical bulbs from a single planted bulb of the same size, the bulbs being separate, not enclosed in a common outer set of scales (de Rougemont, 1989). Different cultivars are recognisable by the colour of the skins - grey, pinkish, or golden brown, with subtly different flavours.

Allium cepa var. *aggregatum* G. Don (Potato or Multiplier Onion, *n=8*): The potato onion is quite distinct in that it forms fairly large round bulbs which are wider than they are high, are aggregated, and have brown outer scales; they also have numerous laterals enclosed by the outer scales. These laterals produce separate tops and bulbs in their second year. The number of bulbs formed from a single bulb varies.

Allium fistulosum L. (Japanese Bunching Onion or Welsh Onion, *n=8*): The latter name is misleading, as this plant has no connection with Wales. This has been the main garden onion in China and Japan since prehistoric times. It is very similar to the common onion and, like the latter, is not known in the wild. The bulbs are small, narrowly oblong, slightly asymmetrical like those of shallot, and white or pink in colour. They are milder than those of the common onion and so require less cooking.

Allium sativum L. (Garlic, *n=8*): The leaves of this plant are flat, linear, somewhat longitudinally folded and keeled abaxially. The inflorescence is not usually produced but when present it is small and bears small aerial bulbs which displace the flowers. The bulbs (Fig. 42.10) are expanded axillary buds from within the leaf bases. Garlic is an important flavouring agent.

Allium tuberosum Rottl. ex Spr. (Chinese Chives, *n=16*): These are small plants, growing in dense clumps from a prominent rhizome. Bulbs are insignificant. Leaves linear, flat, grass-like. Umbels small and white. The leaves and scapes are used as a green vegetable.

Allium porrum (ampeloprasum) L. (Leek, Kurrat and Great-headed Garlic, *n=16*): This is a variable species with flat leaves and large dense umbels. Leek has v-shaped, keeled, thin leaves; blanched bases form the edible portion. Bulbs are insignificant (Fig. 42.13). Kurrat is similar but smaller, though the bulbs are often better developed. Prior to flowering, the great-headed garlic resembles a large garlic plant. The 2-lobed bulbs are always well developed; axillary bulbils are not enclosed in a papery membrane.

Allium schoenoprasum L. (Chives, *n=8,12 or 16*): This plant (Fig. 42.12) is grown for its very narrow, thick, hollow, tubular leaves which are chopped and used raw to garnish some foods. The bulbs are insignificant. The flowers occur in pinkish-mauve umbels which are borne on rather slender hollow scapes.

Allium chinense L. (Rakkyo, *n=8, 12 or 16*): Similar to chives but leaves are 3-5-angled and not as stiffly erect. Scape solid. Bulbs are ovoid and well developed. Rakkyo is mostly used for pickles.

The Genus *Asparagus*

Members of this genus differ strikingly from those of *Allium*. It contains about 150 species. Some are grown as ornamental plants but only 1, *A. officinalis* L. (asparagus), is cultivated. Asparagus, a diploid with a basic chromosome number $x=10$, occurs throughout central and southern Europe and in North Africa and western and central Asia. It has been in cultivation since ancient Greek times. It is a rhizomatous plant (Fig. 42.14) with annual stems and very extensive root-systems. The tall stems are strongly branched and bear small, appressed, triangular scale-leaves. Photosynthesis is carried out by cladodes, *i.e.* thread-like branches which resemble feathery leaves. Flowers are small and inconspicuous, yellowish or green, and borne singly or in bundles of 2 or 3 in the axils of the cladodes. Asparagus is a dioecious plant and red berries are produced on the stems of female plants.

The crop, which may be established from seed or cuttings, consists of newly emerging succulent 'spears' which are cut below ground level. Establishment from seed is rather slow. Asparagus is looked upon as a luxury vegetable. Large quantities are canned, but there is an inevitable loss of flavour.

BIBLIOGRAPHY

Aberg, E. (1940). The taxonomy and phylogeny of *Hordeum* L. Sect. Cerealia Ands. *Symb. Bot.* Upsaliens, 4, (2), 1.

Allen, O. N. and E. K. Allen (1981). *Leguminosae.* University of Wisconsin Press, Wisconsin, USA.

Appelqvist, L. A. and R. Ohlson (1972). *Rapeseed; cultivation, composition, processing and utilisation.* Elsevier, Amsterdam.

Baker, M. J. and W. M. Williams, eds, (1987). White clover. CAB International, Wallingford.

Banga, O. (1976). Carrot, *Daucus carota* (Umbelliferae). In *Evolution of Crop Plants* (ed. N. W. Simmonds), pp. 291-293. Longman, London and New York.

Barton, L.V. (1961). *Seed Preservation and Longevity.* Leonard Hill (Books) Ltd., London.

Baum, B.R. (1977). *Oats: Wild and Cultivated.* A monograph of the genus *Avena* L. (Poaceae). Can. Min. of Agric.

Bell, G. D. H. (1965). The Comparative Phylogeny of the Temperate Cereals. In *Crop Plant Evolution* (ed. Sir J. Hutchinson), p.70. Cambridge University Press, London.

Bergal, P. and L. Friedberg (1940). *Ann. Epipyht,* 6:157-306.

Beaven, E. S. (1947). *Barley.* Duckworth, London.

Berrie, A. M. M. (1977). *Introduction to the Botany of major Crop Plants.* Heyden, London.

Bond, D.A. (1976). Field bean. *Vicia faba* (Leguminosae/Papilionatae). In *Evolution of Crop Plants* (ed. N. W. Simmonds), pp. 179-182. Longman, London and New York.

Booth, W. E. (1964). *Agrostology.* Edward Brothers, Michigan, USA.

Bowden, B.N. (1965). Modern grass taxonomy. *Outlook on agric.* 4:244.

Bowden, W.M. (1959). The taxonomy and nomenclature of the Wheats, Barleys and Rye and their Wild Relatives. *Canad. J. Bot.* 37, 657.

Bowden, W. M. (1966). Chromosome numbers in seven genera of the tribe Triticeae. *Can. J. Genet. Cytol.* 11:130-136.

Briggle, L. W. (1967). Morphology of the Wheat plant. In *Wheat and Wheat Improvement* (eds Quisenberry, K. S. and L. P. Reitz), pp. 89-116. *Agronomy,* 13. Wisconsin, USA.

Briggs, D. E. (1978). *Barley.* Chapman and Hall, London.

Brown, H. F. (1958). Leaf anatomy in grass systematics. *Bot. Gaz.,* 119:170.

Campbell, G. K. G. (1976). Sugar beet - *Beta vulgaris* (Chenopodiaceae). In *Evolution of Crop Plants* (ed. N. W. Simmonds), pp. 25-28. Longman, London and New York.

Catchside, D. G. (1934). The chromosomal relationships in swede and turnip groups of *Brassica. Ann. Bot.,* 48:601-632.

Clapham, T. G., Tutin, T. G. and E. E. Warburg (1962). *Flora of the British Isles.* Cambridge University Press, London.

Clifford, H. T. and D. W. Goodall (1967). A numerical contribution to the classification of the Poaceae. *Aust. J. Bot.,* 15:499.

Cobley, L. S. (1976). *An Introduction to the Botany of Tropical Crops.* 2nd Edit. Longman, New York and London.

Coffman, F.A. (1946). Origin of cultivated Oats. *J. Am. Soc. Agron.,* 38:983-1002. Coffman, F.A., (ed). (1961). Oats and Oats improvement. *Agronomy,* 8. 15-40. Wisconsin.

Coffman, F.A. and J.W. Taylor (1936). Widespread occurrence and origin of fatuoids in Fulghum Oats. *J. Agr. Res.* 52:123-131.

Coons, G. H., Owen, F. V. and S, Dewey (1955). Improvement of Sugar Beet in the United States. *Advances in Agronomy,* 7:89-139.

Cooper, M. R. and A. W. Johnson (1984). *Poisonous Plants in Britain* and the effects on animals and man. HMSO, London.

Correll, D. S. (1962). *The Potato and its Wild Relatives*. Renner. Texas Research Foundation.

Crane, M. B. (1943). The origin and relationships of *Brassica* crops. *J. Roy. Hort. Soc.* 68:172-174.

Curran, P.L. (1962). The Nature of our *Brassica* crop. Part 1. Nomenclature and Cytology. *Sci. Proc.* of the *Roy. Dub. Soc., Vol.* 1 (12), pp. 319-335.

Curran, P.L. (1966). *Avena fatua:* The common Wild-oat. *J. Dept. Agric. Republ.,* Vol. 64 : 3-15.

Curran, P.L. (1970). Aneuploids of *Brassica oleracea. Sci. Proc.* of the *Roy. Dub. Soc.,* 2:217-219.

Curran, P.L. and C.M. Brickell (1969). Desynapsis in *Brassica oleracea* spp. *balerarica. Sci. Proc.* of the *Roy. Dub. Soc.,* 2:163-169.

Davis, D.R. (1976). Peas. *Pisum sativum* (Leguminosae - Papilionatae). In *Evolution of Crop Plants* (ed. N. W. Simmonds), pp. 172-174. Longman, London and New York.

de Candolle, A. (1882). *Origin of Cultivated Plants*. The Int. Sci. Ser. New York. D. Appleton and Co.

de Rougemont, G.M. (1989). *A Field guide to the Crops of Britain and Europe*. Collins, Grafton Street, London.

de Wit, J. M. J., Harlan, J. R. and D. E. Brink (1986). In *Infraspecific Classification of Wild and Cultivated Plants* (ed. B. T. Styles), pp. 211-212. Clarendon Press, London.

Dodds, K.S. (1965). The history and relationships of Cultivated Potatoes. In *Essay on Crop Evolution* (ed. S. J. Hutchinson), pp. 279-283.

Dony, J.D., Jury, S.L. and F.H. Perring (1986). *English Names of Wild Flowers*. The Botanical Society of the British Isles.

Encyclopaedia Britannica (1984). 5th Edit. The University of Chicago.

Engledew, F. L. (1920). *J. Genet.* 10:93-108.

Evans, G. M. (1976). Rye: *Secale cereale* (Gramineae - Triticinae). In *Evolution of Crop Plants* (ed. N. W. Simmonds), pp. 108-11. Longman, London and New York.

Evans, L. T. and W. J. Peacock (1981). *Wheat Science* - today and tomorrow. Cambridge University Press, Cambridge.

Farragher, M.A. (1964). The use of chemicals for determining the germination capacity of farm and garden seeds. *Agric. Rec.* 20:15-33.

Farragher, M.A. (1968). Yield and longevity of some American varieties of lucerne under Irish climatic conditions. *J. Dept. Agric. Republ.* 66:3-8.

Farragher, M.A. (1969). Lucerne investigations. II. Yield and persistency of some Canadian produced varieties under Irish climatic conditions. *Ir. J. agric. Res.* 8:213-220.

Farragher, M.A. (1970). *Taxonomy and Ecology of Irish Grasses*. Ph.D. Thesis. National University of Ireland Library, Dublin.

Feekes, W. (1941). Verslagen van de Technische Tarwe Commissie XVII De Taewe en haar milieu. pp. 523-588. Groeningen : Gebroeders Hoitsema.

Feldman, M. (1976). Wheats: *Triticum* spp. (Gramineae - Triticinae). In *Evolution of Crop Plants* (ed. N. W. Simmonds), pp. 120-128. Longman, London and New York.

Feldman, M and E. R. Sears (1981). The wild gene resource of Wheat. *Sci. Amer.,* Vol. 204:98-107.

Forsyth, A. A. (1954). *British Poisonous Plants*. HMSO Bull. No. 161. London.

Frick, G.W., Loomis, R. S. and W. A. Williams (1975). Sugar Beet. In *Crop Physiology*, some case histories (ed. L. T. Evans). Cambridge University Press.

Gill, N. W. Vear, K. C. and D. J. Barnard (1980). *Agricultural Botany*, Vols 1 & 2. Duckworth, London.

Godwin, H. (1956). *The History of the British Flora*. Cambridge.

Goodman, Major M. (1976). Maize - *Zea mays* (Gramineae-Maydeae). In *Evolution of Crop Plants* (ed. N. W. Simmonds), pp. 165-168. Longman, London and New York.

Goodsell, H. F. (1948). Triphenyltetrazolium chloride for viability determination of frozen seed corn. *J. Amer. Agron.,* 40 (5). 432-442.

Gould, F. W. (1968). *Grass Systematics*. Mc-Graw-Hill, New York.

Hackel, E. (1890). The true grasses. (Translated from Die Nathurlichen Pflanzenfamilien by F. Lamson-Scribner and Effie A. Southworth), New York.

Hanlet, P. (1990). Taxonomy, evolution and history. In *Onions and Allied Crops*. Vol. 1, Botany, physiology and genetics (ed. H. D. Rabinowitch and J. L. Brewster), pp 1-24. CRC Press, Boca Raton, Florida.

Harlan, H. V. (1957). One man's life with Barley. The memories and observations of H. V. Harlan (ed. J. R. Harlan). New York.

Harlan, J. R. (1976). Barley - *Hordeum vulgare* (Gramineae - Triticinae). In *Evolution of Crop Plants* (ed. N.W. Simmonds), pp. 93-98. Longman, London and New York.

Harlan, H. V. and M. L. Martini (1936). Problems and results in Barley breeding. In *USDA Yearbook*, 303-346.

Harlan, H. V. and M. L. Martini (1938). The effect of natural selection in a mixture of Barley varieties. *J. Ag. Res.* 57:189-200.

Harlan, J. R. and D. Zohary (1966). Distribution of Wild Wheats and Barley. *Science*, 153:1074-1080.

Harris, P. M. (ed., 1978). *The Potato Crop*. Chapman and Hall, London.

Harrison, S. G., Masefield, G. B. and M. Wallis (1985). *The Illustrated Book of Food Plants*. Peerage Books, London.

Haussknecht, C. (1885). Über die Abstammung des Saathabers. Mitt. Georgr. Gesell. 3:231-242. The biology and taxonomy of the Solanaceae. Linnean Society of London. Academic Press, London.

Hawkes, J.G. (1990). *The Potato*: evolution, biodiversity and genetic resources. Belhaven Press, London.

Hawkes, J.G., Lester, R.N. and A.D. Skelding (eds,1979). The biology and taxonomy of the Solanaceae. Linnean Society of London. Academic Press, London.

Hector, J. M. (1936). *Introduction to the Botany of Field Crops*, Vols 1 & 2. Central News Agency, Ltd., Johannesburg, South Africa.

Heiser, C. B., Jr. (1984). Peppers - *Capsicum* (Solanaceae). In *Evolution of Crop Plants* (ed. N.W. Simmonds), pp. 77-78. Longman, London and New York.

Helbaek, H. (1959). Domestication of food plants in the old world. *Science*, 130:365-372.

Helbaek, H. (1966). Commentary on the phylogenesis of *Triticum* and *Hordeum*. *Economic Botany*, 20:350-360.

Helbaek, H. (1960). Paleoethnobotany of the Near East and Europe, in prehistoric Investigations in Iraq Kurdistan, by R.J. Braidwood and B. Howe (No. 311 in series studios in ancient Oriental civilivation, pp. 99-18. University of Chicago Press.

Hemingway, J. S. (1976). Mustards. *Brassica spp.* and *Sinapis alba* (Cruciferae). In *Evolution of Crop Plants* (ed. N.W. Simmonds), pp. 56-59. Longman, London and New York.

Heywood, V.H. (1968). *Modern methods in Plant Taxonomy*. Academic Press, London and New York.

Heywood, V.H., Harborne, J. B. and B. L. Turner, eds, (1977). The biology and chemistry of the *Compositae*. Academic Press, London, New York and San Francisco.

Holden, J.H.W. (1976). Oats: *Avena* ssp. (Gramineae - Aveneae). In *Evolution of Crop Plants* (ed. N.W. Simmonds), pp. 86-90. Longman, London and New York.

Holmes, Sandra (1983). *Outline of Plant Classification*. London and New York.

Hubbard, C. E. (1984). *Grasses*. Penguin Books, Harmondsworth.

Hutchinson, J. (1955). *British Wild Flowers*. Vols 1 & 2. Penguin Books, Richard Clay & Co., Suffolk.

Hyde, E. O. C. (1955). Methods for determining the viability of seeds by tetrazolium staining. III. Italian Rye-grass and short rotation Rye-grass. *N. Z. J. Sci. Tech.* 37:36-39.

Hymowitz, T. (1976). Soybeans: *Glycine max* (Leguminosae - Papilionatae). In *Evolution of Crop Plants* (ed. N. S. Simmonds), pp. 159-162. Longman, London and New York.

International Code of Botanical Nomenclature (1972). International Association for Plant Nomenclature, Utrecht, Netherlands.

Jain, S. K. (1960). The genetics of Rye (*Secale* spp.). *Bibliogr. Genet.* 19:1.

Janick, J., Schery, R. W. and V. W. Ruttan (1969). *Plant Science*. Freeman and Co., San Francisco.

Jeffrey, Charles (1977). *Biological Plant Nomenclature*. Edward Arnold, London.

Jones, J. K. (1984). Strawberry - *Fragaria ananassa* (Rosaceae). In evolution of crop plants (ed. N. S. Simmonds), pp. 237-242. Longman, London and New York.

Jones, A. J. and L. K. Mann (1963). *Onions and their Allies*. Leonard Hill (Books) Limited, London.

Jones, R. N. (1990). Cytogenetics. In *Onions and Allied Crops*, Vol. 1, Botany, physiology and genetics (ed. H. D. Rabinowitch and J. L. Brewster), pp 199-214. CRC Press, Inc., Boca Raton, Florida.

Jones, M. B. and A. Lazenby, eds, (1988). *The Grass Crop*. Chapman and Hall, London.

Kamm, A. (1954). The discovery of wild six-row Barley and wild *Hordeum intermedium* in Israel. *Ann. Roy. Agric. Coll., Sweden*. 21:287.

Kimber, G. (1974). The relationships of the S-genome diploids to polyploid Wheats. *Proc. 4th. Internl Wheat Genet. Symp.*, Columbia. 81-85.

Klush, G. S. (1963). Cytogenetic and evolutionary studies in *Secale* III. Cytogenetics of weedy Ryes and origin of Cultivated Rye. *Econ. Bot.,* 17 : 60-71.

Klush, G. S. and G. L. Stebbins (1961). Cytogenetic and evolutionary studies in *Secale.* I. Some new data on the ancestry of *Secale. Amer. J. Bot.,* 48:721-730.

Korniche, F. and H. Werner (1885). Handbuch des Getreidebaues. Berlin. Paul Parey. Vols 1 & 2.

Langer, R. H. M. and G. D. Hill (1991). *Agricultural Plants.* Cambridge University Press, Cambridge, Port Chester, Melbourne and Sydney.

Large, E. C. (1954). Growth stages in cereals. *Plant. Path.* 3:128-129.

Lafferty, H.A. (1934). Seventh International Seed Testing Congress, July, 1934. *J. of the Dept. of Agric.,* Vol. 33, p. 53.

Lawrence, G. H. M. (1951). *Taxonomy of Vascular Plants.* Macmillan, London.

Leonard, W. H. and J. H. Martin (1963). *Cereal Crops.* Macmillan, London.

Lesins, K. (1976). Alfalfa, Lucerne - *Medicago sativa* (Leguminosae-Papilionatae). In evolution of crop plants (ed. N. W. Simmonds), pp. 165-168. Longman, London and New York.

Li, C. W. (1981). The origin, evolution, taxonomy, and hybridization of Chinese cabbage. In Chinese cabbage, Proceedings of the *First Int. Symp.* 1981 (eds N. S. Talakar and T. D. Griggs), pp. 1-10. AVRDC, Taiwan.

Lisinska, G. and W. Leszczynski (1989). *Potato Science and Ttechnology.* Elsevier, London.

Lowson, J. M. (1977). *Textbook of Botany.* 15th Edit. London University Tutorial Press.

Mackay, J. T. (1836). *Flora Hibernica.* W. Curry and Co., Dublin.

Malzew, A. I. (1930). Wild and cultivated Oats (Sectio Euavena. Griseb). *Bull. appl. Bot. Genet.* Plant Breeding (Lennigrad), Suppl., 38:1-522.

Mangeldorf, P. C. (1950). The mystery of corn. *Sci. Am.,* 183 (i) : 2-6.

Mangeldorf, P. C. (1953). Wheat. *Sci. Am.,* 189 (i) : 50-59.

Mangeldorf, P. C., MacNeish, R. S. and W. C. Galinat (1964). The domestication of corn. *Science,* 143: 538-545.

Martin, J. H. and W. H. Leonard (1967). *Principles of Field Crop Production.* Macmillan, New York.

Matz, Samuel A. (1969). *Cereal Science* The AVI Publishing Company, Inc., Westport, Connecticut.

McCarthy, H.V. (1988). The Testing of Farm and Garden Seeds (unpublished data). Official Seed Testing Station, Department of Agriculture, Abbotstown, Castlenock, County Dublin.

McFadden, E. S. and E. R. Sears (1946). The origin of *Triticum spelta* and its free-threshing hexaploid relatives. *J. Hered.,* 37:81-89 and 107-116.

McNaughton, I.H. (1976a). Turnip and relatives *Brassica campestris* (Cruciferae). In *Evolution of Crop Plants* (ed. N. W. Simmonds), pp. 45-48. Longman, London and New York.

McNaughton, I.H. (1976b). Swedes and rapes. *Brassica napus* (Cruciferae). In *Evolution of Crop Plants* (ed. N. W. Simmonds), pp. 53-56. Longman, London and New York.

Moore, D. and A. G. Moore (1886). *Cybele Hibernica.* Hodges, Smith & Co., Dublin.

Morris, R. and E. R. Sears (1967). The cytogenetics of Wheat and its relatives. In *Wheat and Wheat Improvement* (eds Quisenberry, K. S. and L. P. Reitz), 19-88. *Agronomy,* 13. Wisconsin, USA.

Mott, G. O. (1983). Potential productivity of temperate and tropical grassland systems. In *Proc.* of the XIV *Intn. Grass. Congr,* (eds Smith, S. A. and V. W. Hays), pp. 35-41. Colorado, USA.

Muntzing, A. (1943). Genetical effects of duplicated fragment chromosomes in Rye. *Hered.,* 29:91-112.

Muntzing, A. (1944). Cytological studies of extra fragment chromosomes in Rye. I: Iso-fragments produced by misdivision. *Hered.,* 30:231-248.

Muntzing, A. (1945). Cytological studies of extra fragment chromosomes in Rye. II: Transmission and multiplication of standard fragments and iso-fragments. *Hered.,* 31:457-477.

Muntzing, A. (1946). Cytological studies of extra fragment chromosomes in Rye. III: The mechanism of non-disjunction at the pollen mitosis. *Hered.,* 32:97-119.

Nevski, S. A. (1934). Herbarium Florae Asiae Mediae ab Universitate Asiae Mediae editum. Fasc. XXI, No. 5036. *Avena ludoviciana* Dur.

Nicolaisen, W. (1940). Hafer, *Avena sativa* L. Paul Brey, Berlin. *Handbuck der Pflanzen-Zuchtung,* 2 (13-18):224-288.

Nishi, S. (1980). Differentiation of *Brassica* crops in Asia and the breeding of 'Hakuran', a newly synthesized leafy vegetable. In *Brassica* Crops and Wild Allies (eds S. Tsunoda, K. Hinata and C. Gomez-Campo), pp. 133-150. *Jap. Scient. Soc.* Press, Tokyo.

Norman, A.G. (1978). *Soybean: Physiology, Agronomy and Utilisation.* Academic Press, New York.

Ollson, G. (1954). Crosses within the *campestris* group of the genus *Brassica*. *Hereditas*, 40:398-418.

Oost, E.H. (1986). A proposal for an interspecific classification of *Brassica rapa* L. In *Interaspecific Classification of Wild and Cultivated Plants* (ed. B. T. Styles), 309-315. Clarendon Press, Oxford.

Peterson, R. F. (1965). *Wheat*. Leonard Hill (Books) Ltd., London.

Philip, J. (1933). Genetics and cytology of some interspecific hybrids of *Avena*. *J. Genet.*, 27:133-179.

Philipson, W. R. (1937). A Revision of the British species of *Agrostis*. *J. Linn. Soc. L.* 1:73-263.

Poehlman, J. M. (1959). *Breeding Field Crops*. Holt, Rinehart and Winston, Inc., New York.

Porter, C.L. (1967). *Taxonomy of Flowering Plants*. Freeman and Company, San Francisco and London.

Prakash, S. and K. Hinta (1980). Taxonomy, cytogenetics and origin of crop *Brassicas*. A review. *Opera Bot.* 55, 1-57.

Prat, H. (1960). Vers une classification naturelle des Graminees. *Bull. Soc. Bot.* Fr., 107:32.

Purseglove, J. W. (1968). *Tropical Crops*. Dicotyledons. Longman, London.

Purseglove, J. W. (1972). *Tropical crops*. Monocotyledons. Longman, London.

Quisenberry, K. S. (1938). Survival of Wheat varieties in the Great Plains winter-hardiness nursery. *J. Amer. Soc. Agron.*, 30:399-405.

Renfrew, J. M. (1969). The Archaeological evidence for the domestication of plants; methods and problems. The domestication and exploitation of plants and animals (P. J. Ucko and G. W. Dimbleby, eds). Chicago, 149-172.

Rick, M. (1976). Tomato. *Lycopersicon esculentum* (Solanaceae). In *Evolution of Crop Plants* (ed. N. W. Simmonds), pp. 268-272. Longman, London and New York.

Riley, R. (1965). Cytogenetics and evolution of Wheat. In *Essays on Crop Plant Evolution* (ed. J. Hutchinson), pp. 103-122. Cambridge University Press, London.

Robinson, D. H. (1937). *Leguminous Forage Plants*. Edward Arnold, London.

Row, H. C. and J. R. Reader (1957). Root-hair development as evidence of relationship among genera of *Gramineae: Amer. J. Bot.* 44:596.

Ryder, R. J. (1979). *Leafy Salad Vegetables*. AVI Publishing Company, Westport, Connecticut.

Sampson, D. R. (1954). On the origin of Oats. *Bot.* Mus. Leaf., Harvard University 16 (10): 265-303. Cambridge, Mass.

Sakamura, T. (1918). *Kurze Mitteilung über die chromosomenzahlen und die Verwandtschafts-verhaltnisse der Triticum* Arten. *Bot. Mag.* (Tokyo), 32:151-154.

Savitsky, H. (1950). A method of determining self-fertility and self-sterility in sugar beet. *Proc. Soc.* of Beet Technologist. 6:198-201.

Savitsky, H. (1950). A method of determining self-fertility and self-sterility in sugar beet. *Proc. Amer. Soc.* of Sugar Beet Technologist. 6:198-201.

Sax, K. (1922). Sterility in Wheat hybrids. II Chromosoma behaviour in partially sterile hybrids. *Genetics*, 7:513-552.

Scannell, M.J P. and D.M. Synnott (1987). *Census Catalogue of the Flora of Ireland*. Gov. Publication, Molesworth St., Dublin.

Schaben, L. J. (1948). Rye, a source of daily bread. *For. Agr.*, 12:163-168.

Scholz, F. and C. O. Lehmann (1961). *Kulturpfl.*, 9:230-272.

Shands, H. L. and J. G. Dickson (1953). Barley - botany, production, harvesting, processing, utilization and economics. *Econ. Bot.*, 7:3-26.

Shantz, H. L. (1954). The place of grasslands in the earth's cover of vegetation. *Ecol.* 35:143.

Shantz, H. L. and L. N. Piemeisel (1927). The water requirements of plants at Akron, Colorado. *J. Agr. Res.* 34:1093-1190.

Sikka, S. M. (1940). Cytogenetics of *Brassica* hybrids and species. *J. Genet.*, 40:441-506.

Simmonds, N.W. (1976a). *Quinoa* and relatives: *Chenopodium* spp. (Chenopodiaceae). In *Evolution of Crop Plants* (ed. N. W. Simmonds), pp. 29-30. Longman, London and New York.

Simmonds, N.W. (1976b). *Solanum tuberosum* (Solanaceae). In *Evolution of Crop Plants* (ed. N. W. Simmonds), pp. 279-283. Longman, London and New York.

Simpson, B. B. and M. C. Ogorzaly (1986). *Economic Botany. Plants in our World*. McGraw-Hill, N.Y.

Solbrig, O.T. (1970). *Principles and Methods of Plant Biosystematics*. Collier-Macmillan Ltd., London.

Stace, C. A. (1980). *Plant Taxonomy and Biosystematics*. Edward Arnold, London.

Stace, C. A. (1992). *New Flora of the British Isles* (with illustrations by Hilli Thompson). Cambridge University Press, Cambridge.

Stanton, T. R. (1953). Production, harvesting, processing, utilization and economic importance of Oats. *Econ. Bot.*, 7 (1) : 43-64.

Stanton, T.R. (1961). Classification of *Avena*. In *Oats and Oat Improvement* (ed. F.A. Coffman). Agronomy, 8. *Amer. Soc.* of *Agron.*, Wisconsin, USA.

Stanton, T. R. Coffman, F. A. and G. A. Weibe (1926). Fatuoid or false wild forms in Fulghum and other oat varieties. *J. Heredity*, 17:152-226.

Stebbins, G. L. and B. Crampton (1961). A suggested revision of the grass genera of N America. *Rec. Adv. in Bot.*, 1:41.

Stewart, S. A. and T. H. Corry (1888). Flora of the North-East of Ireland. Macmillan and Bowes, Cambridge.

Takahashi, R. (1955). The origin and evolution of cultivated Barley. *Advanc. Genet.* 7:227-266.

Taylor, A. E. (1928). Rye in its relation to Wheat. Standford U. Food Res. Inst. Wheat Studies, 4 (5) : 181-234.

Thomas, T. (1984). Returns from cereal production in Ireland. In *Cereal Production* (ed. E. J. Gallagher), pp. 297-304. Butterworth, London.

Thomas, P. T. and M. B. Crane (1942). Genetic classification of *Brassica* crops. *Nature*, 150 : 431

Thompson, K. F. (1956). Production of haploid plants of marrow-stem kale. *Nature*, London, 178:748.

Thompson, K. F. (1976). Cabbages, Kales, etc. *Brassica oleracea* (Cruciferae). In *Evolution of Crop Plants* (ed. N. W. Simmonds), pp. 49-52. Longman, London and New York.

Thurston, J.M. (1954). A Survey of Wild-oats (*Avena fatua* and *A. ludoviciana*) in England and Wales - 1951. *Ann. App. Biol.*, 41:619-636.

Thurston, J. M. and A. Philipson (1976). Distribution. In *Wild-oats in World Agriculture* (ed. D. Price-Jones), pp. 19-64.

Tottman, D. R., Makepeace, R. J. and H. Broad (1979). An explanation of the decimal code for the growth stages of cereals; with illustrations. *Ann. App. Biol.*, 93:221-234.

Tsunoda, S., Hinata, K. and C. Gomez-Campo, eds, (1980). *Brassica* Crops and their Allies. Biology and breeding. Japan Scientific Societies Press, Tokyo.

U, N. (1935). Genomic analysis in *Brassica* with special reference to the experimental formation of *B. napus* and peculiar mode of fertilization. *Jap. J. Bot.*, 7:389-452.

Vaughan, J. G., A. J. Macleod and B. M. G. Jones (1976). *The Biology and Chemistry of the Cruciferae*. Academic Press, London, New York and San Francisco.

Vavilov, N. I. (1951). *The Origin, Variation, Immunity and Breeding of Cultivated Plants. Chronica Botanica*, Vol. 13 (No. 1-6).

Walkins, R. (1984). Apples and Pears - *Malus* and *Pyrus* spp. In *Evolution of Crop Plants* (ed. N. W. Simmonds), pp. 247-250.

Webb, D. A. (1977). *An Irish Flora*. Dundalgan Press (W. Tempest) Ltd., Dundalk, Ireland.

Wellington, P. S. and C. E. Quartly (1972). A Practical system for classifying, naming and identifying some cultivated *Brassicas*. *J. Natn. Inst. Agric. Bot.*, 12:413-432.

Whitaker, T. W. and G. N. Davis (1962). *Cucurbits*. Leonard Hill (Books) Ltd., New York.

Whitaker, T. W. and W. P. Bemis (1984). Cucurbits: *Cucumis, Citrillus, Cucurbita, Lagenaria* (Cucurbitaceae). In *Evolution of Crop Plants* (ed. N. W. Simmonds), pp. 64-69. Longman. London and New York.

Weiss, A. N. (1962). *Oilseed Crops*. Longman, London.

Wilsie, C. P. (1951). *Crop Adaptation and Distribution*. Freeman and Co., San Francisco and London.

Yarnell, S. H. (1956). Cytogenetics of vegetable crops II. Cruficers. *Bot. Rev.*, 22 (2) : 81-166.

Yasuda, A. (1903). On the comparative anatomy of the Cucurbitaceae, Wild and Cultivated. *J. Coll. Sci. Imp. Univ.*, Tokyo, 18:156.

Zadoks, J. C., Chang, T. T. and G. F. Konzak (1974). A decimal code for the growth stages of cereals. *Weed Res.*, 14:415-421.

Ziegler, A. (1911). Z. ges. Brauw., 34:513-568.

Zohary, D. (1960). Studies on the origin of cultivated Barley. *Bull. Res. Counc.*, Israel, 9D, 21.

Zohary, D. (1971). Origin of south-west Asiatic Cereals: Wheat, Barley, Oats and Rye. In plant life of southwest Asia. (ed. P. M. Davis, *et al.*,) p. 235. Edinburgh.

Zohary, D. and M. Feldman (1961). Hybridization between amphidiploids and the evolution of polyploids in the Wheat (*Aegilops - Triticum*) group. *Evolution*, 16:44-61.

GLOSSARY OF SCIENTIFIC TERMS

abaxially .... on the surface directed away from the axis (lower surface).
accumbent .... lying against and face to face.
acicular .... like a needle in shape; sharp-pointed.
actinomorphic .... having 2 or more planes of symmetry.
acuminate .... tapered to a fine point.
adaxially .... on the surface directed towards the axis (upper surface).
adnate .... joined to another organ of a different kind.
aggregate .... refers to distinct fruits or florets which are grouped together.
allopolyploid .... polyploid derived through hybridisation between 2 different species.
amplexicaul .... base of blade extending around stem to form 2 ears.
anastomosing .... refers to leaves where the veins come together to form a network.
andromonoecious .... male and hermaphrodite flowers on the same plant.
anthesis .... time of opening of a grass floret when pollination takes place.
apocarpous .... refers to a pistil with free carpels.
appressed .... pressed closely and flatly against.
arcuate .... shaped like an arch or bow.
aristate .... with an awn or bristle.
ascending .... rising up; produced somewhat obliquely.
asexual .... without stamens and carpels.
attachment scar .... mark or scar left on a fruit when it separates from the pedicel.
atuberculate .... lacking tubercles or wart-like structures.
auricles .... small projections at base of a leaf-blade (see ear).
awn .... a stiff bristle-like projection from the flowering pale or lemma.
axillary .... arising in the axil of a leaf or bract.

beak .... a mostly prominent point, particularly of fruits.
bifid .... split deeply in 2.
bipinnate .... refers to a compound leaf where the pinnae are also pinnate.
bisexual .... with male and female parts in the same flower.
biternate .... leaf type where the 3 main divisions are themselves ternate.
bract .... a modified leaf at the base of an inflorescence or flower.
bracteole .... a secondary bract at the base of an umbel.
bulbil .... a small bulb, usually axillary, on an aerial part of a plant.
bulbous .... swollen and bulb-like.

c. .... about (from the latin *circa*).
caducous .... falling off early.
caespitose .... see cespitose.
calcicole .... a plant which thrives on or confined to soils with free calcium carbonate.
calcifuge .... a plant which grows to advantage on acid soils.
campanulate .... bell-shaped.
campylotropous ovule .... ovules with the micropyle projecting laterally.
capitate .... head-like inflorescence or knob-like stigma.
carpel .... 1 female reproductive organ.
caruncle .... refers to a small fleshy body attached to the apex of a seed.
cauline .... borne on aerial part of stem.

cespitose .... tufted growth-habit.

ciliate .... margins fringed with hairs.

cladode .... a green leaf-like shoot arising from the axil of a leaf.

clavate .... club-shaped.

claw .... the narrow lower part of some petals (*e.g.* in *Brassica*).

cleistogamous .... of flowers which never open and are self-pollinated.

collenchyma .... thick-walled, angular supporting tissue.

commissure .... the place of joining or meeting (see Apiaceae).

connate .... growing together and becoming joined, though distinct in origin.

connivent .... coming together or converging.

convolute .... rolled together or margins rolled inwards.

cordate .... somewhat heart-shaped.

coriaceous .... leathery texture.

corm .... short, usually erect, swollen underground stem.

cortex .... the collective term for the epidermis, hypodermis and endodermis.

corymb .... a flat-topped raceme.

corymbose .... like a corymb.

costa .... longitudinal rib or mid-vein of a leaf; raised vascular bundle.

cotyledon .... the first leaf formed from a seed.

crenate .... blunt teeth projecting towards leaf apex.

crisped .... curled.

culm .... the stem of grasses which carries the leaves and inflorescence.

cuneate .... wedge-shaped.

cusp .... point; projection; a prominence.

cuspidate .... terminating in a point.

deciduous .... falling after one season's growth.

decumbent .... refers to stems which lie on the ground and tend to rise at the apex.

decurrent .... having leaf base prolonged down the stem.

deflexed .... pointing downwards.

dehiscent .... opening to shed seeds.

deltoid .... shaped like the fourth letter of the greek alphabet.

dentate .... with rounded or blunt teeth.

dermatogenic .... relating to the epidermis or skin.

diadelphous .... in 2 bundles.

diarch .... vascular bundle with a central xylem and 2 lateral strands of phloem.

didymus .... 2-lobed.

digitate .... radiating from a basal point like the fingers on a hand.

dimorphic .... 2 different forms.

dioecious .... having the sexes on different plants.

distichous .... arranged in 2 rows.

dorso-ventrally .... from dorsal surface to ventral surface.

ear .... small projection from the base of a leaf-blade.

emarginate .... notched apically.

epicotyl .... the area of stem immediately above the cotyledons.

epigeal .... above ground (germination).

epigynous .... perianth and stamens at a higher level than the consequent inferior ovary.

epipetalous .... stamens inserted on petals.

expeliolate .... without a petiole.

exstipulate .... lacking stipules.

face .... flat, curved or rounded surface.

falcate .... sickle-shaped; hooked.

fascicle .... a bundle or cluster, especially of stamens.

filament	the thread-like structure of a stamen.
filiform	thread-like.
flexuous	wavy.
floret	an alternative name for a flower (principally grass flowers).
foliaceous	leaf-like.
follicle	a dry, dehiscent fruit, splitting along 1 margin.
funicle	the stalk of the ovule.
fusiform	spindle-shaped.
gamopetalous	having the petals joined into a tube, at least at the base.
gamosepalous	having the sepals joined into a tube, at least at the base.
geniculate	bent and knee-like.
glabrescent	becoming glabrous or hairless.
glabrous	hairless.
glandular	with glands.
glaucescent	slightly glaucous.
glaucous	a somewhat bluish-green colour.
glycoside	a compound which on hydrolysis gives a sugar and a non-sugar residue.
gynaeceum	a collective term for all the carpels of 1 flower.
gynoecium	a collective term for all the carpels of 1 flower.
habitat	the external environment in which a plant grows.
halophyte	a plant which thrives on salt-impregnated soils.
haploid	chromosome number characteristic of a gamete, denoted by x or n.
hastate	spear-like.
herbaceous	soft, pliable, lacking lignin.
hermaphrodite	male and female part in same flower.
hesperidium	fruit with epicarp and mesocarp fused and endocarp projecting inwards.
heterospiculate	spikelets not alike.
hilum	scar left on seed by the ovule stalk.
hirsute	clothed with short, rather rough or coarse hairs.
hispid	covered with short stiff hairs.
hoary	covered with a close white or whitish pubescence.
homogeneous	having the same kind of constituent units throughout.
homologous	similar in structure and origin.
husk	refers to the presence of the lemma and palea on the fruit of oat or barley.
hyaline	thin and translucent.
hygroscopic	responsive to the presence or absence moisture, *e.g.* lodicules.
hypanthium	the cup-like or saucer-like expansion of the receptacle.
hypocotyl	embryo axis below the point of attachment of the cotyledons.
hypogeal	below ground (germination).
hypogynous	sepals, petals and stamens arranged at a lower level than the consequent superior ovary.
imparipinnate	pinnate leaf with an unpaired terminal leaflet.
indehiscent	not opening to shed seeds.
indumentum	hairy covering as a whole.
inferior	ovary apparently at a lower level than the other floral parts.
infra	used in combination to signify below (see next term).
infraspecific	refers to any unit of classification below the level of species.
intercostal	between ribs, ridges, veins, etc.
inverted	turned over, *e.g.* upside down.
involucel	small bracts at the base of a secondary umbel.
involucre	a whorl of bracts resembling a calyx.
irregular	refers to a flower, especially its corolla, in which there is only 1 plane.
involute	with margins rolled inwards.

isospiculate	spikelets all alike.
lamina	blade portion of leaf.
lanate	woolly, with long intermingled hairs, *e.g.* as in *Holcus lanatus*.
lanceolate	widest in the lower half.
latex	milky or yellowish juice, *e.g.* as in *Euphorbia*.
lenticel	air pore of stem or root (*c.f.* stomata).
lenticular	convex on both surfaces and more or less circular in outline.
lepidote	covered with small scurfy scales.
limb	the flattened or expanded portion of a petal or sepal.
ligulate	strap-shaped.
limb	expanded flattened apical part of a petal or sepal, *e.g.* as in *Brassica*.
linear	long and narrow with more or less parallel sides.
lodicule	small hygroscopic scales of a grass floret.
lyrate	pinnatifid, but with a large terminal lobe; lobes projecting basally.
median	middle or central.
membranous	thin, dry, semi-transparent tissue.
monadelphous	united a single bundle.
monoecious	plant having separate male and female flowers.
monostichous	arranged in 1 row.
mucronate	mid-vein extended as a short projection or point.
muricate	covered with short sharp outgrowths.
naked	refers to flowers with no petals or sepals, or grains with no pales.
nerve	vascular bundle or vein.
obcordate	deeply lobed apically.
oblanceolate	widest in the upper half.
oblique	slanting or unequal-sided.
oblong	longer than broad, and mostly parallel-sided.
obovate	inverted egg-shaped.
obtuse	blunt.
ocrea	tubular sheath-like expansion at petiole-base (sheathing stipule).
orbicular	rounded, with length and width about the same.
orthotropous ovule	ovule where the micropyle projects upwards.
ovate	egg-shaped.
ovoid	somewhat egg-shaped.
ovule	a minute ovoid structure inside the ovary.
palea	inner flowering pale of a grass floret.
palmate	leaf with leaflets radiating from a basal point.
palmatifid	resembling a palmate leaf.
panduriform	with a sagittate outline but with blunt or rounded base and apex.
panicle	a branched racemose inflorescence.
papilla	a glandular hair, mostly with a swollen apex.
pappus	the hairs representing the calyx in a flower of the daisy family.
pedicel	stalk of a single flower.
peduncle	stalk of an inflorescence.
perianth	collectively, the sepals and petals.
pericarp	the fruit wall.
perigynous	of flowers in which there is an annular region, flat or concave, between the base of the gynoecium and the insertion of the other floral parts.
petaloid	resembling a petal.
petiole	the leaf-stalk.

pericycle	the layer of tissue immediately beneath the endodermis.
petiolate	with a petiole.
pilose	covered with rather long soft hairs.
pinna	a leaflet of a pinnate leaf.
pinnate	refers to a compound leaf in which the leaflets are arranged in more or less opposite ranks on either side of the stem.
pinnatifid	resembling a pinnate leaf.
pinnatisect	leaves divided almost to the base.
pinnule	a secondary leaflet of a bipinnate leaf.
pistil	the carpel or carpels of a flower, whether free or united.
pistillate	containing pistil only.
plicate	folded, or nearly so.
polygamous	having unisexual and hermaphrodite flowers on the same plant.
polypetalous	having free petals.
polysepalous	having free sepals.
pome	fruit of apple and related plants.
prickle	stout outgrowth with a sharp apex.
procumbent	lying loosely on the surface but not usually rooting.
proliferous	bearing other similar structures on itself.
protandrous	refers to a flower whose anthers ripen and and shed their pollen before the stigma of the same flower is receptive.
protandry	see protandrous.
protogynous	refers to a flower whose stigma ripens before the pollen is shed from anthers of the same flower.
protogyny	see protogynous.
pruinose	having a whitish 'bloom'.
pubescent	hairy.
pulvinus	petiole-base, particularly when swollen.
punctate	dotted or pitted.
raceme	inflorescence with primary branches only.
rachilla	central branch of a spikelet; also seed-stalk.
rachis	central stem of a grass inflorescence.
receptacle	the apex of the peduncle to which the floral parts are attached.
regular	refers to a flower, especially its corolla, which has 2 or more planes of symmetry.
reniform	kidney-shaped.
revolute	rolled backwards from margin upon under surface.
reticulate	marked with a raised network.
rhizome	an underground horizontal stem.
rhomboid	more or less diamond-shaped.
rind	the outer layer of a fruit of any member of the Cucurbitaceae.
rootstock	subterranean vertical swollen stem, *e.g.* as in the Apiaceae.
rugose	wrinkled.
runner	a slender trailing shoot which roots at the nodes.
runcinate	pinnately lobed, with the lobes directed basally.
sagittate	shaped like the head of an arrow.
seed-stalk	rachilla internode attached to some grass 'seeds'.
sepaloid	green and sepal-like.
septa	singular *septum*, a partition separating 2 cavities, etc.
serrate	sharp-toothed like the teeth of a saw.
sessile	without a petiole or stalk.
setaceous	bristle-like.
sinuate	having a wavy indented margin.

sinus	depression between 2 lobes, teeth or ridges.
spadix	a fleshy spike bearing unisexual flowers enveloped with a spathe.
spathe	a large leaf-like structure protecting an inflorescence.
spatulate	spoon-shaped.
spike	inflorescence with flowers or spikelets attached directly to main stem.
spikelet	one or more grass or sedge florets subtended by 1 or 2 glumes.
spur	a tubular or sac-like projection from a flower.
staminate	having stamens only.
staminode	a sterile stamen, sometimes modified to perform some other function.
stellate	star-shaped.
stipel	outgrowth from leaflet-base.
stipulate	with a stipule.
stipule	outgrowth, mostly leaflet-like, from petiole-base.
stolon	an overground creeping stem.
stomata	minute openings with guard cells.
striate	marked with long narrow, shallow depressions or ridges.
strict	forming a small angle to the vertical.
stricted	straight, narrow and erect.
strigose	with stiff appressed hairs.
stylopodium	a disk-like enlargement at the base.
subtend	attached below and close to.
subulate	awl-shaped.
suffrutescent	semi-shrubby, or the lowermost part of stem becoming woody.
superior	ovary apparently at a higher level than the other floral parts.
syncarpous	carpels united into a single structure.
taxon	any unit of classification, *e.g.* family, genus, species, etc.
tendril	a thread-like extension of the leaf mid-rib.
tepal	a perianth segment not clearly differentiated into sepals or petals.
terete	circular in cross-section.
ternate	divided, from near the base, into 3 principal parts.
tertiary	refers to an organ which arises on a secondary organ.
tomentum	dense covering of white cottony hairs.
triad	occurring in groups of 3.
trifoliate	divided into 3 parts.
trigonous	solid but more or less triangular in cross-section.
triploid	plant with 3 times the basic chromosome number.
triquetrous	of a solid body triangular in section and acutely angled.
truncate	flat-topped.
tuberculate	with small blunt or wart-like projections.
tubercle	a small wart-like projection.
umbel	an inflorescence with branches arising from a common point.
unisexual	of 1 or other sex, i.e. male or female.
vein	vascular bundle or nerve.
ventral	surface facing the axis.
versatile	attached near the middle and usually moving freely.
vitta	resinous gland.
web	tuft of silky hairs.
whorl	3 or more organs of the same kind attached at the same level.
woolly	clothed with shaggy hairs.
zygomorphic	having only 1 plane of symmetry.

SUBJECT AND NAME INDEX